# THE HANDY SCIENCE ANSWER BOOK

# The Handy Answer Book™ Series

The Handy Answer Book for Kids (and Parents)
The Handy Bug Answer Book
The Handy Dinosaur Answer Book
The Handy Geography Answer Book
The Handy History Answer Book
The Handy Ocean Answer Book
The Handy Physics Answer Book
The Handy Politics Answer Book
The Handy Religion Answer Book
The Handy Science Answer Book
The Handy Space Answer Book
The Handy Sports Answer Book
The Handy Weather Answer Book

**Compiled by the Science and Technology Department
of the Carnegie Library of Pittsburgh**

**Edited by James E. Bobick and Naomi E. Balaban**

**Detroit**

# THE HANDY SCIENCE ANSWER BOOK™

Centennial Edition

Carnegie
Library of
Pittsburgh



Visible Ink Press™
42015 Ford Rd. #208
Canton, MI 48187-3669

Visible Ink Press is a trademark of Visible Ink Press LLC.

Most Visible Ink Press books are available at special quantity discounts when purchased in bulk by corporations, organizations, or groups. Customized printings, special imprints, messages, and excerpts can be produced to meet your needs. For more information, contact Special Markets Director, Visible Ink Press, at www.visibleink.com.

Art Director: Mary Claire Krzewinski
Typesetter: Graphix Group

ISBN 1-57859-140-6

**Library of Congress Cataloging-in-Publication Data**

The handy science answer book / compiled by the Science and Technology Department of the Carnegie Library of Pittsburgh ; edited by James E. Bobick and Naomi E. Balaban. -- Centennial ed.
p. cm.
Includes bibliographical references and index.
ISBN 1-57859-140-6 (pbk. : alk. paper)
1. Science--Miscellanea. I. Bobick, James E. II. Balaban, Naomi E. III. Carnegie Library of Pittsburgh. Science and Technology Dept.
Q173 .H24 2002
500--dc21 2002015562

Printed in the United States of America


10 9 8 7 6 5 4 3 2 1

# Contents


INTRODUCTION *XI*
ACKNOWLEDGMENTS *XV*
CREDITS *XVI*

## PHYSICS AND CHEMISTRY

Energy, Motion, Force, and Heat *1*
Light, Sound, and Other Waves *7*
Matter *12*
Chemical Elements, etc. *17*
Measurement, Methodology, etc. *27*


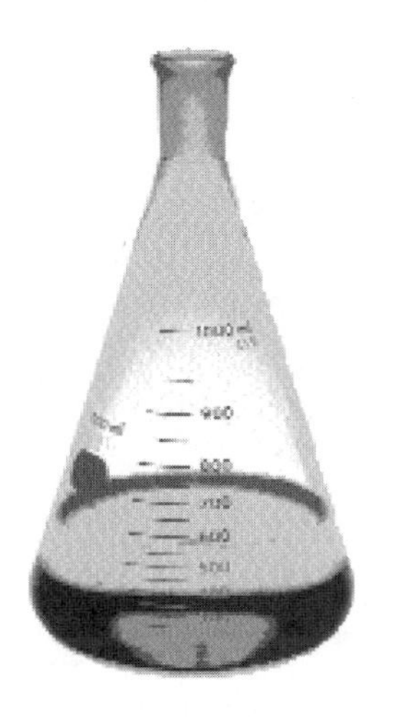

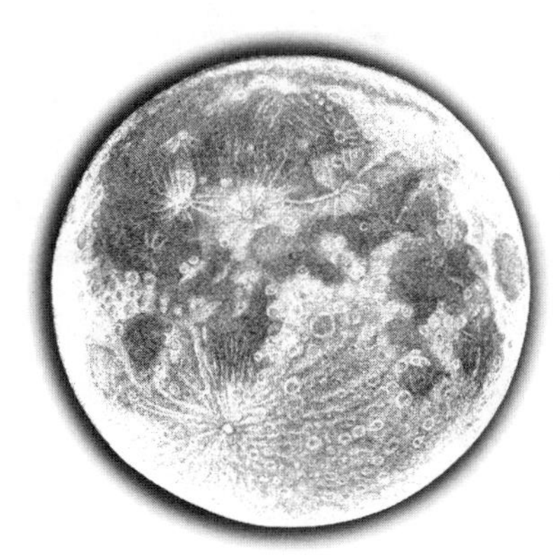


## SPACE

Universe *33*
Stars *36*
Planets and Moons *48*
Comets, Meteorites, etc. *60*
Observation and Measurement *64*
Exploration *68*

# EARTH


Air *81*

Physical Characteristics, etc. *83*

Water *87*

Land *95*

Volcanoes and Earthquakes *105*

Observation and Measurement *113*

# CLIMATE AND WEATHER


Temperature *119*

Air Phenomena *122*

Wind *127*

Precipitation *138*

Weather Prediction *143*


# MINERALS AND OTHER MATERIALS


Rocks and Minerals *147*

Metals *156*

Natural Substances *162*

Man-made Products *168*

# ENERGY

Non-nuclear Fuels *181*
Nuclear Power *190*
Measures and Measurement *195*
Consumption and Conservation *198*

# ENVIRONMENT

Ecology, Resources, etc. *205*
Extinct and Endangered Plants and Animals *216*
Pollution *224*
Recycling, Conservation, and Water *235*

# BIOLOGY

Cells *245*
Evolution and Genetics *249*
Life Processes, Structures, etc. *263*
Classification, Measurement, and Terms *264*
Fungi, Bacteria, Algae, etc. *269*

## PLANT WORLD

Physical Characteristics, Functions, etc. *273*

Trees and Shrubs *275*

Flowers and Other Plants *280*

Gardening, Farming, etc. *289*

## ANIMAL WORLD

Physical Characteristics, etc. *305*

Names *313*

Insects, Spiders, etc. *317*

Aquatic Life *328*

Reptiles and Amphibians *333*

Birds *335*

Mammals *344*

Pets *356*

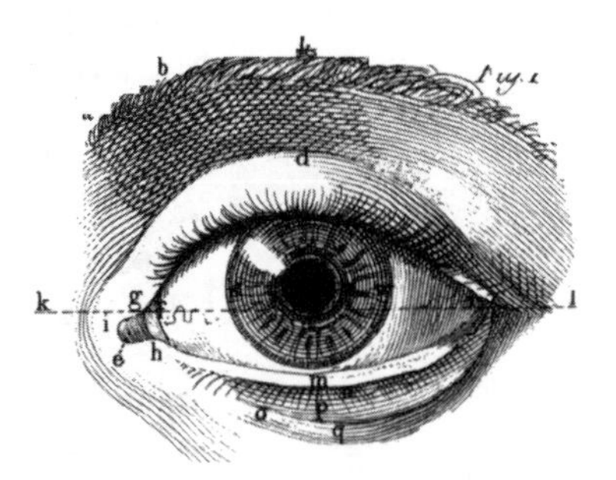

## HUMAN BODY

Functions, Processes, and Characteristics *367*

Bones, Muscles, and Nerves *383*

Organs and Glands *386*

Body Fluids *392*

Skin, Hair, and Nails *397*

Senses and Sense Organs *400*

# HEALTH AND MEDICINE

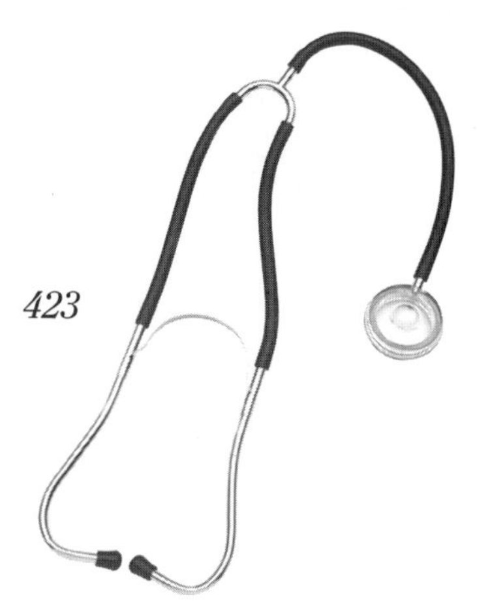


Health Hazards, Risks, etc. *407*

First Aid, Poisons, etc. *416*

Diseases, Disorders, and Other Health Problems *423*

Health Care *440*

Diagnostic Equipment, Tests, etc. *445*

Drugs, Medicines, etc. *447*

Surgery and Other Non-drug Treatments *459*

# WEIGHTS, MEASURES, TIME, TOOLS, AND WEAPONS


Weights, Measures, and Measurement *465*

Time *474*

Tools, Machines, and Processes *489*

Weapons *496*


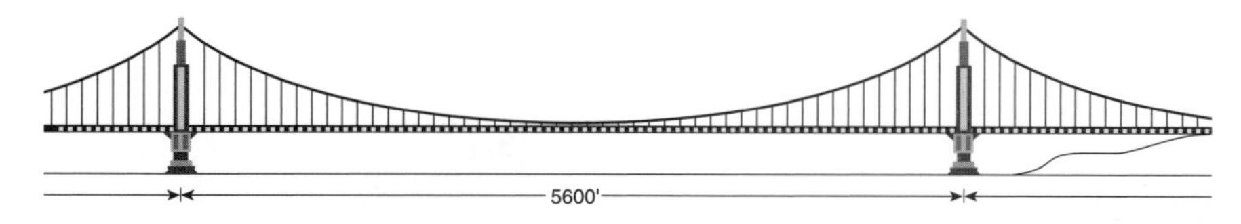


# BUILDINGS, BRIDGES, AND OTHER STRUCTURES


Buildings and Building Parts *505*

Roads, Bridges, and Tunnels *513*

Miscellaneous Structures *519*

## BOATS, TRAINS, CARS, AND PLANES

Boats and Ships *525*

Trains and Trolleys *530*

Motor Vehicles *533*

Aircraft *546*

Military Vehicles *546*


## COMMUNICATIONS

Symbols, Writing, and Codes *557*

Radio and Television *565*

Telecommunications, Recording, etc. *571*

Computers *577*


## GENERAL SCIENCE AND TECHNOLOGY

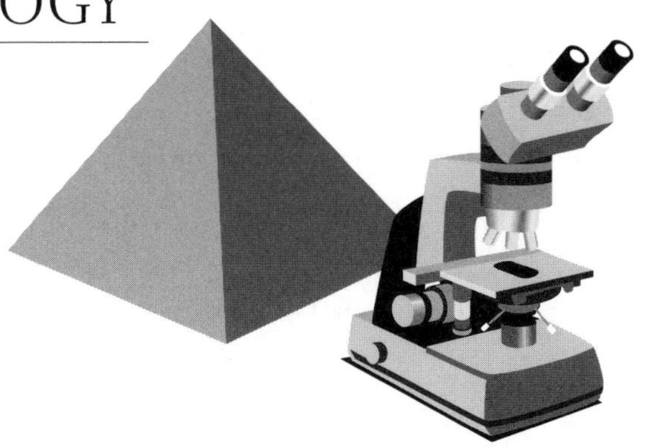


Numbers *595*

Mathematics *600*

Terms and Theories *613*

FURTHER READING *617*

INDEX *627*

# Introduction

The revolution is happening, all the time, everywhere. The rapidity of advances in science and technology seems to rival that of the speed of light (186,282 miles per second). How do we keep up, and where can we find answers to all our daily questions—from the mundane to the esoteric? How much data can a 3.5-inch floppy disk hold? (From 400 kilobytes to more than two megabytes.) I'd like a dog, but I don't want one that sheds; which kind should I get? (A poodle, a Kerry blue terrier, or a schnauzer.) When will the sun die? (In about five billion years.) How much waste paper does my daily newspaper subscription generate? (550 pounds each year.) Is there life on Mars? (Haven't found any yet.)

Science and technology have become the cornerstones of modern life. Imagine a world without computers. Less than twenty years ago, the overwhelming popular assumption was that the computer would remain a highly specialized tool of big business. A basic personal computer now houses more calculating power than the giant mainframes of not so long ago. Although the general public is largely using this awesome power to surf the Web, e-shop, make greeting cards, view digital photos, download music, and drop Kerouac-inspired stream-of-consciousness e-mail on unsuspecting friends, networks of home computers have been used to do complex scientific equations, and scores of home businesses depend on computers. The machine is now part of our daily ritual, a general utility that has transformed the behavior of its operators.

Scientists utilizing computer mapping and analyses are unraveling the mysteries of the genetic code. Manipulation at the gene level may be the key to finding cures for cancer and other major diseases and extending human life. Scientists can now clone animals, while politicians huff and puff about the morality and prevention of human cloning (perhaps as a public service, recognizing that a cloned politician would certainly be a menace to society). The President declares stem cell research more or less off limits, maybe. Cellular phones link everyone to everyone else, all the time, everywhere. Our understanding of the universe is expanding at a revolutionary rate, thanks to orbiting telescopes and computer analysis of signals from deep space. Perhaps someday, in the not-too-distant future, we'll witness the big bang itself. The beginning of everything. It boggles the mind. We are dependent upon, and take for granted, these giant leaps in science and technology. But as life has gotten far more complex and sophisticated, as our specialized knowledge of the world and universe has increased, our common understanding of basic science and technology has diminished. We have questions, we are puzzled, but we can't find answers. The modern age is fast upon us, and we are confused. What we need is *The Handy Science Answer Book.*

Concise and easy to understand, *The Handy Science Answer Book* covers hundreds of intriguing science and technology topics, from the inner workings of the human body to outer space, and from math and computers to planes, trains, and automobiles. In 1902, the Carnegie Library of Pittsburgh (which opened in 1895, underwritten by steel magnate Andrew Carnegie) became the first major public library in the United States to establish a separate Science and Technology Department. Since then, the Department has been patiently answering the reference questions of customers, at a rate of more than 60,000 per year via personal visits, fax, e-mail, regular mail, or through the newly implemented Web-based virtual reference service. The bathroom is down the hall. The bus stops at the corner. There are 42 gallons in a barrel of oil. Groundhogs have accurately predicted the weather only 28 percent of the time on Groundhog Day. The ice that covers Antarctica is 15,700 feet in depth at its thickest point. From astronomy to zoology, the Department has accumulated an immense, authoritative reference file. *Handy Science* celebrates the Department's centennial with a collection of 1,700 of the most asked, most interesting, or most unusual questions and answers in the areas of science, pseudo science, and technology. Edited by James Bobick, the Department head, and Naomi Balaban, this edition has been thoroughly revised, with nearly 400 completely new questions added. In addition, 125 illustrations and many tables augment the text.

In some ways, science touches so much of our lives—whether it be our environment, our homes, our workplaces, or our physical bodies themselves—that it can become difficult to categorize what actually constitutes science. *Handy Science* makes no particular effort to restrict the questions to pure science, but focuses on those questions that have achieved noteworthiness either through their popularity, the time-consuming nature of their research, or their uniqueness. How is the glass used in movie stunts made? When do the swallows come back to Capistrano? Why do dogs howl at sirens? What do the different colors and varieties of roses symbolize? Does the familiar phrase "open sesame" have anything to do with sesame seeds? What are the top ten dog names? What is the funny bone? When was the parking meter introduced? Are there trees that predict the weather and tell time? How does driving speed affect gas mileage?

The Carnegie staff has verified figures and dates to the best of their ability. Keep in mind that even in science, figures can seem to be in conflict; many times such discrepancies may be attributable to the authority perspective, or more commonly, to the results of simple mathematical rounding of figures. Occasionally, the figure or date listed is a consensus of the consulted sources; other times, the discrepancy is noted and an alternative given. *Handy Science* rounds off figures whenever it seems that such precision is unnecessary. When designating eras, *Handy Science* uses the abbreviation C.E. ("of the common era") instead of the more familiar A.D. (*anno Domini,* "in the year of the Lord"), and B.C.E. ("before the common era") in place of B.C. ("before Christ").

Designed as a family reference, *Handy Science* is kid friendly, helping satisfy that inspired curiosity about the world. The answers are written in non-technical language and provide either a succinct response or a more elaborate explanation, depending on the nature of the question. Definitions of scientific terminology are given within the answer itself, and both metric and U.S. customary measurements are listed. Following the main Q&A section are suggestions for further reading (most of which were used to answer various questions), including an all-new list of helpful Web sites, and the index.

Since the first edition of *Handy Science* in 1994, the Department has received many favorable comments about its interesting content. Apparently, people simply like having all this information in one, well, handy book. Andrew Carnegie would have been proud of the Department and this publication!

# Acknowledgments

**W**riting a book is a lot of work, and you need a lot of help from a lot of individuals to get the job done. Many people have made significant contributions to the third edition of *The Handy Science Answer Book.* Naomi Balaban, who served as project manager, completed this revision with speed, accuracy, resourcefulness, enthusiasm, and dependability. She is the consummate professional librarian! I'm sure she appreciated the questions that her husband, Carey, and daughters asked and, subsequently, answered. I want to thank the librarians in the Science and Technology Department for their individual and collective work in gathering, reviewing, answering, verifying, and revising many more questions than the number that are included in this volume. Thanks go to Grace Alba, Joan Anderson, Gregg Carter, John Doncevic, Mary Fry, Diane Gerber, Terry Lamperski, Judy Lesso, Matt Marsteller, Dave Murdock, and Donna Strawbridge. These librarians were remarkable at balancing the never-ending needs of our library customers with the frequent deadlines required for submitting chapter questions. All of them know how much I've appreciated their efforts. Students in my "Resources and Services in Science and Technology" classes at the University of Pittsburgh's School of Information Studies contributed some interesting and challenging questions over the past several years, and I am thankful for all of them.

At Visible Ink Press, thanks to Marty Connors, publisher; Christa Brelin, managing editor; Kevin Hile, copyeditor; Marie MacNee and Susan Salter, proofreaders; Larry Baker, indexer; Chad Woolums, photo researcher; Bob Huffman, photo processor; P. J. Butland, copywriter; Mary Claire Krzewinski, designer; and Marco Di Vita of the Graphix Group, typesetter.

This new edition comes at an opportune time. One hundred years ago, in 1902, the Carnegie Library of Pittsburgh became the first major public library in the United States to establish a separate Science and Technology Department. I'm pleased that this book will be published as part of the Department's 100th anniversary.

Finally, thanks to my wife, Sandi, and sons, Andrew and Michael, for their encouragement, patience, and understanding.

*James E. Bobick*
*Head, Science and Technology Department*
*Carnegie Library of Pittsburgh*

# Credits

**Associated Press:** back cover photo of Hindenburg.

**Corbis:** photos on pages 3, 5, 13, 18, 27, 35, 40, 62, 67, 71, 73, 76, 82, 99, 101, 103, 109, 111, 123, 141, 142, 154, 159, 184, 213, 215, 252, 260, 265, 266, 271, 298, 302, 309, 318, 323, 338, 357, 376, 381, 453, 461, 499, 512, 534, 536, 547, 551, 561, 565, 566, 567, 578, 601, 602, 608.

**Electronic Illustrators Group:** all line art.

**Robert J. Huffman/Field Mark Publications:** cover photo of ladybugs; photos on pages 47, 100, 148, 149, 165, 175, 192, 206, 219, 239, 254 (both), 276, 282, 325, 340, 341, 350, 355, 416, 436, 511, 571, 575.

# THE HANDY SCIENCE ANSWER BOOK

# PHYSICS AND CHEMISTRY

## ENERGY, MOTION, FORCE, AND HEAT

*See also: Energy*

### How is **"absolute zero"** defined?

Absolute zero is the theoretical temperature at which all substances have zero thermal energy. Originally conceived as the temperature at which an ideal gas at constant pressure would contract to zero volume, absolute zero is of great significance in thermodynamics and is used as the fixed point for absolute temperature scales. Absolute zero is equivalent to 0 K, –459.67°F, or –273.15°C.

The velocity of a substance's molecules determines its temperature; the faster the molecules move, the more volume they require, and the higher the temperature becomes. The lowest actual temperature ever reached was two-billionth of a degree above absolute zero ($2 \times 10^{-9}$K) by a team at the Low Temperature Laboratory in the Helsinki University of Technology, Finland, in October 1989.

### Does **hot water freeze faster** than cold?

A bucket of hot water will not freeze faster than a bucket of cold water. However, a bucket of water that has been heated or boiled, then allowed to cool to the same temperature as the bucket of cold water, may freeze faster. Heating or boiling drives out some of the air bubbles in water; because air bubbles cut down thermal conductivity, they can inhibit freezing. For the same reason, previously heated water forms denser ice than unheated water, which is why hot-water pipes tend to burst before cold-water pipes.

## What is **superconductivity**?

Superconductivity is a condition in which many metals, alloys, organic compounds, and ceramics conduct electricity without resistance, usually at low temperatures. Heinke Kamerlingh Omnes, a Dutch physicist, discovered superconductivity in 1911. The modern theory regarding the phenomenon was developed by three American physicists—John Bardeen, Leon N. Cooper, and John Robert Schrieffer. Known as the BCS theory after the three scientists, it postulates that superconductivity occurs in certain materials because the electrons in them, rather than remaining free to collide with imperfections and scatter, form pairs that can flow easily around imperfections and do not lose their energy. Bardeen, Cooper, and Schrieffer received the Nobel Prize in Physics for their work in 1972. A further breakthrough in superconductivity was made in 1986 by J. Georg Bednorz and K. Alex Müller. Bednorz and Müller discovered a ceramic material consisting of lanthanum, barium, copper and oxygen which became superconductive at 35 K (–238°C)—much higher than any other material. Bednorz and Müller won the Nobel Prize in Physics in 1987. This was a significant accomplishment since in most situations the Nobel Prize is awarded for discoveries made as many as 20 to 40 years earlier.

## What are some **practical applications** of superconductivity?

A variety of uses have been proposed for superconductivity in fields as diverse as electronics, transportation, and power. Research continues to develop more powerful, more efficient electric motors and devices that measure extremely small magnetic fields for medical diagnosis. The field of electric power transmission has much to gain by developing superconducting materials since 15 percent of the electricity generated must be used to overcome the resistance of traditional copper wire. More powerful electromagnets will be utilized to build high-speed magnetically levitated trains, known as "maglevs."

## What is the **string theory**?

A relatively recent theory in particle physics, the string theory conceives elementary particles not as points but as lines or loops. The idea of these "strings" is purely theoretical since no string has ever been detected experimentally. The ultimate expression of string theory may potentially require a new kind of geometry—perhaps one involving an infinity of dimensions.

## What is **inertia**?

Inertia is a tendency of all objects and matter in the universe to stay still, or, if moving, to continue moving in the same direction, unless acted on by some outside force. This forms the first law of motion formulated by Isaac Newton (1642–1727). To move a body at rest, enough external force must be used to overcome the object's inertia; the larger the object is, the more force is required to move it. In his *Philosophae Naturalis Principia Mathematica,* published in 1687, Newton sets forth all three laws of motion. Newton's second law is that the force to move a body is equal to its mass times its acceleration (F = MA), and the third law states that for every action there is an equal and opposite reaction.

In 1687, Isaac Newton published his *Philosophae Naturalis Principia Mathematica,* laying the foundation for the science of mechanics.

## Why do **golf balls** have dimples?

The dimples minimize the drag (a force that makes a body lose energy as it moves through a fluid or gas), allowing the ball to travel farther than a smooth ball would travel. The air, as it passes over a dimpled ball, tends to cling to the ball longer, reducing the eddies or wake effects that drain the ball's energy. A dimpled ball can travel up to 300 yards (275 meters), but a smooth ball only goes 70 yards (65 meters). A ball can have 300 to 500 dimples that can be 0.01 inch (0.25 millimeter) deep. Another effect to get distance is to give the ball a backspin. With a backspin there is less air pressure on the top of the ball, so the ball stays aloft longer (much like an airplane).

## Why does a **curve ball** curve?

For many years it was debated whether curve balls actually curved or if the apparent change in course was merely an optical illusion. In 1959, Lyman Briggs demonstrated that a ball can curve up to 17.5 inches (44.45 centimeters) over the 60 feet 6 inches (18.4 meters) the ball travels between pitcher and batter. A rapidly spinning baseball experiences two lift forces that cause it to curve in flight. One is the Magnus force named after H. G. Magnus (1802–1870), the German physicist who discovered it, and the other is the wake deflection force. The Magnus force causes the curve ball to move sideways because the pressure forces on the ball's sides do not balance each other. The stitches on a baseball cause the pressure on one side of the ball to be less than on its opposite side. This forces the ball to move faster on one side than the other and forces the ball to "curve." The wake deflection force also causes the ball to curve to one side.

### Why does a boomerang return to its thrower?

Two well-known scientific principles dictate the characteristic flight of a boomerang: (1) the force of lift on a curved surface caused by air flowing over it; and (2) the unwillingness of a spinning gyroscope to move from its position.

When a person throws a boomerang properly, he or she causes it to spin vertically. As a result, the boomerang will generate lift, but it will be to one side rather than upwards. As the boomerang spins vertically and moves forward, air flows faster over the top arm at a particular moment than over the bottom arm. Accordingly, the top arm produces more lift than the bottom arm and the boomerang tries to twist itself, but because it is spinning fast it acts like a gyroscope and turns to the side in an arc. If the boomerang stays in the air long enough, it will turn a full circle and return to the thrower. Every boomerang has a built-in orbit diameter, which is not affected by a person throwing the boomerang harder or spinning it faster.

It occurs because the air flowing around the ball in the direction of its rotation remains attached to the ball longer and the ball's wake is deflected.

## What is **Maxwell's demon**?

An imaginary creature who, by opening and shutting a tiny door between two volumes of gases, could, in principle, concentrate slower molecules in one (making it colder) and faster molecules in the other (making it hotter), thus breaking the second law of thermodynamics. Essentially this law states that heat does not naturally flow from a colder body to a hotter body; work must be expended to make it do so. This hypothesis was formulated in 1871 by James C. Maxwell (1831–1879), who is considered to be the greatest theoretical physicist of the 19th century. The demon would bring about an effective flow of molecular kinetic energy. This excess energy would be useful to perform work and the system would be a perpetual motion machine. About 1950, the French physicist Léon Brillouin disproved Maxwell's hypothesis by demonstrating that the decrease in entropy resulting from the demon's actions would be exceeded by the increase in entropy in choosing between the fast and slow molecules.

## Who is the founder of the science of **magnetism**?

The English scientist William Gilbert (1544–1603) regarded the Earth as a giant magnet and investigated its magnetic field terms of dip and variation. He explored many other magnetic and electrostatic phenomena. The Gilbert (symbol Gb), a unit of magnetism, is named for him.

John H. Van Vleck (1899–1980), an American physicist, made significant contributions to modern magnetic theory. He explained the magnetic, electrical, and optical properties of many elements and compounds with the ligand field theory, demonstrated the effect of temperature on paramagnetic materials (called Van Vleck paramagnetism), and developed a theory on the magnetic properties of atoms and their components.

William Gilbert first explained the connection between magnetism and electricity.

## When was **spontaneous combustion** first recognized?

Spontaneous combustion is the ignition of materials stored in bulk. This is due to internal heat build-up caused by oxidation (generally a reaction in which electrons are lost, specifically when oxygen is combined with a substance, or when hydrogen is removed from a compound). Because this oxidation heat cannot be dissipated into the surrounding air, the temperature of the material rises until the material reaches its ignition point and bursts into flame.

A Chinese text written before 290 C.E. recognized this phenomenon in a description of the ignition of stored oiled cloth. The first Western acknowledgment of spontaneous combustion was by J. P. F. Duhamel in 1757, when he discussed the gigantic conflagration of a stack of oil-soaked canvas sails drying in the July sun. Before spontaneous combustion was recognized, such events were usually blamed on arsonists.

## What is **phlogiston**?

Phlogiston was a name used in the 18th century to identify a supposed substance given off during the process of combustion. The phlogiston theory was developed in the early 1700s by the German chemist and physicist Georg Ernst Stahl (1660–1734).

In essence, Stahl held that combustible material such as coal or wood was rich in a material substance called "phlogiston." What remained after combustion was without phlogiston and could no longer burn. The rusting of metals also involved a transfer of phlogiston. This accepted theory explained a great deal previously unknown to chemists. For instance, metal smelting was consistent with the phlogiston theory, as was the fact that charcoal lost weight when burned. Thus the loss of phlogiston either decreased or increased weight.

The French chemist Antoine Laurent Lavoisier (1743–1794) demonstrated that the gain of weight when a metal turned to a calx was just equal to the loss of weight of the air in the vessel. Lavoisier also showed that part of the air (oxygen) was indispensable to combustion, and that no material would burn in the absence of oxygen. The transition from Stahl's phlogiston theory to Lavoisier's oxygen theory marks the birth of modern chemistry at the end of the 18th century.

## What is the **kindling point of paper**?

Paper ignites at 450°F (230°C).

## What is an **adiabatic process**?

It is any thermodynamic process in which no heat transfer takes place between a system and its surrounding environment.

## Does **water running down a drain** rotate in a different direction in the Northern versus the Southern Hemisphere?

If water runs out from a perfectly symmetrical bathtub, basin, or toilet bowl in the Northern Hemisphere, it would swirl counterclockwise; in the Southern Hemisphere, the water would run out clockwise. This is due to the Coriolis effect (the Earth's rotation influencing any moving body of air or water). However, some scientists think that the effect does not work on small bodies of water. Exactly on the equator, the water would run straight down.

## Who invented the **cyclotron**?

The cyclotron was invented by Ernest Lawrence (1901–1958) at the University of California, Berkeley, in 1934 to study the nuclear structure of the atom. The cyclotron produced high energy particles that were accelerated outwards in a spiral rather than through an extremely long, linear accelerator.

## What is a **Leyden jar**?

A Leyden jar, the earliest form of capacitor, is a device for storing an electrical charge. First described in 1745 by E. Georg van Kleist (c. 1700–1748), it was also used by Pieter van Musschenbroek (1692–1761), a professor of physics at the University of Leyden. The device came to be known as a Leyden jar and was the first device that could

store large amounts of electric charge. The jars contained an inner wire electrode in contact with water, mercury, or wire. The outer electrode was a human hand holding the jar. An improved version coated the jar inside and outside with separate metal foils with the inner foil connected to a conducting rod and terminated in a conducting sphere. This eliminated the need for the liquid electrolyte. In use, the jar was normally charged from an electrostatic generator. The Leyden jar is still used for classroom demonstrations of static electricity.

# LIGHT, SOUND, AND OTHER WAVES

## What is the **speed of light**?

The figure is 186,282 miles (299,792 kilometers) per second.

## What are the **primary colors** in light?

Color is determined by the wavelength of visible light (the distance between one crest of the light wave and the next). Those colors that blend to form "white light" are, from shortest wave length to longest: red, orange, yellow, green, blue, indigo, and violet. All these monochromatic colors, except indigo, occupy large areas of the spectrum (the entire range of wavelengths produced when a beam of electromagnetic radiation is broken up). These colors can be seen when a light beam is refracted through a prism. Some consider the primary colors to be six monochromatic colors that occupy large areas of the spectrum: red, orange, yellow, green, blue, and violet. Many physicists recognize three primary colors: red, yellow, and blue; or red, green, and blue. All other colors can be made from these by adding two primary colors in various proportions. Within the spectrum, scientists have discovered 55 distinct hues. Infra-red and ultraviolet rays at each end of the spectrum are invisible to the human eye.

## How do **polarized sunglasses** reduce glare?

Sunlight reflected from the horizontal surface of water, glass, and snow is partially polarized, with the direction of polarization chiefly in the horizontal plane. Such reflected light may be so intense as to cause glare. The Polaroid material in sunglasses will block light that is polarized in a direction perpendicular to the transmission axes; Polaroid sunglasses are made with the transmission axes of the lenses oriented vertically.

## Why does the color of clothing appear different in **sunlight** than it does in a store under **fluorescent light**?

White light is a blend of all the colors, and each color has a different wavelength. Although sunlight and fluorescent light both appear as "white light," they each contain slightly different mixtures of these varying wavelengths. When sunlight and fluorescent light (white light) are absorbed by a piece of clothing, only some of the wavelengths (composing white light) reflect from the clothing. When the retina of the eye perceives the "color" of the clothing, it is really perceiving these reflected wavelengths. The mixture of wavelengths determines the color perceived. This is why an article of clothing sometimes appears to be a different color in the store than it does on the street.

## What were **Anders Ångström's** contributions to the development of **spectroscopy**?

Swedish physicist and astronomer Anders Jonas Ångström (1814–1874) was one of the founders of spectroscopy. His early work provided the foundation for spectrum analysis (analysis of the ranges of electromagnetic radiation emitted or absorbed). He investigated the sun spectra as well as that of the Aurora Borealis. In 1868, he established measurements for wavelengths of greater than 100 Frauenhofer. In 1907, the angstrom (Å, equal to $10^{-10}$m), a unit of wavelength measurement, was officially adopted.

## Why was the **Michelson-Morley** experiment important?

This experiment on light waves, first carried out in 1881 by physicists Albert A. Michelson (1852–1931) and E. W. Morley (1838–1923) in the United States, is one of the most historically significant experiments in physics and led to the development of Einstein's theory of relativity. The original experiment, using the Michelson interferometer, attempted to detect the velocity of the Earth with respect to the hypothetical "luminiferous ether," a medium in space proposed to carry light waves. The procedure measured the speed of light in the direction of the Earth and the speed of light at right angles to the Earth's motion. No difference was found. This result discredited the ether theory and ultimately led to the proposal by Albert Einstein (1879–1955) that the speed of light is a universal constant.

## Who was the first person to break the **sound barrier**?

On October 14, 1947, Charles E. (Chuck) Yeager (b. 1923) was the first pilot to break the sound barrier. He flew a Bell X-1, attaining a speed of 750 miles (1,207 kilometers) per hour (Mach 1.06) and an altitude of 70,140 feet (21,379 meters) over the town of

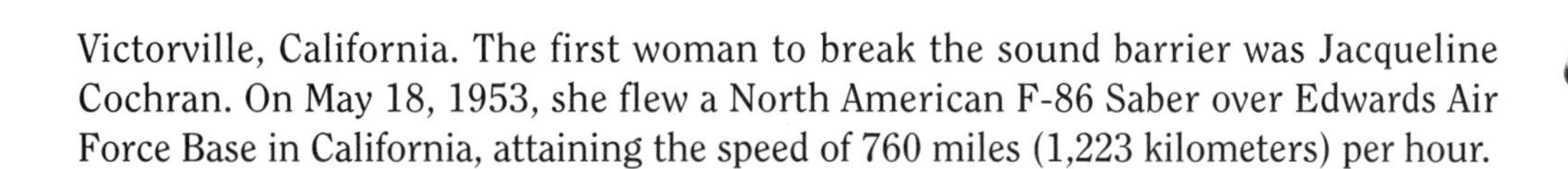
Victorville, California. The first woman to break the sound barrier was Jacqueline Cochran. On May 18, 1953, she flew a North American F-86 Saber over Edwards Air Force Base in California, attaining the speed of 760 miles (1,223 kilometers) per hour.



## Why does a **double sonic boom** occur when the space shuttle enters the atmosphere?

As long as an airborne object, such as a plane, is moving below the speed of sound (called Mach 1), the disturbed air remains well in front of the craft. But as the craft passes Mach 1 and is flying at supersonic speeds, a sharp air pressure rise occurs in front of the craft. In a sense, the air molecules are crowded together and collectively impact. What is heard is a claplike thunder called a sonic boom or a supersonic bang. There are many shocks coming from a supersonic aircraft, but these shocks usually combine to form two main shocks, one coming from the nose and one from the aft end of the aircraft. Each of the shocks moves at a different velocity. If the time difference between the two shock waves is greater than 0.10 seconds apart, two sonic booms will be heard. This usually occurs when an aircraft ascends or descends quickly. If the aircraft moves more slowly, the two booms will sound like only one boom to the observer.

## What causes the sounds that are heard in a **seashell**?

When a seashell is held to an ear, the sounds heard are ambient, soft sounds that have been resonated and thereby amplified by the seashell's cavity. The extreme sensitivity of the human ear to sound is illustrated by the seashell resonance effect.

## What is the **Doppler effect**?

The Austrian physicist Christian Doppler (1803–1853) in 1842 explained the phenomenon of the apparent change in wavelength of radiation—such as sound or light—emitted either by a moving body (source) or by the moving receiver. The frequency of the wavelengths increases and the wavelength becomes shorter as the moving source approaches, producing high-pitched sounds and bluish light (called blue shift). Likewise, as the source recedes from the receiver the frequency of the wavelengths decreases, the sound is pitched lower, and light appears reddish (called red shift). This Doppler effect is commonly demonstrated by the whistle of an approaching train or the roar of a jet aircraft.

There are three differences between acoustical (sound) and optical (light) Doppler effects: The optical frequency change is not dependent on which is moving—the source or observer—nor is it affected by the medium through which the waves are moving, but acoustical frequency is affected by such conditions. Optical frequency changes are affected if the source or observer moves at right angles to the

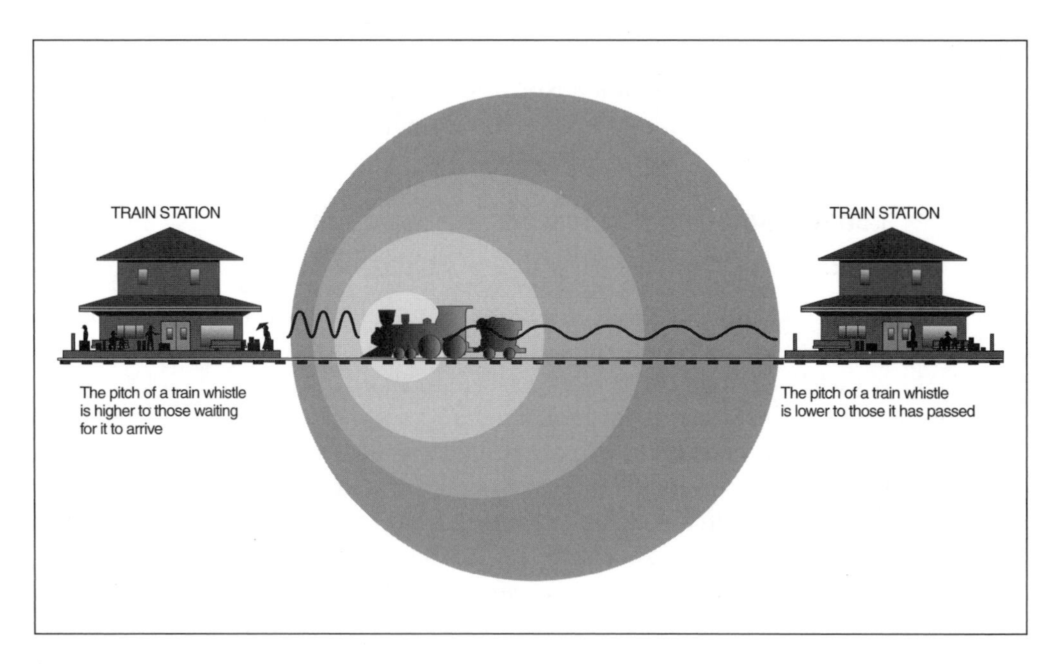


**The Doppler effect.**

line connecting the source and observer. Observed acoustical changes are not affected in such a situation. Applications of the Doppler phenomenon include the Doppler radar and the measurement by astronomers of the motion and direction of celestial bodies.

## What is a **decibel**?

A decibel is a measure of the relative loudness or intensity of sound. A 20 decibel sound is 10 times louder than a 10 decibel sound; 30 decibels is 100 times louder, etc. One decibel is the smallest difference between sounds detectable by the human ear.

| Decibel Level | Equivalent |
|---|---|
| 10 | Light whisper |
| 20 | Quiet conversation |
| 30 | Normal conversation |
| 40 | Light traffic |
| 50 | Loud conversation |
| 60 | Noisy office |
| 70 | Normal traffic, quiet train |
| 80 | Rock music, subway |
| 90 | Heavy traffic, thunder |
| 100 | Jet plane at takeoff |

## What is the sound frequency of the **musical scale**?

**EQUAL TEMPERED SCALE**

| Note | Frequency | Note | Frequency |
|---|---|---|---|
| C♭ | 261.63 | G | 392.00 |
| C# | 277.18 | G# | 415.31 |
| D | 293.67 | A | 440.00 |
| D# | 311.13 | A# | 466.16 |
| E | 329.63 | B | 493.88 |
| F | 349.23 | Cn | 523.25 |
| F# | 369.99 | | |

*Notes:* ♭ indicates flat; # indicates sharp; n indicates return to natural.

The lowest frequency distinguishable as a note is about 20 hertz. The highest audible frequency is about 20,000 hertz. A hertz (symbol Hz) is a unit of frequency that measures the number of the wave cycles per second frequency of a periodic phenomenon whose periodic time is one second (cycles per second).

## What is the **speed of sound**?

The speed of sound is not a constant; it varies depending on the medium in which it travels. The measurement of sound velocity in the medium of air must take into account many factors, including air temperature, pressure, and purity. At sea level and 32°F (0°C), scientists do not agree on a standard figure; estimates range from 740 to 741.5 miles (1,191.6 to 1,193.22 kilometers) per hour. As air temperature rises, sound velocity increases. Sound travels faster in water than in air and even faster in iron and steel. Sounds traveling a mile in air for five seconds will travel the same distance in one second underwater and one-third of a second in steel.

## What are the characteristics of **alpha, beta**, and **gamma radiation**?

Radiation is a term that describes all the ways energy is emitted by the atom as X-rays, gamma rays, neutrons, or as charged particles. Most atoms, being stable, are nonradioactive; others are unstable and give off either particles or gamma radiation. Substances bombarded by radioactive particles can become radioactive and yield alpha particles, beta particles, and gamma rays.

*Alpha particles*, first identified by Antoine Henri Becquerel (1852–1908), have a positive electrical charge and consist of two protons and two neutrons. Because of their great mass, alpha particles can travel only a short distance, around two inches (five centimeters) in air, and can be stopped by a sheet of paper.

*Beta particles*, identified by Ernest Rutherford (1871–1937), are extremely high-speed electrons that move at the speed of light. They can travel far in air and can pass through solid matter several millimeters thick.

*Gamma rays*, identified by Marie (1867–1934) and Pierre Curie (1859–1906), are similar to X-rays, but they usually have a shorter wave length. These rays, which are bursts of photons, or very short-wave electromagnetic radiation, travel at the speed of light. They are much more penetrating than either the alpha or beta particles and can go through seven inches (18 centimeters) of lead.

# MATTER

## Who proposed the **theory of the atom**?

The modern theory of atomic structure was first proposed by the Japanese physicist Hantaro Nagaoka (1865–1950) in 1904. In his model, electrons rotated in rings around a small central nucleus. In 1911, Ernest Rutherford (1871–1937) discovered further evidence to prove that the nucleus of the atom is extremely small and dense and is surrounded by a much larger and less dense cloud of electrons. In 1913, the Danish physicist Niels Bohr (1885–1962) suggested that electrons orbit the nucleus in concentric quantum shells that correspond to the electron's energy levels.

## What is the **fourth state of matter**?

Plasma, a mixture of free electrons and ions or atomic nuclei, is sometimes referred to as a "fourth state of matter." Plasmas occur in thermonuclear reactions as in the sun, in fluorescent lights, and in stars. When the temperature of gas is raised high enough, the collision of atoms becomes so violent that electrons are knocked loose from their nuclei. The result of a gas having loose, negatively charged electrons and heavier, positively charged nuclei is called a plasma.

All matter is made up of atoms. Animals and plants are organic matter; minerals and water are inorganic matter. Whether matter appears as a solid, liquid, or gas depends on how the molecules are held together in their chemical bonds. Solids have a rigid structure in the atoms of the molecules; in liquids the molecules are close together but not packed; in a gas, the molecules are widely spaced and move around, occasionally colliding but usually not interacting. These states—solid, liquid, and gas—are the first three states of matter.

## What is the difference between **nuclear fission** and **nuclear fusion**?

Nuclear fission is the splitting of an atomic nucleus into at least two fragments. Nuclear fusion is a nuclear reaction in which the nuclei of atoms of low atomic number, such as hydrogen and helium, fuse to form a heavier nucleus. In both nuclear fission and nuclear fusion substantial amounts of energy are produced.

## Who is generally regarded as the discoverer of the **electron**, the **proton**, and the **neutron**?

The British physicist Sir Joseph John Thomson (1856–1940) in 1897 researched electrical conduction in gases, which led to the important discovery that cathode rays consist of negatively charged particles called electrons. The discovery of the electron inaugurated the electrical theory of the atom, and this, along with other work, entitled Thomson to be regarded as the founder of modern atomic physics.

Ernest Rutherford (1871–1937) discovered the proton in 1919. He also predicted the existence of the neutron, later discovered by his colleague, James Chadwick (1891–1974). Chadwick was awarded the 1935 Nobel Prize for physics for this discovery.

## How did a total **solar eclipse** confirm Einstein's **theory of general relativity**?

When formulating his theory of general relativity, Albert Einstein (1879–1955) proposed that the curvature of space near a massive object like the sun would bend light that passed close by. For example, a star seen near the edge of the sun during an eclipse would appear to have shifted by 1.75 arc seconds from its usual place. The British astronomer Arthur Eddington (1882–1944) confirmed Einstein's hypothesis during an eclipse on May 29, 1919. The subsequent attention given Eddington's findings helped establish Einstein's reputation as one of science's greatest figures.

Murray Gell-Mann theorized the existence of fundamental particles he called "quarks."

## How did the **quark** get its name?

This theoretical particle, considered to be the fundamental unit of matter, was named by Murray Gell-Mann (b. 1929) an American theoretical physicist and Nobel Prize winner. Its name was initially a playful tag that Gell-Mann invented, sounding something like "kwork." Later, Gell-Mann came across the line "Three quarks for Master Marks" in James Joyce's *Finnegan's Wake*, and the tag became known as a quark. There are six kinds or "flavors" (up, down, strange, charm, bottom, and top) of quarks, and each "flavor" has three varieties or "colors" (red, blue, and green). All eighteen types have different electric charges (a basic characteristic of all elementary particles). Three quarks form a proton (having one unit of positive electric charge) or a neutron (zero charge), and two quarks (a quark and an antiquark) form a meson. Like all known particles, a quark has its anti-matter opposite, known as an antiquark (having the same mass but opposite charge).

## What was **Richard Feynman's** contribution to physics?

Richard Feynman (1918–1988) developed a theory of quantum electrodynamics that described the interaction of electrons, positrons, and photons, providing physicists a new way to work with electrons. He reconstructed quantum mechanics and electrodynamics in his own terms, formulating a matrix of measurable quantities visually represented by a series of graphs knows as the Feynman diagrams. Feynman was awarded the Nobel Prize in Physics in 1965.

## What are the **subatomic particles**?

Subatomic particles are particles that are smaller than atoms. Historically, subatomic particles were considered to be electrons, protons, and neutrons. However, the definition of subatomic particles has now been expanded to include elementary particles, which are the particles so small that they do not appear to be made of more minute units. The physical study of such particles became possible only during the twentieth century with the development of increasingly sophisticated apparatus. Many new particles have been discovered in the last half of the twentieth century.

A number of proposals have been made to organize the particles by their spin, their mass, or their common properties. One system is now commonly known as the Standard Model. This system recognizes two basic types of fundamental particles: quarks and leptons. Other force-carrying particles are called bosons. Photons, gluons, and weakons are bosons. Leptons include electrons, muons, taus, and three kinds of neutrinos. Quarks never occur alone in nature. They always combine to form particles called hadrons. According to the Standard Model, all other subatomic particles consist of some combination of quarks and their antiparticles. A proton consists of three quarks.

## What are **colligative properties**?

Colligative properties are properties of solutions that depend on the number of particles present in the solution and not on characteristics of the particles themselves. Colligative properties include depression of freezing point and elevation of boiling point. For living systems, perhaps the most important colligative property is osmotic pressure.

### What substance, other than water, is **less dense as a solid** than as a liquid?

Only bismuth and water share this characteristic. Density (the mass per unit volume or mass/volume) refers to how compact or crowded a substance is. For instance, the density of water is 1 $g/cm^3$ (gram per cubic centimeter) or 1 kg/l (kilogram per liter); the density of a rock is 3.3 $g/cm^3$; pure iron is 7.9 $g/cm^3$; and the Earth (as a whole) is 5.5 $g/cm^3$ (average). Water as a solid (i.e., ice) floats, which is a good thing; otherwise, ice would sink to the bottom of every lake or stream.

### Why is **liquid water** more dense than ice?

Pure liquid water is most dense at 39.2°F (3.98°C) and decreases in density as it freezes. The water molecules in ice are held in a relatively rigid geometric pattern by their hydrogen bonds, producing an open, porous structure. Liquid water has fewer bonds; therefore, more molecules can occupy the same space, making liquid water more dense than ice.

### What does **half-life** mean?

Half-life is the time it takes for the number of radioactive nuclei originally present in a sample to decrease to one-half of their original number. Thus, if a sample has a half-life of one year, its radioactivity will be reduced to half its original amount at the end of a year and to one quarter at the end of two years. The half-life of a particular radionuclide is always the same, independent of temperature, chemical combination, or any other condition. Natural radiation was discovered in 1896 by the French physicist Antoine Henri Becquerel (1852–1908). His discovery initiated the science of nuclear physics.

### Who made the **first organic compound to be synthesized** from inorganic ingredients?

In 1828, Friedrich Wöhler (1800–1882) synthesized urea from ammonia and cyanic acid. This synthesis dealt a deathblow to the vital-force theory, which held that definite and fundamental differences existed between organic and inorganic compounds. The Swedish chemist Jöns Jakob Berzelius (1779–1848) proposed that the two classes of compounds were produced from their elements by entirely different laws. Organic compounds were produced under the influence of a vital force and so were incapable of being prepared artificially. This distinction ended with Wöhler's synthesis.

### Who is known as the **founder** of **crystallography**?

The French priest and mineralogist René-Just Haüy (1743–1822) is called the father of crystallography. In 1781 Haüy had a fortunate accident when he dropped a piece of

## What is a chemical garden and how is one made?

Mix four tablespoons of bluing, four tablespoons of salt, and one tablespoon household ammonia. Pour this mixture over pieces of coal or brick in a suitable dish or bowl. Put several drops of red or green ink or mercurochrome on various parts of the coal and leave undisturbed for several days.

A chemical garden—a dishful of crystals that grow like plants and look like coral—will begin to appear. How soon the crystals will begin to appear depends on the temperatures and humidity in the room. Before long, crystals will be growing all over the briquettes, on the side of the dish, and down onto the plate. The crystals will be pure white with a snow-like texture.

calcite and it broke into small fragments. He noticed that the fragments broke along straight planes that met at constant angles. Haüy hypothesized that each crystal was built from successive additions of what is now called a unit cell to form a simple geometric shape with constant angles. An identity or difference in crystalline form implied an identity or difference in chemical composition. This was the beginning of the science of crystallography.

By the early 1800s many physicists were experimenting with crystals; in particular, they were fascinated by their ability to bend light and separate it into its component colors. An important member of the emerging field of optical mineralogy was the British scientist David Brewster (1871–1868), who succeeded in classifying most known crystals according to their optical properties.

The work of French chemist Louis Pasteur (1822–1895) during the mid 1800s became the foundation for crystal polarimetry—a method by which light is polarized, or aligned to a single plane. Pierre Curie (1859–1906) and his brother Jacques (1855–1941) discovered another phenomenon displayed by certain crystals called piezoelectricity. It is the creation of an electrical potential by squeezing certain crystals.

Perhaps the most important application of crystals is in the science of X-ray crystallography. Experiments in this field were first conducted by the German physicist Max von Laue (1879–1960). This work was perfected by William Henry Bragg (1862–1942) and William Lawrence Bragg (1890–1971), who were awarded the Nobel Prize in physics for their work. The synthesis of penicillin and insulin were made possible by the use of X-ray crystallography.

# CHEMICAL ELEMENTS, ETC.

*See also: Metals and Other Materials*



## Who are some of the **founders** of **modern chemistry**?

Several contenders share this honor:

Swedish chemist Jöns Jakob Berzelius (1779–1848) devised chemical symbols, determined atomic weights, contributed to the atomic theory, and discovered several new elements. Between 1810 and 1816, he described the preparation, purification, and analysis of 2,000 chemical compounds. Then he determined atomic weights for 40 elements. He simplified chemical symbols, introducing a notation—letters with numbers—that replaced the pictorial symbols his predecessors used, and that is still used today. He discovered cerium (in 1803, with Wilhelm Hisinger), selenium (1818), silicon (1824), and thorium (1829).

Robert Boyle (1627–1691), a British natural philosopher, is considered one of the founders of modern chemistry. Best known for his discovery of Boyle's Law (volume of a gas is inversely proportional to its pressure at constant temperature), he was a pioneer in the use of experiments and the scientific method. A founder of the Royal Society, he worked to remove the mystique of alchemy from chemistry to make it a pure science.

The French chemist Antoine-Laurent Lavoisier (1743–1794) is regarded as another founder of modern chemistry. His wide-ranging contributions include the discrediting of the phlogiston theory of combustion, which had been for so long a stumbling block to a true understanding of chemistry. He established modern terminology for chemical substances and did the first experiments in quantitative organic analysis. He is sometimes credited with having discovered or established the law of conservation of mass in chemical reactions.

John Dalton (1766–1844), an English chemist, proposed an atomic theory of matter that became a basic theory of modern chemistry. His theory, first proposed in 1803, states that each chemical element is composed of its own kind of atoms, all with the same relative weight.

## Who developed the **periodic table**?

Dmitry Ivanovich Mendeleyev (1834–1907) was a Russian chemist whose name will always be linked with the development of the periodic table. He was the first chemist really to understand that all elements are related members of a single ordered system. He changed what had been a highly fragmented and speculative branch of chemistry into a true, logical science. His nomination for the 1906 Nobel Prize for chemistry failed by one vote, but his name became recorded in perpetuity 50 years later when element 101 was called mendelevium.

Dmitry Ivanovich Mendeleyev, the Russian chemist renowned for developing the periodic table of elements.

According to Mendeleyev, the properties of the elements, as well as those of their compounds, are periodic functions of their atomic weights (in the 1920s, it was discovered that atomic number was the key rather than weight). Mendeleyev compiled the first true periodic table listing all the 63 (then-known) elements. In order to make the table work, Mendeleyev had to leave gaps, and he predicted that further elements would eventually be discovered to fill them. Three were discovered in Mendeleyev's lifetime: gallium, scandium, and germanium.

There are 95 naturally occurring elements; of the remaining elements (elements 96 to 109), 10 are undisputed. There are aproximately 17 million chemical compounds registered with *Chemical Abstracts* that have been produced from these elements.

## What was the **first element** to be discovered?

Phosphorus was first discovered by German chemist Hennig Brand in 1669 when he extracted a waxy white substance from urine that glowed in the dark. But Brand did not publish his findings. In 1680, phosphorus was rediscovered by the English chemist Robert Boyle.

## What is the **sweetest chemical compound**?

The sweetest chemical compound is sucronic acid.

| Sweetener | Relative sweetness (sucrose=1) |
|---|---|
| sucronic acid | 200,000 |
| saccharin | 300 |
| aspartame | 180 |
| cyclamate | 30 |
| sugar (sucrose) | 1 |

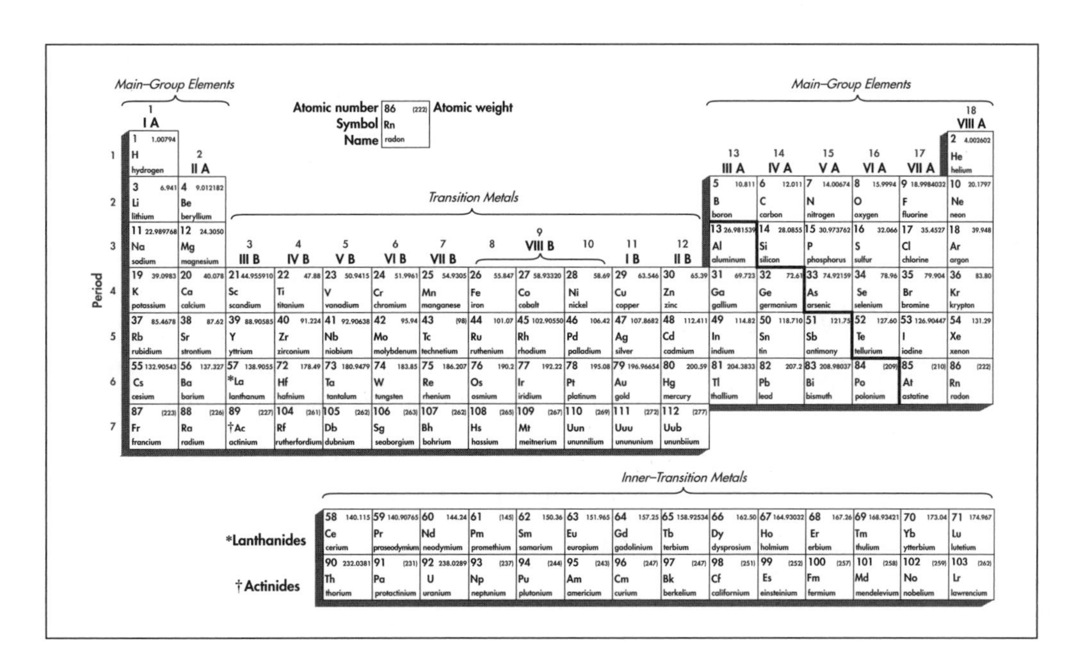


The periodic table.

## What are the **alkali metals**?

These are the elements at the left of the periodic table: lithium (Li, element 3), potassium (K, element 19), rubidium (Rb, element 37), cesium (Cs, element 55), francium (Fr, element 87), and sodium (Na, element 11). The alkali metals are sometimes called the sodium family of elements, or Group I elements. Because of their great chemical reactivity (they easily form positive ions), none exist in nature in the elemental state.

## What are the **alkaline Earth metals**?

These are beryllium (Be, element 4), magnesium (Mg, element 12), calcium (Ca, element 20), strontium (Sr, element 38), barium (Ba, element 56), and radium (Ra, element 88). The alkaline Earth metals are also called Group II elements. Like the alkali metals, they are never found as free elements in nature and are moderately reactive metals. Harder and less volatile than the alkali metals, these elements all burn in air.

## What are the **transition elements**?

The transition elements are the 10 subgroups of elements between Group II and Group XIII, starting with period 4. They include gold (Au, element 79), silver (Ag, element 47), platinum (Pt, element 78), iron (Fe, element 26), copper (Cu, element 29), and other metals. All transition elements are metals. Compared to alkali and alkaline Earth metals, they are usually harder and more brittle and have higher melting points. Transition metals are also good conductors of heat and electricity. They have variable valences, and compounds of transition elements are often colored. Transition elements are so named because they comprise a gradual shift from the strongly electropositive elements of Groups I and II to the electronegative elements of Groups VI and VII.

## What are the **transuranic chemical elements** and the names for elements 102–109?

Transuranic elements are those elements in the periodic system with atomic numbers greater than 92. Many of these elements are ephemeral, do not exist naturally outside the laboratory, and are not stable.

Elements 93–109

| Element Number | Name | Symbol |
|---|---|---|
| 93 | Neptunium | Np |
| 94 | Plutonium | Pu |
| 95 | Americum | Am |
| 96 | Curium | Cm |
| 97 | Berkelium | Bk |
| 98 | Californium | Cf |
| 99 | Einsteinium | Es |
| 100 | Fermium | Fm |
| 101 | Mendelevium | Md |
| 102 | Nobelium | No |
| 103 | Lawrencium | Lr |
| 104 | Rutherfordium | Rf |
| 105 | Dubnium | Db |
| 106 | Seaborgium | Sg |
| 107 | Bohrium | Bh |
| 108 | Hassium | Hs |
| 109 | Meitnerium | Mt |

The names for elements 110–116 are under review by the International Union of Pure and Applied Chemistry.

## Which are the only two **elements** in the periodic table **named after women**?

Curium, atomic number 96, was named after the pioneers of radioactive research Marie (1867–1934) and Pierre Curie (1859–1906). Meitnerium, atomic number 109, was named after Lise Meitner (1878–1968), one of the founders of nuclear fission.

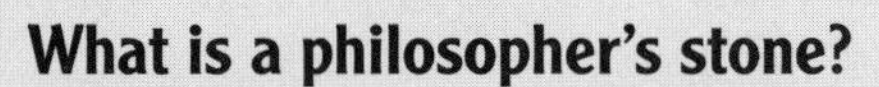

### What is a philosopher's stone?

A philosopher's stone was the name of a substance believed by medieval alchemists to have the power to change baser metals into gold or silver. It had, according to some, the power of prolonging life and of curing all injuries and diseases. The pursuit of it by alchemists led to the discovery of several chemical substances; however, the magical philosopher's stone has since proved fictitious.

## Which elements are the **"noble metals"**?

The noble metals are gold (Au, element 79), silver (Ag, element 47), mercury (Hg, element 80), and the platinum group, which includes platinum (Pt, element 78), palladium (Pd, element 46), iridium (Ir, element 77), rhodium (Rh, element 45), ruthenium (Ru, element 44), and osmium (Os, element 76). The term refers to those metals highly resistant to chemical reaction or oxidation (resistant to corrosion) and is contrasted to "base" metals, which are not so resistant. The term has its origins in ancient alchemy whose goals of transformation and perfection were pursued through the different properties of metals and chemicals. The term is not synonymous with "precious metals," although a metal, like platinum, may be both.

The platinum group metals have a variety of uses. In the United States more than 95 percent of all platinum group metals are used for industrial purposes. While platinum is a coveted material for jewelry making, it is also used in the catalytic converters of automobiles to control exhaust emissions, as are rhodium and palladium. Rhodium can also be alloyed with platinum and palladium for use in furnace windings, thermocouple elements, and in aircraft spark-plug electrodes. Osmium is used in the manufacture of pharmaceuticals and in alloys for instrument pivots and long-life phonograph needles.

## What distinguishes **gold** and **silver** as elements?

Besides their use as precious metals, gold and silver have properties that distinguish them from other chemical elements. Gold is the most ductile and malleable metal—the thinnest gold leaf is 0.0001mm thick. Silver is the most reflective of all metals; thus, it is used in mirrors.

## What is **Harkin's rule**?

Atoms having even atomic numbers are more abundant in the universe than are atoms having odd atomic numbers. Chemical properties of an element are determined by its atomic number, which is the number of protons in the atom's nucleus.

## What are some chemical elements whose **symbols** are **not derived from their English names**?

| Modern Name | Symbol | Older Name |
|---|---|---|
| antimony | Sb | stibium |
| copper | Cu | cuprum |
| gold | Au | aurum |
| iron | Fe | ferrum |
| lead | Pb | plumbum |
| mercury | Hg | hydrargyrum |
| potassium | K | kalium |
| silver | Ag | argentum |
| sodium | Na | natrium |
| tin | Sn | stannum |
| tungsten | W | wolfram |

## Which elements are **liquid at room temperature**?

Mercury ("liquid silver," Hg, element 80) and bromine (Br, element 35) are liquid at room temperature 68° to 70°F (20° to 21°C). Gallium (Ga, element 31) with a melting point of 85.6°F (29.8°C) and cesium (Cs, element 55) with a melting point of 83°F (28.4°C), are liquid at slightly above room temperature and pressure.

## Which chemical element is the **most abundant in the universe**?

Hydrogen (H, element 1) makes up about 75 percent of the mass of the universe. It is estimated that more than 90 percent of all atoms in the universe are hydrogen atoms. Most of the rest are helium (He, element 2) atoms.

## Which chemical elements are the **most abundant on Earth**?

Oxygen (O, element 8) is the most abundant element in the Earth's crust, waters, and atmosphere. It composes 49.5 percent of the total mass of these compounds. Silicon (Si, element 14) is the second most abundant element. Silicon dioxide and silicates make up about 87 percent of the materials in the Earth's crust.

## Why are the **rare gases** and **rare Earth elements** called "rare"?

Rare gases refers to the elements helium, neon, argon, krypton, and xenon. They are rare in that they are gases of very low density ("rarified") at ordinary temperatures and are found only scattered in minute quantities in the atmosphere and in some sub-

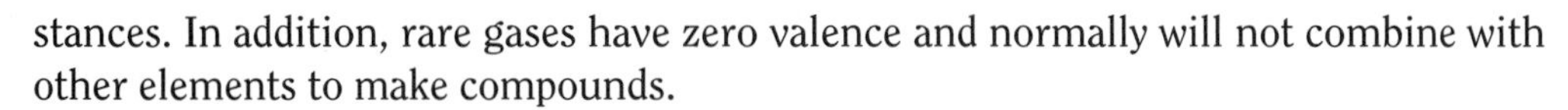

stances. In addition, rare gases have zero valence and normally will not combine with other elements to make compounds.

Rare Earth elements are elements numbered 58 through 71 in the periodic table plus yttrium (Y, element 39) and thorium (Th, element 90). They are called "rare Earths" because they are difficult to extract from monazite ore, where they occur. The term has nothing to do with scarcity or rarity in nature.

## Which elements are the **best** and **worst conductors** of electricity?

The element with the lowest electrical resistance (and thus the highest electrical conductivity) under standard conditions is silver, followed by copper, gold, and aluminum. The poorest conductors of electricity among the metals are manganese, gadolinium, and terbium.

## How do **lead-acid batteries** work?

Lead-acid batteries consist of positive and negative lead plates suspended in a diluted sulfuric acid solution called an electrolyte. Everything is contained in a chemically and electrically inert case. As the cell discharges, sulfur molecules from the electrolyte bond with the lead plates, releasing excess electrons. The flow of electrons is called electricity.

## Which elements have the **most isotopes**?

The elements with the most isotopes, with 36 each, are xenon (Xe) with nine stable isotopes (identified from 1920 to 1922) and 27 radioactive isotopes (identified from 1939 to 1981), and cesium (Cs) with one stable isotope (identified in 1921) and 35 radioactive isotopes (identified from 1935 to 1983).

The element with the fewest number of isotopes is hydrogen (H), with three isotopes, including two stable ones—protium (identified in 1920) and deuterium (identified in 1931)—and one radioactive isotope—tritium (first identified in 1934, but later considered a radioactive isotope in 1939).

## Are there more chemical compounds with **even numbers of carbon atoms** than with **odd**?

A team of chemists recently noted that the database of the Beilstein Information System, containing approximately seven million organic compounds, includes significantly more substances with an even number of carbon atoms than with an odd number. Statistical analyses of smaller sets of organic compounds, such as the *Cambridge Crystallographic Database* or the *CRC Handbook of Chemistry and Physics* led to the same results. A possible explanation for the observed asymmetry might be that

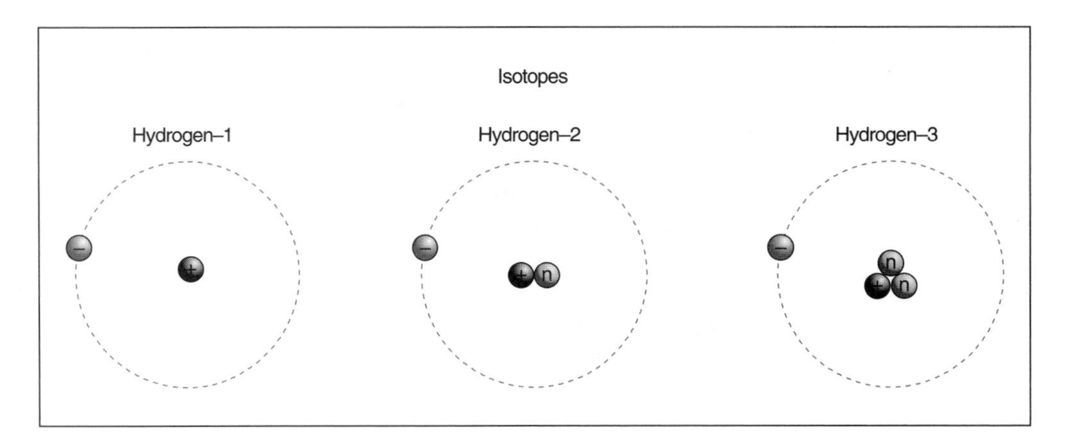


**Hydrogen isotopes.**

organic compounds are ultimately derived from biological sources, and nature frequently utilized acetate, a $C_2$ building block, in its syntheses of organic compounds. It may therefore be that the manufacturers' and synthetic chemists' preferential use of relatively economical starting materials derived from natural sources has left permanent traces in chemical publications and databases.

## Which elements have the **highest** and **lowest boiling points**?

Helium has the lowest boiling point of all the elements at –452.074°F (–268.93°C) followed by hydrogen –423.16°F (–252.87°C). The highest boiling point for an element is that of rhenium 10,104.8°F (5,596°C) followed by tungsten 10,031°F (5,555°C).

## What is the **density of air**?

The density of dry air is 1.29 grams per liter at 32°F (0°C) at average sea level and a barometric pressure of 29.92 inches of mercury (760 millimeters).

The weight of one cubic foot of dry air at one atmosphere of barometric pressure is:

| Temperature (Fahrenheit) | Weight per cubic foot (pounds) |
|---|---|
| 50° | 0.07788 |
| 60° | 0.07640 |
| 70° | 0.07495 |

## Which element has the **highest density**?

Either osmium or iridium is the element with the highest density; however, scientists have yet to gather enough conclusive data to choose between the two. When tradi-

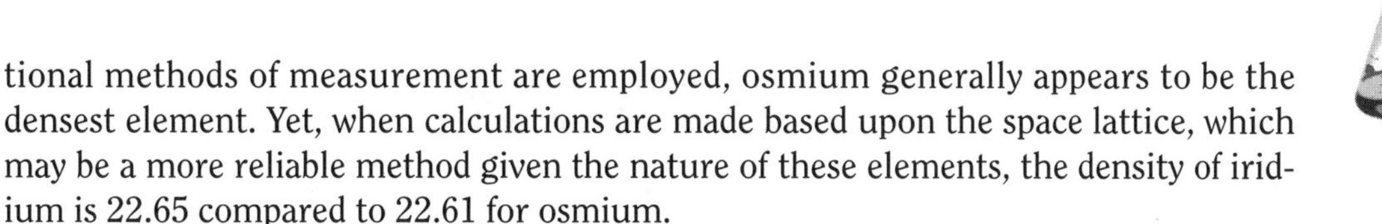

tional methods of measurement are employed, osmium generally appears to be the densest element. Yet, when calculations are made based upon the space lattice, which may be a more reliable method given the nature of these elements, the density of iridium is 22.65 compared to 22.61 for osmium.



## Which elements are the **hardest** and **softest**?

Carbon is both the hardest and softest element occurring in two different forms as graphite and diamond. A single crystal of diamond scores the absolute maximum value on the Knoop hardness scale of 90. Based on the somewhat less informative abrasive hardness scale of Mohs, diamond has a hardness of 10. Graphite is an extremely soft material with a Mohs hardness of only 0.5 and a Knoop hardness of 0.12.

## What are **isomers**?

Isomers are compounds with the same molecular formula but different structures due to the different arrangement of the atoms within the molecules. Structural isomers have atoms connected in different ways. Geometric isomers differ in their symmetry about a double bond. Optical isomers are mirror images of each other.

## What are the **gas laws**?

The gas laws are physical laws concerning the behavior of gases. They include Boyle's law, which states that the volume of a given mass of gas at a constant temperature is inversely proportional to its pressure; and Charles's law, which states that the volume of a given mass of gas at constant pressure is directly proportional to its absolute temperature. These two laws can be combined to give the General or Universal gas law, which may be expressed as:

$$(\text{pressure} \times \text{volume})/\text{temperature} = \text{constant}$$

Avogadro's law states that equal volumes of all gases contain the same number of particles if they all have the same pressure and temperature.

The laws are not obeyed exactly by any real gas, but many common gases obey them under certain conditions, particularly at high temperatures and low pressures.

## What is **heavy water**?

Heavy water, also called deuterium oxide ($D_20$), is composed of oxygen and two hydrogen atoms in the form of deuterium, which has about twice the mass of normal hydrogen. As a result, heavy water has a molecular weight of about 20, while ordinary water has a molecular weight of about 18. Approximately one part heavy water can be found in 6,500 parts of ordinary water, and it may be extracted by fractional distillation. It is used in thermonuclear weapons and nuclear reactors and as an isotopic tracer in studies of chemical and biochemical processes. Heavy water was discovered by Harold C. Urey in 1931.

## What is a **Lewis acid**?

Named after the American chemist Gilbert Newton Lewis (1875–1946), the Lewis theory defines an acid as a species that can accept an electron pair from another atom, and a base as a species that can donate an electron pair to complete the valence shell of another atom. Hydrogen ion (proton) is the simplest substance that will do this, but Lewis acids include many compounds—such as boron trifluoride ($BF_3$) and aluminum chloride ($AlCl_3$)—that can react with ammonia, for example, to form an addition compound or Lewis salt.

## Which chemical is **used in greater quantities** than any other?

Sodium chloride (NaCl), or salt, has over 14,000 uses, and is probably used in greater quantities and for more applications than any other chemical.

## Which chemicals are used today as **embalming fluids**?

During the 19th century embalming became a common practice in the United States, and generally salts of heavy metals such as arsenic, antimony, lead, mercury, and copper were used to preserve the corpse and inhibit bacterial growth. By the early 1900s, however, laws were passed prohibiting the use of metal salts in embalming. Formaldehyde soon became the compound of choice and continues to be the most common preservative in embalming fluids. Its continued popularity is due to low cost, availability in usable form, and simplicity of use. It also provides good cell preservation under a variety of pH conditions. Generally, a mortician, after draining the corpse of bodily fluids, injects it with a solution of formaldehyde in water. Various buffers are also present in the solution to counteract the formation of "formaldehyde pigments" on the body. Concerns over the possible carcinogenic effects of formaldehyde, as well as its tendency to turn a corpse's skin an ashen gray, have instigated numerous attempts at finding a replacement. Glutaraldehyde was first used in 1955, but in spite of some of its distinct advantages, formaldehyde is still the embalming fluid of choice.

## What was the first **national physics society** organized in the United States?

The first national physics society in the United States was the American Physical Society, organized on May 20, 1899, at Columbia University in New York City. The first president was Henry Augustus Rowland.

## What was the first **national chemical society** organized in the United States?

The first national chemical society in the United States was the American Chemical Society, organized in New York City on April 20, 1876. The first president was John William Draper (1811–1882).

## What are the **four major divisions** of chemistry?

Chemistry has traditionally been divided into organic, inorganic, analytical, and physical chemistry. Organic chemistry is the study of compounds that contain carbon. More than 90 percent of all known chemicals are organic. Inorganic chemistry is the study of compounds of all elements except carbon. Analytical chemists determine the structure and composition of compounds and mixtures. They also develop and operate instruments and techniques for carrying out the analyses. Physical chemists use the principles of physics to understand chemical phenomena.

## What were some of the leading contributions of **Albert Einstein**?

Albert Einstein (1879–1955) was the principal founder of modern theoretical physics; his theory of relativity (speed of light is a constant and not relative to the observer or source of light), and the relationship of mass and energy ($e = mc^2$), fundamentally changed human understanding of the physical world.

Albert Einstein revolutionized twentieth-century human understanding of the physical world.

During a single year in 1905, he produced three landmark papers. These papers dealt with the nature of particle movement known as Brownian motion, the quantum nature of electromagnetic radiation as demonstrated by the photoelectric effect, and the special theory of relativity. Although Einstein is probably best known for the last of these works, it was for his quantum explanation of the photoelectric effect that he was awarded the 1921 Nobel Prize in physics.

His stature as a scientist, together with his strong humanitarian stance on major political and social issues, made him one of the outstanding men of the twentieth century.

## What is **antimatter**?

Antimatter is the exact opposite of normal matter. Antimatter was predicted in a series of equations derived by Paul Dirac (1902–1984). He was attempting to combine the Theory of Relativity with equations governing the behavior of electrons. In order to make his equations work, he had to predict the existence of a particle that would be similar to the electron, but opposite in charge. This particle, discovered in 1932, was the antimatter equivalent of the electron and called the positron (electrons with a positive charge). Other antimatter particles would not be discovered until 1955 when particle accelerators were finally able to confirm the existence of the antineutron and antiproton (protons with a negative charge). Antiatoms (pairing of positrons and antiprotons) are other examples of antimatter.

## Who is generally regarded as the founder of **quantum mechanics**?

The German mathematical physicist Werner Karl Heisenberg (1901–1976) is regarded as the father of quantum mechanics (the theory of small-scale physical phenomena). His theory of uncertainty in 1927 overturned traditional classical mechanics and electromagnetic theory regarding energy and motions when applied to subatomic particles such as electrons and parts of atomic nuclei. The theory states that while it is impossible to specify precisely both the position and the simultaneous momentum (mass $\times$ velocity) of a particle, they can only be predicted. This means that the result of an action can be expressed only in terms of probability that a certain effect will occur.

## Who invented the **thermometer**?

The Greeks of Alexandria knew that air expanded as it was heated. Hero of Alexandria (first century C.E.) and Philo of Byzantium made simple "thermoscopes," but they were not real thermometers. In 1592, Galileo (1564–1642) made a kind of thermometer that also functioned as a barometer, and in 1612 his friend Santorio Santorio (1561–1636) adapted the air thermometer (a device in which a colored liquid was driven down by the expansion of air) to measure the body's temperature change during illness and recovery. Still, it was not until 1713 that Daniel Fahrenheit (1686–1736) began developing a thermometer with a fixed scale. He worked out his scale from two "fixed" points: the melting point of ice and the heat of the healthy human body. He realized that the melting point of ice was a constant temperature, whereas the freezing point of water varied.

Fahrenheit put his thermometer into a mixture of ice, water, and salt (which he marked off as 0°) and using this as a starting point, marked off melting ice at 32° and blood heat at 96°. In 1835, it was discovered that normal blood measured 98.6°F. Sometimes Fahrenheit used spirit of wine as the liquid in the thermometer tube, but more often he used specially purified mercury. Later, the boiling point of water (212°F) became the upper fixed point.

## What is the **pH scale**?

The pH scale is the measurement of the $H^+$ concentration (hydrogen ions) in a solution. It is used to measure the acidity or alkalinity of a solution. The pH scale ranges from 0 to 14. A neutral solution has a pH of 7, one with a pH greater than 7 is basic; or alkaline, and one with a pH less than 7 is acidic. The lower the pH below 7, the more acidic the solution. Each whole-number drop in pH represents a tenfold increase in acidity.

| pH Value | Examples of Solutions |
|---|---|
| 0 | Hydrochloric acid (HCl), battery acid |
| 1 | Stomach acid (1.0–3.0) |
| 2 | Lemon juice (2.3) |
| 3 | Vinegar, wine soft drinks, beer, orange juice, some acid rain |
| 4 | Tomatoes, grapes, banana (4.6) |
| 5 | Black coffee, most shaving lotions, bread, normal rainwater |
| 6 | Urine (5–7), milk (6.6), saliva (6.2–7.4) |
| 7 | Pure water, blood (7.3–7.5) |
| 8 | Egg white (8.0), seawater (7.8–8.3) |
| 9 | Baking soda, phosphate detergents, Clorox, Tums |
| 10 | Soap solutions, milk of magnesia |
| 11 | Household ammonia (10.5–11.9), nonphosphate detergents |
| 12 | Washing soda (sodium carbonate) |
| 13 | Hair remover, oven cleaner |
| 14 | Sodium hydroxide (NaOH) |

## What is the **Kelvin** temperature scale?

Temperature is the level of heat in a gas, liquid, or solid. The freezing and boiling points of water are used as standard reference levels in both the metric (centigrade or Celsius) and the English system (Fahrenheit). In the metric system, the difference between freezing and boiling is divided into 100 equal intervals called degree Celsius or degree centigrade (°C). In the English system, the intervals are divided into 180 units, with one unit called degree Fahrenheit (°F). But temperature can be measured from absolute zero (no heat, no motion); this principle defines thermodynamic temperature and establishes a method to measure it upward. This scale of temperature is called the Kelvin temperature scale, after its inventor, William Thomson, Lord Kelvin (1824–1907), who devised it in 1848. The Kelvin (symbol K) has the same magnitude as the degree Celsius (the difference between freezing and boiling water is 100 degrees), but the two temperatures differ by 273.15 degrees (absolute zero, which is –273.15°C on the Celsius scale). Below is a comparison of the three temperatures:

| Characteristic | K | °C | °F |
|---|---|---|---|
| Absolute zero | 0 | –273.15 | –459.67 |
| Freezing point of water | 273.15 | 0 | 32 |
| Normal human body temperature | 310.15 | 37 | 98.6 |
| Boiling point of water | 373.15 | 100 | 212 |

To convert Celsius to Kelvin: Add 273.15 to the temperature($K = C + 273.15$). To convert Fahrenheit to Celsius: Subtract 32 from the temperature and multiply the difference by 5; then divide the product by 9 ($C = \frac{5}{9}[F - 32]$). To convert Celsius to Fahrenheit: Multiply the temperature by 1.8, then add 32 ($F = \frac{9}{5}C + 32$ or $F = 1.8C + 32$).

## What was unusual about the original **Celsius** temperature scale?

In 1742, the Swedish astronomer Anders Celsius (1701–1744) set the freezing point of water at 100°C and the boiling point of water at 0°C. It was Carolus Linnaeus (1707–1778), who reversed the scale, but a later textbook attributed the modified scale to Celsius and the name has remained.

## How are **Celsius** temperatures converted into **Fahrenheit** temperatures?

The formulas for converting Celsius temperatures into Fahrenheit (and the reverse) are as follows:

$$F = (C \times \frac{9}{5}) + 32$$
$$C = (F - 32) \times \frac{5}{9}$$

Some useful comparisons of the two scales:

| Temperature | Fahrenheit | Celsius |
|---|---|---|
| Absolute zero | –459.67 | –273.15 |
| Point of equality | –40.0 | –40.0 |
| Zero Fahrenheit | 0.0 | –17.8 |
| Freezing point of water | 32.0 | 0.0 |
| Normal human blood temperature | 98.4 | 36.9 |
| 100 degrees F | 100.0 | 37.8 |
| Boiling point of water (at standard pressure) | 212.0 | 100.0 |

## Who invented **chromatography**?

Chromatography was invented by the Russian botanist Mikhail Tswett (1872–1919) in the early 1900s. The technique was first used to separate different plant pigments from one another. Chromatography has developed into a widely used method to separate various components of a substance from one another. Three types of chromatography are high-performance liquid chromatography (HPLC), gas chromatography, and paper chromatography. Different chromatography techniques are used in forensic science and analytical laboratories.

## What is **nuclear magnetic resonance**?

Nuclear magnetic resonance (NMR) is a process in which the nuclei of certain atoms absorb energy from an external magnetic field. Analytical chemists use NMR spectroscopy to identify unknown compounds, check for impurities, and study the shapes of molecules. They use the knowledge that different atoms will absorb electromagnetic energy at slightly different frequencies.

## What is **STP**?

The abbreviation STP is often used for standard temperature and pressure. As a matter of convenience, scientists have chosen a specific temperature and pressure as standards for comparing gas volumes. The standard temperature is 0°C (273 K) and the standard pressure is 760 torr (one atmosphere).

## How did the electrical term **ampere** originate?

It was named for André Marie Ampère (1775–1836), the physicist who formulated the basic laws of the science of electrodynamics. The ampere (symbol A), often abbreviated as "amp," is the unit of electric current, defined at the constant current, that, maintained in two straight parallel infinite conductors placed one meter apart in a vacuum, would produce a force between the conductors of $2 \times 10^{-7}$ newton per meter. For example, the amount of current flowing through a 100-watt light bulb is 1 amp; through a toaster, 10 amps; a TV set, 3 amps; a car battery, 50 amps (while cranking). A newton (symbol N) is defined as a unit of force needed to accelerate one kilogram by one meter second$^{-2}$, or $1 \text{ N} = 1\text{Kg}^{\text{MS-2}}$.

## How did the electrical unit **volt** originate?

The unit of voltage is the volt, named after Alessandro Volta (1745–1827), the Italian scientist who built the first modern battery. (A battery, operating with a lead rod and vinegar, was also manufactured in ancient Egypt.) Voltage measures the force or "oomph" with which electrical charges are pushed through a material. Some common voltages are 1.5 volts for a flashlight battery; 12 volts for a car battery; 115 volts for ordinary household receptacles; and 230 volts for a heavy-duty household receptacle.

## How did the electrical unit **watt** originate?

Named for the Scottish engineer and inventor James Watt (1736–1819), the watt is used to measure electric power. An electric device uses one watt when one volt of electric current drives one ampere of current through it.

## What is a **mole** in chemistry?

A mole (symbol mol), a fundamental measuring unit for the amount of a substance, refers to either a gram atomic weight or a gram molecular weight of a substance. It is the quantity of a substance that contains $6.02 \times 10^{23}$ atoms, molecules, or formula units of that substance. This number is called Avogadro's number or constant after Amadeo Avogadro (1776–1856), who is considered to be one of the founders of physical science.

## What is **Mole Day**?

Mole Day was organized by the National Mole Day Foundation to promote an awareness and enthusiasm for chemistry. It is celebrated each year on October 23.

## How does **gram atomic weight** differ from **gram formula weight**?

Gram atomic weight is the amount of an element (substance made up of atoms having the same atomic number) equal to its atomic weight in grams. Gram formula weight is an amount of a compound (a combination of elements) equal to its formula weight in grams.

# SPACE

## UNIVERSE

### What was the **Big Bang**?

The Big Bang theory is the explanation most commonly accepted by astronomers for the origin of the universe. It proposes that the universe began as the result of an explosion—the Big Bang—15 to 20 billion years ago. Two observations form the basis of this cosmology. First, as Edwin Hubble (1889–1953) demonstrated, the universe is expanding uniformly, with objects at greater distances receding at greater velocities. Secondly, the Earth is bathed in a glow of radiation that has the characteristics expected from a remnant of a hot primeval fireball. This radiation was discovered by Arno A. Penzias (b. 1933) and Robert W. Wilson (b. 1936) of Bell Telephone Laboratories. In time, the matter created by the Big Bang came together in huge clumps to form the galaxies. Smaller clumps within the galaxies formed stars. Parts of at least one clump became a group of planets—our solar system.

### What is the **Big Crunch** theory?

According to the Big Crunch theory, at some point in the very distant future, all matter will reverse direction and crunch back into the single point from which it began. Two other theories predict the future of the universe: the Big Bore theory and the Plateau theory. The Big Bore theory, named because it has nothing exciting to describe, claims that all matter will continue to move away from all other matter and the universe will expand forever. According to the Plateau theory, expansion of the universe will slow to the point where it will nearly cease, at which time the universe will reach a plateau and remain essentially the same.

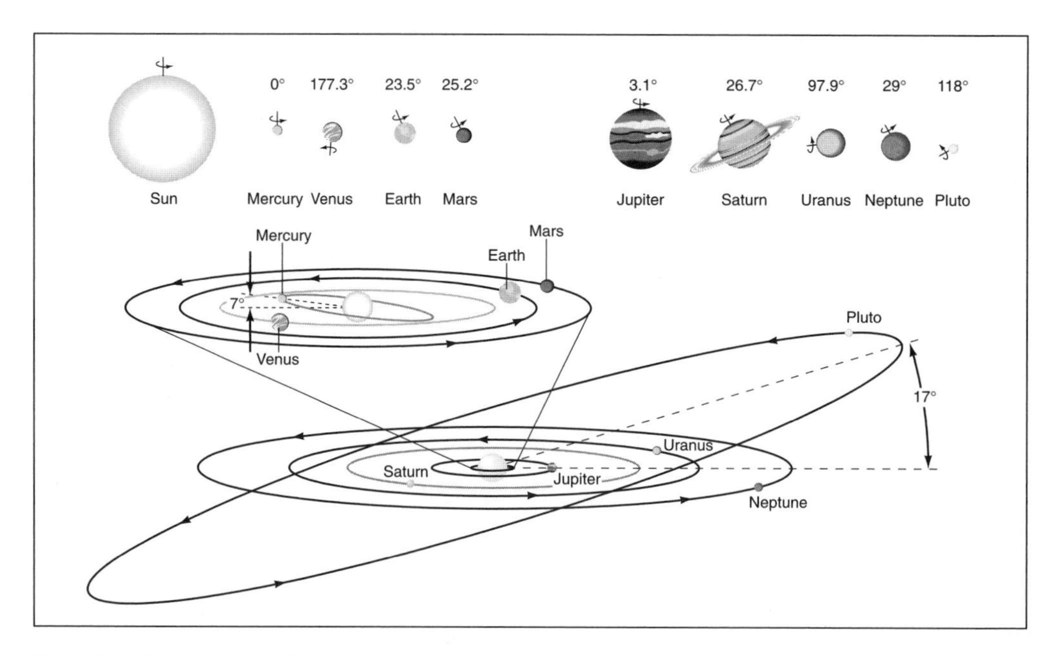


Schematic of the present-day solar system.

## How **old** is the **universe**?

Recent data collected by the Hubble Space Telescope suggests that the universe may only be eight billion years old. This contradicts the previous belief the universe was somewhere between 13 billion and 20 billion years old. The earlier figure was derived from the concept that the universe has been expanding at the same rate since its birth at the Big Bang. The rate of expansion is a ratio known as Hubble's constant. It is calculated by dividing the speed at which the galaxy is moving away from the Earth by its distance from the Earth. By inverting Hubble's Constant—that is, dividing the distance of a galaxy by its recessional speed—the age of the universe can be calculated. The estimates of both the velocity and distance of galaxies from the Earth are subject to uncertainties, and not all scientists accept that the universe has always expanded at the same rate. Therefore, many still hold that the age of the universe is open to question.

## How **large** is the **solar system**?

The size of the solar system can be visualized by imagining the sun (864,000 miles in diameter; 1,380,000 km) shrunk to a diameter of one inch (about the size of a ping-pong ball). Using the same size scale, the Earth would be a speck 0.01 inches (0.25 mm) in diameter and about nine feet (2.7 meters) away from the ping-pong ball sized sun. Our moon would have a diameter of 0.0025 inches (0.06 mm; the thickness of a human hair) and only a little over one-quarter inch (6.3 mm) from the Earth. Jupiter, the largest planet in the solar system, appears as the size of a small pea (0.1 inches

[2.5 mm] in diameter) and 46 feet (14 meters) from the sun. Pluto, the smallest planet in the solar system, appears as a nearly invisible speck (0.00083 inches [0.02 mm] in diameter) and 355 feet (108 meters) from the sun.

## Who is **Stephen Hawking**?

Stephen Hawking, often recognized as the greatest theoretical physicist of the late 20th century.

Hawking (b. 1943), a British physicist and mathematician, is considered to be the greatest theoretical physicist of the late 20th century. In spite of being severely handicapped by amyotrophic lateral sclerosis (ALS), he has made major contributions to scientific knowledge about black holes and the origin and evolution of the universe though his research into the nature of space-time and its anomalies. For instance, Hawking proposed that a black hole could emit thermal radiation, and he predicted that a black hole would disappear after all its mass has been converted into radiation (called "Hawking's radiation"). A current objective of Hawking is to synthesize quantum mechanics and relativity theory into a theory of quantum gravity. He is also the author of several books, including the popular best-selling work *A Brief History of Time*.

## What are **quasars**?

The name quasar originated as a contraction of "quasi-stellar" radio source. Quasars appear to be star like, but they have large red shifts in their spectra indicating that they are receding from the Earth at great speeds, some at up to 90 percent of the speed of light. Their exact nature is still unknown, but many believe quasars to be the cores of distant galaxies, the most distant objects yet seen. Quasars, also called quasi-stellar objects or QSOs, were first identified in 1963 by astronomers at the Palomar Observatory in California.

## What is a **syzygy**?

A syzygy (*sizz*-eh-jee) is a configuration that occurs when three celestial bodies lie in a straight line, such as the sun, Earth, and Moon during a solar or lunar eclipse. The particular syzygy when a planet is on the opposite side of the Earth from the sun is called an opposition.

## Which **galaxy** is **closest** to us?

The Andromeda Galaxy is the galaxy closest to the Milky Way galaxy, where Earth is located. It is estimated to be 2.2 million light years away from Earth. Bigger than the Milky Way, Andromeda is a spiral-shaped galaxy that is also the brightest in Earth's sky.

# STARS

## What is a **supernova**?

A supernova is the death explosion of a massive star. Immediately after the explosion, the brightness of the star can outshine the entire galaxy, followed by a gradual fading. A supernova is a fairly rare event. The last supernova observed in our galaxy was in 1604. In February 1987, Supernova 1987A appeared in the Large Magellanic Cloud, a nearby galaxy.

## What are the **four types** of **nebulae**?

The four types of nebulae are: emission, reflection, dark, and planetary. Primarily the birth place of stars, nebulae are clouds of gas and dust in space. Emission nebulae and reflection nebulae are bright nebulae. Emission nebulae are colorful and self-luminous. The Orion nebula, visible with the naked eye, is an example of an emission nebula. Reflection nebulae are cool clouds of dust and gas. They are illuminated by the light from nearby stars rather than by their own energy. Dark nebulae, also known as absorption nebulae, are not illuminated and appear as holes in the sky. The Horsehead nebula in the constellation Orion is an example of a dark nebula. Planetary nebulae are the remnants of the death of a star.

## Why do **stars twinkle**?

Stars actually shine with a more or less constant light. They appear to twinkle to those of us observing them from the Earth due mostly to atmospheric interference. Mole-

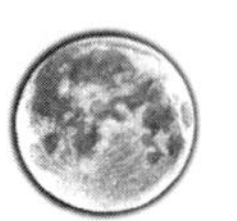



**What is stardust and are we really made of it?**

The heavy elements that comprise the Earth—such as iron, silicon, oxygen, and carbon—originally formed in distant stars that exploded, sending their heavy elements into space. The Earth and all life on it are, in fact, recycled stellar debris.

cules and dust particles float at random in the Earth's covering of gases. When such floating particles pass between a star and a person observing it, there is a brief interruption in the stream of light. Added together, these brief interruptions give rise to twinkling stars.

## What is a **binary star**?

A binary star is a pair of stars revolving around a common center of gravity. About half of all stars are members of either binary star systems or multiple star systems, which contain more than two stars.

The bright star Sirius, about 8.6 light years away, is composed of two stars: one about 2.3 times the mass of the sun, the other a white dwarf star about 980 times the mass of Jupiter. Alpha Centauri, the nearest star to Earth after the sun, is actually three stars: Alpha Centauri A and Alpha Centauri B, two sunlike stars, orbit each other, and Alpha Centauri C, a low-mass red star, orbits around them.

## What is a **black hole**?

When a star with a mass greater than about four times that of the sun collapses even the neutrons cannot stop the force of gravity. There is nothing to stop the contraction, and the star collapses forever. The material is so dense that nothing—not even light—can escape. The American physicist John Wheeler gave this phenomenon the name "black hole" in 1967. Since no light escapes from a black hole, it cannot be observed directly. However, if a black hole existed near another star, it would draw matter from the other star into itself and, in effect, produce X-rays. In the constellation of Cygnus, there is a strong X-ray source named Cygnus X-1. It is near a star, and the two revolve around each other. The unseen X-ray source has the gravitational pull of at least 10 suns and is believed to be a black hole. Another type of black hole, a primordial black hole, may also exist dating from the time of the Big Bang, when regions of gas and dust were highly compressed. Recently, astronomers observed a brief pulse of X-rays from Sagittarius A, a region near the center of the Milky Way Galaxy. The origin of this pulse and its behavior led scientists to conclude that there is probably a black hole in the center of our galaxy.

There are four other possible black holes: a Schwarzschild black hole has no charge and no angular momentum; a Reissner-Nordstrom black hole has charge but no angular momentum; a Kerr black hole has angular momentum but no charge; and a Kerr-Newman black hole has charge and angular momentum.

## What is a **pulsar**?

A pulsar is a rotating neutron star that gives off sharp regular pulses of radio waves at rates ranging from 0.001 to four seconds. Stars burn by fusing hydrogen into helium. When they use up their hydrogen, their interiors begin to contract. During this contraction, energy is released and the outer layers of the star are pushed out. These layers are large and cool; the star is now a red giant. A star with more than twice the mass of the sun will continue to expand, becoming a supergiant. At that point, it may blow up in an explosion called a supernova. After a supernova, the remaining material of the star's core may be so compressed that the electrons and protons become neutrons. A star 1.4 to four times the mass of the sun can be compressed into a neutron star only about 12 miles (20 kilometers) across. Neutron stars rotate very fast. The neutron star at the center of the Crab Nebula spins 30 times per second.

A pulsar is formed by the collapse of a star with 1.4 to four times the mass of the sun. Some of these neutron stars emit radio signals from their magnetic poles in a direction that reaches Earth. These signals were first detected by Jocelyn Bell (b. 1943) of Cambridge University in 1967. Because of their regularity some people speculated that they were extraterrestrial beacons constructed by alien civilizations. This theory was eventually ruled out and the rotating neutron star came to be accepted as the explanation for these pulsating radio sources, or pulsars.

## What does the **color** of a **star** indicate?

The color of a star gives an indication of its temperature and age. Stars are classified by their spectral type. From oldest to youngest and hottest to coolest, the types of stars are:

| Type | Color | Temperature (°F) | Temperature (°C) |
|---|---|---|---|
| O | Blue | 45,000–75,000 | 25,000–40,000 |
| B | Blue | 20,000–45,000 | 11,000–25,000 |
| A | Blue-White | 13,500–20,000 | 7,500–11,000 |
| F | White | 10,800–13,500 | 6,000–7,500 |
| G | Yellow | 9,000–10,800 | 5,000–6,000 |
| K | Orange | 6,300–9,000 | 3,500–5,000 |
| M | Red | 5,400–6,300 | 3,000–3,500 |



Each type is further subdivided on a scale of 0–9. The sun is a type G2 star.

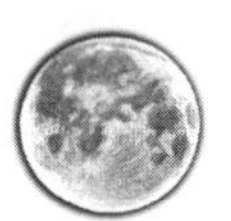



## What is the **most massive** star?

The Pistol Star is both the brightest and most massive star known. Located 25,000 light years away in the area of the constellation Sagittarius, this young (one- to three-million-year-old) star is as bright as ten million suns and may have weighed two hundred times the mass of the sun at one point in its young life.

## Which **stars** are the **brightest**?

The brightness of a star is called its magnitude. Apparent magnitude is how bright a star appears to the naked eye. The lower the magnitude, the brighter the star. On a clear night, stars of about magnitude +6 can be seen with the naked eye. Large telescopes can detect objects as faint as +27. Very bright objects have negative magnitudes; the sun is –26.8.

| Star | Constellation | Apparent Magnitude |
|---|---|---|
| Sirius | Canis Major | –1.47 |
| Canopus | Carina | –0.72 |
| Arcturus | Boötes | –0.06 |
| Rigil Kentaurus | Centaurus | +0.01 |
| Vega | Lyra | +0.04 |
| Capella | Auriga | +0.05 |
| Rigel | Orion | +0.14 |
| Procyon | Canis Minor | +0.37 |
| Betelgeuse | Orion | +0.41 |
| Achernar | Eridanus | +0.51 |

## What is the **Milky Way**?

The Milky Way is a hazy band of light that can be seen encircling the night sky. This light comes from the stars that make up the Milky Way galaxy, the galaxy to which the sun and the Earth belong. Galaxies are huge systems of stars separated from one another by largely empty space. Astronomers estimate that the Milky Way galaxy contains at least 100 billion stars and is about 100,000 light years in diameter. The galaxy is shaped like a compact disk with a central bulge, or nucleus, and spiral arms curving out from the center.

## What is the **Big Dipper**?

The Big Dipper is a group of seven stars that are part of the constellation Ursa Major. They appear to form a sort of spoon with a long handle. The group is known as the Plough in Great Britain. The Big Dipper is almost always visible in the northern hemisphere. It serves as a convenient reference point when locating other stars; for exam-

The constellation Ursa Major, known as the greater bear, contains the cluster of seven stars that make up the Big Dipper.

ple, an imaginary line drawn from the two end stars of the dipper leads to Polaris, the North Star.

## Where is the **North Star**?

If an imaginary line is drawn from the North Pole into space, it will reach a star called Polaris, or the North Star, less than one degree away from the line. As the Earth rotates on its axis, Polaris acts as a pivot-point around which all the stars visible in the northern hemisphere appear to move, while Polaris itself remains motionless.

## Was **Polaris** always the **North Star**?

The Earth has had several North Stars. The Earth slowly wobbles on its axis as it spins. This motion is called precession. The Earth traces a circle in the sky over a period of 26,000 years. In Pharonic times the North Star was Thuban; today it is Polaris; around 14,000 C.E. it will be Vega.

## What is the **summer triangle**?

The summer triangle is the triangle formed by the stars Deneb, Vega, and Altair as seen in the summer Milky Way.

## How many **constellations** are there and how were they named?

Constellations are groups of stars that seem to form some particular shape, such as that of a person, animal, or object. They only appear to form this shape and be close to each other from Earth; in actuality, the stars in a constellation are often very distant from each other. There are 88 recognized constellations whose boundaries were defined in the 1920s by the International Astronomical Union.

Various cultures in all parts of the world have had their own constellations. However, because modern science is predominantly a product of Western culture, many of the constellations represent characters from Greek and Roman mythology. When Europeans began to explore the southern hemisphere in the 16th and 17th centuries, they derived some of the new star patterns from the technological wonders of their time, such as the microscope.

Names of constellations are usually given in Latin. Individual stars in a constellation are usually designated with Greek letters in the order of brightness; the brightest star is alpha, the second brightest is beta, and so on. The genitive, or possessive, form of the constellation name is used, thus Alpha Orionis is the brightest star of the constellation Orion.

| Constellation | Genitive | Abbreviation | Meaning |
|---|---|---|---|
| Andromeda | Andromedae | And | Chained Maiden |
| Antlia | Antliae | Ant | Air Pump |
| Apus | Apodis | Aps | Bird of Paradise |
| Aquarius | Aquarii | Aqr | Water Bearer |
| Aquila | Aquilae | Aql | Eagle |
| Ara | Arae | Ara | Altar |
| Aries | Arietis | Ari | Ram |
| Auriga | Aurigae | Aur | Charioteer |
| Boötes | Boötis | Boo | Herdsman |
| Caelum | Caeli | Cae | Chisel |
| Camelopardalis | Camelopardalis | Cam | Giraffe |
| Cancer | Cancri | Cnc | Crab |
| Canes Venatici | Canum Venaticorum | CVn | Hunting Dogs |
| Canis Major | Canis Majoris | CMa | Big Dog |
| Canis Minor | Canis Minoris | CMi | Little Dog |
| Capricornus | Capricorni | Cap | Goat |
| Carina | Carinae | Car | Ship's Keel |
| Cassiopeia | Cassiopeiae | Cas | Queen of Ethiopia |
| Centaurus | Centauri | Cen | Centaur |
| Cepheus | Cephei | Cep | King of Ethiopia |
| Cetus | Ceti | Cet | Whale |
| Chamaeleon | Chamaeleonis | Cha | Chameleon |
| Circinus | Circini | Cir | Compass |

| Constellation | Genitive | Abbreviation | Meaning |
|---|---|---|---|
| Columba | Columbae | Col | Dove |
| Coma Berenices | Comae Berenices | Com | Berenice's Hair |
| Corona Australis | Coronae Australis | CrA | Southern Crown |
| Corona Borealis | Coronae Borealis | CrB | Northern Crown |
| Corvus | Corvi | Crv | Crow |
| Crater | Crateris | Crt | Cup |
| Crux | Crucis | Cru | Southern Cross |
| Cygnus | Cygni | Cyg | Swan |
| Delphinus | Delphini | Del | Dolphin |
| Dorado | Doradus | Dor | Goldfish |
| Draco | Draconis | Dra | Dragon |
| Equuleus | Equulei | Equ | Little Horse |
| Eridanus | Eridani | Eri | River Eridanus |
| Fornax | Fornacis | For | Furnace |
| Gemini | Geminorum | Gem | Twins |
| Grus | Gruis | Gru | Crane |
| Hercules | Herculis | Her | Hercules |
| Horologium | Horologii | Hor | Clock |
| Hydra | Hydrae | Hya | Hydra, Greek monster |
| Hydrus | Hydri | Hyi | Sea Serpent |
| Indus | Indi | Ind | Indian |
| Lacerta | Lacertae | Lac | Lizard |
| Leo | Leonis | Leo | Lion |
| Leo Minor | Leonis Minoris | LMi | Little Lion |
| Lepus | Leporis | Lep | Hare |
| Libra | Librae | Lib | Scales |
| Lupus | Lupi | Lup | Wolf |
| Lynx | Lyncis | Lyn | Lynx |
| Lyra | Lyrae | Lyr | Lyre or Harp |
| Mensa | Mensae | Men | Table Mountain |
| Microscopium | Microscopii | Mic | Microscope |
| Monoceros | Monocerotis | Mon | Unicorn |
| Musca | Muscae | Mus | Fly |
| Norma | Normae | Nor | Carpenter's Square |
| Octans | Octanis | Oct | Octant |
| Ophiuchus | Ophiuchi | Oph | Serpent Bearer |
| Orion | Orionis | Ori | Orion, the Hunter |
| Pavo | Pavonis | Pav | Peacock |
| Pegasus | Pegasi | Peg | Winged Horse |
| Perseus | Persei | Per | Perseus, a Greek hero |
| Phoenix | Phoenicis | Phe | Phoenix |
| Pictor | Pictoris | Pic | Painter |

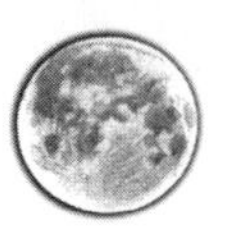



| Constellation | Genitive | Abbreviation | Meaning |
|---|---|---|---|
| Pisces | Piscium | Psc | Fish |
| Piscis Austrinus | Piscis Austrini | PsA | Southern Fish |
| Puppis | Puppis | Pup | Ship's Stern |
| Pyxis | Pyxidis | Pyx | Ship's Compass |
| Reticulum | Reticuli | Ret | Net |
| Sagitta | Sagittae | Sge | Arrow |
| Sagittarius | Sagittarii | Sgr | Archer |
| Scorpius | Scorpii | Sco | Scorpion |
| Sculptor | Sculptoris | Scl | Sculptor |
| Scutum | Scuti | Sct | Shield |
| Serpens | Serpentis | Ser | Serpent |
| Sextans | Sextantis | Sex | Sextant |
| Taurus | Tauri | Tau | Bull |
| Telescopium | Telescopii | Tel | Telescope |
| Triangulum | Trianguli | Tri | Triangle |
| Triangulum | Triangli Australis | TrA | Southern Australe Triangle |
| Tucana | Tucanae | Tuc | Toucan |
| Ursa Major | Ursae Majoris | UMa | Big Bear |
| Ursa Minor | Ursae Minoris | UMi | Little Bear |
| Vela | Velorum | Vel | Ship's Sail |
| Virgo | Virginis | Vir | Virgin |
| Volans | Volantis | Vol | Flying Fish |
| Vulpecula | Vulpeculae | Vul | Little Fox |

## What is the **largest constellation**?

Hydra is the largest constellation, extending from Gemini to the south of Virgo. It has a recognizable long line of stars. The name "hydra" is derived from the watersnake monster killed by Hercules in ancient mythology.

## Which **star** is the **closest to Earth**?

The sun, at a distance of 92,955,900 miles (149,598,000 kilometers), is the closest star to the Earth. After the sun, the closest stars are the members of the triple star system known as Alpha Centauri (Alpha Centauri A, Alpha Centauri B, and Alpha Centauri C, sometimes called Proxima Centauri). They are 4.3 light years away.

## How **hot** is the sun?

The center of the sun is about 27,000,000°F (15,000,000°C). The surface, or photosphere, of the sun is about 10,000°F (5,500°C). Magnetic anomalies in the photosphere cause cooler regions that appear to be darker than the surrounding surface. These sunspots are about 6,700°F (4,000°C). The sun's layer of lower atmosphere, the chromosphere, is only a few thousand miles thick. At the base, the chromosphere is about 7,800°F (4,300°C), but its temperature rises with altitude to the corona, the sun's outer layer of atmosphere, which has a temperature of about 1,800,000°F (1,000,000°C).

## What is the **sun** made of?

The sun is an incandescent ball of gases. Its mass is 1.8 × 1027 tons or 1.8 octillion tons (a mass 330,000 times as great as the Earth).

| Element | % of mass |
|---|---|
| Hydrogen | 73.46 |
| Helium | 24.85 |
| Oxygen | 0.77 |
| Carbon | 0.29 |
| Iron | 0.16 |
| Neon | 0.12 |
| Nitrogen | 0.09 |
| Silicon | 0.07 |
| Magnesium | 0.05 |
| Sulfur | 0.04 |
| Other | 0.10 |

## When will the **sun die**?

The sun is approximately 4.5 billion years old. About five billion years from now, the sun will have burned all of its hydrogen fuel into helium. As this process occurs, the sun will change from the yellow dwarf as we know it to a red giant. Its diameter will extend well beyond the orbit of Venus, and even possibly beyond the orbit of Earth. In either case, the Earth will be burned to a cinder.

## What is the **ecliptic**?

Ecliptic refers to the apparent yearly path of the sun through the sky with respect to the stars. In the spring, the ecliptic in the northern hemisphere is angled high in the evening sky. In fall, the ecliptic lies much closer to the horizon.

## Why does the **color** of the **sun** vary?

Sunlight contains all the colors of the rainbow, which blend to form white light, making sunlight appear white. At times, some of the color wavelengths, especially blue,

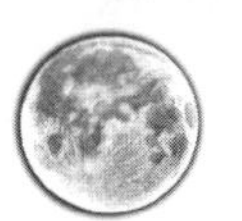



become scattered in the Earth's atmosphere and the sunlight appears colored. When the sun is high in the sky, some of the blue rays are scattered in the Earth's atmosphere. At such times, the sky looks blue and the sun appears to be yellow. At sunrise or sunset, when the light must follow a longer path through the Earth's atmosphere, the sun looks red (red having the longest wavelengths).

## How long does it take **light from the sun** to **reach the Earth**?

Sunlight takes about eight minutes and 20 seconds to reach the Earth, traveling at 186,282 miles (299,792 kilometers) per second, although it varies with the position of the Earth in its orbit. In January the light takes about 495 seconds to reach the Earth; in July the trip takes about 505 seconds.

## How long is a **solar cycle**?

The solar cycle is the periodic change in the number of sunspots. The cycle is taken as the interval between successive minima and is about 11.1 years long. During an entire cycle, solar flares, sunspots, and other magnetic phenomena move from intense activity to relative calm and back again. The solar cycle is one area of study to be carried out by up to 10 ATLAS space missions designed to probe the chemistry and physics of the atmosphere. These studies of the solar cycle will yield a more detailed picture of the Earth's atmosphere and its response to changes in the sun.

## What is the **sunspot cycle**?

It is the fluctuating number of sunspots on the sun during an 11-year period. The variation in the number of sunspots seems to correspond with the increase or decrease in the number of solar flares. An increased number of sunspots means an increased number of solar flares.

## When do **solar eclipses** happen?

A solar eclipse occurs when the moon passes between the Earth and the sun and all three bodies are aligned in the same plane. When the moon completely blocks Earth's view of the sun and the umbra, or dark part of the moon's shadow, reaches the Earth, a total eclipse occurs. A total eclipse happens only along a narrow path 100 to 200 miles (160 to 320 kilometers) wide called the track of totality. Just before totality, the only parts of the sun that are visible are a few points of light called Baily's beads shining through valleys on the moon's surface. Sometimes, a last bright flash of sunlight is seen—the diamond ring effect. During totality, which averages 2.5 minutes but may last up to 7.5 minutes, the sky is dark and stars and other planets are easily seen. The corona, the sun's outer atmosphere, is also visible.

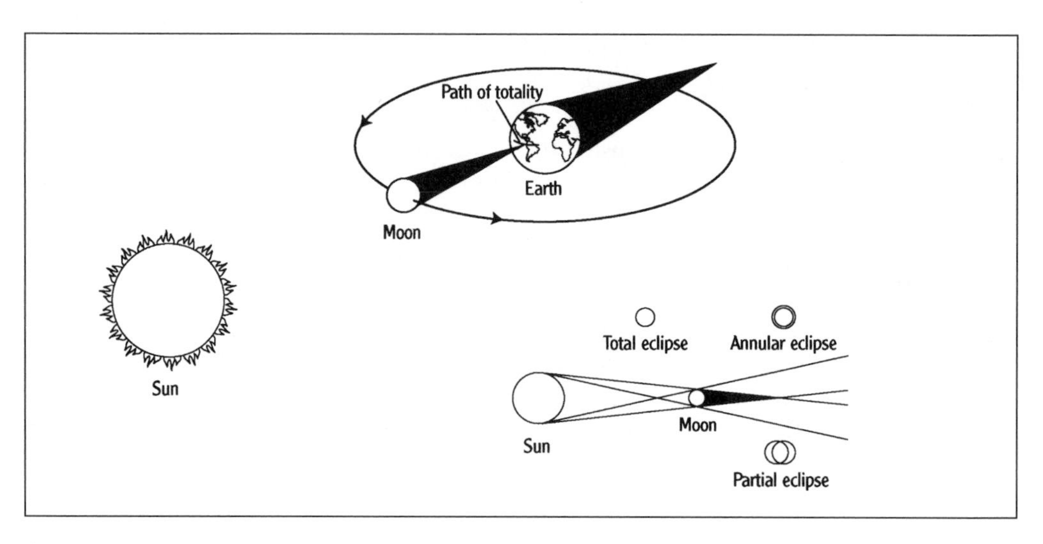


Solar eclipse.

If the moon does not appear large enough in the sky to completely cover the sun, it appears silhouetted against the sun with a ring of sunlight showing around it. This is an annular eclipse. Because the sun is not completely covered, its corona cannot be seen, and although the sky may darken it will not be dark enough to see the stars.

During a partial eclipse of the sun, the penumbra of the moon's shadow strikes the Earth. A partial eclipse can also be seen on either side of the track of totality of an annular or total eclipse. The moon will cover part of the sun and the sky will not darken noticeably during a partial eclipse.

## What is a **sun dog**?

A sun dog is also known as a mock sun, false sun, or the 22° parhelia. It is a bright spot of light that sometimes appears on either side of the sun at the same distance above the horizon as the sun, and is separated from the sun by an angle of 22°.

## What is **solar wind**?

Solar wind is caused by the expansion of gases in the sun's outermost atmosphere, the corona. Because of the corona's extremely high temperature of 1,800,000°F (1,000,000°C), the gases heat up and their atoms start to collide. The atoms lose electrons and become electrically charged ions. These ions create the solar wind. Solar wind has a velocity of 310 miles (500 kilometers) per second, and its density is approximately 82 ions per cubic inch (five ions per cubic centimeter). Because the Earth is surrounded by strong magnetic forces, its magnetosphere, it is protected from the solar wind particles. In 1959, the Soviet spacecraft *Luna 2* acknowledged the existence of solar wind and made the first measurements of its properties.

## What is the **safest way** to **view a solar eclipse**?

Punch a pinhole in an index card and hold it two to three feet in front of another index card. The eclipse can be viewed safely through the hole. Encase the index card contraption in a box, using aluminum foil with a pinhole, and you'll see a sharper image of the eclipse. You may also purchase special glasses with aluminized Mylar lenses. Damage to the retina can occur if the eclipse is viewed with other devices such as photographic filters, exposed film, smoked glass, camera lenses, telescopes, or binoculars.

A simple pinhole camera can be used to focus the image of a solar eclipse for safe viewing.

## When and where will the **next ten total solar eclipses** occur?

| | |
|---|---|
| November 23, 2003 | Antarctica |
| April 8, 2005 | South Pacific |
| March 29, 2006 | Africa, Turkey, Russia |
| August 1, 2008 | Greenland, Siberia, China |
| July 22, 2009 | India, China, Pacific Ocean |
| July 11, 2010 | South Pacific |
| November 13, 2012 | Australia, Pacific Ocean |
| March 20, 2015 | North Atlantic Ocean |
| March 9, 2016 | Indonesia, North Pacific Ocean |
| August 21, 2017 | United States |

The next total solar eclipse in the United States, August 21, 2017, will sweep a path 70 miles (113 kilometers) wide from Salem, Oregon, to Charleston, South Carolina.

# PLANETS AND MOONS

*See also: The Earth*

## Which **planets** are **visible** with the naked eye?

Mercury, Venus, Mars, Jupiter and Saturn are visible with the naked eye at varying times of the year.

## How **old** is the **solar system**?

It is currently believed to be 4.5 billion years old. The Earth and the rest of the solar system formed from an immense cloud of gas and dust. Gravity and rotational forces caused the cloud to flatten into a disc and much of the cloud's mass to drift into the center. This material became the sun. The leftover parts of the cloud formed small bodies called planetesimals. These planetesimals collided with each other, gradually forming larger and larger bodies, some of which became the planets. This process is thought to have taken about 25 million years.

## How **far** are the **planets** from the sun?

The planets revolve around the sun in elliptical orbits, with the sun at one focus of the ellipse. Thus, a planet is at times closer to the sun than at other times. The distances given below are the average distance from the sun, starting with Mercury, the planet closest to the sun, and moving outward.

| Planet | Average distance (miles) | Average distance (km) |
|---|---:|---:|
| Mercury | 35,983,000 | 57,909,100 |
| Venus | 67,237,700 | 108,208,600 |
| Earth | 92,955,900 | 149,598,000 |
| Mars | 141,634,800 | 227,939,200 |
| Jupiter | 483,612,200 | 778,298,400 |
| Saturn | 888,184,000 | 1,427,010,000 |
| Uranus | 1,782,000,000 | 2,869,600,000 |
| Neptune | 2,794,000,000 | 4,496,700,000 |
| Pluto | 3,666,000,000 | 5,913,490,000 |

## Which **planets** have **rings**?

Jupiter, Saturn, Uranus, and Neptune all have rings. Jupiter's rings were discovered by Voyager 1 in March 1979. The rings extend 80,240 miles (129,130 kilometers) from the center of the planet. They are about 4,300 miles (7,000 kilometers) in width and

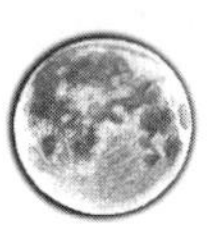



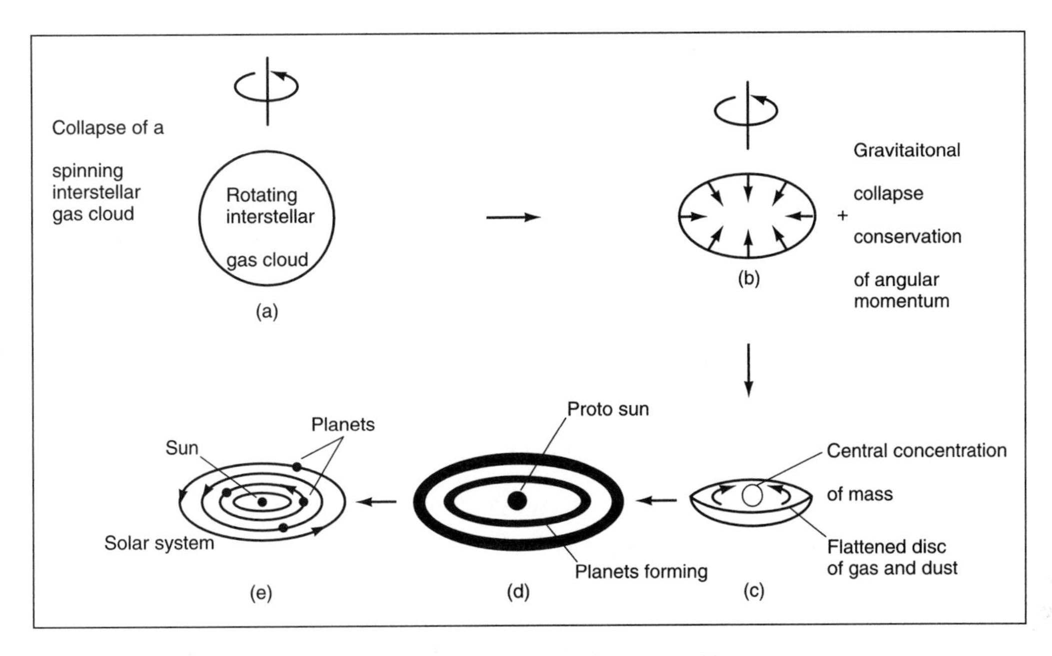


**The evolution of the solar system from a gas cloud (a) to its present-day structure (e).**

less than 20 miles (30 kilometers) thick. A faint inner ring is believed to extend to the edge of Jupiter's atmosphere. Saturn has the largest, most spectacular set of rings in the solar system. That the planet is surrounded by a ring system was first recognized by the Dutch astronomer Christiaan Huygens (1629–1695) in 1659. Saturn's rings are 169,800 miles (273,200 kilometers) in diameter, but less than 10 miles (16 kilometers) thick. There are six different rings, the largest of which appear to be divided into thousands of ringlets. The rings appear to be composed of pieces of water ice ranging in size from tiny grains to blocks several tens of yards in diameter.

In 1977 when Uranus occulted (passed in front of) a star, scientists observed that the light from the star flickered or winked several times before the planet itself covered the star. The same flickering occurred in reverse order after the occultation. The reason for this was determined to be a ring around Uranus. Nine rings were initially identified, and Voyager 2 observed two more in 1986. The rings are thin, narrow, and very dark.

Voyager 2 also discovered a series of at least four rings around Neptune in 1989. Some of the rings appear to have arcs, areas where there is a higher density of material than at other parts of the ring.

## How **thick** are **Saturn's rings**?

Saturn's large rings are actually composed of thousands of smaller rings. The thickness of the rings varies, with some rings less than one hundred meters thick.

## How long do the planets take **to go around the sun**?

| Planet | Period of revolution Earth days | Earth years |
|---|---|---|
| Mercury | 88 | 0.24 |
| Venus | 224.7 | 0.62 |
| Earth | 365.26 | 1.00 |
| Mars | 687 | 1.88 |
| Jupiter | 4,332.6 | 11.86 |
| Saturn | 10,759.2 | 29.46 |
| Uranus | 30,685.4 | 84.01 |
| Neptune | 60,189 | 164.8 |
| Pluto | 90,777.6 | 248.53 |

## What are the **diameters** of the planets?

| Planet | Diameter Miles | Kilometers |
|---|---|---|
| Mercury | 3,031 | 4,878 |
| Venus | 7,520 | 12,104 |
| Earth | 7,926 | 12,756 |
| Mars | 4,221 | 6,794 |
| Jupiter | 88,846 | 142,984 |
| Saturn | 74,898 | 120,536 |
| Uranus | 31,763 | 51,118 |
| Neptune | 31,329 | 50,530 |
| Pluto | 1,423 | 2,290 |

*Note:* All diameters are as measured at the planet's equator.

## What are the **colors** of the **planets**?

| Planet | Color |
|---|---|
| Mercury | Orange |
| Venus | Yellow |
| Earth | Blue, brown, green |
| Mars | Red |
| Jupiter | Yellow, red, brown, white |
| Saturn | Yellow |
| Uranus | Green |
| Neptune | Blue |
| Pluto | Yellow |

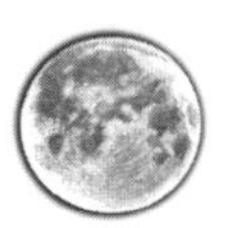



## Who discovered Saturn's rings?

In 1610, Galileo was probably the first to discover Saturn's rings. Because his telescope was small, Galileo could not see the rings properly and assumed they were satellites. In 1656, Christiaan Huygens discovered a ring around Saturn with a more powerful telescope. Later, in 1675, Jean Domenique Cassini distinguished two rings around Saturn. Still later, more rings were discovered and, as recently as 1980, ringlets were observed.

## What is the **gravitational force** on each of the planets, the moon, and the sun relative to the Earth?

If the gravitational force on the Earth is taken as 1, the comparative forces are:

| | |
|---|---|
| Sun | 27.9 |
| Mercury | 0.37 |
| Venus | 0.88 |
| Earth | 1.00 |
| Moon | 0.16 |
| Mars | 0.38 |
| Jupiter | 2.64 |
| Saturn | 1.15 |
| Uranus | 0.93 |
| Neptune | 1.22 |
| Pluto | 0.06 |

Weight comparisons can be made by using this table. If a person weighed 100 pounds (45.36 kilograms) on Earth, then the weight of the person on the Moon would be 16 pounds (7.26 kilograms) or $100 \times 0.16$.

## What is **sidereal time**?

Sidereal time is measured by considering the rotation of the Earth relative to the distant stars (rather than the sun, which is the basis of civil time). A sidereal day is 23 hours, 56 minutes, and 4 seconds, nearly 4 minutes shorter than mean solar time.

## Which planets are called **"inferior" planets** and which are **"superior" planets**?

An inferior planet is one whose orbit is nearer to the sun than Earth's orbit is. Mercury and Venus are the inferior planets. Superior planets are those whose orbits around the sun lie beyond that of Earth. Mars, Jupiter, Saturn, Uranus, Neptune, and

Pluto are the superior planets. The terms have nothing to do with the quality of an individual planet.

## Is a **day** the same on all the planets?

No. A day, the period of time it takes for a planet to make one complete turn on its axis, varies from planet to planet. Venus, Uranus, and Pluto display retrograde motion; that is to say, they rotate in the opposite direction from the other planets. The table below lists the length of the day for each planet.

| Planet | Length of day<br>Earth days | <br>Hours | <br>Minutes |
|---|---|---|---|
| Mercury | 58 | 15 | 30 |
| Venus | 243 | | 32 |
| Earth | | 23 | 56 |
| Mars | | 24 | 37 |
| Jupiter | | 9 | 50 |
| Saturn | | 10 | 39 |
| Uranus | | 17 | 14 |
| Neptune | | 16 | 3 |
| Pluto | 6 | 9 | 18 |

## What are the **Jovian** and **terrestrial planets**?

Jupiter, Saturn, Uranus, and Neptune are the Jovian (the adjectival form for the word "Jupiter"), or Jupiter-like, planets. They are giant planets, composed primarily of light elements such as hydrogen and helium.

Mercury, Venus, Earth, and Mars are the terrestrial (derived from "terra," the Latin word for "earth"), or Earth-like, planets. They are small in size, have solid surfaces, and are composed of rocks and iron. Pluto appears to be a terrestrial-type planet as well, but it may have a different origin from the other planets.

## What is unique about the **rotation** of the **planet Venus**?

Unlike Earth and most of the other planets, Venus rotates in a retrograde, or opposite, direction with relation to its orbital motion about the sun. It rotates so slowly that only two sunrises and sunsets occur each Venusian year. Uranus and Pluto's rotation are also retrograde.

## Is it true that the **rotation speed of the Earth** varies?

The rotation speed is at its maximum in late July and early August and at its minimum in April; the difference in the length of the day is about 0.0012 second. Since

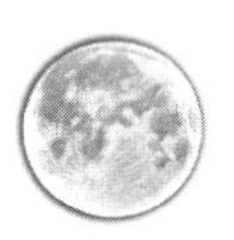

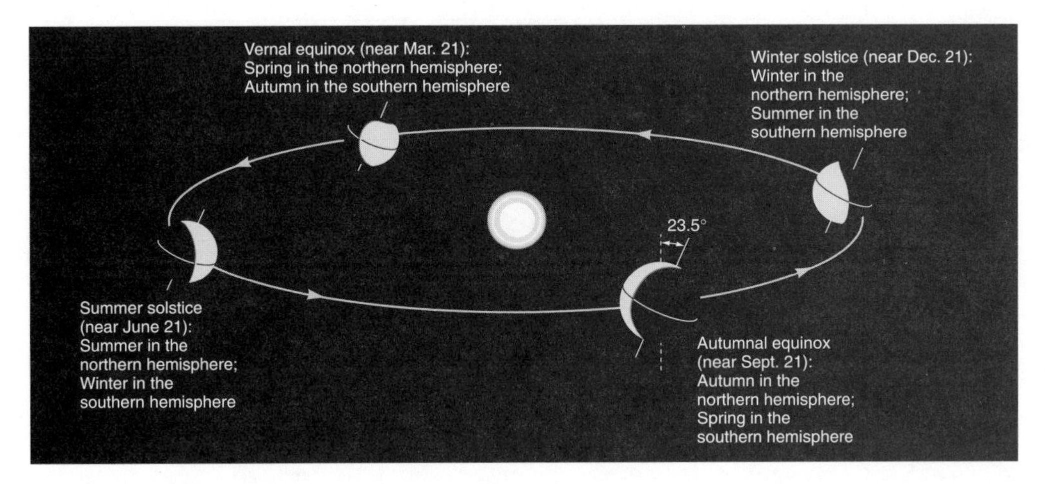


The seasons.

about 1900 the Earth's rotation has been slowing at a rate of approximately 1.7 seconds per year. In the geologic past the Earth's rotational period was much faster; days were shorter and there were more days in the year. About 350 million years ago, the year had 400 to 410 days; 280 million years ago, a year was 390 days long.

## Is it true that the **Earth is closer to the sun in winter** than in summer in the northern hemisphere?

Yes. However, the Earth's axis, the line around which the planet rotates, is tipped 23.5° with respect to the plane of revolution around the sun. When the Earth is closest to the sun (its perihelion, about January 3), the northern hemisphere is tilted away from the sun. This causes winter in the northern hemisphere while the southern hemisphere is having summer. When the Earth is farthest from the sun (its aphelion, around July 4), the situation is reversed, with the northern hemisphere tilted towards the sun. At this time, it is summer in the northern hemisphere and winter in the southern hemisphere.

## What is the **circumference** of the **Earth**?

The Earth is an oblate ellipsoid—a sphere slightly flattened at the poles and bulging at the equator. The distance around the Earth at the equator is 24,902 miles (40,075 kilometers). The distance around the Earth through the poles is 24,860 miles (40,008 kilometers).

## What is the **precession** of the **equinoxes**?

The "precession of the equinoxes" is the 26,000-year circular movement of the Earth's axis. It is caused by the bulging at the equator, which makes the Earth's axis twist in

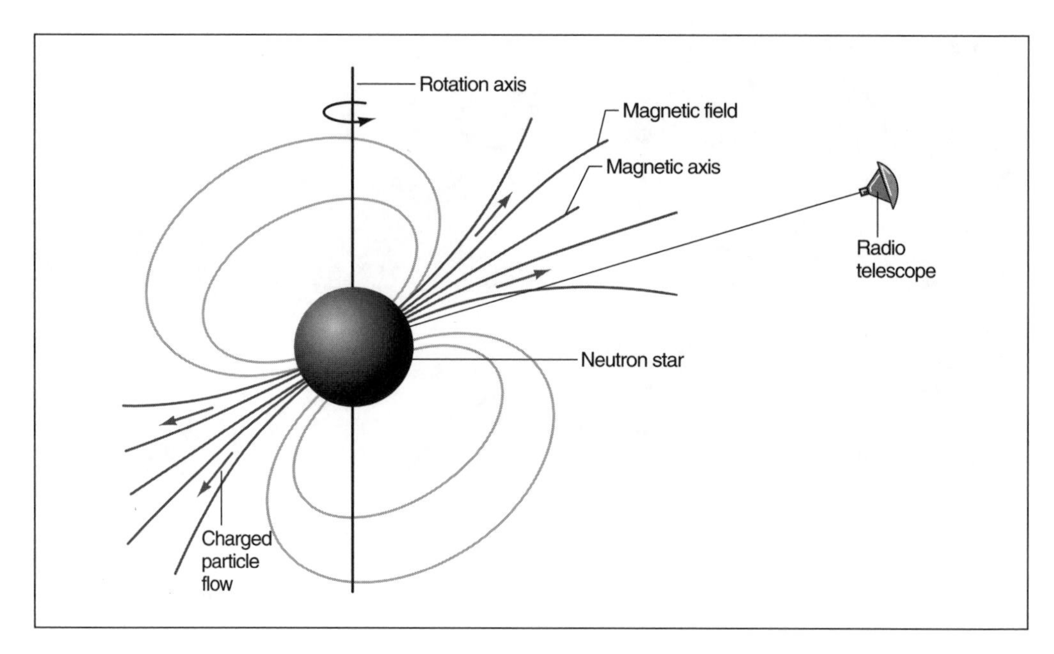


The precessional motion of the Earth.

such a way that the North and South Poles complete a circle every 26,000 years. Every year when the sun crosses the equator at the time of the equinox, it is in a slightly different position than the previous year. This movement proceeds eastward until a circle is completed.

## Is there **life on Mars**?

The answer to this question still remains inconclusive. There is evidence of frozen water on Mars. Results of the Viking soil sample data have been disputed. Microfossil-like imprints contained in meteorites that originated from Mars may indicate early forms of life. More sophisticated investigations will be needed to settle this question.

## Is it true that **Pluto** is not always the **outermost planet** in the solar system?

Pluto's very eccentric orbit carried it inside Neptune's orbit on January 23, 1979. It remained there until March 15, 1999. During that time, Neptune was the outermost planet in the solar system. However, because they are so far apart, the planets are in no danger of colliding with one another.

Pluto, discovered in 1930 by American astronomer Clyde Tombaugh (b. 1906), is the smallest planet in the solar system. It is composed of rock and ice, with methane ice on the surface and a thin methane atmosphere. Pluto's single moon, Charon, discovered by James Christy in 1978, has a diameter of 741 miles (1,192 kilometers). This

**What is the music of the spheres?**

Inaudible to human beings, the "music of the spheres" is a theoretic harmony or music created by the movements of the planets and heavenly bodies. Its existence was proclaimed by Pythagoras and other ancient mathematicians.

makes Charon, at half the size of Pluto, a very large moon relative to the planet. Some astronomers consider Pluto and Charon to be a double planet system.

## What is **Planet X**?

Astronomers have observed perturbations, or disturbances, in the orbits of Uranus and Neptune since the discoveries of both planets. They speculated that Uranus and Neptune were being influenced by the gravity of another celestial body. Pluto, discovered in 1930, does not appear to be large enough to cause these disturbances. The existence of another planet, known as Planet X, orbiting beyond Pluto, has been proposed. As yet there have been no sightings of this tenth planet, but the search continues. There is a possibility that the unmanned space probes Pioneer 10 & 11 and Voyager 1 & 2, now heading out of the solar system, will be able to locate this elusive object.

## What does it mean when a **planet** is said to be **in opposition**?

A body in the solar system is in opposition when its longitude differs from the sun by 180°. In that position, it is exactly opposite the sun in the sky and it crosses the meridian at midnight.

## How can an observer distinguish **planets** from **stars**?

In general, planets emit a constant light or shine, whereas stars appear to twinkle. The twinkling effect is caused by the combination of the distance between the stars and Earth and the refractive effect Earth's atmosphere has on a star's light. Planets are relatively closer to Earth than stars and their disklike shapes average out the twinkling effect, except when they're observed near the Earth's horizon.

## How **far** is the **moon** from the Earth?

Since the moon's orbit is elliptical, its distance varies from about 221,463 miles (356,334 kilometers) at perigee (closest approach to Earth), to 251,968 miles (405,503 kilometers) at apogee (farthest point), with the average distance being 238,857 miles (384,392 kilometers).

## How many **moons** does each planet have?

| Planet | Number of moons | Names of some of the moons |
|---|---|---|
| Mercury | 0 | |
| Venus | 0 | |
| Earth | 1 | The Moon (sometimes called Luna) |
| Mars | 2 | Phobos, Deimos |
| Jupiter | 39 | Metis, Adrastea, Amalthea, Thebe, Io, Europa, Ganymede, Callisto, Leda, Himalia, Lysithia, Elara, Ananke, Carme, Pasiphae, Sinope |
| Saturn | 30 | Epimetheus, Janus, Mimas, Enceladus, Tethys, Telesto, Calypso, Dione, Helene, Rhea, Titan, Hyper-ion, Iapetus, Phoebe |
| Uranus | 20 | Cordelia, Ophelia, Bianca, Cressida, Desdemona, Juliet, Portia, Rosalind, Belinda, Puck, Miranda, Ariel, Umbriel, Titania, Oberon |
| Neptune | 8 | Naiad, Thalassa, Despina, Galatea, Larissa, Proteus, Triton, Nereid |
| Pluto | 1 | Charon |

Jupiter has the greatest number of moons. As scientists continue to explore Jupiter they expect to find even more moons. Its four largest moons—Io, Europa, Calisto, and Ganymede—were discovered by Galileo in 1610.

## Does the **moon** have an **atmosphere**?

The moon does have an atmosphere; however it is very slight, having a density of about 50 atoms per cubic centimeter.

## What are the **diameter** and **circumference** of the **moon**?

The moon's diameter is 2,159 miles (3,475 kilometers) and its circumference is 6,790 miles (10,864 kilometers). The moon is 27 percent the size of the Earth.

## What are the **phases of the moon**?

The phases of the moon are changes in the moon's appearance during the month, which are caused by the moon's turning different portions of its illuminated hemisphere towards the Earth. When the moon is between the Earth and the sun, its daylight side is turned away from the Earth, so it is not seen. This is called the new moon. As the moon continues its revolution around the Earth, more and more of its surface becomes visible. This is called the waxing crescent phase. About a week after the new moon, half the moon is visible—the first quarter phase. During the next week, more than half of the moon is seen; this is called the waxing gibbous phase. Finally, about two weeks after the new moon, the moon and sun are on opposite sides of the Earth. The side of the moon facing the sun is also facing the Earth, and all the moon's illuminated side is seen as a full moon. In the next two weeks the moon goes through the same phases, but in reverse from a waning gibbous to third or last quarter to waning crescent phase. Gradually, less and less of the moon is visible until a new moon occurs again.

## Why does the **moon** always keep the **same face toward the Earth**?

Only one side of the moon is seen because it always rotates in exactly the same length of time that it takes to revolve about the Earth. This combination of motions (called "captured rotation") means that it always keeps the same side toward the Earth.

## What are **moonquakes**?

Similar to earthquakes, moonquakes are a result of the constant shifting of molten or partly molten material in the interior of the moon. These moonquakes are usually very weak. Other moonquakes may be caused by the impact of meteorites on the moon's surface. Still others occur at regular intervals during a lunar cycle, suggesting that gravitational forces from the Earth have an effect on the moon similar to ocean tides.

## What are the **names** of the **full moon** during each month?

| Month | American Folk Name |
|---|---|
| January | Wolf Moon |
| February | Snow Moon |
| March | Sap Moon |
| April | Pink Moon |
| May | Flower Moon |
| June | Strawberry Moon |
| July | Buck Moon |
| August | Sturgeon Moon |
| September | Harvest Moon |
| October | Hunter Moon |
| November | Beaver Moon |
| December | Cold Moon |

## Is the moon really blue during a **blue moon**?

The term "blue moon," the second full moon in a single month, does not refer to the color of the moon. A blue moon occurs, on average, every 2.72 years. Since 29.53 days pass between full moons (a synodial month), there is never a blue moon in February. On rare occasions, a blue moon can be seen twice in one year, but only in certain parts of the world. Blue moons will next occur:

**Upcoming Blue Moons**

| | |
|---|---|
| July 31, 2004 | March 31, 2018 |
| June 30, 2007 | October 31, 2020 |
| December 31, 2009 | August 31, 2023 |
| August 31, 2012 | May 31, 2026 |
| July 31, 2015 | December 31, 2028 |
| January 31, 2018 | |

A bluish-looking moon can result from effects of the Earth's atmosphere. For example, the phenomenon was widely observed in North America on September 26, 1950, due to Canadian forest fires that had scattered high-altitude dust.

## What is the difference between a **hunter's moon** and a **harvest moon**?

The harvest moon is the full moon nearest the autumnal equinox (on or about September 23). It is followed by a period of several successive days when the moon rises soon after sunset. In the southern hemisphere the harvest moon is the full moon closest to the vernal equinox (on or about March 21). This gives farmers extra hours of light for harvesting crops. The next full moon after the harvest moon is called the hunter's moon.

## Why do **lunar eclipses** happen?

A lunar eclipse occurs only during a full moon when the moon is on one side of the Earth, the sun is on the opposite side, and all three bodies are aligned in the same plane. In this alignment the Earth blocks the sun's rays to cast a shadow on the moon. In a total lunar eclipse the moon seems to disappear from the sky when the whole moon passes through the umbra, or total shadow, created by the Earth. A total lunar eclipse may last up to one hour and 40 minutes. If only part of the moon enters the umbra, a partial eclipse occurs. A penumbral eclipse takes place if all or part of the moon passes through the penumbra (partial shadow or "shade") without touching the umbra. It is difficult to detect this type of eclipse from Earth. From the moon one could see that the Earth blocked only part of the sun.

## What is the **moon's tail** that astronomers have discovered?

A glowing 15,000 mile (24,000 kilometer) long tail of sodium atoms streams from the moon. The faint, orange glow of sodium cannot be seen by the naked eye but it is

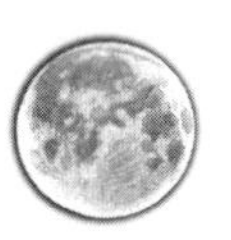



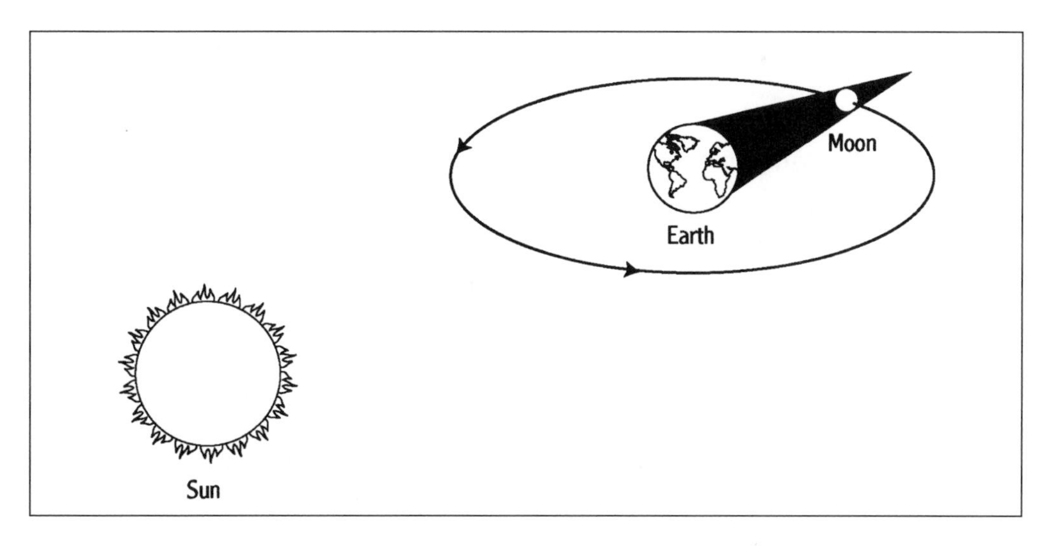


Lunar eclipse.

detectable with the use of instruments. Astronomers are not certain of the source of these sodium atoms.

## What is the **largest crater** on the moon?

The largest crater on the moon is Bailly. Its diameter is 184 miles (296 km).

## What are the **craters on the moon** that are named for the famous Curie family?

*Curie*—named for Pierre Curie (1859–1906), French chemist and Nobel prize winner.

*Sklodowska*—the maiden name of Marie Curie (1867–1934), French physical chemist and Nobel prize winner.

*Joliot*—named for physicist Frederic Joliot-Curie (1900–1958), Pierre and Marie's son-in-law and Nobel prize winner.

## What is the **Genesis rock**?

The Genesis rock is a lunar rock brought to Earth by Apollo 15. It is approximately 4.15 billion years old, which is only 0.5 billion years younger than the generally accepted age of the moon.

# COMETS, METEORITES, ETC.

## Where are **asteroids** found?

The asteroids, also called the minor planets, are smaller than any of the nine major planets in the solar system and are not satellites of any major planet. The term asteroid means "starlike" because asteroids appear to be points of light when seen through a telescope.

Most asteroids are located between Mars and Jupiter, between 2.1 and 3.3 AUs (astronomical units) from the sun. Ceres, the largest and first to be discovered, was found by Giuseppe Piazzi on January 1, 1801, and has a diameter of 582 miles (936 kilometers). A second asteroid, Pallas, was discovered in 1802. Since then, astronomers have identified more than 18,000 asteroids and have established orbits for about 5,000 of them. Some of these have diameters of only 0.62 mile (one kilometer). Originally, astronomers thought the asteroids were remnants of a planet that had been destroyed; now they believe asteroids to be material that never became a planet, possibly because it was affected by Jupiter's strong gravity.

Not all asteroids are in this main asteroid belt. Three groups reside in the inner solar system. The Aten asteroids have orbits that lie primarily inside Earth's orbit. However, at their farthest point from the sun, these asteroids may cross Earth's orbit. The Apollo asteroids cross Earth's orbit; some come even closer than the moon. The Amor asteroids cross the orbit of Mars, and some come close to Earth's orbit. The Trojan asteroids move in virtually the same orbit as Jupiter but at points 60° ahead or 60° behind the planet. In 1977 Charles Kowal discovered an object now known as Chiron orbiting between Saturn and Uranus. Originally cataloged as an asteroid, Chiron was later observed to have a coma (a gaseous halo), and it may be reclassified as a comet.

## What was the **Tunguska Event**?

On June 30, 1908, a violent explosion occurred in the atmosphere over the Podkamennaya Tunguska River in a remote part of central Siberia. The blast's consequences were similar to an H-bomb going off, leveling thousands of square miles of forest. The shock of the explosion was heard more than 600 miles (960 kilometers) away. A number of theories have been proposed to account for this event.

Some people thought that a large meteorite or a piece of anti-matter had fallen to Earth. But a meteorite, composed of rock and metal, would have created a crater and none was found at the impact site. There are no high radiation levels in the area that would have resulted from the collision of anti-matter and matter. Two other theories include a mini-black hole striking the Earth or the crash of an extraterrestrial spaceship. However, a mini-black hole would have passed through the Earth and there is no record of a corresponding explosion on the other side of the world. As for the spaceship, no wreckage of such a craft was ever found.

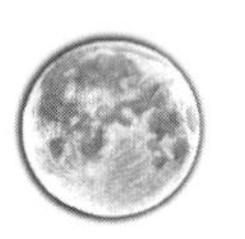



**An asteroid came close to hitting the Earth sometime in 2002. How much damage might it have done?**

Asteroid 2002 EM7, a 70-meter-long rock, is estimated to have the capability of releasing the energy equivalent of a four megaton nuclear bomb.

The most likely cause of the explosion was the entry into the atmosphere of a piece of a comet, which would have produced a large fireball and blast wave. Since a comet is composed primarily of ice, the fragment would have melted during its passage through the Earth's atmosphere, leaving no impact crater and no debris. Since the Tunguska Event coincided with the Earth's passage through the orbit of Comet Encke, the explosion could have been caused by a piece of that comet.

## From where do **comets** originate?

According to a theory developed by Dutch astronomer Jan Oort, there is a large cloud of gas, dust, and comets orbiting beyond Pluto out to perhaps 100,000 astronomical units (AU). Occasional stars passing close to this cloud disturb some of the comets from their orbits. Some fall inwards towards the sun.

Comets, sometimes called "dirty snowballs," are made up mostly of ice, with some dust mixed in. When a comet moves closer to the sun, the dust and ice of the core, or nucleus, heats up, producing a tail of material that trails along behind it. The tail is pushed out by the solar wind and almost always points away from the sun.

Most comets have highly elliptical orbits that carry them around the sun and then fling them back out to the outer reaches of the solar system, never to return. Occasionally, however, a close passage by a comet near one of the planets can alter a comet's orbit, making it stay in the middle or inner solar system. Such a comet is called a short-period comet because it passes close to the sun at regular intervals. The most famous short-period comet is Comet Halley, which reaches perihelion (the point in its orbit that is closest to the sun) about every 76 years. Comet Encke, with an orbital period of 3.3 years, is another short-period comet.

## When will **Halley's comet** return?

Halley's comet returns about every 76 years. It was most recently seen in 1985/1986 and is predicted to appear again in 2061, then in 2134. Every appearance of what is now known as Comet Halley has been noted by astronomers since the year 239 B.C.E.

The comet is named for Edmund Halley (1656–1742), England's second Astronomer Royal. In 1682 he observed a bright comet and noted that it was moving

Edmund Halley observed the comet that bears his name in 1682 and predicted its return every 76 years based upon earlier recorded sightings.

in an orbit similar to comets seen in 1531 and 1607. He concluded that the three comets were actually one and the same and that the comet had an orbit of 76 years. In 1705 Halley published *A Synopsis of the Astronomy of Comets,* in which he predicted that the comet seen in 1531, 1607, and 1682 would return in 1758. On Christmas night, 1758, a German farmer and amateur astronomer named Johann Palitzsch spotted the comet in just the area of the sky that Halley had foretold.

Prior to Halley, comets appeared at irregular intervals and were often thought to be harbingers of disaster and signs of divine wrath. Halley proved that they are natural objects subject to the laws of gravity.

## How does a **meteorite** differ from a **meteoroid**?

A meteorite is a natural object of extraterrestrial origin that survives passage through the Earth's atmosphere and hits the Earth's surface. A meteorite is often confused with a meteoroid or a meteor. A meteoroid is a small object in outer space, generally less than 30 feet (10 meters) in diameter. A meteor (sometimes called a shooting star) is the flash of light seen when an object passes through Earth's atmosphere and burns as a result of heating caused by friction. A meteoroid becomes a meteor when it enters the Earth's atmosphere; if any portion of a meteoroid lands on Earth, it is a meteorite.

There are three kinds of meteorites. Irons contain 85 percent to 95 percent iron; the rest of their mass is mostly nickel. Stony irons are relatively rare meteorites composed of about 50 percent iron and 50 percent silicates. Stones are made up mostly of silicates and other stony materials.

## When do **meteor showers** occur?

There are a number of groups of meteoroids orbiting the sun just as the Earth is. When Earth's orbit intercepts the path of one of these swarms of meteoroids, some of them enter Earth's atmosphere. When friction with the air causes a meteoroid to burn up, the streak, or shooting star, that is produced is called a meteor. Large numbers of meteors can produce a spectacular shower of light in the night sky. Meteor showers are named for the constellation that occupies the area of the sky from which they

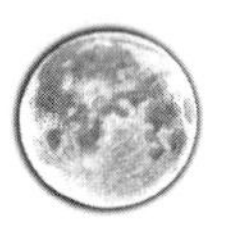



originate. Listed below are 10 meteor showers and the dates during the year during which they can be seen.

| Name of Shower | Dates |
|---|---|
| Quadrantids | January 1–6 |
| Lyrids | April 19–24 |
| Eta Aquarids | May 1–8 |
| Perseids | July 25–August 18 |
| Orionids | October 16–26 |
| Taurids | October 20–November 20 |
| Leonids | November 13–17 |
| Phoenicids | December 4–5 |
| Geminids | December 7–15 |
| Ursids | December 17–24 |

## How many **meteorites land on the Earth** in a given year?

Approximately 26,000 meteorites, each weighing over 3.5 ounces (99.2 grams) land on the Earth during a given year. Three thousand of these meteorites weigh more than 2.2 pounds (1 kg). This figure is compiled from the number of fireballs visually observed by the Canadian Camera Network. Of that number, only five or six falls are witnessed or cause property damage. The majority fall in the oceans, which cover over 70 percent of the Earth's surface.

## What are the **largest meteorites** that have been found in the world?

The famous Willamette (Oregon) iron, displayed at the American Museum of Natural History in New York, is the largest specimen found in the United States. It is 10 feet (3.048 meters) long and five feet (1.524 meters) high.

| Name | Location | Weight | |
|---|---|---|---|
| | | Tons | Tonnes |
| Hoba West | Namibia | 66.1 | 60 |
| Ahnighito (The Tent) | Greenland | 33.5 | 30.4 |
| Bacuberito | Mexico | 29.8 | 27 |
| Mbosi | Tanzania | 28.7 | 26 |
| Agpalik | Greenland | 22.2 | 20.1 |
| Armanty | Outer Mongolia | 22 | 20 |
| Willamette | Oregon, USA | 15.4 | 14 |
| Chupaderos | Mexico | 15.4 | 14 |
| Campo del Cielo | Argentina | 14.3 | 13 |
| Mundrabilla | Western Australia | 13.2 | 12 |
| Morito | Mexico | 12.1 | 11 |

## How do scientists know that some **meteorites** that were found in Antarctica came **from the Moon**?

Because of the high-quality reference collection of lunar rocks collected during space flights to the moon, the original 1979 meteorite find in Antarctica and the 10 subsequent findings were verified as lunar in origin.

# OBSERVATION AND MEASUREMENT

## Who was the first **Astronomer Royal**?

The first Astronomer Royal was John Flamsteed (1646–1719). He was appointed Astronomer Royal in 1675 when the Royal Greenwich Observatory was founded. Until 1972, the Astronomer Royal also served as the Director of the Royal Greenwich Observatory.

## Who is considered the founder of **systematic astronomy**?

The Greek scientist Hipparchus (fl. 146–127 B.C.E.) is considered to be the father of systematic astronomy. He measured as accurately as possible the directions of objects in the sky. He compiled the first catalog of stars, containing about 850 entries, and designated each star's celestial coordinates, indicating its position in the sky. Hipparchus also divided the stars according to their apparent brightness or magnitudes.

## What is a **light year**?

A light year is a measure of distance, not time. It is the distance that light, which travels in a vacuum at the rate of 186,282 miles (299,792 kilometers) per second, can travel in a year (365.25 days). This is equal to 5.87 trillion miles (9.46 trillion kilometers).

## Besides the light year, what other units are used to measure **distances in astronomy**?

The astronomical unit (AU) is often used to measure distances within the solar system. One AU is equal to the average distance between the Earth and the sun, or 92,955,630 miles (149,597,870 kilometers). The parsec is equal to 3.26 light years, or about 19.18 trillion miles (30.82 trillion kilometers).

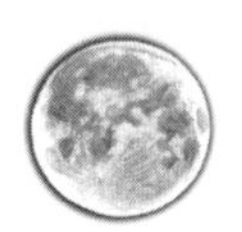



## How are **new celestial objects named**?

Many stars and planets have names that date back to antiquity. The International Astronomical Union (IAU), the professional astronomers organization, has attempted, in this century, to standardize names given to newly discovered celestial objects and their surface features.

Stars are generally called by their traditional names, most of which are of Greek, Roman, or Arabic origin. They are also identified by the constellation in which they appear, designated in order of brightness by Greek letters. Thus Sirius is also called alpha Canis Majoris, which means it is the brightest star in the constellation Canis Major. Other stars are called by catalog numbers, which include the star's coordinates. To the horror of many astronomers, several commercial star registries exist, and for a fee you can submit a star name to them. These names are not officially recognized by the IAU.

The IAU has made some recommendations for naming the surface features of the planets and their satellites. For example, features on Mercury are named for composers, poets, and writers; features of Venus for women; and features on Saturn's moon Mimas for people and places in Arthurian legend.

Comets are named for their discoverers. Newly discovered asteroids are first given a temporary designation consisting of the year of discovery plus two letters. The first letter indicates the half-month of discovery (A=first half of January, B=second half of January, etc.) and the second the order of discovery in that half-month. Thus asteroid 2002EM was the 13th (M) asteroid discovered in the first half of March (E) in 2002. After an asteroid's orbit is determined, it is given a permanent number and its discoverer is given the honor of naming it. Asteroids have been named after such diverse things as mythological figures (Ceres, Vesta), an airline (Swissair), and the Beatles (Lennon, McCartney, Harrison, Starr).

## What is an **astrolabe**?

Invented by the Greeks or Alexandrians around 100 B.C.E. or before, an astrolabe is a two-dimensional working model of the heavens, with sights for observations. It consists of two concentric, flat disks, one fixed, representing the observer on Earth, the other moving, which can be rotated to represent the appearance of the celestial sphere at a given moment. Given latitude, date, and time, the observer can read off the altitude and azimuth of the sun, the brightest stars, and the planets. By measuring the altitude of a particular body, one can find the time. The astrolabe can also be used to find times of sunrise, sunset, twilight, or the height of a tower or depth of a well. After 1600, it was replaced by the sextant and other more accurate instruments.

## Who invented the **telescope**?

Hans Lippershey (ca. 1570–1619), a German-Dutch lens grinder and spectacle maker, is generally credited with inventing the telescope in 1608 because he was the first sci-

## What is widely considered to be one of the earliest celestial observatories?

Built in England over a period of years between 2500 and 1700 B.C.E., Stonehenge is one of the earliest observatories or observatory-temples. It is widely believed that its primary function was to observe the mid-summer and mid-winter solstices.

entist to apply for a patent. Two other inventors, Zacharias Janssen and Jacob Metius, also developed telescopes. Modern historians consider Lippershey and Janssen as the two likely candidates for the title of inventor of the telescope, with Lippershey possessing the strongest claim. Lippershey used his telescope for observing grounded objects from a distance.

In 1609, Galileo also developed his own refractor telescope for astronomical studies. Although small by today's standards, the telescope enabled Galileo to observe the Milky Way and to identify blemishes on the moon's surface as craters.

## What are the differences between **reflecting** and **refracting telescopes**?

Reflecting telescopes capture light using a mirror while refracting telescopes capture light with a lens. The advantages of reflecting telescopes are: 1) they collect light with a mirror so there is no color fringing, and 2) since a mirror can be supported at the back there is no size limit. In an effort to alleviate the problem of color fringing always associated with lenses, Newton built a reflecting telescope in 1668 that collected light with mirrors.

## What is the **Very Large Array** (VLA) and what information have we learned from it?

The Very Large Array (VLA) is one of the world's premier astronomical radio observatories. The VLA consists of 27 antennas arranged in a huge Y pattern up to 36km (22 miles) across—roughly one-and-a-half times the size of Washington, D.C. Each antenna is 25 meters (81 feet) in diameter; they are combined electronically to give the resolution of an antenna 36km (22 miles) across, with the sensitivity of a dish 130 meters (422 feet) in diameter. Each of the 27 radio telescopes in the VLA is the size of a house and can be moved on train tracks. In its twenty-second year of operation, the VLA has been one of the most productive observatories with more than 2,200 scientists using it for more than 10,000 separate observing projects. The VLA has been used to discover water on the planet Mercury, radio-bright coronae around ordinary stars, micro-quasars in our galaxy, gravitationally induced Einstein rings around distant galaxies, and radio counterparts to cosmologically distant gamma-ray bursts. The vast

A series of radio telescope dishes make up the Very Large Array in New Mexico.

size of the VLA has allowed astronomers to study the details of super-fast cosmic jets, and even map the center of our galaxy.

## Who is the **Hubble** for whom the **space telescope** is named?

Edwin Powell Hubble (1889–1953) was an American astronomer known for his studies of galaxies. His study of nebulae, or clouds—the faint, unresolved luminous patches in the sky—showed that some of them were large groups of many stars. Hubble classified galaxies by their shapes as being spiral, elliptical, or irregular.

Hubble's Law establishes a relationship between the velocity of recession of a galaxy and its distance. The speed at which a galaxy is moving away from our solar system (measured by its redshift, the shift of its light to longer wavelengths, presumed to be caused by the Doppler effect) is directly proportional to the galaxy's distance from it.

The Hubble Space Telescope was deployed by the space shuttle Discovery on April 25, 1990. The telescope, which would be free of distortions caused by the Earth's atmosphere, was designed to see deeper into space than any telescope on land. However, on June 27, 1990, the National Aeronautics and Space Administration announced that the telescope had a defect in one of its mirrors that prevented it from properly focusing. Although other instruments, including one designed to make observations in ultraviolet light, were still operating, nearly 40 percent of the telescope's experiments had to be postponed until repairs were made. On December 2, 1993, astronauts were able to make the necessary repairs. Four of Hubble's six gyroscopes were replaced as well as two solar panels. Hubble's primary camera, which had a flawed mirror, was also replaced. Since that mission four other servicing missions have been conducted, dramatically improving the HST's capabilities.

# EXPLORATION

## Who was the first person to propose **space rockets**?

In 1903, Konstantin E. Tsiolkovsky, a Russian high school teacher, completed the first scientific paper on the use of rockets for space travel. Several years later, Robert H. Goddard of the United States and Herman Oberth of Germany awakened wider scientific interest in space travel. These three men worked individually on many of the technical problems of rocketry and space travel. They are known, therefore, as the "fathers of space flight." In 1919, Goddard wrote the paper, "A Method of Reaching Extreme Altitudes," which explained how rockets could be used to explore the upper atmosphere and described a way to send a rocket to the moon. During the 1920s Tsiolkovsky wrote a series of new studies that included detailed descriptions of multi-stage rockets. In 1923, Oberth wrote "The Rocket into Interplanetary Space," which discussed the technical problems of space flight and also described what a spaceship would be like.

## What is the difference between **zero-gravity** and **microgravity**?

Zero-gravity is the absence of gravity; a condition in which the effects of gravity are not felt; weightlessness. Microgravity is a condition of very low gravity, especially approaching weightlessness. On a spaceship, while in zero- or microgravity, objects would fall freely and float weightlessly. Both terms, however, are technically incorrect. The gravitation in orbit is only slightly less than the gravitation on the earth. A spacecraft and its contents continuously fall toward earth. It is the spacecraft's immense forward speed that appears to make the earth's surface curve away as the vehicle falls toward it. The continuous falling seems to eliminate the weight of everything inside the spacecraft. For this reason, the condition is sometimes referred to as weightlessness or zero-gravity.

## How probable is it that **intelligent life** exists on other planets?

The possibility of intelligent life depends on several factors. An estimation can be calculated by using an equation developed originally by American astronomer Frank Drake (b. 1930). Drake's equation reads $N=N_*f_pn_ef_1f_if_cf_L$. This means that the number of advanced civilizations (N) is equal to

$N_*$, the number of stars in the Milky Way galaxy, times

$f_p$, the fraction of those stars that have planets, times

$n_e$, the number of planets capable of supporting life, times

$f_l$, the fraction of planets suitable for life on which life actually arises, times

$f_i$, the fraction of planets where intelligent life evolves, times

$f_c$, the fraction of planets with intelligent life that develops a technically advanced civilization, times

$f_L$, the fraction of time that a technical civilization lasts.

The equation is obviously subjective, and the answer depends on whether optimistic or pessimistic numbers are assigned to the various factors. However, the galaxy is so large that the possibility of life elsewhere cannot be ruled out.

## What is meant by the phrase **"greening of the galaxy"**?

The expression means the spreading of human life, technology, and culture through interstellar space and eventually across the entire Milky Way galaxy, the Earth's home galaxy.

## When was the **Outer Space Treaty** signed?

The United Nations Outer Space Treaty was signed on January 23, 1967. The treaty provides a framework for the exploration and sharing of outer space. It governs the outer space activities of nations that wish to exploit and make use of space, the moon, and other celestial bodies. It is based on a humanist and pacifist philosophy and on the principle of the nonappropriation of space and the freedom that all nations have to explore and use space. A very large number of countries have signed this agreement, including those from the Western alliance, the former Eastern bloc, and nonaligned countries.

Space law, or those rules governing the space activities of various countries, international organizations, and private industries, has been evolving since 1957, when the General Assembly of the United Nations created the Committee on the Peaceful Uses of Outer Space (COPUOS). One of its subcommittees was instrumental in drawing up the 1967 Outer Space Treaty.

## What is a **"close encounter of the third kind"**?

UFO expert J. Allen Hynek (1910–1986) developed the following scale to describe encounters with extraterrestrial beings or vessels:

*Close Encounter of the First Kind*—sighting of a UFO at close range with no other physical evidence.

### Is anyone looking for extraterrestrial life?

A program called SETI (the Search for Extraterrestrial Intelligence) began in 1960, when American astronomer Frank Drake (b. 1930) spent three months at the National Radio Astronomy Observatory in Green Bank, West Virginia, searching for radio signals coming from the nearby stars Tau Ceti and Epsilon Eridani. Although no signals were detected and scientists interested in SETI have often been ridiculed, support for the idea of seeking out intelligent life in the universe has grown.

Project Sentinel, which used a radio dish at Harvard University's Oak Ridge Observatory in Massachusetts, could monitor 128,000 channels at a time. This project was upgraded in 1985 to META (Megachannel Extraterrestrial Assay), thanks in part to a donation by filmmaker Steven Spielberg. Project META is capable of receiving 8.4 million channels. NASA began a 10-year search in 1992 using radio telescopes in Arecibo, Puerto Rico, and Barstow, California.

Scientists are searching for radio signals that stand out from the random noises caused by natural objects. Such signals might repeat at regular intervals or contain mathematical sequences. There are millions of radio channels and a lot of sky to be examined. As of October 1995, Project BETA (Billion-channel Extraterrestrial Assay) has been scanning a quarter of a billion channels. This new design improves upon Project META 300-fold, making the challenge of scanning millions of radio channels seem less daunting. SETI has since developed other projects, some "piggybacking" on radio telescopes while engaged in regular uses. A program launched in 1999, SETI@HOME, uses the power of home computers while they are at rest.

*Close Encounter of the Second Kind*—sighting of a UFO at close range, but with some kind of proof, such as a photograph, or an artifact from a UFO.

*Close Encounter of the Third Kind*—sighting of an actual extraterrestrial being.

*Close Encounter of the Fourth Kind*—abduction by an extraterrestrial spacecraft.

## Who was the **first man in space**?

Yuri Gagarin (1934–1968), a Soviet cosmonaut, became the first man in space when he made a full orbit of the Earth in *Vostok I* on April 12, 1961. Gagarin's flight lasted only one hour and 48 minutes, but as the first man in space, he became an international hero. Partly because of this Soviet success, U.S. President John F. Kennedy (1917–1963) announced on May 25, 1961, that the United States would land a man on the moon before the end of the decade. The United States took its first step toward that goal when it launched the first American into orbit on February 20, 1962. Astronaut John H. Glenn Jr. (b. 1921) completed three orbits in *Friendship 7* and traveled about 81,000 miles (130,329 kilometers). Prior to this, on May 5, 1961, Alan B. Shepard Jr. (b. 1923) became the first American to pilot a spaceflight, aboard *Freedom 7*. This sub-orbital flight reached an altitude of 116.5 miles (187.45 kilometers).

Russian Air Force Major Yuri Gagarin became the first man in space on April 12, 1961.

## What did NASA mean when it said ***Voyager 1*** and ***2*** would take a "grand tour" of the planets?

Once every 176 years the giant outer planets—Jupiter, Saturn, Uranus, and Neptune—align themselves in such a pattern that a spacecraft launched from Earth to Jupiter at just the right time might be able to visit the other three planets on the same mission. A technique called "gravity assist" used each planet's gravity as a power boost to point *Voyager* toward the next planet. The first opportune year for the "grand tour" was 1977.

## What is the **message** attached to the ***Voyager*** spacecraft?

*Voyager 1* (launched September 5, 1977) and *Voyager 2* (launched August 20, 1977) were unmanned space probes designed to explore the outer planets and then travel out of the solar system. A gold-coated copper phonograph record containing a message to any possible extraterrestrial civilization that they might encounter is attached to each spacecraft. The record contains both video and audio images of Earth and the civilization that sent this message to the stars.

The record begins with 118 pictures. These show the Earth's position in the galaxy; a key to the mathematical notation used in other pictures; the sun; other planets in the solar system; human anatomy and reproduction; various types of terrain (seashore, desert, mountains); examples of vegetation and animal life; people of both sexes and of all ages and ethnic types engaged in a number of activities; structures (from grass huts to the Taj Mahal to the Sydney Opera House) showing diverse architectural styles; and means of transportation, including roads, bridges, cars, planes, and space vehicles.

The pictures are followed by greetings from Jimmy Carter, who was then president of the United States, and Kurt Waldheim, then Secretary General of the United Nations. Brief messages in 54 languages, ranging from ancient Sumerian to English, are included, as is a "song" of the humpback whales.

The next section is a series of sounds common to the Earth. These include thunder, rain, wind, fire, barking dogs, footsteps, laughter, human speech, the cry of an infant, and the sounds of a human heartbeat and human brainwaves.

The record concludes with approximately 90 minutes of music: "Earth's Greatest Hits." These musical selections were drawn from a broad spectrum of cultures and include such diverse pieces as a Pygmy girl's initiation song; bagpipe music from Azerbaijan; the Fifth Symphony, First Movement by Ludwig von Beethoven; and "Johnny B. Goode" by Chuck Berry.

It will be tens, or even hundreds of thousands of years before either Voyager comes close to another star, and perhaps the message will never be heard; but it is a sign of humanity's hope to encounter life elsewhere in the universe.

## Which astronauts have **walked on the moon**?

Twelve astronauts have walked on the moon. Each Apollo flight had a crew of three. One crew member remained in orbit in the command service module (CSM) while the other two actually landed on the moon.

*Apollo 11,* July 16–24, 1969
Neil A. Armstrong
Edwin E. Aldrin, Jr.
Michael Collins (CSM pilot, did not walk on the moon)

*Apollo 12,* November 14–24, 1969
Charles P. Conrad
Alan L. Bean
Richard F. Gordon, Jr. (CSM pilot, did not walk on the moon)

*Apollo 14,* January 31–February 9, 1971
Alan B. Shepard, Jr.
Edgar D. Mitchell
Stuart A. Roosa (CSM pilot, did not walk on the moon)

*Apollo 15,* July 26–August 7, 1971
David R. Scott
James B. Irwin
Alfred M. Worden (CSM pilot, did not walk on the moon)

*Apollo 16,* April 16–27, 1972
John W. Young
Charles M. Duke, Jr.
Thomas K. Mattingly, II (CSM pilot, did not walk on the moon)

*Apollo 17,* December 7–19, 1972
Eugene A. Cernan
Harrison H. Schmitt
Ronald E. Evans (CSM pilot, did not walk on the moon)

**Laika made history as the first living creature to orbit the Earth aboard the Soviet *Sputnik 2.***

## Which **manned space flight** was the longest?

Dr. Valerij Polyakov manned a flight to the space station *Mir* on January 8, 1994. He returned aboard *Soyuz TM-20* on March 22, 1995, making the total time in space equal 438 days and 18 hours.

## When and what was the **first animal** sent into orbit?

A small female dog named Laika, aboard the Soviet *Sputnik 2,* launched November 3, 1957, was the first animal sent into orbit. This event followed the successful Soviet launch on October 4, 1957, of *Sputnik 1,* the first man-made satellite ever placed in orbit. Laika was placed in a pressurized compartment within a capsule that weighed 1,103 pounds (500 kilograms). After a few days in orbit, she died, and *Sputnik 2* reentered the Earth's atmosphere on April 14, 1958. Some sources list the dog as a Russian samoyed laika named "Kudyavka" or "Limonchik."

## What were the **first monkeys** and **chimpanzees** in space?

On a United States *Jupiter* flight on December 12, 1958, a squirrel monkey named Old Reliable was sent into space, but not into orbit. The monkey drowned during recovery.

On another *Jupiter* flight, on May 28, 1959, two female monkeys were sent 300 miles (482.7 kilometers) high. Able was a six-pound (2.7-kilogram) rhesus monkey and Baker was an 11-ounce (0.3-kilogram) squirrel monkey. Both were recovered alive.

A chimpanzee named Ham was used on a *Mercury* flight on January 31, 1961. Ham was launched to a height of 157 miles (253 kilometers) into space but did not go into orbit. His capsule reached a maximum speed of 5,857 miles (9,426 kilometers) per hour and landed 422 miles (679 kilometers) downrange in the Atlantic Ocean, where he was recovered unharmed.

On November 29, 1961, the United States placed a chimpanzee named Enos into orbit and recovered him alive after two complete orbits around the Earth. Like the Soviets, who usually used dogs, the United States had to obtain information on the effects of space flight on living beings before they could actually launch a human into space.

## Who were the first man and woman to **walk in space**?

On March 18, 1965, the Soviet cosmonaut Alexei Leonov (b. 1934) became the first person to walk in space when he spent 10 minutes outside his *Voskhod 2* spacecraft. The first woman to walk in space was Soviet cosmonaut Svetlana Savitskaya (b. 1947), who, during her second flight aboard the *Soyuz T-12* (July 17, 1984), performed 3.5 hours of extravehicular activity.

The first American to walk in space was Edward White II (1930–1967) from the spacecraft *Gemini 4* on June 3, 1965. White spent 22 minutes floating free attached to the *Gemini* by a lifeline. The photos of White floating in space are perhaps some of the most familiar of all space shots. Kathryn D. Sullivan (b. 1951) became the first American woman to walk in space when she spent 3.5 hours outside the *Challenger* orbiter during the space shuttle mission 41G on October 11, 1984.

American astronaut Bruce McCandless II (b. 1937) performed the first untethered space walk from the space shuttle *Challenger* on February 7, 1984, using an MMU (manual maneuvering unit) backpack.

## Who was the **first woman** in space?

Valentina V. Tereshkova-Nikolaeva (b. 1937), a Soviet cosmonaut, was the first woman in space. She was aboard the *Vostok 6,* launched June 16, 1963. She spent three days circling the Earth, completing 48 orbits. Although she had little cosmonaut training, she was an accomplished parachutist and was especially fit for the rigors of space travel.

The United States space program did not put a woman in space until 20 years later when, on June 18, 1983, Sally K. Ride (b. 1951) flew aboard the space shuttle *Challenger* mission STS-7. In 1987, she moved to the administrative side of NASA and was instrumental in issuing the "Ride Report," which recommended future missions and direction for NASA. She retired from NASA in August 1987 to become a research fellow at Stanford University after serving on the Presidential Commission that investigated the *Challenger* disaster. At present, she is the director of the California Space Institute at the University of California San Diego.

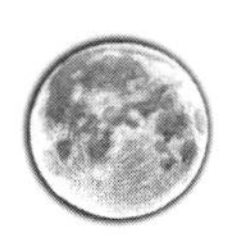



## What were the **first words spoken** by an astronaut after touchdown of the lunar module on the ***Apollo 11*** flight, and by an astronaut standing on the moon?

On July 20, 1969, at 4:17:43 p.m. Eastern Daylight Time (20:17:43 Greenwich Mean Time), Neil A. Armstrong (b. 1930) and Edwin E. Aldrin Jr. (b. 1930) landed the lunar module *Eagle* on the moon's Sea of Tranquility, and Armstrong radioed: "Houston, Tranquility Base here. The *Eagle* has landed." Several hours later, when Armstrong descended the lunar module ladder and made the small jump between the *Eagle* and the lunar surface, he announced: "That's one small step for man, one giant leap for mankind." The article "a" was missing in the live voice transmission, and was later inserted in the record to amend the message to "one small step for a man."

## What material was used in the **United States flag** planted on the moon by astronauts Neil Armstrong and Edwin Aldrin Jr.?

The astronauts erected a three-by-five foot nylon U.S. flag, its top edge braced by a spring wire to keep it extended.

## What was the **first meal** on the moon?

American astronauts Neil A. Armstrong (b. 1930) and Edwin E. Aldrin, Jr. (b. 1930) ate four bacon squares, three sugar cookies, peaches, pineapple-grapefruit drink, and coffee before their historic moonwalk on July 20, 1969.

## Who made the **first golf shot** on the moon?

Alan B. Shepard Jr. (b. 1923), commander of *Apollo 14*, launched on January 31, 1971, made the first golf shot. He attached a six iron to the handle of the contingency sample return container, dropped a golf ball on the moon, and took a couple of one-handed swings. He missed with the first, but connected with the second. The ball, he reported, sailed for miles and miles.

## What are some of the accomplishments of **female astronauts**?

First American woman in space: Sally K. Ride—June 18, 1983, aboard *Challenger* STS-7.

First American woman to walk in space: Kathryn D. Sullivan—October 11, 1984, aboard *Challenger* STS 41G.

First woman to make three spaceflights: Shannon W. Lucid—June 17, 1985; October 18, 1989; and August 2, 1991.

First African American woman in space: Mae Carol Jemison—September 12, 1992, aboard *Endeavour*.

Astronaut Sally K. Ride became the first U.S. woman in space on June 18, 1983.

First American woman space shuttle pilot: Eileen M. Collins—February 3, 1995, aboard *Discovery*.

## Who was the **first African American** in space?

Guion S. Bluford, Jr. (b. 1942), became the first African American to fly in space during the space shuttle *Challenger* mission STS-8 (August 30–September 5, 1983). Astronaut Bluford, who holds a Ph.D. in aerospace engineering, made a second shuttle flight aboard *Challenger* mission STS-61-A/Spacelab D1 (October 30–November 6, 1985). The first black man to fly in space was Cuban cosmonaut Arnaldo Tamayo-Mendez, who was aboard *Soyuz 38* and spent eight days aboard the Soviet space station *Salyut 6* during September 1980. Dr. Mae C. Jemison became the first African American woman in space on September 12, 1992 aboard the space shuttle *Endeavour* mission Spacelab-J.

## Who were the **first married couple** to go into space together?

Astronauts Jan Davis and Mark Lee were the first married couple in space. They flew aboard the space shuttle *Endeavor* on an eight-day mission that began on September 12, 1992. Ordinarily, NASA bars married couples from flying together. An exception was made for Davis and Lee because they had no children and had begun training for the mission long before they got married.

## How many successful space flights were **launched in 2000**, and by what countries?

A total of 82 flights that achieved Earth orbit or beyond were made in 2000:

| Country or Organization | Number of Launches |
|---|---|
| U.S.S.R. | 35 |
| United States | 28 |
| European Space Agency | 12 |
| People's Republic of China | 5 |
| Ukraine | 2 |

In 1990, there were 116 such launches, with 75 by the former U.S.S.R.; 27 (including seven commercial launches) by the United States; five by the European Space Agency; five by the People's Republic of China; and one by Israel.

## Who has spent the **most time in space**?

As of December 31, 2001, cosmonaut Sergei Vasilyevich Avdeyev has accumulated the most spaceflight time—747.6 days in three flights.

## When was the **first United States satellite** launched?

*Explorer 1,* launched January 31, 1958, by the U.S. Army, was the first United States satellite launched into orbit. This 31-pound (14.06-kilogram) satellite carried instrumentation that led to the discovery of the Earth's radiation belts, which would be named after University of Iowa scientist James A. Van Allen. It followed four months after the launching of the world's first satellite, the Soviet Union's *Sputnik 1.* On October 3, 1957, the Soviet Union placed the large 184-pound (83.5-kilogram) satellite into low Earth orbit. It carried instrumentation to study the density and temperature of the upper atmosphere, and its launch was the event that opened the age of space.

## What is the mission of the ***Galileo*** spacecraft?

*Galileo,* launched October 18, 1989, required almost six years to reach Jupiter after looping past Venus once and the Earth twice. The *Galileo* spacecraft was designed to make a detailed study of Jupiter and its rings and moons over a period of years. On December 7, 1995, it released a probe to analyze the different layers of Jupiter's atmosphere. *Galileo* recorded a multitude of measurements of the planet, its four largest moons, and its mammoth magnetic field. The mission was originally scheduled to continue until the end of 1997, but, since it has continued to operate successfully, missions exploring Jupiter's moons were added in 1997, 1999, and 2001. *Galileo* is scheduled to plunge into Jupiter's atmosphere in September 2003.

## Who was the **founder** of the **Soviet space program**?

Sergei P. Korolev (1907–1966) made enormous contributions to the development of Soviet manned space flight, and his name is linked with their most significant space achievements. Trained as an aeronautical engineer, he directed the Moscow group studying the principles of rocket propulsion, and in 1946 took over the Soviet program to develop long-range ballistic rockets. Under Korolev, the Soviets used these rockets for space projects and launched the world's first satellite on October 4, 1957. Besides a vigorous, unmanned, interplanetary research program, Korolev's goal was to place men in space, and following tests with animals his manned space flight program was initiated when Yuri Gagarin (1934–1968) was successfully launched into Earth's orbit.

## How many **fatalities** have occurred during space-related missions?

The 14 astronauts and cosmonauts listed below died in space-related accidents.

| Date | Astronaut/Cosmonaut | Mission |
|---|---|---|
| January 27, 1967 | Roger Chaffee (U.S.) | Apollo 1 |
| January 27, 1967 | Edward White II (U.S.) | Apollo 1 |
| January 27, 1967 | Virgil "Gus" Grissom (U.S.) | Apollo 1 |
| April 24, 1967 | Vladimir Komarov (U.S.S.R.) | Soyuz 1 |
| June 29, 1971 | Viktor Patsayev (U.S.S.R.) | Soyuz 11 |
| June 29, 1971 | Vladislav Volkov (U.S.S.R.) | Soyuz 11 |
| June 29, 1971 | Georgi Dobrovolsky (U.S.S.R.) | Soyuz 11 |
| January 28, 1986 | Gregory Jarvis (U.S.) | STS 51L |
| January 28, 1986 | Christa McAuliffe (U.S.) | STS 51L |
| January 28, 1986 | Ronald McNair (U.S.) | STS 51L |
| January 28, 1986 | Ellison Onizuka (U.S.) | STS 51L |
| January 28, 1986 | Judith Resnik (U.S.) | STS 51L |
| January 28, 1986 | Francis Scobee (U.S.) | STS 51L |
| January 28, 1986 | Michael Smith (U.S.) | STS 51L |

Chaffee, Grissom, and White died in a cabin fire during a test firing of the *Apollo 1* rocket. Komarov was killed in *Soyuz 1* when the capsule's parachute failed. Dobrovolsky, Patsayev, and Volkov were killed during the *Soyuz II*'s re-entry when a valve accidently opened and released their capsule's atmosphere. Jarvis, McAuliffe, McNair, Onizuka, Resnik, Scobee, and Smith died when the space shuttle *Challenger* STS 51L exploded 73 seconds after lift-off.

In addition, 19 other astronauts and cosmonauts have died of non-space related causes. Fourteen of these died in air crashes, four died of natural causes, and one died in an auto crash.

## What was the **worst disaster** in the U.S. space program and what caused it?

*Challenger* mission STS 51L was launched on January 28, 1986, but exploded 73 seconds after lift-off. The entire crew of seven was killed, and the *Challenger* was completely destroyed. The investigation of the *Challenger* tragedy was performed by the Rogers Commission, established and named for its chairman, former Secretary of State William Rogers.

The consensus of the Rogers Commission (which studied the accident for several months) and participating investigative agencies was that the accident was caused by a failure in the joint between the two lower segments of the right solid rocket motor. The specific failure was the destruction of the seals that are intended to prevent hot gases from leaking through the joint during the propellant burn of the rocket motor.

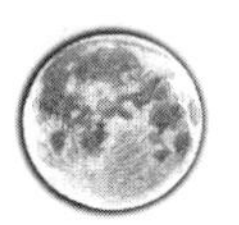



The evidence assembled by the commission indicated that no other element of the space shuttle system contributed to this failure.

Although the commission did not affix blame to any individuals, the public record made clear that the launch should not have been made that day. The weather was unusually cold at Cape Canaveral, and temperatures had dipped below freezing during the night. Test data had suggested that the seals (called O-rings) around the solid rocket booster joints lost much of their effectiveness in very cold weather.

## What were some of the accomplishments of the **first nine *Challenger* spaceflights**?

First American woman in space—Sally Ride

First African American man in space—Guion S. Bluford Jr.

First American woman to spacewalk—Kathryn Sullivan

First shuttle spacewalk—Donald Peterson and Story Musgrave

First untethered spacewalk—Robert Stewart and Bruce McCandless

First satellite repair in orbit—Pinky Nelson and Ox Van Hoften

First Coke and Pepsi in orbit—1985

## What is the composition of the **tiles** on the **underside of the space shuttle** and how hot do they get?

The 20,000 tiles are composed of a low-density, high purity silica fiber insulator hardened by ceramic bonding. Bonded to a Nomex fiber felt pad, each tile is directly bonded to the shuttle exterior. The maximum surface temperature can reach up to 922K to 978K (649°C to 704°C or 1,200°F to 1,300°F).

## What are the **liquid fuels** used by the space shuttles?

Liquid hydrogen is used as a fuel, with liquid oxygen used to burn it. These two fuels are stored in chambers separately and then mixed to combust the two. Because oxygen must be kept below –183°C to remain a liquid, and hydrogen must be at –253°C to remain a liquid, they are both difficult to handle, but make useful rocket fuel.

# EARTH

## AIR

*See also: Climate and Weather*

### What is the **composition** of the **Earth's atmosphere**?

The Earth's atmosphere, apart from water vapor and pollutants, is composed of 78 percent nitrogen, 21 percent oxygen, and less than 1 percent each of argon and carbon dioxide. There are also traces of hydrogen, neon, helium, krypton, xenon, methane, and ozone. The Earth's original atmosphere was probably composed of ammonia and methane; 20 million years ago the air started to contain a broader variety of elements.

### How many **layers** does the **Earth's atmosphere** contain?

The atmosphere, the "skin" of gas that surrounds the Earth, consists of five layers that are differentiated by temperature:

*The troposphere* is the lowest level; it averages about seven miles (11 kilometers) in thickness, varying from five miles (eight kilometers) at the poles to 10 miles (16 kilometers) at the equator. Most clouds and weather form in this layer. Temperature decreases with altitude in the troposphere.

*The stratosphere* ranges between seven and 30 miles (11 to 48 kilometers) above the Earth's surface. The ozone layer, important because it absorbs most of the sun's harmful ultraviolet radiation, is located in this band. Temperatures rise slightly with altitude to a maximum of about 32°F (0°C).

*The mesosphere* (above the stratosphere) extends from 30 to 55 miles (48 to 85 kilometers) above the Earth. Temperatures decrease with altitude to –130°F (–90°C).

James Van Allen discovered two regions of highly charged particles above the Earth's equator. Van Allen appears here (center) with William Pickering and Wernher von Braun, holding a model of the first successfully launched U.S. satellite, *Explorer.*

*The thermosphere* (also known as the hetereosphere) is between 55 to 435 miles (85 to 700 kilometers). Temperatures in this layer range to 2696°F (1475°C).

*The exosphere,* beyond the thermosphere, applies to anything above 435 miles (700 kilometers). In this layer, temperature no longer has any meaning.

*The ionosphere* is a region of the atmosphere that overlaps the others, reaching from 30 to 250 miles (48 to 402 kilometers). In this region, the air becomes ionized (electrified) from the sun's ultraviolet rays, etc. This area affects the transmission and reflection of radio waves. It is divided into three regions: the D region (at 35 to 55 miles [56 to 88 kilometers]), the E Region (Heaviside-Kennelly Layer, 55 to 95 miles [88 to 153 kilometers]), and the F Region (Appleton Layer, 95 to 250 miles [153 to 402 kilometers]).

## What are the **Van Allen belts**?

The Van Allen belts (or zones) are two regions of highly charged particles above the Earth's equator trapped by the magnetic field that surrounds the Earth. Also called the magnetosphere, the first belt extends from a few hundred to about 2,000 miles (3,200 kilometers) above the Earth's surface and the second is between 9,000 and 12,000 miles (14,500 to 19,000 kilometers). The particles, mainly protons and electrons, come from the solar wind and cosmic rays. The belts are named in honor of James Van Allen (b. 1914), the American physicist who discovered them in 1958 and 1959 with

the aid of radiation counters carried aboard the artificial satellites, *Explorer I* (1958) and *Pioneer 3* (1959).

In May 1998 there were a series of large, solar disturbances that caused a new Van Allen belt to form in the so-called "slot region" between the inner and outer Van Allen belts. The new belt eventually disappeared once the solar activity subsided. There were also a number of satellite upsets around the same time involving the *Galaxy IV* satellite, Iridium satellites, and others. This is not the first time that a temporary new belt has been observed to form in the same region, but it takes a prolonged period of solar storm activity to populate this region with particles.

### Why is the **sky blue**?

The sunlight interacting with the Earth's atmosphere makes the sky blue. In outer space the astronauts see blackness because outer space has no atmosphere. Sunlight consists of light waves of varying wavelengths, each of which is seen as a different color. The minute particles of matter and molecules of air in the atmosphere intercept and scatter the white light of the sun. A larger portion of the blue color in white light is scattered, more so than any other color because the blue wavelengths are the shortest. When the size of atmospheric particles are smaller than the wavelengths of the colors, selective scattering occurs—the particles only scatter one color and the atmosphere will appear to be that color. Blue wavelengths especially are affected, bouncing off the air particles to become visible. This is why the sun looks yellow (yellow equals white minus blue). At sunset, the sky changes color because as the sun drops to the horizon, sunlight has more atmosphere to pass through and loses more of its blue wavelengths. The orange and red, having the longer wavelengths and making up more of sunlight at this distance, are most likely to be scattered by the air particles.

## PHYSICAL CHARACTERISTICS, ETC.

*See also: Space—Planets and Moons*

### What is the **mass** of the **Earth**?

The mass of the Earth is estimated to be 6 sextillion, 588 quintillion short tons (6.6 sextillion short tons) or $5.97 \times 10^{24}$ kilograms, with the Earth's mean density being 5.515 times that of water (the standard). This is calculated from using the parameters of an ellipsoid adopted by the International Astronomical Union in 1964 and recognized by the International Union of Geodesy and Geophysics in 1967.

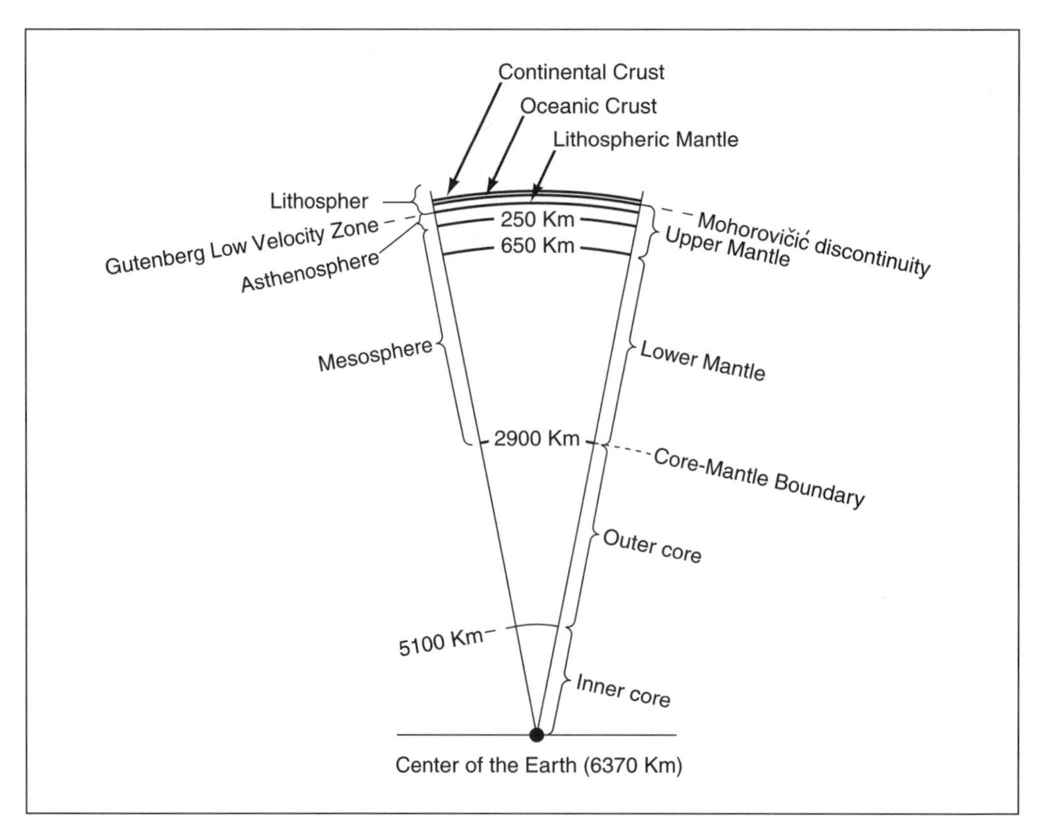


**A diagram of the interior of the Earth.**

## What is the **interior of the Earth** like?

The Earth is divided into a number of layers. The topmost layer is the crust, which contains about 0.6 percent of the Earth's volume. The depth of the crust varies from 3.5 to five miles (five to nine kilometers) beneath the oceans to 50 miles (80 kilometers) beneath some mountain ranges. The crust is formed primarily of rocks such as granite and basalt.

Between the crust and the mantle is a boundary known as the Mohorovičić discontinuity (or Moho for short), named for the Croatian seismologist, Andrija Mohorovičić (1857–1936), who discovered it in 1909. Below the Moho is the mantle, extending down about 1,800 miles (2,900 kilometers). The mantle is composed mostly of oxygen, iron, silicon, and magnesium, and accounts for about 82 percent of the Earth's volume. Although mostly solid, the upper part of the mantle, called the asthenosphere, is partially liquid.

The core-mantle boundary, also called the Gutenberg discontinuity for the German-American seismologist, Beno Gutenberg (1889–1960), separates the mantle from the core. Made up primarily of nickel and iron, the core contains about 17 percent of

the Earth's volume. The outer core is liquid and extends from the base of the mantle to a depth of about 3,200 miles (5,155 kilometers). The solid inner core reaches from the bottom of the outer core to the center of the Earth, about 3,956 miles (6,371 kilometers) deep. The temperature of the inner core is estimated to be about 7,000°F (3,850°C).

## What causes **sinkholes**?

A sinkhole is a depression shaped like a well or funnel that occurs in a land surface. Most common in limestone regions, sinkholes are usually formed by the dissolving action of groundwater or the seepage of above-ground streams into the limestone below, causing cracks or fractures in subterranean rock. The collapse of cave roofs can also cause large sinkholes. The resulting depression may be several miles in diameter.

## What is at the **center of the Earth**?

Geophysicists have held since the 1940s that the Earth's interior core is a partly crystallized sphere of iron and nickel that is gradually cooling and expanding. As it cools, this inner core releases energy to an outer core called the fluid core, which is composed of iron, nickel, and lighter elements, including sulfur and oxygen. Another model called the "nuclear earth model" holds that there is a small core, perhaps five miles wide, of uranium and plutonium surrounded by a nickel-silicon compound. The uranium and plutonium work as a natural nuclear reactor, generating radiating energy in the form of heat, which in turn drives charged particles to create the Earth's magnetic field. The traditional model of the Earth's core is still dominant; however, scientists have yet to disprove the nuclear earth model.

## How does the **temperature of the Earth** change as one goes deeper underground?

The Earth's temperature increases with depth. Measurements made in deep mines and drill-holes indicate that the rate of temperature increase varies from place to place in the world, ranging from 59° to 167°F (15 to 75°C) per kilometer in depth. Actual temperature measurements cannot be made beyond the deepest drill-holes, which are a little more than 6.2 miles (10 kilometers) deep. Estimates suggest that the temperatures at the Earth's center can reach values of 5,000°F (2,760°C) or higher.

## What are the **highest** and **lowest** points on Earth?

The highest point on land is the top of Mt. Everest (in the Himalayas on the Nepal-Tibet border) at 29,028 feet (8,848 meters) above sea level, plus or minus 10 feet (three meters) because of snow. This height was established by the Surveyor General of India in 1954 and accepted by the National Geographic Society. Prior to that the height was taken to be 29,002 feet (8,840 meters). Satellite measurements taken in 1987 indicate that Mt. Everest is 29,864 feet (9,102 meters) high, but this measurement has not been adopted by the National Geographic Society.

The lowest point on land is the Dead Sea between Israel and Jordan, which is 1,312 feet (399 meters) below sea level. The lowest point on the Earth's surface is thought to be in the Marianas Trench in the western Pacific Ocean, extending from southeast of Guam to northwest of the Marianas Islands. It has been measured as 36,198 feet (11,034 meters) below sea level.

## Which elements are contained in the **Earth's crust**?

The most abundant elements in the Earth's crust are listed in the table below. In addition, nickel, copper, lead, zinc, tin, and silver account for less than 0.02 percent with all other elements comprising 0.48 percent.

| Element | Percentage |
|---|---|
| Oxygen | 47.0 |
| Silicon | 28.0 |
| Aluminum | 8.0 |
| Iron | 4.5 |
| Calcium | 3.5 |
| Magnesium | 2.5 |
| Sodium | 2.5 |
| Potassium | 2.5 |
| Titanium | 0.4 |
| Hydrogen | 0.2 |
| Carbon | 0.2 |
| Phosphorus | 0.1 |
| Sulfur | 0.1 |

## What are the **highest** and **lowest elevations** in the **United States**?

Named in honor of U.S. president William McKinley (1843–1901), Mt. McKinley, Alaska, at 20,320 feet (6,194 meters), is the highest point in the United States and North America. Located in central Alaska, it is part of the Alaska Range. Its south peak measures 20,320 feet (6,194 meters) high and the north peak is 19,470 feet (5,931 meters) high. It boasts one of the world's largest unbroken precipices and is the main scenic attraction at Denali National Park. Denali means the "high one" or the "great one" and is a native American name sometimes used for Mt. McKinley. Mt. Whitney, California, at 14,494 feet (4,421 meters), is the highest point in the continental United

**How much of the Earth's surface is land and how much is water?**

Approximately 30 percent of the Earth's surface is land. This is about 57,259,000 square miles (148,300,000 square kilometers). The area of the Earth's water surface is approximately 139,692,000 square miles (361,800,000 square kilometers), or 70 percent of the total surface area.

States. Death Valley, California, at 282 feet (86 meters) below sea level, is the lowest point in the United States and in the western hemisphere.

# WATER

## Does **ocean water circulate**?

Ocean water is in a constant state of movement. Horizontal movements are called currents while vertical movements are called upwelling and downwelling. Wind, tidal motion, and differences in density due to temperature or salinity are the main causes of ocean circulation. Temperature differentials arise from the equatorial water being warmer than water in the polar regions. In the Northern Hemisphere the currents circulate in a clockwise direction while in the Southern Hemisphere the currents circulate in a counter-clockwise direction. In the equatorial regions the currents move in opposite directions—from left to right in the north and from right to left in the south. Currents moving north and south from equatorial regions carry warm water while those in polar regions carry cold water.

| Major Cold Currents | Major Warm Currents |
|---|---|
| California | North Atlantic (Gulf Stream) |
| Humboldt | South Atlantic |
| Labrador | South Indian Ocean |
| Canaries | South Pacific |
| Benguela | North Pacific |
| Falkland | Monsoons |
| West Australian | Okhotsk |

## Is there **gold** in **seawater**?

There are very minute amounts of gold in seawater. Searching all of the planet's seawater one could find nine pounds (four kilograms) of gold for every person on Earth.

## If the Earth were a uniform sphere, **how much water** would cover the surface?

It is estimated that 97 percent of all water in the world or over one quadrillion acre-feet (1,234 $\times$ $10^{15}$ cubic meters) is contained within the oceans. If the Earth were a uniform sphere, this volume of water would cover the Earth to a depth of 800 feet (244 meters).

## If you **melted all the ice** in the world, how high would the oceans rise?

If you melted all the ice in the world, some 5.5 million cubic miles (23 million cubic kilometers) in all, the oceans would rise 1.7 percent or about 180 feet (60 meters), which is enough, for example, for 20 stories of the Empire State Building to be underwater.

## What fraction of an **iceberg** shows above water?

Only one seventh to one-tenth of an iceberg's mass shows above water.

## What **color** is an **iceberg**?

Most icebergs are blue and white. However, one in a thousand icebergs in Antarctica is emerald green. They are only found in Antarctica because the Northern Hemisphere is not cold enough. These icebergs form when seawater freezes to the bottom of floating ice shelves. The ice looks green from the combination of yellow and blue. Yellow is from the yellowish-brown remains of dead plankton dissolved in seawater and trapped in the ice. Blue is present because although ices reflects virtually all the wavelengths of visible light, it absorbs slightly more red wavelengths than blue.

## What is an **aquifer**?

Some rocks of the upper part of the Earth's crust contain many small holes, or pores. When these holes are large or are joined together so that water can flow through them easily, the rock is considered to be permeable. A large body of permeable rock in which water is stored and flows through is called an aquifer (from the Latin for "water" and "to bear"). Sandstones and gravels are excellent examples of permeable rock.

As water reservoirs, aquifers provide about 60 percent of American drinking water. The huge Ogallala Aquifer, underlying about two million acres of the Great Plains, is a major source of water for the central United States. It has been estimated that after oceans (containing 850 million cubic miles [1,370 million cubic kilometers] of water), aquifers, with an estimated 31 million cubic miles (50 million cubic kilometers), are the second largest store of water. Water is purified as it is filtered through the rock, but it can be polluted by spills, dumps, acid rain, and other causes. In addition, recharging of water by rainfall often cannot keep up with the volume removed by

heavy pumping. The Ogallala Aquifer's supply of water could be depleted by 25 percent in the year 2020.

## What is the **chemical composition** of the **ocean**?

The ocean contains every known naturally occurring element plus various gases, chemical compounds, and minerals. Below is a sampling of the most abundant chemicals.

| Constituent | Concentration (parts per million) |
|---|---|
| Chloride | 18,980 |
| Sodium | 10,560 |
| Sulfate | 2,560 |
| Magnesium | 1,272 |
| Calcium | 400 |
| Potassium | 380 |
| Bicarbonate | 142 |
| Bromide | 65 |
| Strontium | 13 |
| Boron | 4.6 |
| Fluoride | 1.4 |

## Why is the **sea blue**?

There is no single cause for the colors of the sea. What is seen depends in part on when and from where the sea is observed. Eminent authority can be found to support almost any explanation. Some explanations include absorption and scattering of light by pure water; suspended matter in sea water; the atmosphere; and color and brightness variations of the sky. For example, one theory is that when sunlight hits seawater, part of the white light, composed of different wavelengths of various colors, is absorbed, and some of the wavelengths are scattered after colliding with the water molecules. In clear water, red and infrared light are greatly absorbed but blue is least absorbed, so that the blue wavelengths are reflected out of the water. The blue effect requires a minimum depth of 10 feet (three meters) of water.

## What causes **waves** in the ocean?

The most common cause of surface waves is air movement (the wind). Waves within the ocean can be caused by tides, interactions among waves, submarine earthquakes or volcanic activity, and atmospheric disturbances. Wave size depends on wind speed, wind duration, and the distance of water over which the wind blows. The longer the distance the wind travels over water, or the harder it blows, the higher the waves. As the wind blows over water it tries to drag the surface of the water with it. The surface cannot move as fast as air, so it rises. When it rises, gravity pulls the water back, carrying the falling water's momentum below the surface. Water pressure from below pushes this swell back up again. The tug of war between gravity and water pressure constitutes wave motion. Capillary waves are caused by breezes of less than two knots. At 13 knots the waves grow taller and faster than they grow longer, and their steepness cause them to break, forming whitecaps. For a whitecap to form, the wave height must be one-seventh the distance between wave crests.

## How **deep** is the ocean?

The average depth of the ocean floor is 13,124 feet (4,000 meters). The average depth of the four major oceans is given below:

| | Average depth | |
|---|---|---|
| **Ocean** | **Feet** | **Meters** |
| Pacific | 13,740 | 4,188 |
| Atlantic | 12,254 | 3,735 |
| Indian | 12,740 | 3,872 |
| Arctic | 3,407 | 1,038 |

There are great variations in depth because the ocean floor is often very rugged. The greatest depth variations occur in deep, narrow depressions known as trenches along the margins of the continental plates. The deepest measurements made—36,198 feet (11,034 meters), deeper than the height of the world's tallest mountains—was taken in Mariana Trench east of the Mariana Islands. In January 1960, the French oceanographer Jacques Piccard, together with the United States Navy Lieutenant David Walsh, took the bathyscaphe *Trieste* to the bottom of the Mariana Trench.

| | | Depth | |
|---|---|---|---|
| **Ocean** | **Deepest point** | **Feet** | **Meters** |
| Pacific | Mariana Trench | 36,200 | 11,033 |
| Atlantic | Puerto Rico Trench | 28,374 | 8,648 |
| Indian | Java Trench | 25,344 | 7,725 |
| Arctic | Eurasia Basin | 17,881 | 5,450 |

## How far can **sunlight penetrate** into the ocean?

Because seawater is relatively transparent, approximately 5 percent of sunlight can penetrate clean ocean water to a depth of 262 feet (80 meters). When the water is turbid (cloudy) due to currents, mixing of silt, increased growth of algae, or other factors, the depth of penetration is reduced to less than 164 feet (50 meters).

## What is a **tidal bore**?

A tidal bore is a large, turbulent, wall-like wave of water that moves inland or upriver as an incoming tidal current surges against the flow of a more narrow and shallow river, bay, or estuary. It can be 10 to 16 feet (three to five meters) high and move rapidly (10 to 15 knots) upstream with and faster than the rising tide.

## Where are the **world's highest tides**?

The Bay of Fundy (New Brunswick, Canada) has the world's highest tides. They average about 45 feet (14 meters) high in the northern part of the bay, far surpassing the world average of 2.5 feet (0.8 meter).

## Is there one numerical value for **sea level**?

Sea level is the average height of the sea surface. Scientists have calculated a mean sea level based on observations around the world. The mean average sea level takes into account all stages of ocean tides over a nineteen-year period. Numerous irregularities and slopes on the sea surface make it difficult to calculate an accurate measure of sea level.

## What is the difference between an **ocean** and a **sea**?

There is no neatly defined distinction between ocean and sea. One definition says the ocean is a great body of interconnecting salt water that covers 71 percent of the Earth's surface. There are four major oceans—the Arctic, Atlantic, Indian, and Pacific—but some sources do not include the Arctic Ocean, calling it a marginal sea. The terms "ocean" and "sea" are often used interchangeably but a sea is generally considered to be smaller than an ocean. The name is often given to saltwater areas on the margins of an ocean, such as the Mediterranean Sea.

## How much salt is in **brackish water**?

Brackish water has a saline (salt) content between that of fresh water and sea water. It is neither fresh nor salty, but somewhere in between. Brackish waters are usually regarded as those containing 0.5 to 30 parts per thousand salt, while the average saltiness of seawater is 35 parts per thousand.

## How **salty** is seawater?

Seawater is, on average, 3.3 to 3.7 percent salt. The amount of salt varies from place to place. In areas where large quantities of fresh water are supplied by melting ice, rivers, or rainfall, such as the Arctic or Antarctic, the level of salinity is lower. Areas such as the Persian Gulf and the Red Sea have salt contents over 4.2 percent. If all the salt in

### What are rip tides and why are they so dangerous?

At points along a coast where waves are high, a substantial buildup of water is created near the shore. This mass of water moves along the shore until it reaches an area of lower waves. At this point, it may burst through the low waves and move out from shore as a strong surface current moving at an abnormally rapid speed known as a rip current. Swimmers who become exhausted in a rip current may drown unless they swim parallel to the shore. Rip currents are sometimes incorrectly called rip tides.

the ocean were dried, it would form a mass of solid salt the size of Africa. Most of the ocean salt comes from processes of dissolving and leaking from the solid Earth over hundreds of millions of years. Some is the result of salty volcanic rock that flows up from a giant rift that runs through all the ocean's basins.

## Is the **Dead Sea** really dead?

Because the Dead Sea, on the boundary between Israel and Jordan, is the lowest body of water on the Earth's surface, any water that flows into it has no outflow. It is called "dead" because its extreme salinity makes impossible any animal or vegetable life except bacteria. Fish introduced into the sea by the Jordan River or by smaller streams die instantly. The only plant life consists primarily of halophytes (plants that grow in salty or alkaline soil). The concentration of salt increases toward the bottom of the lake. The water also has such a high density that bathers float on the surface easily.

## What is the **bearing capacity of ice** on a lake?

The following chart indicates the maximum safe load. It applies only to clear lake ice that has not been heavily traveled. For early winter slush ice, ice thickness should be doubled for safety.

| Ice thickness | | | Maximum safe load | |
|---|---|---|---|---|
| **Inches** | **Centimeters** | **Examples** | **Tons** | **Kilograms** |
| 2 | 5 | One person on foot | | |
| 3 | 7.6 | Group in single file | | |
| 7.5 | 19 | Car or snowmobile | 2 | 907.2 |
| 8 | 20.3 | Light truck | 2.5 | 1,361 |
| 10 | 25.4 | Medium truck | 3.5 | 1,814.4 |
| 12 | 30.5 | Heavy truck | 9 | 7,257.6 |
| 15 | 38 | | 10 | 9,072 |
| 20 | 50.8 | | 25 | 22,680 |

## Where is the world's **deepest lake**?

Lake Baikal, located in southeast Siberia, Russia, is approximately 5,371 feet (1,638 meters) deep at its maximum depth, Olkhon Crevice, making it the deepest lake in the world. Lake Tanganyika in Tanzania and Zaire is the second deepest lake, with a depth of 4,708 feet (1,435 meters).

## Where are the **five largest lakes** in the world located?

| | Area | | Length | | Depth | |
|---|---|---|---|---|---|---|
| **Location** | **Square miles** | **Square Km[a]** | **Miles** | **Km[a]** | **Feet** | **Meters** |
| Caspian Sea,[b] Asia-Europe | 143,244 | 370,922 | 760 | 1,225 | 3,363 | 1,025 |
| Superior, North America | 31,700 | 82,103 | 350 | 560 | 1,330 | 406 |
| Victoria, Africa | 26,828 | 69,464 | 250 | 360 | 270 | 85 |
| Aralb, Asia | 24,904 | 64,501 | 280 | 450 | 220 | 67 |
| Huron, North America | 23,010 | 59,600 | 206 | 330 | 750 | 229 |

[a] = Kilometers — [b] = Salt water lake

## What is a **yazoo**?

A yazoo is a tributary of a river that runs parallel to the river, being prevented from joining the river because the river has built up high banks. The name is derived from the Yazoo River, a tributary of the Mississippi River, which demonstrates this effect.

## Which of the **Great Lakes** is the largest?

| | Surface area | | Maximum depth | |
|---|---|---|---|---|
| **Lake** | **Square miles** | **Square kilometers** | **Feet** | **Meters** |
| Superior | 31,700 | 82,103 | 1,333 | 406 |
| Huron | 23,010 | 59,600 | 750 | 229 |
| Michigan | 22,300 | 57,757 | 923 | 281 |
| Erie | 9,910 | 25,667 | 210 | 64 |
| Ontario | 7,540 | 9,529 | 802 | 244 |

Lake Superior is the largest of the Great Lakes. The North American Great Lakes form a single watershed with one common outlet to the sea—the St. Lawrence Seaway. The total volume of all five basins is 6,000 trillion gallons (22.7 trillion liters) equivalent to about 20 percent of the world's freshwater. Only Lake Michigan lies wholly within the United States' borders; the others share their boundaries with Canada. Some believe that Lake Huron and Lake Michigan are two lobes of one lake, since they are the same elevation and are connected by the 120-foot (36.5-meter) deep

Strait of Mackinac, which is 3.6 to five miles (six to eight kilometers) wide. Gage records indicate that they both have similar water level regimes and mean long-term behavior, so that hydrologically they act as one lake. Historically they were considered two by the explorers who named them, but this is considered a misnomer by some.

## What are the **longest rivers** in the world?

The two longest rivers in the world are the Nile in Africa and the Amazon in South America. However, which is the longest is a matter of some debate. The Amazon has several mouths that widen toward the South Atlantic, so the exact point where the river ends is uncertain. If the Pará estuary (the most distant mouth) is counted, its length is approximately 4,195 miles (6,750 kilometers). The length of the Nile as surveyed before the loss of a few miles of meanders due to the formation of Lake Nasser behind the Aswan Dam was 4,145 miles (6,670 kilometers). The table below lists the five longest river systems in the world.

| River | Length | |
|---|---|---|
| | Miles | Kilometers |
| Nile (Africa) | 4,145 | 6,670 |
| *Amazon (South America) | 4,000 | 6,404 |
| Chang jiang-Yangtze (Asia) | 3,964 | 6,378 |
| Mississippi-Missouri river system (North America) | 3,740 | 6,021 |
| Yenisei-Angara river system (Asia) | 3,442 | 5,540 |

*excluding Pará estuary

## What is the world's **highest waterfall**?

Angel Falls, named after the explorer and bush pilot Jimmy Angel, on the Carrao tributary in Venezuela is the highest waterfall in the world. It has a total height of 3,212 feet (979 meters) with its longest unbroken drop being 2,648 feet (807 meters).

It is difficult to determine the height of a waterfall because many are composed of several sections rather than one straight drop. The highest waterfall in the United States is Yosemite Falls on a tributary of the Merced River in Yosemite National Park, California, with a total drop of 2,425 feet (739 meters). There are three sections to the Yosemite Falls: Upper Yosemite is 1,430 feet (435 meters), Cascades (middle portion) is 675 feet (205 meters), and Lower Yosemite is 320 feet (97 meters).

## Why was **Niagara Falls shut down** for thirty hours in 1848?

The volume of Niagara waters depends on the height of Lake Erie at Buffalo, a factor that varies with the direction and intensity of the wind. Changes of as much as eight feet (2.5

### When will Niagara Falls disappear?

The water dropping over Niagara Falls digs great plunge pools at the base, undermining the shale cliff and causing the hard limestone cap to cave in. Niagara has eaten itself seven miles (11 kilometers) upstream since it was formed 10,000 years ago. At this rate, it will disappear into Lake Erie in 22,800 years. The Niagara River connects Lake Erie with Lake Ontario, and marks the U.S.-Canada boundary (New York-Ontario).

meters) in the level of Lake Erie at the Niagara River source have been recorded. On March 29, 1848, a gale drove the floating ice in Lake Erie to the lake outlet, quickly blocking that narrow channel and shutting off a large proportion of the river's flow. Eyewitness accounts stated that the American falls were passable on foot, but for that day only.

# LAND

*See also: Space—Planets and Moons*

## Are there **tides** in the **solid part** of the Earth as well as in its waters?

The solid Earth is distorted about 4.5 to 14 inches (11.4 to 35.6 centimeters) by the gravitational pull of the sun and moon. It is the same gravitational pull that creates the tides of the waters. When the moon's gravity pulls water on the side of the Earth near to it, it pulls the solid body of the Earth on the opposite side away from the water to create bulges on both sides, and causing high tides. These occur every 12.5 hours. Low tides occur in those places from which the water is drained to flow into the two high-tide bulges. The sun causes tides on the Earth that are about 33 to 46 percent as high as those due to the moon. During a new moon or a full moon when the sun and moon are in a straight line, the tides of the moon and the sun reinforce each other to make high tides higher; these are called spring tides. At the quarter moons, the sun and moon are out of step (at right angles), the tides are less extreme than usual; these are called neap tides. Smaller bodies of water, such as lakes, have no tides because the whole body of water is raised all at once, along with the land beneath it.

## Do the **continents move**?

In 1912, a German geologist, Alfred Lothar Wegener (1880–1930), theorized that the continents had drifted or floated apart to their present locations and that once all the

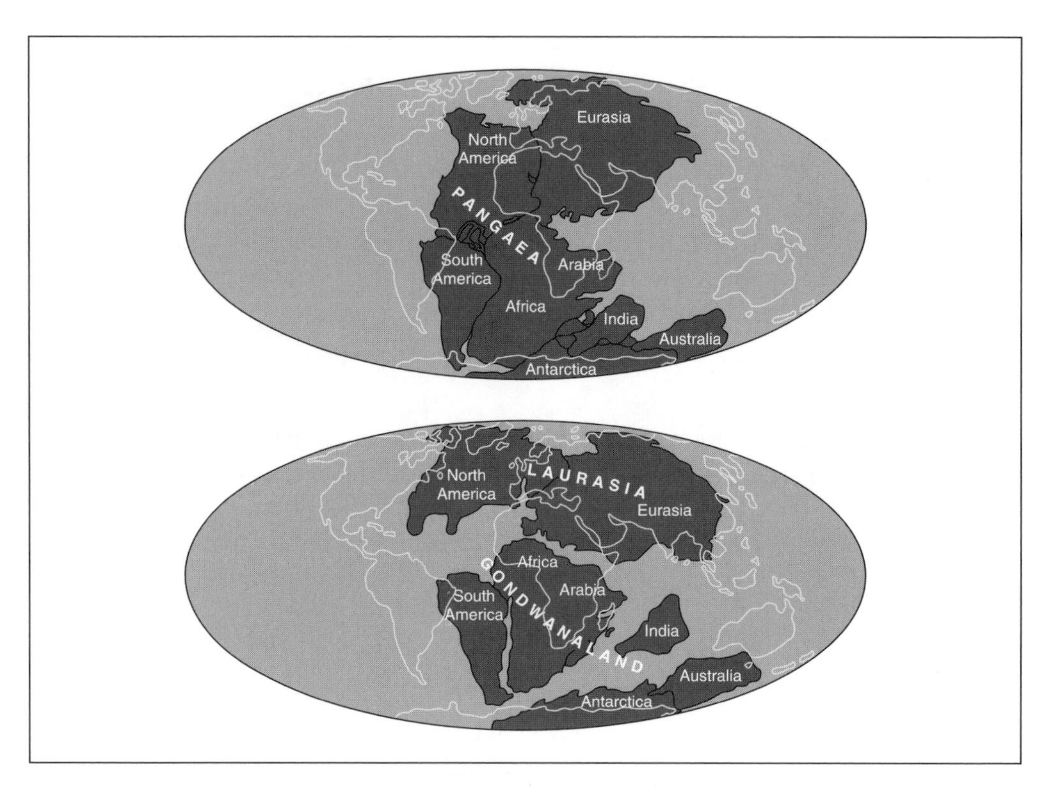


**The Pangaea supercontinent (top) and its breakup into Laurasia and Gondwanaland.**

continents had been a single land mass near Antarctica, which is called Pangaea (from the Greek word meaning *all-earth*). Pangaea then broke apart some 200 million years ago into two major continents called Laurasia and Gondwanaland. These two continents continued drifting and separating until the continents evolved their present shapes and positions. Wegener's theory was discounted, but it has since been found that the continents do move sideways (not drift) at an estimated 0.75 inch (19 millimeters) annually because of the action of plate tectonics. American geologist William Maurice Ewing (1906–1974) and Harry Hammond Hess (1906–1969) proposed that the Earth's crust is not a solid mass, but composed of eight major and seven minor plates that can move apart, slide by each other, collide, or override each other. Where these plates meet are major areas of mountain-building, earthquakes, and volcanoes.

## How much of the **Earth's surface** is covered with **ice**?

About 10.4 percent of the world's land surface is glaciated, or permanently covered with ice. Approximately 6,020,000 square miles (15,600,000 square kilometers) are covered by ice in the form of ice sheets, ice caps, or glaciers. An ice sheet is a body of ice that blankets an area of land, completely covering its mountains and valleys. Ice

sheets have an area of over 19,000 square miles (50,000 square kilometers); ice caps are smaller. Glaciers are larger masses of ice that flow, under the force of gravity, at a rate of between 10 and 1,000 feet (three to 300 meters) per year. Glaciers on steep slopes flow faster. For example, the Quarayoq Glacier in Greenland averages 65 to 80 feet (20 to 24 meters) per day. The areas of glaciation in some parts of the world are:

| Place | Area | |
|---|---|---|
| | **Square miles** | **Square kilometers** |
| Antarctica | 5,250,000 | 12, 588, 000 |
| North Polar Regions (Greenland, Northern Canada, Arctic Ocean islands) | 799,000 | 2,070,000 |
| Asia | 44,000 | 115,800 |
| Alaska & Rocky Mountains | 29,700 | 76,900 |
| South America | 10,200 | 26,500 |
| Iceland | 4,699 | 12,170 |
| Alpine Europe | 3,580 | 9,280 |
| New Zealand | 391 | 1,015 |
| Africa | 5 | 12 |

## Which is purer—**glacier ice** or regular ice?

Impurities found in the snow grains of glaciers have mostly moved to the boundaries of the grains and been flushed out. Glacier ice is like triple-distilled water and hence purer than regular ice.

## How much of the **Earth's surface** is **permanently frozen**?

About one-fifth of the Earth's land is permafrost, or ground that is permanently frozen. This classification is based entirely on temperature and disregards the composition of the land. It can include bedrock, sod, ice, sand, gravel, or any other type of material in which the temperature has been below freezing for over two years. Nearly all permafrost is thousands of years old.

## Where are the **northernmost** and **southernmost points of land**?

The most northern point of land is Cape Morris K. Jesup on the northeastern extremity of Greenland. It is at 83 degrees, 39 minutes north latitude and is 440 miles (708 kilometers) from the North Pole. However, the *Guinness Book of Records* reports that an islet of 100 feet (30 meters) across, called Oodaq, is more northerly at 83 degrees, 40 minutes north latitude and 438.9 miles (706 kilometers) from the North Pole. The southernmost point of land is the South Pole (since the South Pole, unlike the North Pole, is on land).

In the United States, the northernmost point of land is Point Barrow, Alaska (71 degrees, 23 minutes north latitude), and the southernmost point of land is Ka Lae or South Cape (18 degrees, 55 minutes north latitude) on the island of Hawaii. In the 48 contiguous states, the northernmost point is Northwest Angle, Minnesota (49 degrees, 23 minutes north latitude); the southernmost point is Key West, Florida (24 degrees, 33 minutes north latitude).

## How **thick** is the ice that covers **Antarctica**?

The ice that covers Antarctica is 15,700 feet (4,785 meters) in depth at its thickest point. This is about ten times taller than the Sears Tower in Chicago. However, the average thickness is 7,100 feet (2,164 meters).

## Who was the **first person on Antarctica**?

Historians are unsure who first set foot on Antarctica, the fifth largest continent covering 10 percent of the Earth's surface with its area of 5.4 million square miles (14 million square kilometers). In 1773–1775 British Captain James Cook (1728–1779) circumnavigated the continent. American explorer Nathaniel Palmer (1799–1877) discovered Palmer Peninsula in 1820, without realizing that this was a continent. That same year, Fabian Gottlieb von Bellingshausen (1779–1852) sighted the Antarctic continent. American sealer John Davis went ashore at Hughes Bay on February 7, 1821. In 1823, sealer James Weddell (1787–1834) traveled the farthest south (74 degrees south) that anyone had until that time and entered what is now called the Weddell Sea. In 1840, American Charles Wilkes (1798–1877), who followed the coast for 1,500 miles, announced the existence of Antarctica as a continent. In 1841, Sir James Clark Ross (1800–1862) discovered Victoria Land, Ross Island, Mount Erebus, and the Ross Ice Shelf. In 1895, the whaler Henryk Bull landed on the Antarctic continent. Norwegian explorer Roald Amundsen (1872–1928) was the first to reach the South Pole on December 14, 1911. Thirty-four days later, Amundsen's rival Robert Falcon Scott (1868–1912) stood at the South Pole, the second to do so, but he and his companions died upon their return trip.

## When was the **Ice Age**?

Ice ages, or glacial periods, have occurred at irregular intervals for over 2.3 billion years. During an ice age, sheets of ice cover large portions of the continents. The exact reasons for the changes in the Earth's climate are not known, although some think they are caused by changes in the Earth's orbit around the sun.

The Great Ice Age occurred during the Pleistocene Epoch, which began about two million years ago and lasted until 11,000 years ago. At its height, about 27 percent of the world's present land area was covered by ice. In North America, the ice covered Canada and moved southward to New Jersey; in the Midwest, it reached as far south as

The 200-foot oceanic face of Antarctica's Ross Ice Shelf stretches straight across the horizon, dwarfed in the midst of Mount Erebus.

St. Louis. Small glaciers and ice caps also covered the western mountains. Greenland was covered in ice as it is today. In Europe, ice moved down from Scandinavia into Germany and Poland; the British Isles and the Alps also had ice caps. Glaciers also covered the northern plains of Russia, the plateaus of Central Asia, Siberia, and the Kamchetka Peninsula.

The glaciers' effect on the United States can still be seen. The drainage of the Ohio River and the position of the Great Lakes were influenced by the glaciers. The rich soil of the Midwest is mostly glacial in origin. Rainfall in areas south of the glaciers formed large lakes in Utah, Nevada, and California. The Great Salt Lake in Utah is a remnant of one of these lakes. The large ice sheets locked up a lot of water; sea level fell about 450 feet (137 meters) below what it is today. As a result, some states, such as Florida, were much larger during the ice age.

The glaciers of the last ice age retreated about 11,000 years ago. Some believe that the ice age is not over yet; the glaciers follow a cycle of advance and retreat many times. There are still areas of the Earth covered by ice, and this may be a time in between glacial advances.

## What is a **moraine**?

A moraine is a mound, ridge, or any other distinct accumulation of unsorted, unstratified material or drift, deposited chiefly by direct action of glacier ice.

Hoodoo formations at Chirichua National Monument in Arizona.

## What is a **hoodoo**?

A hoodoo is a fanciful name for a grotesque rock pinnacle or pedestal, usually of sandstone, that is the result of weathering in a semi-arid region. An outstanding example of hoodoos occurs in the Wasatch Formation at Bryce Canyon, Utah.

## Do loud noises cause **avalanches**?

Avalanches are not triggered by noise, but they are the only natural hazard triggered by individuals. Avalanches are great masses of snow that slide down mountainsides. The snow may be powdery, sliding over compacted older snow; slabs of snow that roll down the slope; or a combination of ice and snow from the slope, including rocks and other debris. Dry slab avalanches are the most dangerous, traveling at speeds of 60–80 miles per hour.

## What are **sand dunes** and how are they formed?

Mounds of wind-blown sand in deserts and coastal areas are called dunes. Winds transport grains of sand until it accumulates around obstacles to form ridges and mounds. Wind direction, the type of sand, and the amount of vegetation determine the type of dune. Dunes are named either for their shape (e.g. star dunes and parabolic dunes) or according to their alignment with the wind (e.g. longitudinal dunes and transverse dunes).

## Which are the world's **largest deserts**?

A desert is an area that receives little precipitation and has little plant cover. Many deserts form a band north and south of the equator at about 20 degrees latitude

These sand dunes in California present a landscape of varying shapes, sizes and textures.

because moisture-bearing winds do not release their rain over these areas. As the moisture-bearing winds from the higher latitudes approach the equator, their temperatures increase and they rise higher and higher in the atmosphere. When the winds arrive over the equatorial areas and come in contact with the colder parts of the Earth's atmosphere, they cool down and release all their water to create the tropical rain forests near the equator.

The Sahara Desert, the world's largest, is three times the size of the Mediterranean Sea. In the United States, the largest desert is the Mojave Desert in southern California with an area of 15,000 square miles (38,900 square kilometers).

| Desert | Location | Area | |
|---|---|---|---|
| | | Square miles | Square kilometers |
| Sahara | North Africa | 3,500,000 | 9,065,000 |
| Arabian | Arabian Peninsula | 900,000 | 2,330,000 |
| Australian | Australia | 600,000 | 1,554,000 |
| Gobi | Mongolia & China | 500,000 | 1,295,000 |
| Libyan | Libya, SW Egypt, Sudan | 450,000 | 1,165,500 |

## What is **quicksand**?

Quicksand is a mass of sand and mud that contains a large amount of water. A thin film separates individual grains of sand so the mixture has the characteristics of a liq-

### What is the difference between spelunking and speleology?

Spelunking, or sport caving, is exploring caves as a hobby or for recreation. Speleology is the scientific study of caves and related phenomena, such as the world's deepest cave, Réseau Jean Bernard, in Haute Savoie, France, with a depth of 5,256 feet (1,602 meters), or the world's longest cave system, Mammoth Cave in Kentucky, with a length of 348 miles (560 kilometers).

uid. Quicksand is found at the mouths of large rivers or other areas that have a constant source of water. Heavy objects, including humans, can sink when encountering quicksand and the mixture collapses. However, since the density of the sand/water mix is slightly greater than the density of the human body, most humans can actually float on quicksand.

## Are all **craters** part of a volcano?

No, not all craters are of volcanic origin. A crater is a nearly circular area of deformed sedimentary rocks, with a central, ventlike depression. Some craters are caused by the collapse of the surface when underground salt or limestone dissolves. The withdrawal of groundwater and the melting of glacial ice can also cause the surface to collapse, forming a crater.

Craters are also caused by large meteorites, comets, and asteroids that hit the Earth. A notable impact crater is Meteor Crater near Winslow, Arizona. It is 4,000 feet (1,219 meters) in diameter, 600 feet (183 meters) deep and is estimated to have been formed 30,000 to 50,000 years ago.

## How are **caves** formed?

Water erosion creates most caves found along coastal areas. Waves crashing against the rock over years wears away part of the rock forming a cave. Inland caves are also formed by water erosion—in particular, groundwater eroding limestone. As the limestone dissolves, underground passageways and caverns are formed.

## How is **speleothem** defined?

Speleothem is a term given to those cave features that form after a cave itself has formed. They are secondary mineral deposits that are created by the solidification of fluids or from chemical solutions. These mineral deposits usually contain calcium carbonate ($CaCO_3$) or limestone, but gypsum or silica may also be found. Stalactites, stalagmites, soda straws, cave coral, boxwork and cave pearls are all types of speleothems.

A massive stalactite and stalagmite column takes center stage in South Africa's Cango Caves.

## What is a **tufa**?

It is a general name for calcium carbonate ($CaCO_3$) deposits or spongy porous limestone found at springs in limestone areas, or in caves as massive stalactite or stalagmite deposits. Tufa, derived from the Italian word for "soft rock," is formed by the precipitation of calcite from the water of streams and springs.

## Which is the **deepest cave** in the United States?

Lechuguilla Cave, in Carlsbad Caverns National Park, New Mexico is the deepest cave in the United States. Its depth is 1,565 feet (477 meters). Unlike most caves in which carbon dioxide mixes with rainwater to produce carbonic acid, Carlsbad Caverns was shaped by sulfuric acid. The sulfuric acid was a result of a reaction between oxygen that was dissolved in groundwater and hydrogen sulfide that emanated from far below the cave's surface.

## How does a **stalactite** differ from a **stalagmite**?

A stalactite is a conical or cylindrical calcite ($CaCO_3$) formation hanging from a cave roof. It forms from the centuries-long buildup of mineral deposits resulting from the seepage of water from the limestone rock above the cave. This water containing calcium bicarbonate evaporates, losing some carbon dioxide, to deposit small quantities of calcium carbonate (carbonate of lime), which eventually forms a stalactite.

A stalagmite is a stone formation that develops upward from the cave floor and resembles an icicle upside down. Formed from water containing calcite that drips from the limestone walls and roof of the cave, it sometimes joins a stalactite to form a column.

## What and where is the **continental divide** of North America?

The Continental Divide, also known as the Great Divide, is a continuous ridge of peaks in the Rocky Mountains that marks the watershed separating easterly flowing waters from westerly flowing waters in North America. To the east of the Continental Divide, water drains into Hudson Bay or the Mississippi River before reaching the Atlantic Ocean. To the west, water generally flows through the Columbia River or the Colorado River on its way to the Pacific Ocean.

## Which **national parks** in the United States are the most popular?

| Park/Location | Visitors (2001) |
| --- | --- |
| Great Smoky Mountains National Park (N. Carolina & Tennessee) | 9,457,323 |
| Grand Canyon National Park (Arizona) | 4,219,726 |
| Yosemite National Park (California) | 3,453,345 |
| Olympic National Park (Washington) | 3,401,245 |
| Rocky Mountain National Park (Colorado) | 3,211,689 |
| Yellowstone National Park (Wyoming) | 2,769,775 |
| Grand Teton National Park (Wyoming) | 2,531,844 |
| Acadia National Park (Maine) | 2,504,708 |
| Zion National Park (Utah) | 2,086,264 |
| Mammoth Cave National Park (Kentucky) | 1,889,096 |

## How long is the **Grand Canyon**?

The Grand Canyon, cut out by the Colorado River over a period of 15 million years in the northwest corner of Arizona, is the largest land gorge in the world. It is four to 13 miles (6.4 to 21 kilometers) wide at its brim, 4,000 to 5,500 feet (1,219 to 1,676 meters) deep, and 217 miles (349 kilometers) long, extending from the mouth of the Little Colorado River to Grand Wash Cliffs (and 277 miles, 600 feet [445.88 kilometers] if Marble Canyon is included).

However, it is not the deepest canyon in the United States; that distinction belongs to Kings Canyon, which runs through the Sierra and Sequoia National Forests near East Fresno, California, with its deepest point being 8,200 feet (2,500 meters). Hell's Canyon of the Snake River between Idaho and Oregon is the deepest United States canyon in low-relief territory. Also called the Grand Canyon of the Snake, it plunges 7,900 feet (2,408 meters) down from Devil Mountain to the Snake River.

## What are the **LaBrea Tar Pits**?

The tar pits are located in an area of Los Angeles, California, formerly known as Rancho LaBrea. Heavy, sticky tar oozed out of the Earth there, the scum from great petroleum reservoirs far underground. The pools were cruel traps for uncounted numbers of animals. Today, the tar pits are a part of Hancock Park, where many fossil remains are displayed along with life-sized reconstructions of these prehistoric species.

The tar pits were first recognized as a fossil site in 1875. However, scientists did not systematically excavate the area until 1901. By comparing Rancho La Brea's fossil specimens with their nearest living relatives, paleontologists have a greater understanding of the climate, vegetation, and animal life in the area during the Ice Age. Perhaps the most impressive fossil bones recovered belong to such large extinct mammals as the imperial mammoth and the saber-toothed cat. Paleontologists have even found the remains of the western horse and the camel, which originated in North America, migrated to other parts of the world, and became extinct in North America at the end of the Ice Age.

## From what type of stone was **Mount Rushmore National Monument** carved?

Granite. The monument, in the Black Hills of southwestern South Dakota, depicts the 60-foot-high (18-meter-high) faces of four United States presidents: George Washington, Thomas Jefferson, Abraham Lincoln, and Theodore Roosevelt. Sculptor Gutzon Borglum (1867–1941) designed the monument, but died before the completion of the project; his son, Lincoln, finished it. From 1927 to 1941, 360 people, mostly construction workers, drillers, and miners, "carved" the figures using dynamite.

## What is the composition of the **Rock of Gibraltar**?

It is composed of gray limestone, with a dark shale overlay on parts of its western slopes. Located on a peninsula at the southern extremity of Spain, the Rock of Gibraltar is a mountain at the east end of the Strait of Gibraltar, the narrow passage between the Atlantic Ocean and the Mediterranean Sea. "The Rock" is 1,398 feet (425 meters) tall at its highest point.

# VOLCANOES AND EARTHQUAKES

## Which is the **most famous volcano**?

The erruption of Mt. Vesuvius in Italy on August 79 C.E. is perhaps the most famous historical volcanic eruption. Vesuvius had been dormant for generations. When it erupted, entire cities were destroyed, including Pompeii, Stabiae, and Herculaneum.

Pompeii and Stabiae were buried under ashes while Herculaneum was buried under a mud flow.

## How many **kinds of volcanoes** are there?

Volcanoes are usually cone-shaped hills or mountains built around a vent connecting to reservoirs of molten rock, or magma, below the surface of the Earth. At times the molten rock is forced upwards by gas pressure until it breaks through weak spots in the Earth's crust. The magma erupts forth as lava flows or shoots into the air as clouds of lava fragments, ash, and dust. The accumulation of debris from eruptions cause the volcano to grow in size. There are four kinds of volcanoes:

*Cinder cones* are built of lava fragments. They have slopes of 30 degrees to 40 degrees and seldom exceed 1,640 feet (500 meters) in height. Sunset Crater in Arizona and Paricutin in Mexico are examples of cinder cones.

*Composite cones* are made of alternating layers of lava and ash. They are characterized by slopes of up to 30 degrees at the summit, tapering off to five degrees at the base. Mount Fuji in Japan and Mount St. Helens in Washington are composite cone volcanoes.

*Shield volcanoes* are built primarily of lava flows. Their slopes are seldom more than 10 degrees at the summit and two degrees at the base. The Hawaiian Islands are clusters of shield volcanoes. Mauna Loa is the world's largest active volcano, rising 13,653 feet (4,161 meters) above sea level.

*Lava domes* are made of viscous, pasty lava squeezed like toothpaste from a tube. Examples of lava domes are Lassen Peak and Mono Dome in California.

## How do we determine when **ancient volcanic eruptions** occurred?

The most common method used to date ancient volcanic eruptions is carbon dating. Carbon dating relies on the rate of radioactive decay of carbon 14. It is used to date eruptions that took place more than 200 years ago. Charcoal from trees burned during a volcanic eruption is almost pure carbon and is ideal for tracing the tiny amounts of carbon 14.

## Where is the **Circle of Fire**?

The belt of volcanoes bordering the Pacific Ocean is often called the "Circle of Fire" or the "Ring of Fire." The Earth's crust is composed of 15 pieces, called plates, which "float" on the partially molten layer below them. Most volcanoes, earthquakes, and mountain building occur along the unstable plate boundaries. The Circle of Fire marks the boundary between the plate underlying the Pacific Ocean and the surrounding plates. It runs up the west coast of the Americas from Chile to Alaska (through the Andes Mountains, Central America, Mexico, California, the Cascade Mountains, and the Aleutian Islands) then down the east coast of Asia from Siberia to

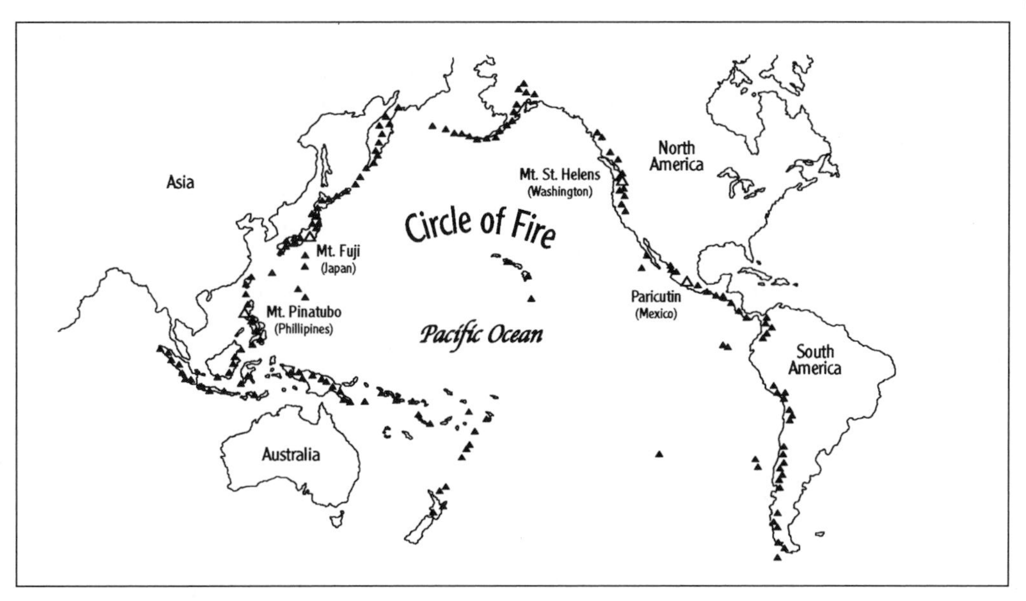


The Circle of Fire is a belt of volcanoes bordering the Pacific Ocean.

New Zealand (through Kamchatka, the Kurile Islands, Japan, the Philippines, Celebes, New Guinea, the Solomon Islands, New Caledonia, and New Zealand). Of the 850 active volcanoes in the world, over 75 percent of them are part of the Circle of Fire.

## Which **island region** has the **greatest concentration** of active volcanoes?

In the early 1990s 1,133 seamounts (submerged mountains) and volcanic cones were discovered in and around Easter Island. Many of the volcanoes rise more than a mile above the ocean floor. Some are close to 7,000 feet (2,134 meters) tall, although their peaks are still 2,500 to 5,000 feet (760 to 1,500 meters) below the surface of the sea.

## Which **volcanoes** in the **contiguous 48 states** are considered **active**?

Seven major volcanoes in the contiguous 48 states are considered active: three in California—Cinder Cone, Lassen Peak, and Mt. Shasta; three in Washington—Mt. Baker, Mt. Rainier, and Mt. St. Helens; and one in Oregon—Mt. Hood.

## When did **Mount St. Helens** erupt?

Mount St. Helens, located in southwestern Washington state in the Cascades mountain range, erupted on May 18, 1980. Sixty-one people died as a result of the eruption. This was the first known eruption in the 48 contiguous United States to claim a human life. Geologists call Mount St. Helens a composite volcano (a steep-sided, often symmetrical cone constructed of alternating layers of lava flows, ash, and other volcanic debris). Composite volcanoes tend to erupt explosively. Mount St. Helens and the other active volcanoes in the Cascade Mountains are a part of the "Ring of Fire," the Pacific zone having frequent and destructive volcanic activity.

Volcanoes have not only been active in Washington, but also in three other U.S. states: California, Alaska, and Hawaii. Lassen Peak is one of several volcanoes in the Cascade Range. It last erupted in 1921. Mount Katmai in Alaska had an eruption in 1912 in which the flood of hot ash formed the Valley of Ten Thousand Smokes 15 miles (24 kilometers) away. And Hawaii has its famed Mauna Loa, which is the world's largest volcano, being 60 miles (97 kilometers) in width at its base.

## Which **volcanoes** have been the **most destructive**?

The five most destructive eruptions from volcanoes since 1700 are as follows:

| Volcano | Date of eruption | Number killed | Lethal agent |
|---|---|---|---|
| Mt. Tambora, Indonesia | April 5, 1815 | 92,000 | 10,000 directly by the volcano, 82,000 from starvation afterwards |
| Karkatoa, Indonesia | Aug. 26, 1883 | 36,417 | 90% killed by a tsunami |
| Mt. Pelee, Martinque | Aug. 30, 1902 | 29,025 | Pyroclastic flows |
| Nevada del Ruiz, Colombia | Nov. 13, 1985 | 23,000 | Mud flow |
| Unzen, Japan | 1792 | 14,300 | 70% killed by cone collapse; 30% by a tsunami |

## What is a **tsunami**?

A tsunami is a giant wave set in motion by a large earthquake occurring under or near the ocean that causes the ocean floor to shift vertically. This vertical shift pushes the water ahead of it, starting a tsunami. These are very long waves (100 to 200 miles [161 to 322 kilometers]) with high speeds (500 mph [805 kph]) that, when approaching shallow water, can grow into a 100-foot (30.5-meter) high wave as its wavelength is reduced abruptly. Ocean earthquakes below a magnitude of 6.5 on the Richter scale, and those that shift the sea floor only horizontally, do not produce these destructive waves. The highest recorded tsunami was 1,719 feet (524 meters) high along Lituya Bay, Alaska, on July 9, 1958. Caused by a giant landslip, it moved at 100 miles per hour. This wave would have swamped the Petronas Towers in Kuala Lumpur, Malaysia, which are 1,483 feet (452 meters) high and are recognized as the tallest in the world.

## How many different kinds of **faults** have been identified?

Faults are fractures in the Earth's crust. Faults are classified as normal, reverse or strike-slip. Normal faults occur when the end of one plate slides vertically down the end

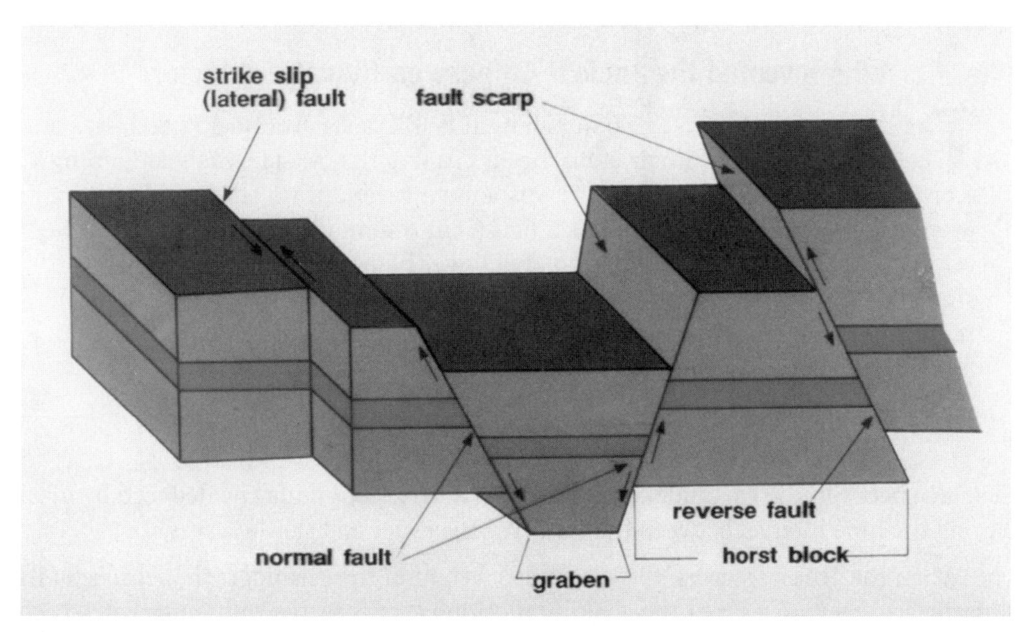


An illustration of three main types of fault motion.

of another. Reverse faults occur when one plate slides vertically up the end of another. A strike-slip fault occurs when two plate ends slide past each other horizontally. In an oblique fault there is plate movement vertically and horizontally at the same time.

## Where is the **San Andreas Fault**?

Perhaps the most famous fault in the world, the San Andreas Fault extends from Mexico north through most of California. The San Andreas Fault is not a single fault, but rather a system of faults. The northern half of the fault, near San Francisco, has reverse faults and is mostly mountainous. The southern half, near Los Angeles, has mostly normal faults. Land development has made it difficult to see the fault except in a few locations, notably near Lake San Andreas south of San Francisco. The fault was named in honor of Andrew Lawson (1861–1952), a geologist who studied the 1906 San Francisco earthquake.

## How does a **seismograph** work?

A seismograph records earthquake waves. When an earthquake occurs, three types of waves are generated. The first two, the P and S waves, are propagated within the Earth, while the third, consisting of Love and Rayleigh waves, is propagated along the planet's surface. The P wave travels about 3.5 miles (5.6 kilometers) per second and is the first wave to reach the surface. The S wave travels at a velocity of a little more than half of the P waves. If the velocities of the different modes of wave propagation are known, the

### Who invented the ancient Chinese earthquake detector?

The detector, invented by Zhang Heng (78–139 C.E.) around 132 C.E., was a copper-domed urn with dragons' heads circling the outside, each containing a bronze ball. Inside the dome was suspended a pendulum that would swing when the ground shook and knock a ball from the mouth of a dragon into the waiting open mouth of a bronze toad below. The ball made a loud noise and signaled the occurrence of an earthquake. Knowing which ball had been released, one could determine the direction of the earthquake's epicenter (the point on the Earth's surface directly above the quake's point of origin).

distance between the earthquake and an observation station may be deduced by measuring the time interval between the arrival of the faster and slower waves.

When the ground shakes, the suspended weight of the seismograph, because of its inertia, scarcely moves, but the shaking motion is transmitted to the marker, which leaves a record on the drum.

## What is the **Richter scale**?

On a machine called a seismograph, the Richter scale measures the magnitude of an earthquake, i.e., the size of the ground waves generated at the earthquake's source. The scale was devised by American geologist Charles W. Richter (1900–1985) in 1935. Every increase of one number means a tenfold increase in magnitude.

**Richter Scale**

| Magnitude | Possible effects |
|---|---|
| 1 | Detectable only by instruments |
| 2 | Barely detectable, even near the epicenter |
| 3 | Felt indoors |
| 4 | Felt by most people; slight damage |
| 5 | Felt by all; damage minor to to moderate |
| 6 | Moderately destructive |
| 7 | Major damage |
| 8 | Total and major damage |

## What is the **modified Mercalli Scale**?

The modified Mercalli Scale is a means of measuring the intesity of an earthquake. Unlike the Richter Scale, which uses mathematical calculation to measure seismic waves, the

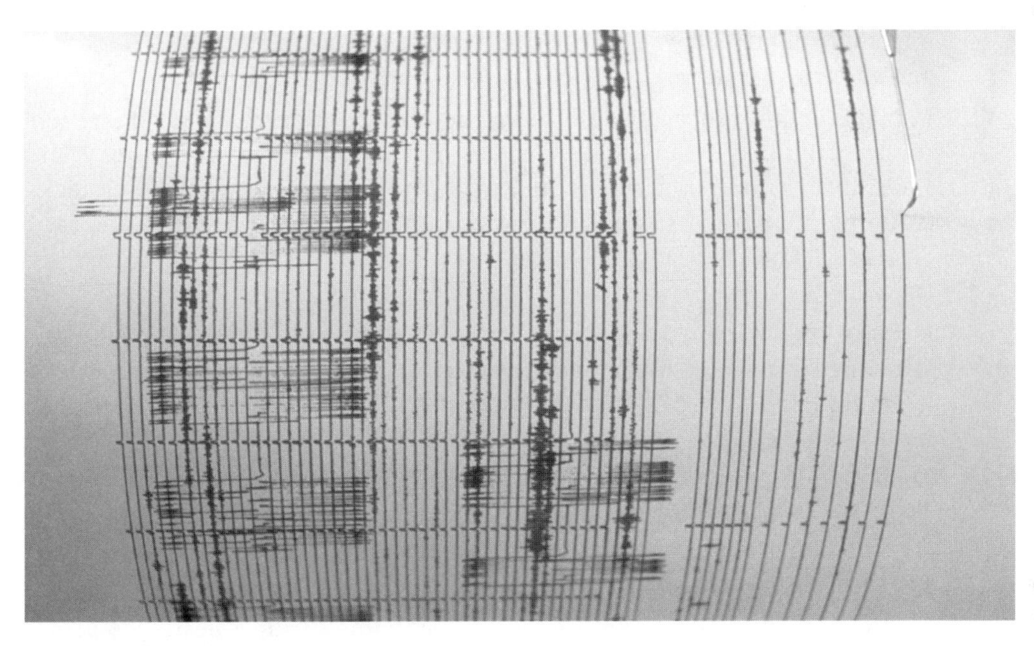

A seismograph records seismic activity in the Philippines.

modified Mercalli Scale uses the effects of an earthquake on the people and structures in a given area to determine its intensity. It was invented by Guiseppe Mercalli (1850–1914) in 1902 and modified by Harry Wood and Frank Neumann in the 1930s to take into consideration such modern inventions as the automobile and the skyscraper.

The Modified Mercalli Scale

I. Only felt by a few under especially favorable circumstances.

II. Felt only by a few sleeping persons, particularly on upper floors of buildings. Some suspended objects may swing.

III. Felt quite noticeably indoors, especially on upper floors of buildings, but may not be recognized as an earthquake. Standing automobiles may rock slightly. Vibration like passing of truck.

IV. During the day felt indoors by many, outdoors by few. At night some awakened. Dishes, windows, doors disturbed; walls make creaking sound. Sensation like heavy truck striking building. Standing automobiles rocked noticeably.

V. Felt by nearly everyone, many awakened. Some dishes, windows, and so on broken; cracked plaster in a few places; unstable objects overturned. Disturbances of trees, poles, and other tall objects sometimes noticed. Pendulum clocks may stop.

VI. Felt by all; many frightened and run outdoors. Some heavy furniture moved; a few instances of fallen plaster and damaged chimneys. Damage slight.

VII. Everybody runs outdoors. Damage negligible in buildings of good design and construction; slight to moderate in well-built ordinary structures; considerable in poorly built or badly designed structures; some chimneys broken. Noticed by persons driving cars.

VIII. Damage slight in specially designed structures; considerable in ordinary substantial buildings with partial collapse; great in poorly built structures. Panel walls thrown out of frame structures. Fall of chimneys, factory stacks, columns, monuments, walls. Heavy furniture overturned. Sand and mud ejected in small amounts. Changes in well water. Persons driving cars disturbed.

IX. Damage considerable in specially designed structures; well-designed frame structures thrown out of plumb; great in substantial buildings, with partial collapse. Building shifted off foundations. Ground cracked conspicuously. Underground pipes broken.

X. Some well-built wooden structures destroyed; most masonry and frame structures destroyed with foundations; ground badly cracked. Rails bent. Landslides considerable from river banks and steel slopes. Shifted sand and mud. Water splashed, slopped over banks.

XI. Few, if any, (masonry) structures remain standing. Bridges destroyed. Broad fissures in ground. Underground pipelines completely out of service. Earth slumps and landslips in soft ground. Rails bent greatly.

XII. Damage total. Waves seen on ground surface. Lines of sight and level distorted. Objects thrown into the air.

## When did the **most severe earthquake** in **American history** occur?

The New Madrid series of earthquakes (a series of quakes starting on December 16, 1811, and lasting until March 1812) is considered to be the most severe earthquake event in United States history. It shook more than two-thirds of the United States and was felt in Canada. It changed the level of land by as much as 20 feet (six meters), altered the course of the Mississippi River, and created new lakes, such as Lake St. Francis west of the Mississippi and Reelfoot Lake in Tennessee. Because the area was so sparsely populated, no known loss of life occurred. Scientists agree that at least three, and possibly five, of the quakes had surface wave magnitudes of 8.0 or greater. The largest was probably a magnitude of 8.8, which is larger than any quake yet experienced in California.

## Of what magnitude was the **earthquake** that hit **San Francisco** on April 18, 1906?

The historic 1906 San Francisco earthquake took a mighty toll on the city and surrounding area. Over 700 people were killed; the newly constructed $6 million city hall was ruined; the Sonoma Wine Company collapsed, destroying 15 million gallons (57 million liters) of wine. The quake registered 8.3 on the Richter scale and lasted 75 seconds total. Many poorly constructed buildings built on landfills were flattened and the

quake destroyed almost all of the gas and water mains. Fires broke out shortly after the quake, and when they were finally eliminated, 3,000 acres of the city, the equivalent of 520 blocks, were charred. Damage was estimated to be $500 million, and many insurance agencies went bankrupt after paying out the claims.

Another large earthquake hit San Francisco on October 17, 1989. It measured 7.1 on the Richter scale, killed 67 people, and caused billions of dollars worth of damage.

## OBSERVATION AND MEASUREMENT

### What is the **Gaia Hypothesis**?

British scientists James Lovelock (1919–) and Lynn Margulis (1938–) proposed the Gaia hypothesis in the 1970s. According to the theory, all living and non-living organisms on the Earth form a single unity that is self-regulated by the organisms themselves. Therefore, the whole planet can be considered a huge single organism. Evidence for this theory is the stability of atmospheric conditions over eons.

### What is **magnetic declination**?

It is the angle between magnetic north and true north at a given point on the Earth's surface. It varies at different points on the Earth's surface and at different times of the year.

### Which direction does a **compass needle point** at the **north pole**?

At the north magnetic pole, the compass needle would be attracted by the ground and point straight down.

### What is a **Foucault pendulum**?

An instrument devised by Jean Foucault (1819–1868) in 1851 to prove that the Earth rotates on an axis, the pendulum consisted of a heavy ball suspended by a very long fine wire. Sand beneath the pendulum recorded the plane of rotation of the pendulum over time.

A reconstruction of Foucault's experiment is located in Portland, Oregon, at its Convention Center. It swings from a cable 90 feet (27.4 meters) long, making it the longest pendulum in the world.

### What is the Piri Re'is map?

In 1929 a map was found in Constantinople that caused great excitement. Painted on parchment and dated in the Muslim year 919 (1541 according to the Christian calendar), it was signed by an admiral of the Turkish navy known as Piri Re'is. This map appears to be one of the earliest maps of America, and it shows South America and Africa in their correct relative longitudes. The mapmaker also indicated that he had used a map drawn by Columbus for the western part. It was an exciting statement because for several centuries geographers had been trying to find a "lost map of Columbus" supposedly drawn by him in the West Indies.

## What are the major eras, periods, and epochs in **geologic time**?

Modern dating techniques have given a range of dates as to when the various geologic time periods have started, as they are listed below:

| Era | Period | Epoch | Beginning date (Est. millions of years) |
|---|---|---|---|
| Cenozoic | Quaternary | Holocene | 10,000 years ago |
| | | Pleistocene | 1.9 |
| | Tertiary | Pliocene | 6 |
| | | Miocene | 25 |
| | | Oligocene | 38 |
| | | Eocene | 55 |
| | | Paleocene | 65 |
| Mesozoic | Cretaceous | | 135 |
| | Jurassic | | 200 |
| | Triassic | | 250 |
| Paleozoic | Permian | | 285 |
| | Carboniferous (divided into Mississippian and Pennsylvanian periods by some in the U.S.) | | 350 |
| | Devonian | | 410 |
| | Silurian | | 425 |
| | Ordovician | | 500 |
| | Cambrian | | 570 |
| Precambrian | Proterozoic | | 2500 |
| | Archeozoic | | 3800 |
| | Azoic | | 4600 |

## What is the **prime meridian**?

The north-south lines on a map run from the North Pole to the South Pole and are called "meridians." The word "meridian" means "noon": When it is noon on one place on the line, it is noon at any other point as well. The lines are used to measure longitudes, or how far east or west a particular place might be, and they are 69 miles (111 kilometers) apart at the equator. The east-west lines are called parallels, and, unlike meridians, are all parallel to each other. They measure latitude, or how far north or south a particular place might be. There are 180 lines circling the Earth, one for each degree of latitude. The degrees of both latitude and longitude are divided into 60 minutes, which are then further divided into 60 seconds each.

The prime meridian is the meridian of 0 degrees longitude, used as the origin for measurement of longitude. The meridian of Greenwich, England, is used almost universally for this purpose.

## What is **Mercator's projection** for maps?

The Mercator projection is a modification of a standard cylindrical projection, a technique used by cartographers to transfer the spherical proportions of the Earth to the flat surface of a map. For correct proportions, the parallels, or lines of latitude, are spaced at increasing distances toward the poles, resulting in severe exaggeration of size in the polar regions. Greenland, for example, appears five times larger than it actually is. Created by Flemish cartographer Gerardus Mercator in 1569, this projection is useful primarily because compass directions appear as straight lines, making it ideal for navigation.

## Who is regarded as the **founder** of **American geology**?

Born in Scotland, the American William Maclure (1763–1840) was a member of a commission set up to settle claims between the United States and France from 1803 through 1807. In 1809 he made a geographical chart of the United States in which the land areas were divided by rock types. In 1817 he revised and enlarged this map. Maclure wrote the first English language articles and books on United States geology.

## When were **relief maps** first used?

The Chinese were the first to use relief maps, in which the contours of the terrain were represented in models. Relief maps in China go back at least to the third century B.C.E. Some early maps were modeled in rice or carved in wood. It is likely that the idea of making relief maps was transmitted from the Chinese to the Arabs and then to Europe. The earliest known relief map in Europe was a map showing part of Austria, made in 1510 by Paul Dox.

## Who was the **first person** to **map the Gulf Stream**?

In his travels to and from France as a diplomat, Benjamin Franklin (1706–1790) noticed a difference in speed in the two directions of travel between France and America. He was the first to study ships' reports seriously to determine the cause of the speed variation. As a result, he found that there was a current of warm water coming from the Gulf of Mexico that crossed the North Atlantic Ocean in the direction of Europe. In 1770, Franklin mapped it.

Franklin thought the current started in the Gulf of Mexico. However, the Gulf Stream actually originates in the western Caribbean Sea and moves through the Gulf of Mexico, the Straits of Florida, then north along the east coast of the United States to Cape Hatteras in North Carolina, where it becomes northeast. The Gulf Stream eventually breaks up near Newfoundland, Canada, to form smaller currents or eddies. Some of these eddies blow toward the British Isles and Norway, causing the climate of these regions to be milder than other areas of northwestern Europe.

## From what distance are **satellite photographs** taken?

U.S. Department of Defense satellites orbit at various distances above the Earth. Some satellites are in low orbit, 100 to 300 miles (160 to 483 kilometers) above the surface, while others are positioned at intermediate altitudes from 500 to 1,000 miles (804 to 1,609 kilometers) high. Some have an altitude of 22,300 miles (35,880 kilometers).

## What are **Landsat maps**?

They are images of the Earth taken at an altitude of 567 miles (912 kilometers) by an orbiting Landsat satellite, or ERTS (Earth Resources Technology Satellite). The Landsats were originally launched in the 1970s. Rather than cameras, the Landsats use multispectral scanners, which detect visible green and blue wavelengths, and four infrared and near-infrared wavelengths. These scanners can detect differences between soil, rock, water, and vegetation; types of vegetation; states of vegetation (e.g., healthy/unhealthy or underwatered/well-watered); and mineral content. The differences are especially accurate when multiple wavelengths are compared using multispectral scanners. Even visible light images have proved useful—some of the earliest Landsat images showed that some small Pacific islands were up to 10 miles (16 kilometers) away from their charted positions.

The results are displayed in "false-color" maps, where the scanner data is represented in shades of easily distinguishable colors—usually, infrared is shown as red, red as green, and green as blue. The maps are used by farmers, oil companies, geologists, foresters, foreign governments, and others interested in land management. Each image covers an area approximately 115 square miles (185 square kilometers). Maps are offered for sale by the United States Geological Survey.

Other systems that produce similar images include the French SPOT satellites, Russian Salyut and Mir manned space stations, and NASA's Airborne Imaging Spectrometer, which senses 128 infrared bands. NASA's Jet Propulsion Laboratories are developing instruments that will sense 224 bands in infrared, which will be able to detect specific minerals absorbed by plants.

## What is **Global Positioning System (GPS)** and how does it work?

The Global Positioning System (GPS) has three parts: the space part, the user part, and the control part. The space part consists of 24 satellites in orbit 11,000 nautical miles (20,300 kilometers) above the Earth. The user part consists of a GPS receiver, which may be hand-held or mounted in a vehicle. The control part consists of five ground stations worldwide to assure the satellites are working properly. Using a GPS receiver an individual can determine his or her location on or above the Earth to within about 300 feet (90 meters).

# CLIMATE AND WEATHER

## TEMPERATURE

### What is **El Niño**?

El Niño is the unusual warming of the surface waters of large parts of the tropical Pacific Ocean. Occurring around Christmastime, it is named after the Christ child. El Niño occurs erratically every three to seven years. It brings heavy rains and flooding to Peru, Ecuador, and southern California, and a milder winter with less snow to the northeastern United States. Studies reveal that El Niño is not an isolated occurrence, but is instead part of a pattern of change in the global circulation of the oceans and atmosphere. The 1982–83 El Niño was one of the most severe climate events of the twentieth century in both its geographical extent as well as in the degree of warming (14°F or 8°C).

### What is **La Niña**?

La Niña is the opposite of El Niño. It refers to a period of cold surface temperatures of the tropical Pacific Ocean. La Niña winters are especially harsh with heavy snowfall in the northeastern United States and rain in the Pacific Northweast.

### Is the **world** actually **getting warmer**?

During the past century, global surface temperatures have increased at a rate near 1.1°F/century (0.6°C/century). This trend has dramatically increased to a rate approaching 3.6°F/century (2.0°C/century) during the past 25 years. There have been two sustained periods of warming, one beginning around 1910 and ending around

1945, and the most recent beginning about 1976. The year 1998 was the warmest year on record, with an average land temperature that was 1.35°F (0.75°C) above the 1880–2000 average land temperature of 47.3°F (8.5°C). The second warmest year on record was 2001, when the average annual global temperature was 57.9°F (14.4°C), which is 0.9°F (0.5°C) above the 1880–2000 long-term average. For the contiguous U.S., the preliminary 2001 average annual temperature was 54.3°F (12.4°C), which was 1.5°F (0.8°C) above the 1895–2001 mean, the sixth warmest on record.

### Which is **colder**—the **North Pole** or the **South Pole**?

The South Pole is considerably colder than the North Pole. The average temperature at the South Pole is –56°F (–49°C), which is approximately 35° lower than the average temperature at the North Pole. Located in the large snowy continent of Antarctica, very little radiation from the sun is retained at the earth's surface on the South Pole. Also, the weather observation station is located at an elevation of 12,000 feet (3,660 meters). At elevations of that height there is less air to hold in the heat from solar radiation.

### What are the **highest** and **lowest recorded temperatures** on Earth?

The highest temperature in the world was recorded as 136°F (58°C) at Al Aziziyah (el-Aziia), Libya, on September 13, 1922; the highest temperature recorded in the United States was 134°F (56.7°C) in Death Valley, California, on July 10, 1913. The temperatures of 140°F (60°C) at Delta, Mexico, in August of 1953 and 136.4°F (58°C) at San Luis, Mexico, on August 11, 1933, are not internationally accepted. The lowest temperature was –128.6°F (–89.6°C) at Vostok Station in Antarctica on July 21, 1983. The record cold temperature for an inhabited area was –90.4°F (–68°C) at Oymyakon, Siberia (population 4,000), on February 6, 1933. This temperature tied with the readings at Verkhoyansk, Siberia, on January 3, 1885, and February 5 and 7, 1892. The lowest temperature reading in the United States was –79.8°F (–62.1°C) on January 23, 1971, in Prospect Creek, Alaska; for the contiguous 48 states, the coldest temperature was –69.7°F (–56.5°C) at Rogers Pass, Montana, on January 20, 1954.

### What is a **heat wave**?

A heat wave is a period of two days in a row when apparent temperatures on the National Weather Service heat index exceed 105°F to 110°F (40°C to 43°C). The temperature standards vary greatly for different locales. Heat waves can be extremely dangerous. According to the National Weather Service, 175–200 Americans die from heat in a normal summer. Between 1936 and 1975, as many as 15,000 Americans died from problems related to heat. In 1980, 1,250 people died during a brutal heat wave in the Midwest. In 1995, more than 500 people died in the city of Chicago from heat related problems. A majority of these individuals were the elderly living in high-rise apart-

ment buildings without proper air conditioning. Large concentrations of buildings, parking lots, and roads create an "urban heat island" in cities.

## What is the **heat index**?

The index is a measure of what hot weather feels like to the average person for various temperatures and relative humidities. Heat exhaustion and sunstroke are inclined to happen when the heat index reaches 105°F (40°C). The chart below provides the heat index for some temperatures and relative humidities.

| | Air Temperature (°F) | | | | | | | | | | |
|---|---|---|---|---|---|---|---|---|---|---|---|
| | 70 | 75 | 80 | 85 | 90 | 95 | 100 | 105 | 110 | 115 | 120 |
| **Relative Humidity** | Feels like (°F) | | | | | | | | | | |
| 0% | 64 | 69 | 73 | 78 | 83 | 87 | 91 | 95 | 99 | 103 | 107 |
| 10% | 65 | 70 | 75 | 80 | 85 | 90 | 95 | 100 | 105 | 111 | 116 |
| 20% | 66 | 72 | 77 | 82 | 87 | 93 | 99 | 105 | 112 | 120 | 130 |
| 30% | 67 | 73 | 78 | 84 | 90 | 96 | 104 | 113 | 123 | 135 | 148 |
| 40% | 68 | 74 | 79 | 86 | 93 | 101 | 110 | 123 | 137 | 151 | |
| 50% | 69 | 75 | 81 | 88 | 96 | 107 | 120 | 135 | 150 | | |
| 60% | 70 | 76 | 82 | 90 | 100 | 114 | 132 | 149 | | | |
| 70% | 70 | 77 | 85 | 93 | 106 | 124 | 144 | | | | |
| 80% | 71 | 78 | 86 | 97 | 113 | 136 | | | | | |
| 90% | 71 | 79 | 88 | 102 | 122 | | | | | | |
| 100% | 72 | 80 | 91 | 108 | | | | | | | |

## Which place has the **maximum amount of sunshine** in the United States?

Yuma, Arizona, has an annual average of 90 percent of sunny days, or over 4,000 sunny hours per year. St. Petersburg, Florida, had 768 consecutive sunny days from February 9, 1967, to March 17, 1969. On the other extreme, the South Pole has no sunshine for 182 days annually, and the North Pole has none for 176 days.

## Why was 1816 known as the **year without a summer**?

The eruption of Mount Tambora, a volcano in Indonesia, in 1815 threw billions of cubic yards of dust over 15 miles (24 kilometers) into the atmosphere. Because the dust penetrated the stratosphere, wind currents spread it throughout the world. As a consequence of this volcanic activity, in 1816 normal weather patterns were greatly altered. Some parts of Europe and the British Isles experienced average temperatures 2.9 to 5.8°F (1.6 to 3.2°C) below normal. In New England heavy snow fell between June 6 and June 11 and frost occurred every month of 1816. Crop failures were experi-

### How can the temperature be determined from the frequency of cricket chirps?

Listen for the chirping of either katydids or crickets, then count the number of chirps you hear in one minute. For the following equations "C" equals the number of chirps per minute:

For katydids, Fahrenheit temperature = 60 + (C–19)/3

For crickets, Fahrenheit temperature = 50 + (C–50)/4

enced in Western Europe and Canada as well as in New England. By 1817 the excess dust had settled and the climate returned to more normal conditions.

### Why are the hot, humid days of summer called **"dog days"**?

This period of extremely hot, humid, sultry weather that traditionally occurs in the northern hemisphere in July and August received its name from the dog star Sirius of the constellation Canis Major. At this time of year, Sirius, the brightest visible star, rises in the east at the same time as the sun. Ancient Egyptians believed that the heat of this brilliant star added to the sun's heat to create this hot weather. Sirius was blamed for the withering droughts, sickness, and discomfort that occurred during this time. Traditional dog days start on July 3 and end on August 11.

### What is **Indian summer**?

The term Indian summer dates back to at least 1778. The term may relate to the way the Native Americans availed themselves of the nice weather to increase their winter food supplies. It refers to a period of pleasant, dry, warm days in the middle to late autumn, usually after the first killing frost.

## AIR PHENOMENA

### What is a **bishop's ring**?

It is a ring around the sun, usually with a reddish outer edge. It is probably due to dust particles in the air, since it is seen after all great volcanic eruptions.

The Aurora Borealis, or Northern Lights, illuminate the sky over a Canadian boreal forest.

## When does the **green flash phenomenon** occur?

On rare occasions, the sun may look bright green for a moment, as the last tip of the sun is setting. This green flash occurs because the red rays of light are hidden below the horizon and the blue are scattered in the atmosphere. The green rays are seldom seen because of dust and pollution in the lower atmosphere. It may best be seen when the air is cloudless and when a distant, well-defined horizon exists, as on an ocean.

## How often does an **aurora** appear?

Because it depends on solar winds (electrical particles generated by the sun) and sunspot activity, the frequency of an aurora cannot be determined. Auroras usually appear two days after a solar flare (a violent eruption of particles on the sun's surface) and reach their peak two years into the 11-year sunspot cycle. The auroras, occurring in the polar regions, are broad displays of usually colored light at night. The northern polar aurora is called Aurora Borealis or Northern Lights and the southern polar aurora is called the Aurora Australis.

## When and by whom were **clouds** first classified?

The French naturalist Jean Lamarck (1744–1829) proposed the first system for classifying clouds in 1802. His work, however, did not receive wide acclaim. A year later the Englishman Luke Howard (1772–1864) developed a cloud classification system that has been generally accepted and is still used today. Clouds are distinguished by their general appearance ("heap clouds" and "layer clouds") and by their height above the ground. Latin names and prefixes are used to describe these characteristics. The shape names are cirrus (curly or fibrous), stratus (layered), and cumulus (lumpy or piled). The prefixes denoting height are cirro (high clouds with bases above 20,000 feet [6,067 meters]) and alto (mid-level clouds from 6,000 to 20,000 feet [1,820 to 6,067 meters]). There is no prefix for low clouds. Nimbo or nimbus is also added as a name or prefix to indicate that the cloud produces precipitation.

## What are the **four major cloud groups** and their types?

1. High Clouds—composed almost entirely of ice crystals. The bases of these clouds start at 16,500 feet (5,000 meters) and reach 45,000 feet (13,650 meters).

   *Cirrus* (from Latin, "lock of hair")—are thin, feather-like crystal clouds in patches or narrow bands. The large ice crystals that often trail downward in well-defined wisps are called "mares tails."

   *Cirrostratus*—is a thin, white cloud layer that resembles a veil or sheet. This layer can be striated or fibrous. Because of the ice content, these clouds are associated with the halos that surround the sun or moon.

   *Cirrocumulus*—are thin clouds that appear as small white flakes or cotton patches and may contain super-cooled water.

2. Middle Clouds—composed primarily of water. The height of the cloud bases range from 6,500 to 23,000 feet (2,000 to 7,000 meters).

   *Altostratus*—appears as a bluish or grayish veil or layer of clouds that can gradually merge into altocumulus clouds. The sun may be dimly visible through it, but flat, thick sheets of this cloud type can obscure the sun.

   *Altocumulus*—is a white or gray layer or patches of solid clouds with rounded shapes.

3. Low Clouds—composed almost entirely of water that may at times be super-cooled; at subfreezing temperatures, snow and ice crystals may be present as well. The bases of these clouds start near the Earth's surface and climb to 6,500 feet (2,000 meters) in the middle latitudes.

   *Stratus*—are gray, uniform, sheet-like clouds with a relatively low base, or they can be patchy, shapeless, low gray clouds. Thin enough for the sun to shine through, these clouds bring drizzle and snow.

   *Stratocumulus*—are globular rounded masses that form at the top of the layer.

   *Nimbostratus*—are seen as a gray or dark, relatively shapeless, massive cloud layer containing rain, snow, and ice pellets.

4. Clouds with Vertical Development—contain super-cooled water above the freezing level and grow to great heights. The cloud bases range from 1,000 feet (300 meters) to 10,000 feet (3,000 meters).

   *Cumulus*—are detached, fair-weather clouds with relatively flat bases and dome-shaped tops. These usually do not have extensive vertical development and do not produce precipitation.

   *Cumulonimbus*—are unstable, large, vertical clouds with dense boiling tops that bring showers, hail, thunder, and lightning.

## What is the **color** of **lightning**?

The atmospheric conditions determine the color of lightning. Blue lightning within a cloud indicates the presence of hail. Red lightning within a cloud indicates the presence of rain. Yellow or orange lightning indicates a large concentration of dust in the air. White lightning is a sign of low humidity in the air.

## How **hot** is **lightning**?

The temperature of the air around a bolt of lightning is about 54,000°F (30,000°C), which is six times hotter than the surface of the sun, yet many times people survive being struck by a bolt of lightning. American park ranger Roy Sullivan, for example, was hit by lightning seven times between 1942 and 1977. In cloud-to-ground lightning, its energy seeks the shortest route to Earth, which could be through a person's shoulder, down the side of the body, through the leg, and to the ground. As long as the lightning does not pass across the heart or spinal column, the victim usually does not die.

## How many **volts** are in **lightning**?

A stroke of lightning discharges from 10 to 100 million volts of electricity. An average lightning stroke has 30,000 amperes.

## How **long** is a **lightning stroke**?

The visible length of the streak of lightning depends on the terrain and can vary greatly. In mountainous areas where clouds are low, the flash can be as short as 300 yards (273 meters); whereas in flat terrain, where clouds are high, the bolt can measure as long as four miles (6.5 kilometers). The usual length is about one mile (1.6 kilometers), but streaks of lightning up to 20 miles (32 kilometers) have been recorded. The stroke channel is very narrow—perhaps as little as half an inch (1.27 centimeters). It is surrounded by a "corona envelope" or a glowing discharge that can be as wide as 10 to 20 feet (three to six meters) in diameter. The speed of lightning can vary from 100 to 1,000 miles (161 to 1,610 kilometers) per second for the downward

### Does lightning ever strike twice in the same place?

It is not true that lightning does not strike twice in the same place. In fact, tall buildings, such as the Empire State Building in New York, can be struck several times during the same storm. During one storm, lightning struck the Empire State Building 12 times.

leader track; the return stroke is 87,000 miles (140,070 kilometers) per second (almost half the speed of light).

## How is the **distance** of a lightning flash calculated?

Count the number of seconds between seeing a flash of lightning and hearing the sound of the thunder. Divide the number by five to determine the number of miles away that the lightning flashed.

## What is **ball lightning**?

Ball lightning is a rare form of lightning in which a persistent and moving luminous white or colored sphere is seen. It can last from a few seconds to several minutes, and it travels at about a walking pace. Spheres have been reported to vanish harmlessly, or to pass into or out of rooms—leaving, in some cases, sign of their passage such as a hole in a window pane. Sphere dimensions vary but are most commonly from four to eight inches (10 to 20 centimeters) in diameter.

Other types of lightning include the common streak lightning (a single or multiple zigzagging line from cloud to ground); forked lightning (lightning that forms two branches simultaneously); sheet lightning (a shapeless flash covering a broad area); ribbon lightning (streak lightning blown sideways by the wind to make it appear like parallel successive strokes); bead or chain lightning (a stroke interrupted or broken into evenly spaced segments or beads); and heat lightning (lightning seen along the horizon during hot weather and believed to be a reflection of lightning occurring beyond the horizon).

## What are **fulgurites**?

Fulgurites (from the Latin word *fulgur,* meaning lightning) are petrified lightning, created when lightning strikes an area of dry sand. The intense heat of the lightning melts the sand surrounding the stroke into a rough, glassy tube forming a fused record of its path. These tubes may be one-half to two inches (1.5 to five centimeters) in diameter, and up to 10 feet (three meters) in length. They are extremely brittle and break easily. The inside walls of the tube are glassy and lustrous while the outside is

rough, with sand particles adhering to it. Fulgurites are usually tan or black in color, but translucent white ones have been found.

### What is **Saint Elmo's fire**?

Saint Elmo's fire has been described as a corona from electric discharge produced on high grounded metal objects, chimney tops, and ship masts. Since it often occurs during thunderstorms, the electrical source may be lightning. Another description refers to this phenomenon as weak static electricity formed when an electrified cloud touches a high exposed point. Molecules of gas in the air around this point become ionized and glow. The name originated with sailors who were among the first to witness the display of spearlike or tufted flames on the tops of their ships' masts. Saint Elmo (which is a corruption of Saint Ermo) is the patron saint of sailors, so they named the fire after him.

### What is the order of **colors** in a **rainbow**?

Red, orange, yellow, green, blue, indigo, and violet are the colors of the rainbow, but these are not necessarily the sequence of colors that an observer might see. Rainbows are formed when raindrops reflect sunlight. As sunlight enters the drops, the different wavelengths of the colors that compose sunlight are refracted at different lengths to produce a spectrum of color. Each observer sees a different set of raindrops at a slightly different angle. Drops at different angles from the observer send different wave lengths (i.e., different color) to the observer's eyes. Since the color sequence of the rainbow is the result of refraction, the color order depends on how the viewer sees this refraction from the viewer's angle of perception.

### What is the origin of the **Brown Mountain Lights** of North Carolina?

For a period of some 30 years the "lights" seen at Brown Mountain could not be explained. In 1922, the United States Geological Survey studied the mystery. The area has extraordinary atmospheric conditions, and automobile, locomotive, and fixed lights from miles away are reflected in the atmosphere.

## WIND

### Where do **haboobs** occur?

Haboobs, derived from the Arabic word "habb" meaning to blow, are violent dust storms with strong winds of sand and dust. They are most common in the Sahara region of Africa and the deserts of southwestern United States, Australia, and Asia.

## What is **wind shear**?

Wind shear refers to rapid changes in wind speed and/or direction over short distances and is usually associated with thunderstorms. It is especially dangerous to aircraft.

## What is the effect of a **microburst** on aircraft?

Microbursts are downbursts with a diameter of 2.5 miles (4 kilometers) or less. Often associated with thunderstorms, they can generate winds of hurricane force that change direction abruptly. Headwinds can become tailwinds in a matter of seconds, resulting in a loss of airspeed and loss of altitude. After microbursts caused several major air catastrophes in the 1970s and 1980s, the Federal Aviation Administration (FAA) installed warning and radar systems at airports to alert pilots to wind shear and microburst conditions.

## What conditions result in **microclimates**?

Microclimates are small-scale regions where the average weather conditions are measurably different from the larger, surrounding region. Differences in temperature, precipitation, wind, or cloud cover can produce microclimates. Frequent causes of microclimates are differences in elevation, mountains that alter wind patterns, shorelines, and man-made structures, such as buildings, that can alter the wind patterns.

## What is the **Coriolis effect**?

The 19th-century French engineer Gaspard C. Coriolis (1792–1843) discovered that the rotation of the Earth deflects streams of air. Because the Earth spins to the east, all moving objects in the Northern Hemisphere tend to turn somewhat to the right of a straight path, while those in the Southern Hemisphere turn slightly left. The Coriolis effect explains the lack of northerly and southerly winds in the tropics and polar regions; the northeast and southeast trade winds and the polar easterlies all owe their westward deflection to the Coriolis effect.

## When was the **jet stream** discovered?

A jet stream is a flat and narrow tube of air that moves more rapidly than the surrounding air. Discovered by World War II bomber pilots flying over Japan and the Mediterranean Sea, jet streams have become important with the advent of airplanes capable of cruising at over 30,000 feet (9,144 meters). The currents of air flow from west to east and are usually a few miles deep, up to 100 miles (160 kilometers) wide, and well over 1,000 miles (1,600 kilometers) in length. The air current must flow at over 57.5 miles (92 kilometers) per hour.

There are two polar jet streams, one in each hemisphere. They meander between 30 and 70 degrees latitude, occur at altitudes of 25,000 to 35,000 feet (7,620 to 10,668

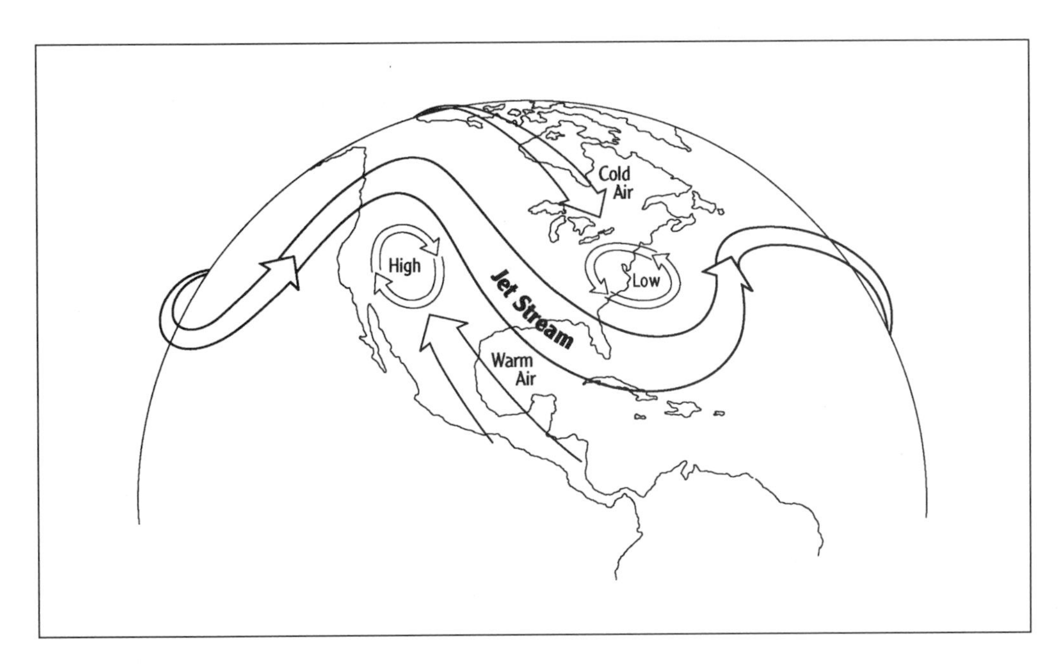


Jet stream

meters), and achieve maximum speeds of over 230 miles (368 kilometers) per hour. The subtropical jet streams (again one per hemisphere) wander between 20 and 50 degrees latitude. They are found at altitudes of 30,000 to 45,000 feet (9,144 to 13,715 meters) and have speeds of over 345 miles (552 kilometers) per hour.

## Why are the **horse latitudes** called by that name?

The horse latitudes are two high pressure belts characterized by low winds about 30 degrees north and south of the equator. Dreaded by early sailors, these areas have undependable winds with periods of calm. In the northern hemisphere, particularly near Bermuda, sailing ships carrying horses from Spain to the New World were often becalmed. When water supplies ran low, these animals were the first to be rationed water. Dying from thirst or tossed overboard, the animals were sacrificed to conserve water for the men. Explorers and sailors reported that the seas were "strewn with bodies of horses," which may be why the areas are called the horse latitudes. The term might also be rooted in complaints by sailors who were paid in advance and received no overtime when the ships slowly traversed this area. During this time they were said to be "working off a dead horse."

## What are **halcyon days**?

This term is often used to refer to a time of peace or prosperity. Among sailors, it is the two-week period of calm weather before and after the shortest day of the year, approximately December 21. The phrase is taken from halcyon, the name the ancient Greeks gave to the kingfisher. According to legend, the halcyon built its nest on the surface of the ocean and was able to quiet the winds while its eggs were hatching.

## What is a **Siberian express**?

This term describes storms that are severely cold and cyclonic; they descend from northern Canada and Alaska to other parts of the United States.

## What is an **Alberta clipper**?

An Alberta Clipper is a little gyrating storm that develops on the Pacific front, usually over the Rocky Mountains of Alberta, Canada. This quick-moving storm moves southeast into the Great Plains, leaving a trail of cold air.

## What is a **Chinook**?

It is a wind that is generally warm and originates from the eastern slope of the Rocky Mountains. It often moves from the southwest in a downslope manner, causing a noticeable rise in temperature that helps to warm the plains just east of the Rocky Mountains.

The Chinook is classified as a katabatic wind. A katabatic wind develops because of cold, heavy air spilling down sloping terrain, moving the lighter, warmer air in front of it. The air is dried and heated as it streams down the slope. At times the falling air becomes warmer than the air it restores. Some katabatic winds have been interestingly named, like Taku, a frigid wind in Alaska, or Santa Ana, a warmer wind from the Sierras.

## Is Chicago the **windiest city**?

Chicago ranks 21st in a list of 68 windy cities with an average wind speed of 10.3 miles (16.6 kilometers) per hour. Cheyenne, Wyoming, with an average wind speed of 12.9 miles (20.8 kilometers) per hour, ranks number one, closely followed by Great Falls, Montana, with an average wind speed of 12.8 miles (20.6 kilometers) per hour. The highest surface wind ever recorded was on Mount Washington, New Hampshire, at an elevation of 6,288 feet (1.9 kilometers). On April 12, 1934, its wind was 231 miles (371.7 kilometers) per hour and its average wind speed was 35 miles (56.3 kilometers) per hour.

## Who is associated with developing the **concept of wind chill**?

The Antarctic explorer Paul A. Siple (1908–1968) coined the term in his 1939 dissertation, "Adaptation of the Explorer to the Climate of Antarctica." Siple was the youngest member of Admiral Richard Byrd's Antarctica expedition in 1928–30, and later made

other trips to the Antarctic as part of Byrd's staff and for the United States Department of the Interior assigned to the United States Antarctic Expedition. He also served in many other endeavors related to the study of cold climates.

## What is meant by the **wind chill factor**?

The wind chill factor, or wind chill index, is a number that expresses the cooling effect of moving air at different temperatures. It indicates in a general way how many calories of heat are carried away from the surface of the body. The National Weather Service began reporting the equivalent wind chill temperature along with the actual air temperature in 1973. For years it was believed that the index overestimated the wind's cooling effect on skin. In 2001–02 a new wind chill index was instituted, and additional corrections are expected in the next few years.

## Is there a **formula** for **computing the wind chill**?

Here is how you calculate the *Old Wind Chill Index*:

$$T_{wc} = 0.081 \times (3.71 \times \sqrt{V} + 5.81 - 0.25 \times V) \times (T - 91.4) + 91.4$$

Here is how you calculate the *New Wind Chill Index* (2001–2002):

$$T_{wc} = 35.74 + 0.6215T - 35.75(V^{0.16}) + 0.4275T(V^{0.16})$$

where $T_{wc}$ is the wind chill in degrees F, V is the wind speed in miles per hour and T is the temperature in degrees F.

## How does a **cyclone** differ from a **hurricane** or a **tornado**?

All three wind phenomena are rotating winds that spiral in toward a low-pressure center as well as upward. Their differences lie in their size, wind velocity, rate of travel, and duration. Generally, the faster the winds spin, the shorter (in time) and smaller (in size) the event becomes.

A cyclone has rotating winds from 10 to 60 miles per hour (16 to 97 kilometers per hour); can be up to 1,000 miles (1,600 kilometers) in diameter, travels about 25 miles per hour (40 kilometers per hour), and lasts from one to several weeks. A hurricane (or typhoon, as it is called in the Pacific Ocean area) has winds that vary from 75 to 200 miles per hour (120 to 320 kilometers per hour), moves between 10 to 20 miles per hour (16 to 32 kilometers per hour), can have a diameter up to 600 miles (960 kilometers), and can exist from several days to more than a week. A tornado can reach a rotating speed of 300 miles per hour (400 kilometers per hour), travels between 25 to 40 miles per hour (40 to 64 kilometers per hour), and generally lasts only minutes, although some have lasted for five to six hours. Its diameter can range from 300 yards (274 meters) to one mile (1.6 kilometers) and its average path length is 16 miles (26 kilometers), with a maximum of 300 miles (483 kilometers).

Typhoons, hurricanes, and cyclones tend to breed in low-altitude belts over the oceans, generally from five degrees to fifteen degrees latitude north or south. A tornado generally forms several thousand feet above the Earth's surface, usually during warm, humid weather; many times it is in conjunction with a thunderstorm. Although a tornado can occur in many places, they mostly appear on the continental plains of North America (i.e., from the Plains States eastward to western New York and the southeastern Atlantic states). Eighty-two percent of tornadoes materialize during the warmest hours of the day (noon to midnight), while 23 percent of all tornado activity occurs between 4 p.m. and 6 p.m.

## What is the **Fujita** and **Pearson Tornado Scale**?

The Fujita and Pearson Tornado Scale, developed by T. Theodore Fujita and Allen Pearson, ranks tornadoes by their wind speed, path, length, and width. Sometimes known simply as the Fujita scale, the ranking ranges from F0 (very weak) to F6 (inconceivable).

F0—Light damage: damage to trees, billboards, and chimneys.

F1—Moderate damage: mobile homes pushed off their foundations and cars pushed off roads.

F2—Considerable damage: roofs torn off, mobile homes demolished, and large trees uprooted.

F3—Severe damage: even well-constructed homes torn apart, trees uprooted, and cars lifted off the ground.

F4—Devastating damage: houses leveled, cars thrown, and objects become flying missiles.

F5—Incredible damage: structures lifted off foundations and carried away; cars become missiles. Less than 2 percent of tornadoes are in this category.

F6—Maximum tornado winds not expected to exceed 318 mph (511 kph).

**Fujita and Pearson Tornado Scale**

| Scale | Speed miles per hour | Path length in miles | Path width |
|---|---|---|---|
| 0 | ≤72 | ≤1.0 | ≤17 yards |
| 1 | 73–112 | 1.0–3.1 | 18–55 yards |
| 2 | 113–157 | 3.2–9.9 | 56–175 yards |
| 3 | 158–206 | 10.0–31.0 | 176–556 yards |
| 4 | 207–260 | 32.0–99.0 | 0.34–0.9 miles |
| 5 | 261–318 | 100–315 | 1.0–3.1 miles |
| 6 | 319–380 | 316–999 | 3.2–9.9 miles |

## What is the **Beaufort scale**?

The Beaufort scale was devised in 1805 by a British Admiral, Sir Francis Beaufort (1774–1857), to help mariners in handling ships. It uses a series of numbers from 0 to 17 to indicate wind speeds and applies to both land and sea.

| Beaufort number | Name | Wind speed | |
|---|---|---|---|
| | | Miles per hour | Kilometers per hour |
| 0 | Calm | less than 1 | less than 1.5 |
| 1 | Light air | 1–3 | 1.5–4.8 |
| 2 | Light breeze | 4–7 | 6.4–11.3 |
| 3 | Gentle breeze | 8–12 | 12.9–19.3 |
| 4 | Moderate breeze | 13–18 | 21–29 |
| 5 | Fresh breeze | 19–24 | 30.6–38.6 |
| 6 | Strong breeze | 25–31 | 40.2–50 |
| 7 | Moderate gale | 32–38 | 51.5–61.1 |
| 8 | Fresh gale | 39–46 | 62.8–74 |
| 9 | Strong gale | 47–54 | 75.6–86.9 |
| 10 | Whole gale | 55–63 | 88.5–101.4 |
| 11 | Storm | 64–73 | 103–117.5 |
| 12–17 | Hurricane | 74 and above | 119.1 and above |

## Which year had the **most tornadoes**?

From the period 1916 (when records started to be kept) to 1999, more tornadoes occurred in 1998 than in any other year. That year 1,424 tornadoes struck, killing 130 people. The largest outbreak of tornadoes occurred on April 3 and 4, 1974; 148 tornadoes were recorded in this "Super Outbreak" in the Great Plains and Midwestern states. Six of these tornadoes had winds greater than 260 miles (420 kilometers) per hour and some of them were the strongest ever recorded. In the 1990s, even greater numbers of tornadoes per year have been reported:

| Year | Number of Tornadoes |
|---|---|
| 1995 | 1,234 |
| 1996 | 1,173 |
| 1997 | 1,148 |
| 1998 | 1,424 |

## Who are **storm chasers**?

Storm chasers are scientists and amateur storm enthusiasts who track and intercept severe thunderstorms and tornadoes. Two reasons for storm chasing are: 1) to gather data to use in researching severe storms and 2) to provide a visual observation of severe storms indicated on remote radar stations. In addition, television personnel will chase storms to produce a dramatic storm video. Storm chasing can be an extremely dangerous activity in which strong winds, heavy rain, hail, and lightning threaten one's safety. Individuals who chase storms are trained in the behavior of severe storms.

## Which **month** is the **most dangerous for tornadoes** in the United States?

According to one study, May is the most dangerous month for tornadoes in the United States, with an average of 329, while February's average is the safest with only three. In another study the months December and January were usually the safest, and the months having the greatest number of tornadoes were April, May, and June. In February, tornado frequency begins to increase. February tornadoes tend to occur in the central Gulf states; in March the center of activity moves eastward to the southeastern Atlantic states, where tornado activity peaks in April. In May the center of activity is in the southern Plains states; in June this moves to the northern Plains and Great Lakes area (into western New York). The most costly outbreak of tornadoes occurred in May 1999, when at least 74 tornadoes touched down in less than 48 hours in Oklahoma and Kansas, including an F5 (see scale above) on the outskirts of Oklahoma City causing $$1.1 billion in damage.

## How are **hurricanes classified**?

The Saffir/Simpson Hurricane Damage-Potential scale assigns numbers 1 through 5 to measure the disaster potential of a hurricane's winds and its accompanying storm surge. The purpose of the scale, developed in 1971 by Herbert Saffir and Robert Simpson, is to help disaster agencies gauge the potential significance of these storms in terms of assistance.

**Saffir/Simpson Hurricane Scale Ranges**

| Scale number (category) | Barometric pressure (in inches) | Winds (miles per hour) | Surge (in feet) | Damage |
|---|---|---|---|---|
| 1 | ⩾28.94 | 74–95 | 4–5 | Minimal |
| 2 | 28.50–28.91 | 96–110 | 6–8 | Moderate |
| 3 | 27.91–28.47 | 111–130 | 9–12 | Extensive |
| 4 | 27.17–27.88 | 131–155 | 13–18 | Extreme |
| 5 | <27.17 | >155 | >18 | Catastrophic |

Damage categories:

*Minimal*—No real damage to building structures. Some tree, shrubbery, and mobile home damage. Coastal road flooding and minor pier damage.

*Moderate*—Some roof, window, and door damage. Considerable damage to vegetation, mobile homes, and piers. Coastal and low-lying escape routes flood two to four hours before center of storm arrives. Small craft can break moorings in unprotected areas.

*Extensive*—Some structural damage to small or residential buildings. Mobile homes destroyed. Flooding near coast destroys structures and floods of homes five feet (1.5 meters) above sea level as far inland as six miles (9.5 kilometers).

*Extreme*—Extensive roof, window, and door damage. Major damage to lower floors of structures near the shore, and some roof failure on small residences. Complete beach erosion. Flooding of terrain 10 feet (three meters) above sea level as far as six miles (9.5 kilometers) inland requiring massive residential evacuation.

*Catastrophic*—Complete roof failure to many buildings; some complete building failure, with small utility buildings blown away. Major damage to lower floors of all structures 19 feet (5.75 meters) above sea level located within 500 yards (547 meters) of the shoreline. Massive evacuation of residential areas on low ground five to 10 miles (eight to 16 kilometers) from shoreline may be required.

## How do **hurricanes** get their **names**?

Since 1950, hurricane names have been officially selected from library sources and are decided on during the international meetings of the World Meteorological Organization (WMO). The names are chosen to reflect the cultures and languages found in the Atlantic, Caribbean, and Hawaiian regions. When a tropical storm with rotary action and wind speeds above 39 miles (63 kilometers) per hour develops, the National Hurricane Center near Miami, Florida, selects a name from one of the six listings for Region 4 (Atlantic and Caribbean area). Letters Q, U, X, Y, and Z are not included because of the scarcity of names beginning with those letters. Once a storm has done great damage, its name is retired from the six-year list cycle.

| 2002 | 2003 | 2004 | 2005 | 2006 | 2007 |
|---|---|---|---|---|---|
| Arthur | Ana | Alex | Arlene | Alberto | Andrea |
| Bertha | Bill | Bonnie | Bret | Beryl | Barry |
| Cristobel | Claudette | Charley | Cindy | Chris | Chantal |
| Dolly | Danny | Danielle | Dennis | Debbie | Dean |
| Edouard | Erika | Earl | Emily | Ernesto | Erin |
| Fay | Fabian | Frances | Franklin | Florence | Felix |
| Gustav | Grace | Gaston | Gert | Gordon | Gabrielle |
| Hanna | Henri | Hermine | Harvey | Helene | Humberto |
| Isidore | Isabel | Ivan | Irene | Isaac | Ingrid |
| Josephine | Juan | Jeanne | Jose | Joyce | Jerry |
| Kyle | Kate | Karl | Katrina | Kirk | Karen |
| Lili | Larry | Lisa | Lee | Leslie | Lorenza |

| | | | | | |
|---|---|---|---|---|---|
| Marco | Mindy | Matthew | Maria | Michael | Melissa |
| Nana | Nicholas | Nicole | Nate | Nadine | Noel |
| Omar | Odette | Otto | Ophelia | Oscar | Olga |
| Paloma | Peter | Paula | Philippe | Patty | Pablo |
| Rene | Rose | Richard | Rita | Rafael | Rebekah |
| Sally | Sam | Shary | Stan | Sandy | Sebastien |
| Teddy | Teresa | Tomas | Tammy | Tony | Tanya |
| Vicky | Victor | Virginie | Vince | Valerie | Van |
| Wilfred | Wanda | Walter | Wilma | William | Wendy |

## Which **hurricane names** have been **retired**?

### Atlantic Storms Retired Into Hurricane History

Agnes (1972): Florida, Northeast U.S.
Alicia (1983): North Texas
Allen (1980): Antilles, Mexico, South Texas
Andrew (1992): Bahamas, South Florida, Louisiana
Anita (1977): Mexico
Audrey (1957): Louisiana, North Texas
Betsy (1965): Bahamas, Southeast Florida, Southeast Louisiana
Beulah (1967): Antilles, Mexico, South Texas
Bob (1991): North Carolina & Northeast U.S.
Camille (1969): Louisiana, Mississippi and Alabama
Carla (1961): Texas
Carmen (1974): Mexico, Central Louisiana
Carol (1954): Northeast U.S.
Cesar (1996): Honduras
Celia (1970): South Texas
Cleo (1964): Lesser Antilles, Haiti, Cuba, Southeast Florida
Connie (1955): North Carolina
David (1979): Lesser Antilles, Hispañola, Florida and Eastern U.S.
Diana (1990): Mexico
Diane (1955): Mid-Atlantic U.S. & Northeast U.S.
Donna (1960): Bahamas, Florida and Eastern U.S.
Dora (1964): Northeast Florida
Edna (1968)
Elena (1985): Mississippi, Alabama, Western Florida
Eloise (1975): Antilles, Northwest Florida, Alabama
Fifi (1974): Yucatan Peninsula, Louisiana

Flora (1963): Haiti, Cuba
Floyd (1999): North Carolina, eastern seaboard
Fran (1996): North Carolina
Frederic (1979): Alabama and Mississippi
Gilbert (1988): Lesser Antilles, Jamaica, Yucatan Peninsula, Mexico
Gloria (1985): North Carolina, Northeast U.S.
Hattie (1961): Belize, Guatemala
Hazel (1954): Antilles, North and South Carolina
Hilda (1964): Louisiana
Hortense (1996)
Hugo (1989): Antilles, South Carolina
Inez (1966): Lesser Antilles, Hispanola, Cuba, Florida Keys, Mexico
Ione (1955): North Carolina
Janet (1955): Lesser Antilles, Belize, Mexico
Joan (1988): Curacao, Venezuela, Colombia, Nicaragua (crossed into the Pacific and became Miriam)
Klaus (1990): Martinique
Lenny (1999): Antilles
Luis (1995)
Marilyn (1995): Bermuda
Mitch (1998): Central America, Nicaragua, Honduras
Opal (1995): Florida Panhandle
Roxanne (1995): Yucatan Peninsula

## Which United States **hurricanes** have **caused the most deaths**?

The ten deadliest United States hurricanes are listed below.

| Hurricane | Year | Deaths |
|---|---|---|
| 1. Texas (Galveston) | 1900 | 6,000 |
| 2. Florida (Lake Okeechobee) | 1928 | 1,836 |
| 3. Florida (Keys/S. Texas) | 1919 | 600–900+ |
| 4. New England | 1938 | 600 |
| 5. Florida (Keys) | 1935 | 408 |
| 6. Louisiana/Texas | 1957 | 390 |
| 7. Northeast U.S. | 1944 | 390 |
| 8. Louisiana (Grand Isle) | 1909 | 350 |
| 9. Louisiana (New Orleans) | 1915 | 275 |
| 10. Texas (Galveston) | 1915 | 275 |

### What was the greatest natural disaster in United States history?

The greatest natural disaster occurred when a hurricane struck Galveston, Texas, on September 8, 1900, and killed over 6,000 people. However, the costliest national disaster to date was Hurricane Andrew, which hit Florida on August 31, 1992, and Louisiana on September 1, 1992. Early warning kept the death toll low, but property damage is estimated at $20 billion.

### Which United States **hurricanes** have been the **most destructive**?

| Hurricane | Year | Category | Damage |
|---|---|---|---|
| 1. Andrew (SE Florida/SE Louisiana) | 1992 | 4 | $26,500,000,000 |
| 2. Hugo (S Carolina) | 1989 | 4 | $7,000,000,000 |
| 3. Floyd (Mid-Atlantic/ NE United States) | 1999 | 2 | $4,500,000,000 |
| 4. Fran (N Carolina) | 1996 | 3 | $3,200,000,000 |
| 5. Opal (NW Florida/Alabama) | 1995 | 3 | $3,000,000,000 |
| 6. Georges (Florida Keys, Mississippi, Alabama) | 1998 | 2 | $2,310,000,000 |
| 7. Frederic (Alabama/Mississippi) | 1979 | 3 | $2,300,000,000 |
| 8. Agnes (NE United States) | 1979 | 3 | $2,100,000,000 |
| 9. Alicia (N Texas) | 1983 | 3 | $2,000,000,000 |
| 10. Bob (N Carolina & NE United States) | 1991 | 2 | $1,500,000,000 |

## PRECIPITATION

### What is a **"white-out"**?

An official definition for "white-out" does not exist. It is a colloquial term that can describe any condition during snowfall that severely restricts visibility. That may mean a blizzard, or snowsquall, etc. If you get some sunlight in the mix, that makes the situation even worse—it's like driving in fog with your headlights on high-beam. The light gets backscattered right into your eyes and you can't see.

### What is the **dew point**?

The dew point is the temperature at which air is full of moisture and cannot store any more. When the relative humidity is 100 percent, the dew point is either the same as

or lower than the air temperature. If a fine film of air contacts a surface and is chilled to below the dew point, then actual dew is formed. This is why dew often forms at night or early morning: as the temperature of the air falls, the amount of water vapor the air can hold also decreases. Excess water vapor then condenses as very small drops on whatever it touches. Fog and clouds develop when sizable volumes of air are cooled to temperatures below the dew point.

## What is the **shape of a raindrop**?

Although a raindrop has been illustrated as being pear-shaped or tear-shaped, high-speed photographs reveal that a large raindrop has a spherical shape with a hole not quite through it (giving it a doughnut-like appearance). Water surface tension pulls the drop into this shape. As a drop larger than 0.08 inch (two millimeters) in diameter falls, it will become distorted. Air pressure flattens its bottom and its sides bulge. If it becomes larger than one-quarter inch (6.4 millimeters) across, it will keep spreading crosswise as it falls and will bulge more at its sides, while at the same time, its middle will thin into a bow-tie shape. Eventually in its path downward, it will divide into two smaller spherical drops.

## What is the **greatest amount of rainfall** ever measured?

| Time duration | Amount | | Place |
|---|---|---|---|
| | Inches | Centimeters | |
| 1 minute | 1.50 | 3.80 | Barst, Guadeloupe, West Indies, November 26, 1970 |
| 24 hours | 73.62 | 187.00 | Cilaos, La Réunion, Indian Ocean, March 15–16, 1952 |
| 24 hours in the United States | 19.00 | 48.30 | Alvin, Texas, July 25–26, 1979 |
| calendar month | 366.14 | 930.00 | Cherrapunji, Meghalaya, India, July 1861 |
| 12 months | 1,041.78 | 2,646.00 | Cherrapunji, Meghalaya, India, between August 1, 1860 and July 31, 1861 |
| 12 months in the United States | 739.00 | 1,877.00 | Kukui, Maui, Hawaii, December 1981–December 1982 |

## Where is the **rainiest place on Earth**?

The wettest place in the world is Tutunendo, Colombia, with an average annual rainfall of 463.4 inches (1,177 centimeters) per year. The place that has the most rainy days per year is Mount Wai-'ale'ale on Kauai, Hawaii. It has up to 350 rainy days annually.

### How fast does rain fall?

The speed of rainfall varies with drop size and wind speed. A typical raindrop in still air falls about 600 feet (182 meters) per minute or about seven miles (11 kilometers) per hour.

In contrast, the longest rainless period in the world was from October 1903 to January 1918 at Arica, Chile—a period of 14 years. In the United States the longest dry spell was 767 days at Bagdad, California, from October 3, 1912, to November 8, 1914.

## When do **thunderstorms** occur?

In the United States thunderstorms usually occur in the summertime, especially from May through August. Thunderstorms tend to occur in late spring and summer when large amounts of tropical maritime air move across the United States. Storms usually develop when the surface air is heated the most from the sun (2 to 4 p.m.). Thunderstorms are relatively rare in the New England area, North Dakota, Montana, and other northern states (latitude 60 degrees), where the air is often too cold. These storms are also rare along the Pacific Ocean because the summers there are too dry for these storms to occur. Florida, the Gulf states, and the southeastern states tend to have the most storms, averaging 70 to 90 annually. The mountainous southwest averages 50 to 70 storms annually. In the world, thunderstorms are most plentiful in the areas between latitude 35 degrees north and 35 degrees south; in these areas there can be as many as 3,200 storms within a 12-hour nighttime period. As many as 1,800 storms can occur at once throughout the world.

Lightning performs a vital function; it returns to the Earth much of the negative charge the Earth loses by leakage into the atmosphere. The annual death toll in the United States from lightning is greater than the annual death toll from tornadoes or hurricanes—150 Americans die annually from lightning-related causes, and 250 are injured.

## How far away can **thunder** be heard?

Thunder is the crash and rumble associated with lightning. It is caused by the explosive expansion and contraction of air heated by the stroke of lightning. This results in sound waves that can be heard easily six to seven miles (9.7 to 11.3 kilometers) away. Occasionally such rumbles can be heard as far away as 20 miles (32.2 kilometers). The sound of great claps of thunder is produced when intense heat and the ionizing effect of repeated lightning occurs in a previously heated air path. This creates a shockwave that moves at the speed of sound.

## How large can **hailstones** become?

The largest hailstone recorded in the U.S. weighed nearly two pounds—many times larger than these golf ball size specimens.

The average hailstone is about one-quarter inch (0.64 centimeter) in diameter. However, hailstones weighing up to 7.5 pounds (3.4 kilograms) are reported to have fallen in Hyderabad state in India in 1939, although scientists think these huge hailstones may be several stones that partly melted and stuck together. On April 14, 1986, hailstones weighing 2.5 pounds (one kilogram) were reported to have fallen in the Gopalgang district of Bangladesh.

The largest hailstone ever recorded in the United States fell in Coffeyville, Kansas, on September 3, 1970. It measured 5.57 inches (14.15 centimeters) in diameter and weighed 1.67 pounds (0.75 kilogram).

Hail is precipitation consisting of balls of ice. Hailstones usually are made of concentric, or onion-like, layers of ice alternating with partially melted and refrozen snow, structured around a tiny central core. It is formed in cumulonimbus or thunderclouds when freezing water and ice cling to small particles in the air, such as dust. The winds in the cloud blow the particles through zones of different temperatures, causing them to accumulate additional layers of ice and melting snow and to increase in size.

**Hail Size Estimates**

| Description | Size | Description | Size |
|---|---|---|---|
| Pea size | 0.25 inch | Golfball | 1.75 inches |
| Penny/Dime | 0.75 inch | Hen Egg | 2.00 inches |
| Nickel | 0.88 inch | Tennis Ball | 2.50 inches |
| Quarter | 1.00 inch | Baseball | 2.75 inches |
| Half Dollar or Susan B. Anthony Dollar | 1.25 inches | Grapefruit | 4.00 inches |
| Ping Pong Ball | 1.50 inches | | |

## What is the difference between **freezing rain** and **sleet**?

Freezing rain is rain that falls as a liquid but turns to ice on contact with a freezing object to form a smooth ice coating called glaze. Usually freezing rain only lasts a short time, because it either turns to rain or to snow. Sleet is frozen or partially frozen rain in the form of ice pellets. Sleet forms when rain falls from a warm layer of air, passes through a freezing air layer near the Earth's surface, and forms hard, clear, tiny ice pellets that can hit the ground so fast that they bounce off with a sharp click.

## How does **snow form**?

Snow is not frozen rain. Snow forms by sublimation of water vapor—the turning of water vapor directly into ice, without going through the liquid stage. High above the ground, chilled water vapor turns to ice when its temperature reaches the dew point. The result of this sublimation is a crystal of ice, usually hexagonal. Snow begins in the form of these tiny hexagonal ice crystals in the high clouds; the young crystals are the seeds from which snowflakes will grow. As water vapor is pumped up into the air by updrafts, more water is deposited on the ice crystals, causing them to grow. Soon some of the larger crystals fall to the ground as snowflakes.

Snowflakes may appear similar in shape but are indeed unique on a molecular level.

## Are all **snowflakes** shaped alike?

Some snowflakes may have strikingly similar shapes, but these twins are probably not molecularly identical. In 1986, cloud physicist Nancy Knight believed she found a uniquely cloned pair of crystals on an oil coated slide that had been hanging from an airplane. This pair may have been the result of breaking off from a star crystal, or were attached side by side, thereby experiencing the same weather conditions simultaneously. Unfortunately the smaller aspects of each of the snow crystals could not be studied because the photograph was unable to capture possible molecular differences. So, even if the human eye may see twins flakes, on a minuscule level these flakes are different.

## Is it ever **too cold** to snow?

No matter how cold the air gets, it still contains some moisture, and this can fall out of the air in the form of very small snow crystals. Very cold air is associated with no snow because these invasions of air from northerly latitudes are associated with clearing conditions behind cold fronts. Heavy snowfalls are associated with relatively mild air in advance of a warm front. The fact that snow piles up, year after year, in Arctic regions illustrates that it is never too cold to snow.

## When does **frost** form?

A frost is a crystalline deposit of small thin ice crystals formed on objects that are at freezing or below freezing temperatures. This phenomenon occurs when atmospheric

### How much water is in an inch of snow?

An average figure is 10 inches (25 centimeters) of snow is equal to one inch (2.5 centimeters) of water. Heavy, wet snow has a high water content; four to five inches (10 to 12 centimeters) may contain one inch (2.5 centimeters) of water. A dry, powdery snow might require 15 inches (38 centimeters) of snow to equal one inch (2.5 centimeters) of water.

water vapor condenses directly into ice without first becoming a liquid; this process is called sublimation. Usually frost appears on clear, calm nights, especially during early autumn when the air above the Earth is quite moist. Permafrost is permanently frozen ground that never thaws out completely.

### What is the **record** for the **greatest snowfall** in the United States?

The record for the most snow in a single storm is 189 inches (480 centimeters) at Mount Shasta Ski Bowl in California from February 13–19, 1959. For the most snow in a 24-hour day, the record goes to Silver Lake, Colorado, on April 14–15, 1921, with 76 inches (193 centimeters) of snow. The year record goes to Paradise, Mount Rainier, in Washington with 1,224.5 inches (3,110 centimeters) from February 19, 1971, to February 18, 1972. The highest average annual snowfall was 241 inches (612 centimeters) for Blue Canyon, California. In March 1911, Tamarack, California, had the deepest snow accumulation—over 37.5 feet (11.4 meters).

## WEATHER PREDICTION

### What is the difference between a National Weather Service **statement, advisory, watch**, and **warning**?

The National Weather Service will issue a *statement* as a "first alert" of the possibility of severe weather. An *advisory* is issued when weather conditions are not life threatening, but individuals need to be alert to weather conditions. A weather *watch* is issued when conditions are more favorable than usual for dangerous weather conditions, e.g. tornadoes and violent thunderstorms. A watch is a recommendation for planning, preparation, and increased awareness (i.e., to be alert for changing weather, listen for further information, and think about what to do if the danger materializes). A *warning* is issued when a particular weather hazard is either imminent or has been reported. A warning indicates the need to take action to protect life and property. The type of hazard is reflected in the type of warning (e.g., tornado warning, blizzard warning).

## What is **Doppler radar**?

Doppler radar measures frequency differences between signals bouncing off objects moving away from or toward it. By measuring the difference between the transmitted and received frequencies, Doppler radar calculates the speed of the air in which the rain, snow, ice crystals, and even insects are moving. It can then be used to predict speed and direction of wind and amount of precipitation associated with a storm. The National Weather Service has installed a series of NEXRAD (Next Generation Radar) Doppler Radar systems throughout the country. They are especially helpful in measuring the speed of tornadoes and other violent thunderstorms.

## When did **modern weather forecasting** begin?

On May 14, 1692, a weekly newspaper, *A Collection for the Improvement of Husbandry and Trade,* gave a seven-day table with pressure and wind readings for the comparable dates of the previous year. Readers were expected to make up their own forecasts from the data. Other journals soon followed with their own weather features. In 1771, a new journal called the *Monthly Weather Paper* was completely devoted to weather prediction. In 1861, the British Meteorological Office began issuing daily weather forecasts. The first broadcast of weather forecasts was done by the University of Wisconsin's station 9XM at Madison, Wisconsin, on January 3, 1921.

## What is **barometric pressure** and what does it mean?

Barometric, or atmospheric, pressure is the force exerted on a surface by the weight of the air above that surface, as measured by an instrument called a barometer. Pressure is greater at lower levels because the air's molecules are squeezed under the weight of the air above. So while the average air pressure at sea level is 14.7 pounds per square inch (1,013.53 hecto Pascals), at 1,000 feet (304 meters) above sea level, the pressure drops to 14.1 pounds per square inch (972.1 hecto Pascals), and at 18,000 feet (5,486 meters) the pressure is 7.3 pounds (503.32 hecto Pascals), about half of the figure at sea level. Changes in air pressure bring weather changes. High pressure areas bring clear skies and fair weather; low pressure areas bring wet or stormy weather. Areas of very low pressure have serious storms, such as hurricanes.

## Can **groundhogs** accurately predict the weather?

Over a 60-year period, groundhogs have accurately predicted the weather (i.e., when spring will start) only 28 percent of the time on Groundhog Day, February 2. Groundhog Day was first celebrated in Germany, where farmers would watch for a badger to

emerge from winter hibernation. If the day was sunny, the sleepy badger would be frightened by his shadow and duck back for another six weeks' nap; if it was cloudy he would stay out, knowing that spring had arrived. German farmers who emigrated to Pennsylvania brought the celebration to America. Finding no badgers in Pennsylvania, they chose the groundhog as a substitute.

## Can weather be predicted from the stripes on a **wooly-bear caterpillar**?

It is an old superstition that the severity of the coming winter can be predicted by the width of the brown bands or stripes around the wooly-bear caterpillar in the autumn. If the brown bands are wide, says the superstition, the winter will be mild, but if the brown bands are narrow, a rough winter is foretold. Studies at the American Museum of Natural History in New York failed to show any connection between the weather and the caterpillar's stripes. This belief is only a superstition; it has no basis in scientific fact.

## Are there **trees** that **predict the weather** and tell time?

Observing the leaves of a tree may be an old-fashioned method of predicting the weather, but farmers have noted that when maple leaves curl and turn bottom up in a blowing wind, rain is sure to follow. Woodsmen claim they can tell how rough a winter is going to be by the density of lichens on a nut tree. Before the katydid awakes, a black gum tree is able to indicate the oncoming winter. Trees can also be extraordinary timekeepers: Griffonia, in tropical west Africa, has two-inch (five-centimeter) inflated pods that burst with a hearty noise, indicating that it is time for farmers of the Accra Plains to plant crops; Trichilia is a 60-foot (18-meter) tree that flowers in February and again in August, signaling that it is time, just before the second rains arrive, for the second planting of corn. In the Fiji Islands, planting yams is cued by the flowering of the coral tree.

## Is a **halo** around the **sun or moon** a sign of rain or snow approaching?

The presence of a ring around the sun or, more commonly, the moon in the night sky, indicates very high ice crystals composing cirrostratus clouds. The brighter the ring, the greater the odds of precipitation and the sooner it may be expected. Rain or snow will not always fall, but two times out of three, precipitation will start to fall within twelve to eighteen hours. These cirroform clouds are a forerunner of an approaching warm front and an associated low pressure system.

## What would cause **frogs and toads** to fall from the sky like a **rain shower**?

Documented cases of showers of frogs have been recorded since 1794, usually during heavy summer rainstorms. Whirlwinds, waterspouts, and tornadoes are given as the conventional explanation. Extensive falls of fish, birds, and other animals have also been reported.

# MINERALS AND OTHER MATERIALS

## ROCKS AND MINERALS

*See also: Energy*

### How do **rocks** differ?

Rocks can be conveniently placed into one of three groups—igneous, sedimentary, and metamorphic.

*Igneous rocks,* such as granite, pegmatite, rhyolite, obsidian, gabbro, and basalt, are formed by the solidification of molten magma that emerges through the Earth's crust via volcanic activity. The nature and properties of the crystals vary greatly, depending in part on the composition of the original magma and partly on the conditions under which the magma solidified. There are thousands of different igneous rock types. For example, granite is formed by slow cooling of molten material (within the Earth). It has large crystals of quartz, feldspars, and mica.

*Sedimentary rocks,* such as brecchia, sandstone, shale, limestone, chert, and coals, are produced by the accumulation of sediments. These are fine rock particles or fragments, skeletons of microscopic organisms, or minerals leached from rocks that have accumulated from weathering. These sediments are then redeposited under water and later compressed in layers over time. The most common sedimentary rock is sandstone, which is predominantly quartz crystals.

*Metamorphic rocks,* such as marble, slate, schist, gneiss, quartzite, and hornsfel, are formed by the alteration of igneous and sedimentary rocks through heat and/or pressure. One example of these physical and chemical changes is the formation of marble from thermal changes in limestone.

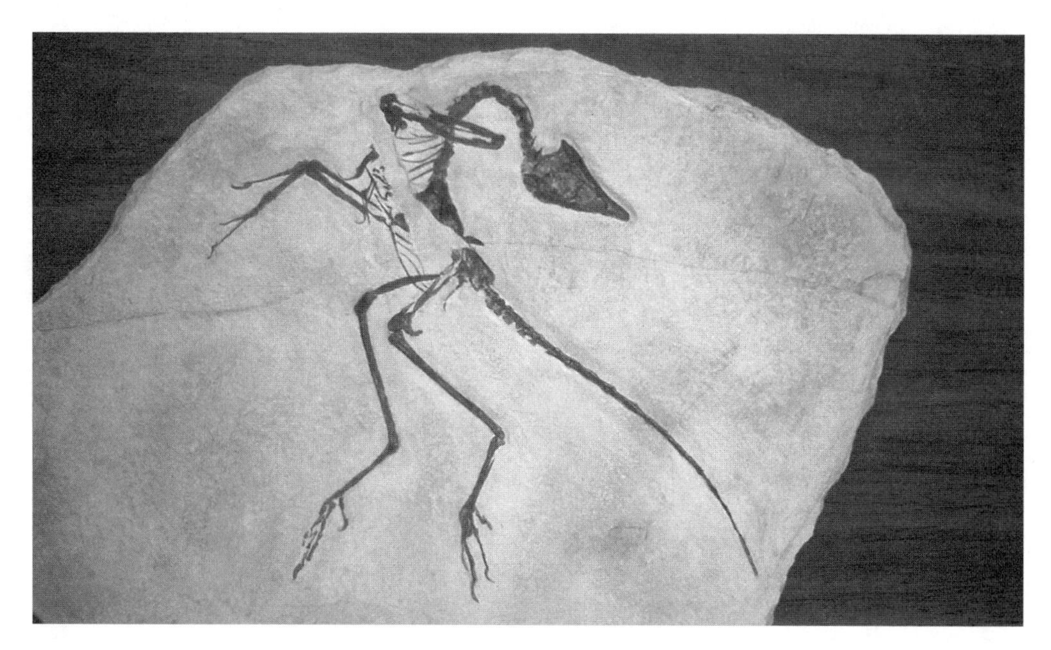

The fossilized remains of an archaeopteryx.

## What is **petrology** and what does a **petrologist** do?

Petrology is the science of rocks. A petrologist is a person who studies the mineralogy of rocks and the record of the geological past contained within rocks. From rocks, a petrologist can learn about past climates and geography, past and present composition of the Earth, and the conditions that prevail within the interior of the Earth.

## How are **fossils formed**?

Fossils are the remains of animals or plants that were preserved in rock before the beginning of recorded history. It is unusual for complete organisms to be preserved; fossils usually represent the hard parts such as bones or shells of animals and leaves, seeds, or woody parts of plants.

Some fossils are simply the bones, teeth, or shells themselves, which can be preserved for a relatively short period of time. Another type of fossil is the imprint of a buried plant or animal that decomposes, leaving a film of carbon that retains the form of the organism.

Some buried material is replaced by silica and other materials that permeate the organism and replace the original material in a process called petrification. Some woods are replaced by agate or opal so completely that even the cellular structure is duplicated. The best examples of this can be found in the Petrified Forest National Park in Arizona.

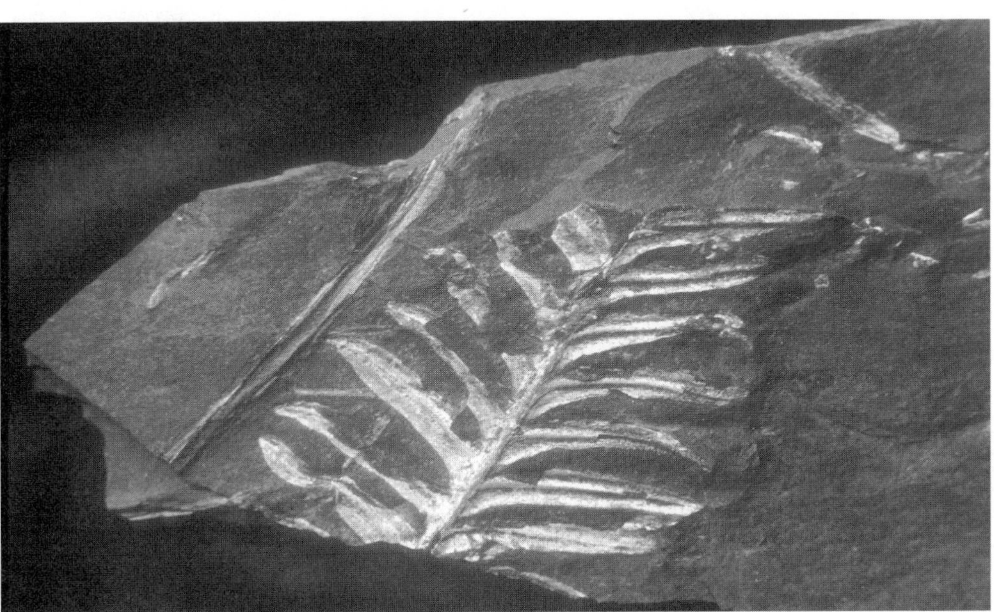

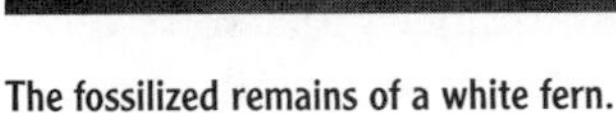

The fossilized remains of a white fern.

Molds and casts are other very common fossils. A mold is made from an imprint, such as a dinosaur footprint, in soft mud or silt. This impression may harden, then be covered with other materials. The original footprint will have formed a mold and the sediments filling it will be a cast of the footprint.

## How **old** are **fossils**?

The oldest known fossils are of bacteria that left their impressions approximately 3.5 billion years ago. The oldest animal fossils are of invertebrates that lived approximately 700 million years ago. The largest number of fossils come from the Cambrian period of 590-505 million years ago, when living organisms began to develop skeletons and hard parts. Since these parts tended to last longer than ordinary tissue, they were more likely to be preserved in clay and become fossilized.

## What is a **tektite**?

Tektites are silica-rich glass objects (rocks) found scattered in selected regions of the Earth's surface. They are generally black, oblong, teardrop or dumbbell-shaped, and several centimeters in length. They are formed from molten rock resulting when a meteorite, asteroid, or comet fragment impacts the Earth's surface. The molten rock is hurled high into atmosphere, where it rapidly cools into its unique shape and physical characteristics. Their mode of formation is considered indisputable evidence of such impacts. Tektites range from 0.7 to 35 million years in age.

## What is **cinnabar**?

Cinnabar is the main ore of the mineral mercury. Its cinnamon to scarlet-red color makes it a colorful mineral. It is produced primarily in the United States (California, Oregon, Texas and Arkansas), Spain, Italy, and Mexico. It is often used as a pigment.

## What are **Indian Dollars**?

They are six-sided, disk-shaped, twin crystals of aragonite ($CaCO_3$), which have altered to calcite but retained their outer form. They occur in large numbers in northern Colorado, where they are known as "Indian Dollars." In New Mexico they are called "Aztec Money" and in western Kansas they are called "Pioneer Dollars."

## How does a **rock** differ from a **mineral**?

Mineralogists use the term "mineral" for a substance that has all four of the following features: it must be found in nature; it must be made up of substances that were never alive (organic); it has the same chemical makeup wherever it is found; and its atoms are arranged in a regular pattern to form solid crystals.

While "rocks" are sometimes described as an aggregate or combination of one or more minerals, geologists extend the definition to include clay, loose sand, and certain limestones.

## What is the **Mohs scale**?

The Mohs scale is a standard of 10 minerals by which the hardness of a mineral is rated. It was introduced in 1812 by the German mineralogist Friedrich Mohs (1773–1839). The minerals are arranged from softest to hardest. Harder minerals, with higher numbers, can scratch those with a lower number.

| Hardness | Mineral | Comment |
|---|---|---|
| 1 | Talc | Hardness 1–2 can be scratched by a fingernail |
| 2 | Gypsum | Hardness 2–3 can be scratched by a copper coin |
| 3 | Calcite | Hardness 3–6 can be scratched by a steel pocket knife |
| 4 | Fluorite | |
| 5 | Apatite | |
| 6 | Orthoclase | Hardness 6–7 will not scratch glass |
| 7 | Quartz | |
| 8 | Topaz | Hardness 8–10 will scratch glass |
| 9 | Corundum | |
| 10 | Diamond | |

**If diamond is the hardest substance, what is the next hardest substance?**

Cubic boron nitride, which is the hardest ceramic, is the second hardest substance in the world.

## Who was the first person to attempt a **color standardization** scheme for minerals?

The German mineralogist Abraham Gottlob Werner (c.1750–1817) devised a method of describing minerals by their external characteristics, including color. He worked out an arrangement of colors and color names, illustrated by an actual set of minerals.

## What is meant by the term **strategic minerals**?

Strategic minerals are minerals essential to national defense—the supply of which a country uses but cannot produce itself. A third to a half of the 80 minerals used by industry could be classed as strategic minerals. Wealthy countries, such as the United States, stockpile these minerals to avoid any crippling effect on their economy or military strength if political circumstances were to cut off their supplies. The United States, for instance, stockpiles bauxite (10.5 million metric dry tons), manganese (1.7 million metric tons), chromium (1.4 million metric tons), tin (59,993 metric tons), cobalt (189 metric tons), tantalum (635 tons), palladium (1.25 million troy ounces), and platinum-group metals (platinum—4,704 kilograms, palladium—16,715 kilograms and iridium—784 kilograms).

## What is **pitchblende**?

Pitchblende is a massive variety of uraninite, or uranium oxide, found in metallic veins. It is a radioactive material and the most important ore of uranium. In 1898, Marie and Pierre Curie discovered that pitchblende contained radium, a rare element that has since been used in medicine and the sciences.

## What is **galena**?

Galena is a lead sulphide (PbS) and the most common ore of lead, containing 86.6 percent lead. Lead-gray in color, with a brilliant metallic luster, galena has a specific gravity of 7.5 and a hardness of 2.5 on the Mohs scale, and usually occurs as cubes or a modification of an octahedral form. Mined in Australia, it is also found in Canada, China, Mexico, Peru, and the United States (Missouri, Kansas, Oklahoma, Colorado, Montana, and Idaho).

## What is **stibnite**?

Stibnite is a lead-gray mineral ($Sb_2S_3$) with a metallic luster. It is the most important ore of antimony, and is also known as antimony glance. One of the few minerals that fuse easily in a match flame (977°F or 525°C), stibnite has a hardness of two on the Mohs scale and a specific gravity of 4.5 to 4.6. It is commonly found in hydrothermal veins or hot springs deposits. Stibnite is mined in Germany, Romania, France, Bolivia, Peru, and Mexico. The Yellow Pine mine at Stibnite, Idaho, is the largest producer in the United States, but California and Nevada also have deposits.

## What are **Cape May diamonds**?

They are pure quartz crystals of many sizes and colors, found in the vicinity of the Coast Guard station in Cape May, New Jersey. When polished and faceted, these crystals have the appearance of real diamonds. Prior to the development of modern gem examination equipment, many people were fooled by these quartz crystals. The possibility of finding a Cape May diamond on one's own, and the availability of already-polished faceted stones, has been a long-standing tourist attractions in the Cape May area.

## Are there any **diamond mines** in the United States?

The United States has no commercial diamond mines. The only significant diamond deposit in North America is Crater of Diamonds State Park near Murfreesboro, Arkansas. It is on government-owned land and has never been systematically developed. For a small fee, tourists can dig there and try to find diamonds. The largest crystal found there weighed 40.23 carats and was named the "Uncle Sam" diamond.

Diamonds crystallize directly from rock melts rich in magnesium and saturated with carbon dioxide gas that has been subjected to high pressures and temperatures exceeding 2,559°F (1400°C). These rock melts originally came from deep in the Earth's mantle at depths of 93 miles (150 kilometers).

Diamonds are minerals composed entirely of the element carbon, with an isometric crystalline structure. The hardest natural substance, gem diamonds have a density of 3.53, though black diamonds (black carbon cokelike aggregates of microscopic crystals) may have a density as low as 3.15. Diamonds have the highest thermal conductivity of any known substance. This property enables diamonds to be used in cutting tools, because they do not become hot.

## How can a **genuine diamond** be identified?

There are several tests that can be performed without the aid of tools. A knowledgeable person can recognize the surface lustre, straightness and flatness of facets, and high light reflectivity. Diamonds become warm in a warm room and cool if the sur-

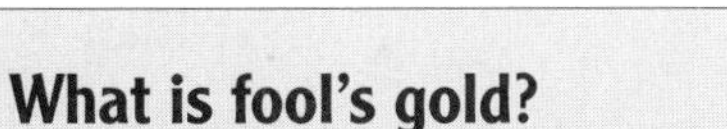

### What is fool's gold?

Iron pyrite ($FeS_2$) is a mineral popularly known as "fool's gold." Because of its metallic luster and pale brass yellow color, it is often mistaken for gold. Real gold is much heavier, softer, not brittle, and not grooved.

roundings are cool. A simple test that can be done is exposing the stones to warmth and cold and then touching them to one's lips to determine their appropriate temperature. This is especially effective when the results of this test are compared to the results of the test done on a diamond known to be genuine. Another test is to pick up the stone with a moistened fingertip. If this can be done, then the stone is likely to be a diamond. The majority of other stones cannot be picked up in this way.

The water test is another simple test. A drop of water is placed on a table. A perfectly clean diamond has the ability to almost "magnetize" water and will keep the water from spreading. An instrument called a diamond probe can detect even the most sophisticated fakes. Gemologists always use this as part of their inspection.

## How is the **value of a diamond** determined?

Demand, beauty, durability, rarity, freedom from defects, and perfection of cutting generally determine the value of a gemstone. But the major factor in establishing the price of gem diamonds is the control over output and price as exercised by the Central Selling Organization's (CSO) Diamond Trading Corporation Ltd. The CSO is a subsidiary of DeBeers Consolidated Mines Ltd.

## What are the **four "C"s of diamonds**?

The four "C"s are cut, color, clarity, and carat. Cut refers to the proportions, finish, symmetry, and polish of the diamond. These factors determine the brilliance of a diamond. Color describes the amount of color the diamond contains. Color ranges from colorless to yellow with tints of yellow, gray, or brown. Colors can also range from intense yellow to the more rare blue, green, pink, and red. Clarity describes the cleanness or purity of a diamond as determined by the number and size of imperfections. Carat is the weight of the diamond.

## How are **diamonds weighed**?

The basic unit is a carat, which is set at 200 milligrams (0.00704 ounces or 1/142 of an avoirdupois ounce). A well-cut, round diamond of one carat measures almost exactly 0.25 inch (6.3 millimeters) in diameter. Another unit commonly used is the point, which is one hundredth of a carat. A stone of one carat weighs 100 points. "Carat" as a unit of weight should not be confused with the term "karat" used to indicate purity of the gold into which gems are mounted.

A miner holds the Cullinan diamond—the largest ever discovered—pieces of which were later used in the British Crown and Sceptre.

## Which **diamond** is the **world's largest**?

The Cullinan Diamond, weighing 3,106 carats, is the world's largest. It was discovered on January 25, 1905, at the Premier Diamond Mine, Transvaal, South Africa. Named for Sir Thomas M. Cullinan, chairman of the Premier Diamond Company, it was cut into nine major stones and 96 smaller brilliants. The total weight of the cut stones was 1,063 carats, only 35 percent of the original weight.

Cullinan I, also known as the "Greater Star of Africa" or the "First Star of Africa," is a pear-shaped diamond weighing 530.2 carats. It is 2.12 inches (5.4 centimeters) long, 1.75 inches (4.4 centimeters) wide, and one inch (2.5 centimeters) thick at its deepest point. It was presented to Britain's King Edward VII in 1907, and was set in the British monarch's sceptre with the cross. It is still the largest cut diamond in the world.

Cullinan II, also know as the "Second Star of Africa," is an oblong stone that weighs 317.4 carats. It is set in the British Imperial State Crown.

## What are the **common cuts** of **gemstones**?

Modern gem cutting uses faceted cutting for most transparent gems. In faceted cutting, numerous facets—geometrically disposed to bring out the beauty of light and color to the best advantage—are cut. The four most common cuts are the brilliant, the rose, the baguette, and the step or trap cut. The step or trap cut is also known as the emerald cut and is used for emeralds. The brilliant and rose cuts are often used for diamonds.

## What is **cubic zirconium**?

Cubic zirconium was discovered in 1937 by two German mineralogists. It became popular with jewelry designers in the 1970s after Soviet scientists learned how to

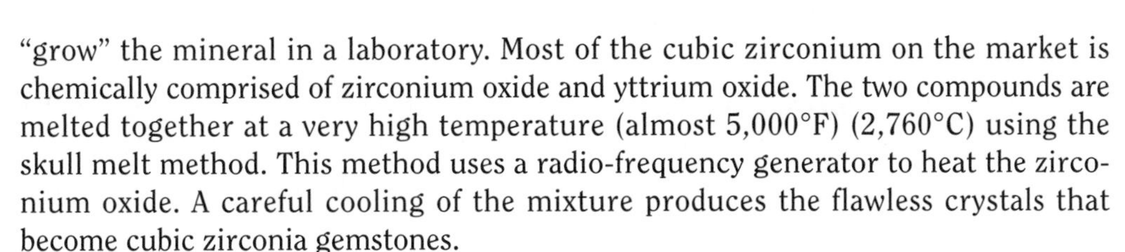

"grow" the mineral in a laboratory. Most of the cubic zirconium on the market is chemically comprised of zirconium oxide and yttrium oxide. The two compounds are melted together at a very high temperature (almost 5,000°F) (2,760°C) using the skull melt method. This method uses a radio-frequency generator to heat the zirconium oxide. A careful cooling of the mixture produces the flawless crystals that become cubic zirconia gemstones.

## What is the difference between **cubic zirconium** and **diamonds**?

Cubic zirconium is a gemstone material that is an imitation of diamonds. The word "imitation" is key. The U.S. Federal Trade Commission defines imitation materials as resembling the natural material in appearance only. Cubic zirconia may be cut the same way as diamonds. It is very dense and solid, weighing 1.7 times more than a diamond of the same millimeter size.

## Besides the Cullinan diamonds, what are the **largest precious stones**?

The largest ruby is a 8,500 carat stone that is 5.5 inches (14 centimeters) tall, carved to resemble the Liberty Bell. The largest star ruby is the 6,465 carat "Eminent Star" from India that has a six-line star. The largest cut emerald was found in Carnaiba, Brazil, in August 1974. It is 86,136 carats. A 2,302 carat sapphire from Anakie, Queensland, Australia, was carved into a 1,318 carat head of Abraham Lincoln, making it the largest carved sapphire. "The Lone Star," at 9,719.5 carats, is the largest star sapphire. The largest natural pearl is the "Pearl of Lao-tze," also called the "Pearl of Allah." Found in May 1934 in the shell of a giant clam at Palawan, Philippines, the pearl weighs 14 pounds, 1 ounce (6.4 kilograms).

## How does the **emerald** get its color?

Emerald is a variety of green beryl ($Be_3Al_2Si_6O_{18}$) that is colored by a trace of chromium (Cr), which replaces the aluminum (Al) in the beryl structure. Other green beryls exist; but if no chromium is present, they are, technically speaking, not emeralds.

## How is the star in **star sapphires** produced?

Sapphires are composed of gem-quality corundum ($Al_2O_3$). Color appears in sapphires when small amounts of iron and titanium are present. Star stapphires contain needles of the mineral rutile that will display as a six-ray star figure when cut in the unfaceted cabochon (dome or convex) form. The most highly prized star sapphires are blue. Black or white star sapphires are less valuble. Since a ruby is simply the red variety of corundum, star rubies also exist.

### What is a **tiger's eye**?

Tiger's eye is a semiprecious quartz gem that has a vertical luminescent band like that of a cat's eye. To achieve the effect of a cat's eye, veins of parallel blue asbestos fibers are first altered to iron oxides and then replaced by silica. The gem has a rich yellow to yellow-brown or brown color.

## METALS

### What is **coltan**?

Coltan is the shortened name for the metallic ore Columbite-tantalite. When refined it becomes a heat-resistant powder, tantalum, which can hold a high electrical charge. These properties make it a vital element in creating capacitors, the electronic elements that control current flow inside miniature circuit boards. Tantalum capacitors are used in almost all cell phones, laptop computers, pagers, and other electrical devices.

### Which **metallic element** is the **most abundant**?

Aluminum is the most abundant metallic element on the surface of the Earth and moon; it comprises more than 8 percent of the Earth's crust. It is never free in nature, combining with oxygen, sand, iron, titanium, etc.; its ores are mainly bauxites (aluminum hydroxide). Nearly all rocks, particularly igneous rocks, contain aluminum as aluminosilicate minerals. Napoleon III (1808–1883) recognized that the physical characteristic of its lightness could revolutionize the arms industry, so he granted a large subsidy to French chemist Sainte-Claire Deville (1818–1881) to develop a method to make its commercial use feasible. In 1854, Deville obtained the first pure aluminum metal through the process of reduction of aluminum chloride. In 1886, the American Charles Martin Hall (1863–1914) and the Frenchman Paul Heroult (1863–1914) independently discovered an electrolytic process to produce aluminum from bauxite. Because of aluminum's resistance to corrosion, low density, and excellent heat-conducting property, the packaging industry uses a large percentage of the aluminum alloy that is produced for drink and food containers and covers, and foil pouches and wraps. It is a good conductor of electricity and is widely used in power and telephone cables, light bulbs, and electrical equipment. Large amounts of aluminum are used in the production of all types of vehicles; alloys have high tensile strengths and are of considerable industrial importance to the aerospace industry. The building construction industry uses aluminum alloys in such items as gutters, panels, siding, window frames, and roofing. Examples of the

numerous other products containing aluminum and aluminum alloys are cookware, golf clubs, air conditioners, lawn furniture, license plates, paints, refrigerators, rocket fuel, and zippers.

## Why are **alchemical symbols for metals** and **astrological symbols for planets** identical?

The ancient Greeks and Romans knew seven metals and also knew seven "planets" (the five nearer planets plus the sun and the moon). They related each planet to a specific metal. Alchemy, originating in about the third century B.C.E., focused on changing base metals, such as lead, into gold. Although at times alchemy bordered on mysticism, it contained centuries of chemical experience, which provided the foundation for the development of modern chemistry.

| English name | Chemical symbol | Latin name | Alchemical symbol |
|---|---|---|---|
| Gold | Au | *aurum* | ☉ (Sun) |
| Silver | Ag | *argentum* | ☽ (Moon) |
| Copper | Cu | *cuprum* | ♀ (Venus) |
| Iron | Fe | *ferrum* | ♂ (Mars) |
| Mercury | Hg | *hydrargyrum* | ☿ (Mercury) |
| Tin | Sn | *stannum* | ♃ (Jupiter) |
| Lead | Pb | *plumbum* | ♄ (Saturn) |

## What are the **noble metals**?

The noble metals are gold, silver, mercury, and the platinum group (including palladium, iridium, rhodium, ruthenium, and osmium). The term refers to those metals highly resistant to chemical reaction or corrosion and is contrasted with "base" metals, which are not so resistant. The term has its origins in ancient alchemy whose goals of transformation and perfection were pursued through the different properties of metals and chemicals. The term is not synonymous with "precious metals," although a metal, like platinium, may be both.

## What are the **precious metals**?

This is a general term for expensive metals that are used for making coins, jewelry, and ornaments. The name is limited to gold, silver, and platinum. Expense or rarity does not make a metal precious, but rather it is a value set by law that states that the object made of these metals has a certain intrinsic value. The term is not synonymous with "noble metals," although a metal (such as platinum) may be both noble and precious.

### Is white gold really gold?

White gold is the name of a class of jeweler's white alloys used as substitutes for platinum. Different grades vary widely in composition, but usual alloys consist of from 20 percent to 50 percent nickel, with the balance gold. A superior class of white gold is made of 90 percent gold and 10 percent palladium. Other elements used include copper and zinc. The main use of these alloys is to give the gold a white color.

## What is **24 karat gold**?

The term "karat" refers to the percentage of gold versus the percentage of an alloy in a piece of jewelry or a decorative object. Gold is too soft to be usable in its purest form and has to be mixed with other metals. One karat is equal to one-24th part fine gold. Thus, 24 karat gold is 100 percent pure and 18 karat gold is 18/24 or 75 percent pure.

| Karatage | Percentage of fine gold |
|---|---|
| 24 | 100 |
| 22 | 91.75 |
| 18 | 75 |
| 14 | 58.5 |
| 12 | 50.25 |
| 10 | 42 |
| 9 | 37.8 |
| 8 | 33.75 |

## How far can a **troy ounce of gold**, if formed into a thin wire, **be stretched** before it breaks?

Ductility is the characteristic of a substance to lend itself to shaping and stretching. A troy ounce of gold (31.1035 grams) can be drawn into a fine wire that is 50 miles (80 kilometers) long.

## How thick is **gold leaf**?

Gold leaf is pure gold that is hammered or rolled into sheets or leaves so extremely thin that it can take 300,000 units to make a stack one inch high. The thickness of a single gold leaf is typically 0.0000035 inch (3.5 millionths of an inch), although this may vary widely according to which manufacturer makes it. Also called gold foil, it is used for architectural coverings and for hot-embossed printing on leather.

Gold leaf is shown here being worked on a smooth stone.

## What are the chief **gold-producing countries**?

The Republic of South Africa leads the world in mine production and gold reserves and holds approximately one-half of all world resources. The United States is the second largest gold-producing nation. Commercial usage in 1998 was estimated as follows: jewelry and arts, 79 percent; industrial (mainly electronic), 4 percent; dental, 2 percent; and other industrial uses, 15 percent. In the United States, Nevada is the leading gold producer, with California a distant second, followed by South Dakota.

World mine production of the top six countries in 2001 was:

| Country | Gold production |
|---|---|
| South Africa | 400 metric tons (400,000 kg) |
| United States | 350 metric tons (350,000 kg) |
| Australia | 290 metric tons (290,000 kg) |
| China | 185 metric tons (185,000 kg) |
| Canada | 160 metric tons (160,000 kg) |
| Russia | 155 metric tons (155,000 kg) |

## What is **sterling silver**?

Sterling silver is a high-grade alloy that contains a minimum of 925 parts in 1,000 of silver (92.5 percent silver and 7.5 percent of another metal—usually copper).

## What is **German silver**?

Nickel silver, sometimes known as German silver or nickel brass, is a silver-white alloy composed of 52 percent to 80 percent copper, 10 percent to 35 percent zinc, and 5 percent to 35 percent nickel. It may also contain a small percent of lead and tin. There are other forms of nickel silver, but the term "German silver" is the name used in the silverware trade.

## Which metal is the main component of **pewter**?

Tin—at least 90 percent. Antimony, copper, and zinc may be added in place of lead to harden and strengthen pewter. Pewter may still contain lead, but high lead content will both tarnish the piece and dissolve into food and drink to become toxic. The alloy used today in fine quality pieces contain 91–95 percent minimum tin, 8 percent maximum antimony, 2.5 percent maximum copper, and 0–5 percent maximum bismuth, as determined by the European Standard for pewter.

## Where were the first successful **ironworks** in America?

Although iron ore in this country was first discovered in North Carolina in 1585, and the manufacture of iron was first undertaken (but never accomplished) in Virginia in 1619, the first successful ironworks in America was established by Thomas Dexter and Robert Bridges near the Saugus River in Lynn, Massachusetts. As the original promoters of the enterprise, they hired John Winthrop, Jr. from England to begin production. By 1645, a blast furnace had begun operations, and by 1648 a forge was working there.

## What is **high speed steel**?

High speed steel is a general name for high alloy steels that retain their hardness at very high temperatures and are used for metal-cutting tools. All high speed steels are based on either tungsten or molybdenum (or both) as the primary heat-resisting alloying element. These steels require a special heat so that their unique properties can be fully realized. The manufacturing process consists of heating the steel to a temperature of 2,150°F to 2,400°F (1,175°C to 1,315°C) to obtain solution of a substantial percentage of the alloy carbides, quenching to room temperature, tempering at 1,000°F to 1,150°F (535°C to 620°C), and again cooling to room temperature.

## Who invented **stainless steel**?

Metallurgists in several countries developed stainless steel, a group of iron-based alloys combined with chromium in order to be resistant to rusting and corrosion.

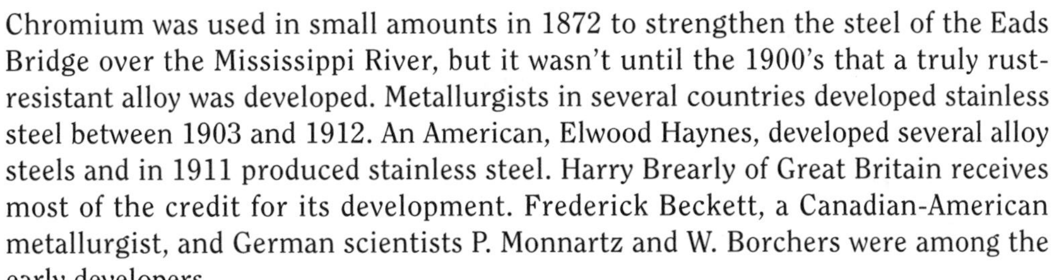

Chromium was used in small amounts in 1872 to strengthen the steel of the Eads Bridge over the Mississippi River, but it wasn't until the 1900's that a truly rust-resistant alloy was developed. Metallurgists in several countries developed stainless steel between 1903 and 1912. An American, Elwood Haynes, developed several alloy steels and in 1911 produced stainless steel. Harry Brearly of Great Britain receives most of the credit for its development. Frederick Beckett, a Canadian-American metallurgist, and German scientists P. Monnartz and W. Borchers were among the early developers.



## What material is used to make a **tuning fork**?

A tuning fork, an instrument that when struck emits a fixed pitch, is generally made of steel. Some tuning forks are made of aluminum, magnesium-alloy, fused quartz, or other elastic material.

## Which countries have **uranium** deposits?

Uranium, a radioactive metallic element, is the only natural material capable of sustaining nuclear fission. But only one isotope, uranium-235, which occurs in one molecule out of 40 of natural uranium, can undergo fission under neutron bombardment. Mined in various parts of the world, it must then be converted during purification to uranium dioxide ($UO_2$). Uranium deposits occur throughout the world; the largest are in the United States (the Colorado Plateau and low-grade reserves in Florida, Tennessee, North Dakota, and South Dakota), Canada (Ontario, Northwest Territories, and west-central), South Africa (the Witwatersrand), and Gabon (Oklo). Other countries and areas having significant low-grade deposits are Brazil, Russia, North Africa, and Sweden. Previously significant reserves in Zaire are almost exhausted.

## What is **technetium**?

Technetium (Tc, element 43) is a radioactive metallic element that does not occur naturally either in its pure form or as compounds; it is produced during nuclear fission. A fission product of molybdenum (Mo, element 42), Tc can also occur as a fission product of uranium (U, element 92). It was the first element to be made artificially in 1937 when it was isolated and extracted by C. Perrier and Emilio Segrè (1905–1989).

Tc has found significant application in diagnostic imaging and nuclear medicine. Ingested soluble technetium compounds tend to concentrate in the liver and are valuable in labeling and in radiological examination of that organ. Also, by technetium labeling of blood serum components, diseases involving the circulatory system can be explored.

# NATURAL SUBSTANCES

*See also: Energy*

## What is **obsidian**?

Obsidian is a volcanic glass that usually forms in the upper parts of lava flows. Embryonic crystal growths, known as crystallites, make the glass an opaque, jet-black color. Red or brown obsidian could result if iron oxide dust is present. There are some well-known formations in existence, including the Obsidian Cliffs in Yellowstone Park and Mount Hekla in Iceland.

## Is **lodestone** a magnet?

Lodestone is a naturally occurring variety of magnetic iron oxide or magnetite. Lodestone is frequently called a natural magnet because it attracts iron objects and possesses polarity. It was used by early mariners to find magnetic north. Other names for lodestone are loadstone, leading stone, and Hercules stone.

## What is **red dog**?

Red dog is the residue from burned coal dumps. The dumps are composed of waste products incidental to coal mining. Under pressure in these waste dumps, the waste frequently ignites from spontaneous combustion, producing a red-colored ash, which is used for driveways, parking lots, and roads.

## What are some uses for **coal** other than as an energy resource?

In the past, many of the aromatic compounds, such as benzene, toluene, and xylene were made from coal. These compounds are now chiefly byproducts of petroleum. Naphthalene and phenanthrene are still obtained from coal tar. Coal tar, a byproduct of coal, is used in roofing.

## What is **diatomite**?

Diatomite (also called diatomaceous earth) is a white- or cream-colored, friable, porous rock composed of the fossil remains of diatoms (small water plants with silica cell walls). These fossils build up on the ocean bottoms to form diatomite, and in some places these areas have become dry land or diatomacceous earth. Chemically inert and having a rough texture and other unusual physical properties, it is suitable for many scientific and industrial purposes, including use as a filtering agent; building material; heat, cold, and sound insulator; catalyst carrier; filler absorbent; abrasive; and ingredi-

ent in pharmaceutical preparations. Dynamite is made from it by soaking it in the liquid explosive nitroglycerin.

## What is **fly ash**?

Fly ash is the very fine portion or ash residue that results from the combustion of coal. The fly ash portion is usually removed electrostatically from the coal combustion gasses before they are released to the atmosphere. About 31 percent of the 57 million metric tons produced annually in the United States are beneficially used; the remainder must be disposed of in ponds or landfills.

## In **coal mining** what is meant by **damp**?

Damp is a poisonous or explosive gas in a mine. The most common type of damp is firedamp, also known as methane. White damp is carbon monoxide. Blackdamp (or chokedamp) is a mixture of nitrogen and carbon dioxide formed by mine fires and explosion of firedamp in mines. Blackdamp extinguishes fire and suffocates its victims.

## What is **fuller's earth**?

It is a naturally occurring white or brown clay containing aluminum magnesium silicate. Fuller's earth acts as a catalyst and was named for a process known as fulling—a process used to clean grease from wool and cloth. It is currently used for lightening the color of oils and fats, as a pigment extender, as a filter, as an absorbent (for example, in litter boxes to absorb animal waste), and in floor sweeping compounds.

## How is **charcoal** made?

Commercial production of charcoal uses wood processing residues, such as sawdust, shavings, milled wood and bark as a raw material. Depending on the material, the residues are placed in kilns or furnaces and heated at low oxygen concentrations. A Herreshoff furnace can produce at least a ton of charcoal per hour.

## How much wood is used to make a ton of **paper**?

In the United States, the wood used for the manufacture of paper is mainly from small diameter bolts and pulpwood. It is usually measured by the cord or by weight. Although the fiber used in making paper is overwhelmingly wood fiber, a large percentage of other ingredients is needed. One ton of a typical paper requires two cords of wood, but also requires 55,000 gallons (208,000 liters) of water, 102 pounds (46 kilograms) of sulfur, 350 pounds (159 kilograms) of lime, 289 pounds (131 kilograms) of clay, 1.2 tons of coal, 112 kilowatt hours of power, 20 pounds (9 kilograms) of dye and pigments, and 108 pounds (49 kilograms) of starch, as well as other ingredients.

## What products come from **tropical forests**?

**Products from Tropical Forests**

| **Woods** | **Houseplants** | **Spices** | **Foods** |
|---|---|---|---|
| Balsa | *Anthurium* | Allspice | Avocado |
| Mahogany | Croton | Black pepper | Banana |
| Rosewood | *Dieffenbachia* | Cardamom | Coconut |
| Sandalwood | *Dracaena* | Cayenne | Grapefruit |
| Teak | Fiddle-leaf fig | Chili | Lemon |
| | Mother-in-law's tongue | Cinnamon | Lime |
| **Fibers** | Parlor ivy | Cloves | Mango |
| Bamboo | *Philodendron* | Ginger | Orange |
| Jute/Kenaf | Rubber tree plant | Mace | Papaya |
| Kapok | *Schefflera* | Nutmeg | Passion fruit |
| Raffia | Silver vase bromeliad | Paprika | Pineapple |
| Ramie | *Spathiphyllum* | Sesame seeds | Plantain |
| Rattan | Swiss cheese plant | Turmeric | Tangerine |
| | Zebra plant | Vanilla bean | Brazil nuts |
| **Gums, resins** | | | Cane sugar |
| Chicle latex | **Oils, etc.** | | Cashew nuts |
| Copaiba | Camphor oil | | Chocolate |
| Copal | Cascarilla oil | | Coffee |
| Gutta percha | Coconut oil | | Cucumber |
| Rubber latex | Eucalyptus oil | | Hearts of palm |
| Tung oil | Oil of star anise | | Macadamia nuts |
| | Palm oil | | Manioc/tapioca |
| | Patchouli oil | | Okra |
| | Rosewood oil | | Peanuts |
| | Tolu balsam oil | | Peppers |
| | Annatto | | Cola beans |
| | Curare | | Tea |
| | Diosgenin | | |
| | Quinine | | |
| | Reserpine | | |
| | Strophanthus | | |
| | Strychnine | | |
| | Yang-Yang | | |

## What wood is the favorite for **butcher's blocks**?

The preferred wood for butcher's blocks is the American sycamore *(Platanus occidentalis)*, also known as American planetree, buttonball, buttonwood, plane-tree, and water beech, because of its toughness. It is also used as a veneer for decorative surfaces as well as for railroad ties, fence posts, and fuel.

## What does **one acre of trees** yield when cut and processed?

There are about 660 trees on one acre in a forest. When cut, one acre of trees may yield approximately 105,000 board feet of lumber or more than 30 tons of paper or 16 cords of firewood.

## Which woods are used for **telephone poles**?

The principal woods used for telephone poles are southern pine, Douglas fir, western red cedar, and lodgepole pine. Ponderosa pine, red pine, jack pine, northern white cedar, other cedars, and western larch are also used.

## Which woods are used for **railroad ties**?

Many species of wood are used for ties. The more common are oaks, gums, Douglas fir, mixed hardwoods, hemlock, southern pine, and mixed softwoods.

Pieces of petrified wood litter the landscape at Painted Desert National Monument, Arizona.

## Does any type of **wood sink** in water?

Ironwood is a name applied to many hard, heavy woods. Some ironwoods are so dense that their specific gravity exceeds 1.0 and they are therefore unable to float in water. North American ironwoods include the American hornbeam, the mesquite, the desert ironwood, and leadwood (*Krugiodendron ferreum*), which has a specific gravity of 1.34-1.42, making it the heaviest in the United States.

The heaviest wood is black ironwood (*Olea laurifolia*), also called South African ironwood. Found in the West Indies, it has a specific gravity of 1.49 and weighs up to 93 pounds (42.18 kilograms) per foot. The lightest wood is *Aeschynomene hispida*,

### How is petrified wood formed?

Petrified wood is formed when water containing dissolved minerals such as calcium carbonate ($CaCO_3$) and silicate infiltrates wood or other structures. The process takes thousands of years. The foreign material either replaces or encloses the organic matter and often retains all the structural details of the original plant material. Botanists find these types of fossils to be very important since they allow for the study of the internal structure of extinct plants. After a time, wood seems to have turned to stone because the original form and structure are retained. The wood itself does not turn to stone.

found in Cuba, with a specific gravity of 0.044 and a weight of 2.5 pounds (1.13 kilograms) per foot. Balsa wood (*Ochroma pyramidale*) varies between 2.5 and 24 pounds (one to 10 kilograms) per foot.

## What is **amber**?

Amber is the fossil resin of trees. The two major deposits of amber are in the Dominican Republic and Baltic. Amber came from a coniferous tree that is now extinct. Amber is usually yellow or orange in color, semitransparent or opaque with a glossy surface. It is used by both artisans and scientists.

## What is **rosin**?

Rosin is the resin produced after the distillation of turpentine, obtained from several varieties of pine trees, especially the longleaf pine (*Pinus palustris*) and the slash pine (*Pinus caribaea*). Rosin has many industrial uses, including the preparation of inks, adhesives, paints, sealants, and chemicals. Rosin is also used by athletes and musicians to make smooth surfaces less slippery.

## What are **naval stores**?

Naval stores are products of such coniferous trees as pine and spruce. These products include pitch, tar, resin, turpentine, pine oil, and terpenes. The term "naval stores" originated in the seventeenth century when these materials were used for building and maintaining wooden sailing ships.

## Why are **essential oils** called "essential"?

Called essential oils because of their ease of solubility in alcohol to form essences, essential oils are used in flavorings, perfumes, disinfectants, medicine, and other products. They are naturally occurring volatile aromatic oils found in uncombined forms within various parts of plants (leaves, pods, etc.). These oils contain as one of their main ingredients a substance belonging to the terpene group. Examples of essential oils include bergamot, eucalyptus, ginger, pine, spearmint, and wintergreen oils. Extracted by distillation or enfleurage (extraction using fat) and mechanical pressing, these oils can now be made synthetically.

### What is **gutta percha**?

Gutta percha is a rubberlike gum obtained from the milky sap of trees of the *Sapotaceae* family, found in Indonesia and Malaysia. Once of great economic value, gutta percha is now being replaced by plastics in many items, although it is still used in some electrical insulation and dental work. The English natural historian John Tradescant (c. 1570-1638) introduced gutta percha to Europe in the 1620s, and its inherent qualities gave it a slow but growing place in world trade. By the end of World War II, however, many manufacturers switched from gutta percha to plastics, which are more versatile and cheaper to produce.

### What is **excelsior**?

Excelsior is a trade name dating from the mid-nineteenth century for the curly, fine wood shavings used as packing material when shipping breakable items. It is also used as a cushioning and stuffing material. Poplar, aspen, basswood or cottonwood are woods that are often made into excelsior.

### What is **ambergris**?

Ambergris, a highly odorous, waxy substance found floating in tropical seas, is a secretion from the sperm whale (*Physeter catodon*). The whale secretes ambergris to protect its stomach from the sharp bone of the cuttlefish, a squid-like sea mollusk, which it ingests. Ambergris is used in perfumery as a fixature to extend the life of a perfume and as a flavoring for food and beverages. Today ambergris is synthesized and used by the perfume trade, which has voluntarily refused to purchase natural ambergris to protect sperm whales from exploitation.

### From where do **frankincense and myrrh** originate?

Frankincense is an aromatic gum resin obtained by tapping the trunks of trees belonging to the genus *Boswellia*. The milky resin hardens when exposed to the air and forms irregular lumps—the form in which it is usually marketed. Also called olibanum, frankincense is used in pharmaceuticals, as a perfume, as a fixative, and in fumigants and incense. Myrrh comes from a tree of the genus *Commiphora*, a native of Arabia and Northeast Africa. It too is a resin obtained from the tree trunk and is used in pharmaceuticals, perfumes, and toothpaste.

### Where does **isinglass** come from?

Isinglass is the purest form of animal gelatin. It is manufactured from the swimming bladder of sturgeon and other fishes. It is used in the clarification of wine and beer as well as in the making of some cements, jams, jellies, and soups.

### What exactly is **cashmere**?

Kashmir goats, which live high in the plateaus of the area from northern China to Mongolia, are covered in a coarse outer hair that helps protect them from the cold, harsh weather. As insulation, these goats have a softer, finer layer of hair or down under the coarse outer hair. This fine hair is shed annually and processed to make cashmere. Each goat produces enough cashmere to make one sweater every four years.

## MAN-MADE PRODUCTS

### What are the major distinguishing characteristics of **ceramics**?

Ceramics are crystalline compounds of metallic and nonmetallic elements. Ceramics are the most rigid of all materials with an almost total absence of ductility. They have the highest known melting points of materials with some being as high as 7,000°F (3,870°C) and many that melt at temperatures of 3,500°F (1,927°C). Glass, brick, cement and plaster, dinnerware, artware, and porcelain enamel are all examples of ceramics.

### Why is **Styrofoam** a good insulator?

Styrofoam insulates well because the foam form increases the length of path for heat flow through the material. It also reduces the effective cross-sectional area across which the heat can flow.

### How is **dry ice** made?

Dry ice a solid form of carbon dioxide ($CO_2$) used primarily to refrigerate perishables that are being transported from one location to another. The carbon dioxide, which at normal temperatures is a gas, is stored and shipped as a liquid in tanks that are pressurized at 1,073 pounds per square inch. To make dry ice, the carbon dioxide liquid is withdrawn from the tank and allowed to evaporate at a normal pressure in a porous bag. This rapid evaporation consumes so much heat that part of the liquid $CO_2$ freezes to a temperature of –109°F (–78°C). The frozen liquid is then compressed by machines into blocks of "dry ice," which will melt into a gas again when set out at room temperature.

It was first made commercially in 1925 by the Prest-Air Devices Company of Long Island City, New York, through the efforts of Thomas Benton Slate. It was used by Schrafft's of New York in July 1925 to keep ice cream from melting. The first large sale of dry ice was made later in that year to Breyer Ice Cream Company of New York. Although used mostly as a refrigerant or coolant, other uses include medical procedures such as freezing warts, blast cleaning, freeze branding animals and creating special effects for live performances and films.

## Why is **sulfuric acid** important?

Sometimes called "oil of vitriol," or vitriolic acid, sulfuric acid ($H_2SO_4$) has become one of the most important of all chemicals. It was little used until it became essential for the manufacture of soda in the eighteenth century. It is prepared industrially by the reaction of water with sulfur trioxide, which in turn is made by chemical combination of sulfur dioxide and oxygen by one of two processes (the contact process or the chamber process). Many manufactured articles in common use depend in some way on sulfuric acid for their production. It is used extensively in petroleum refining and in the production of fertilizers. It is also used in the production of chemicals, automobile batteries, explosives, pigments, iron and other metals, and paper pulp.

## What is **aqua regia**?

"Aqua regia," also known as nitrohydrochloric acid, is a mixture of one part concentrated nitric acid and three parts concentrated hydrochloric acid. The chemical reaction between the acids makes it possible to dissolve all metals except silver. The reaction of metals with nitrohydrochloric acid typically involves oxidation of the metals to a metallic ion and the reduction of the nitric acid to nitric oxide. The term comes from Latin and means "royal water." It was named by the alchemists for its ability to dissolve gold and platinum, which were called the "noble metals."

## Who developed the process for making **ammonia**?

Known since ancient times, ammonia ($NH_3$) has been commercially important for more than 100 years. The first breakthrough in the large-scale synthesis of ammonia resulted from the work of Fritz Haber (1863–1934). In 1913, Haber found that ammonia could be produced by combining nitrogen and hydrogen ($N_2 + 3H_2 \rightleftarrows 2NH_3$) with a catalyst (iron oxide with small quantities of cerium and chromium) at 131°F (55°C) under a pressure of about 200 atmospheres. The process was adapted for industrial-quality production by Karl Bosch (1874–1940). Thereafter, many improved ammonia-synthesis systems, based on the Haber-Bosch process, were commercialized using various operating conditions and synthesis loop-designs. One of the five top inorganic chemicals produced in the United States, it is used in refrigerants, detergents, and other cleaning preparations, explosives, fabrics, and fertilizers. Most ammonia production in the United States is used for fertilizers. It has been shown to produce cancer of the skin in humans in doses of 1,000 milligram per kilogram (2.2 pounds) of body weight.

### What does the symbol $H_2O_2$ stand for?

It is hydrogen peroxide, a syrupy liquid compound used as a strong bleaching, oxidizing, and disinfecting agent. It is usually made in large, strategically located anthrahydroquinone autoxidation processes. It is also made electrolytically. The primary use of hydrogen peroxide is in bleaching wood pulp. A more familiar use is as a 3 percent solution as an antiseptic and germicide. Undiluted, it can cause burns to human skin and mucous membranes, is a fire and explosion risk, and can be highly toxic.

### What percentage of **salt** consumed in the United States is used for **de-icing roads**?

Thirty-five percent of the salt consumed in the United States is used to de-ice roads.

### What is the **lightest known solid**?

The lightest solid is silica aerogels, made of tiny spheres of bonded silicon and oxygen atoms linked together into long strands separated with air pockets. They appear almost like frozen wisps of smoke. They also have the lowest conductivity, lowest solid density, highest porosity, highest surface area, and the highest dielectric constant, giving them the potential of being used in many applications. Understandably, their use is not currently widespread due to the expense to create them, and the difficulty in insulating capabilities will allow their use in place of fiberglass and polyurethane foam, significantly reducing global energy consumption and greenhouse gas emissions.

### What is **buckminsterfullerene**?

It is a large molecule in the shape of a soccer ball, containing 60 carbon atoms, whose structure is the shape of a truncated icosahedron (a hollow, spherical object with 32 faces, 12 of them pentagons and the rest hexagons). This molecule was named buckminsterfullerene because of the structure's resemblance to the geodesic domes designed by American architect R. Buckminster Fuller (1895–1983). The molecule was formed by vaporizing material from a graphite surface with a laser. Large molecules containing only carbon atoms have been known to exist around certain types of carbon-rich stars. Similar molecules are also thought to be present in soot formed during the incomplete combustion of organic materials. Chemist Richard Smalley identified buckminsterfullerene in 1985 and speculated that it may be fairly common throughout the universe. Since that time, other stable, large, even-numbered carbon clusters have been produced. This new class of molecules has been called "fullerenes" since they all seem to have the structure of a geodesic dome. They are also popularly known as "bucky balls." Buckminsterfullerene ($C_{60}$) seems to function as an insulator, conductor, semi-conductor, and superconductor in various compounds. Although no practical application has yet to be developed for it or the other fullerenes, research is expected to result in new types of materials, lubricants, coatings, catalysts, electro-optical devices, and medical applications.

## Is **glass** a solid or a liquid?

Even at room temperature, glass appears to be a solid in the ordinary sense of the word. However, it actually is a fluid with an extremely high viscosity, which refers to the internal friction of fluids. Viscosity is a property of fluids by which the flow motion is gradually damped (slowed) and dissipated by heat. Viscosity is a familiar phenomenon in daily life. An opened bottle of wine can be poured: the wine flows easily under the influence of gravity. Maple syrup, on the other hand, cannot be poured so easily; under the action of gravity, it flows sluggishly. The syrup has a higher viscosity than the wine. It has been documented that century-old windows show signs of flow.

Glass is usually composed of mixed oxides based around the silicon dioxide ($SiO_2$) unit. A very good electrical insulator, and generally inert to chemicals, commercial glass is manufactured by the fusing of sand (silica, $SiO_2$), limestone ($CaCO_2$), and soda (sodium carbonate, $Na_2CO_3$) at temperatures around 2,552°F to 2,732°F (1,400°C to 1,500°C). On cooling, the melt becomes very viscous, and at about 932°F (500°C, known as glass transition temperature), the melt "solidifies" to form soda glass. Small amounts of metal oxides are used to color glass, and its physical properties can be changed by the addition of substances like lead oxide (to increase softness, density, and refractive ability for cutglass and lead crystal), and borax (to significantly lower thermal expansion for cookware and laboratory equipment). Other materials can be used to form glasses if rapidly cooled from the liquid or gaseous phase to prevent an ordered crystalline structure from forming.

Glass objects might have been made as early as 2500 B.C.E. in Egypt and Mesopotamia, and glass blowing developed about 100 B.C.E. in Phoenicia.

## What is **crown glass**?

In the early 1800s, window glass was called crown glass. It was made by blowing a bubble, then spinning it until flat. This left a sheet of glass with a bump, or crown, in

## How is bulletproof glass made?

Bulletproof glass is composed of two sheets of plate glass with a sheet of transparent resin in between, molded together under heat and pressure. When subjected to a severe blow, it will crack without shattering. Today's bulletproof glass is a development of laminated or safety glass, invented by the French chemist Edouard Benedictus. It is basically a multiple lamination of glass and plastic layers.

the center. This blowing method of window-pane making required great skill and was very costly. Still, the finished crown glass produced a distortion through which everything looked curiously wavy, and the glass itself was also faulty and uneven. By the end of the nineteenth century, flat glass was mass-produced and was a common material. The cylinder method replaced the old method, and used compressed air to produce glass that could be slit lengthwise, reheated, and allowed to flatten on an iron table under its own weight. New furnaces and better polishing machines made the production of plate-glass a real industry. Today, almost all flat glass is produced by a float-glass process, which reheats the newly formed ribbon of glass and allows it to cool without touching a solid surface. This produces inexpensive glass that is flat and free from distortion.

### What are the advantages of **tempered glass**?

Tempered glass is glass that is heat-treated. Glass is first heated and then the surfaces are cooled rapidly. The edges cool first, leaving the center relatively hot compared to the surfaces. As the center cools, it forces the surfaces and edges into compression. Tempered glass is approximately four times as strong as annealed glass. It is able to resist temperature differences of 200°F–300°F (90°C–150°C). Since it resists breakage it is favored in many applications for its safety characteristics such as automobiles, doors, tub and shower enclosures, and skylights.

### When were **glass blocks** invented?

Dating back to 1847, glass blocks were originally used as telegraph insulators. They were much smaller and thicker than structural glass blocks and were used mostly in the southeastern Untied States until eventually replaced with porcelain and other types of insulating materials. Glass building bricks were invented in Europe in the early 1900s as thin blocks of glass supported by a grid. Structural glass blocks have been manufactured in the United States since Pittsburgh-based Pittsburgh Corning began producing them in 1938. Blocks made at that time measured approximately eight inches square by nearly five inches in depth, and cast a greenish tint as light transmitted through it. Today's glass blocks can be a square foot in size, much more uniformly shaped, and available in many different sizes, textures, and colors.

## Who invented **thermopane glass**?

Thermopane insulated window glass was invented by C. D. Haven in the United States in 1930. It is two sheets of glass that are bonded together in such a manner that they enclose a captive air space in between. Often this space is filled with an inert gas that increases the insulating quality of the window. Glass is also one of the best transparent materials because it allows the short wavelengths of solar radiation to pass through it, but prohibits nearly all of the long waves of reflected radiation from passing back through it.

## What is the **float glass** process?

Manufacture of high-quality flat glass, needed for large areas and industrial uses, depends on the float glass process, invented by Alastair Pilkington in 1952. The float process departs from all other glass processes where the molten glass flows from the melting chamber into the float chamber, which is a molten tin pool approximately 160 feet (49 meters) long and 12 feet (3.5 meters) wide. During its passage over this molten tin, the hot glass assumes the perfect flatness of the tin surface and develops excellent thickness uniformity. The finished product is as flat and smooth as plate glass without having been ground and polished.

## How is the **glass used in movie stunts** made?

The "glass" might be made of candy (sugar boiled down to a translucent pane) or plastic. This looks like glass and will shatter like glass, but will not cut a performer.

## Who developed **fiberglass**?

Coarse glass fibers were used for decoration by the ancient Egyptians. Other developments were made in Roman times. Parisian craftsman Dubus-Bonnel was granted a patent for the spinning and weaving of drawn glass strands in 1836. In 1893, the Libbey Glass Company exhibited lampshades at the World's Columbian Exposition in Chicago that were made of coarse glass thread woven together with silk. However, this was not a true woven glass. Between 1931 and 1939, the Owens Illinois Glass Company and the Corning Glass Works developed practical methods of making fiberglass commercially. Once the technical problem of drawing out the glass threads to a fraction of their original thinness was solved—basically an endless strand of continuous glass filament as thin as 1/5000 of an inch (0.005 millimeters)—the industry began to produce glass fiber for thermal insulation and air filters, among other uses. When glass fibers were combined with plastics during World War II, a new material was formed. Glass fibers did for plastics what steel did for concrete—gave strength and flexibility. Glass-fiber-reinforced plastics (GFRP) became very important in modern engineering. Fiberglass combined with epoxy resins and thermosetting polyesters are now used extensively in boat and ship construction, sporting goods, automobile bodies, and circuit boards in electronics.

## When was **cement** first used?

Cements are finely ground powders that, when mixed with water, set to a hard mass. The cement used by the Egyptians was calcined gypsum, and both the Greeks and Romans used a cement of calcined limestone. Roman concrete (a mixture of cement, sand, and some other fine aggregate) was made of broken brick embedded in a pozzolanic lime mortar. This mortar consisted of lime putty mixed with brick dust or volcanic ash. Hardening was produced by a prolonged chemical reaction between these components in the presence of moisture. With the decline of the Roman empire, concrete fell into disuse. The first step toward its reintroduction was in 1756, when English engineer John Smeaton (1724–1792) found that when lime containing a certain amount of clay was burned, it would set under water. This cement resembled what had been made by the Romans. Further investigations by James Parker in the same decade led to the commercial production of natural hydraulic cement. In 1824, Englishman Joseph Aspdin (1799–1855) obtained a patent for what he called "portland cement," a material produced from a synthetic mixture of limestone and clay. He called it "portland" because it resembled a building stone that was quarried on the Isle of Portland off the coast of Dorset. The manufacture of this cement spread rapidly to Europe and the United States by 1870. Today, concrete is often reinforced or prestressed, increasing its load-bearing capabilities.

## How was early **macadam** different from modern paved roads?

Macadam roads developed originally in England and France and are named after the Scottish road builder and engineer John Louden MacAdam (1756–1836). The term "macadam" originally designated road surface or base in which clean, broken, or crushed ledge stone was mechanically locked together by rolling with a heavy weight and bonded together by stone dust screenings that were worked into the spaces and then "set" with water. With the beginning of the use of bituminous material (tar or asphalt), the terms "plain macadam," "ordinary macadam," or "waterbound macadam" were used to distinguish the original type from the newer bituminous macadam. Waterbound macadam surfaces are almost never built now in the United States, mainly because they are expensive and the vacuum effect of vehicles loosens them. Many miles of bituminous macadam roads are still in service, but their principal disadvantages are their high crowns and narrowness. Today's roads that carry very heavy traffic are usually surfaced with very durable portland cement.

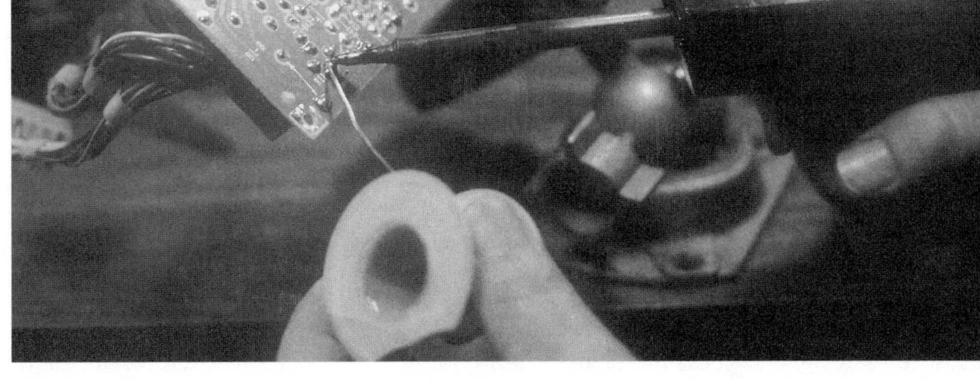

The basic performance of circuit boards relies heavily upon the use of solder.



## What is **Belgian block**?

Belgian block is a road-building material, first used in Brussels, Belgium, and introduced into New York about 1850. Its shape is a truncated pyramid with a base of about five to six inches (13 to 15 centimeters) square and a depth of seven to eight inches (18 to 20.5 centimeters). The bottom of the block is not more than one inch (2.5 centimeters) different from the top. The original blocks were cut from trap-rock from the Palisades of New Jersey.

Belgian blocks replaced cobblestones mainly because their regular shape allowed them to remain in place better than cobblestones. They were not universally adopted, however, because they would wear round and create joints or openings that would then form ruts and hollows. Although they provided a smooth surface compared to the uneven cobblestones, they still made for a rough and noisy ride.

## What is **solder**?

Solder is an alloy of two or more metals used for joining other metals together. One example of solder is half-and-half composed of equal parts of lead and tin. Other metals used in solder are aluminum, cadmium, zinc, nickel, gold, silver, palladium, bismuth, copper, and antimony. Various melting points to suit the work are obtained by varying the proportions of the metals.

Solder is an ancient joining method, mentioned in the Bible (Isaiah 41:7). There is evidence of its use in Mesopotamia some 5,000 years ago, and later in Egypt, Greece, and Rome. Use of numerous types of solder is currently wide and varied, and future use looks bright as well. As long as circuitry based on electrical and magnetic impulses and composed of a combination of conductors, semiconductors, and insulators continues to be in use solder will remain indispensable.

## What is **slag**?

Slag is a non-metallic by-product of iron production that is drawn from the surface of pig iron in the blast furnace. Slag can also be produced in smelting copper, lead, and other metals. Slag from steel blast furnaces contains lime, iron oxide, and silica. The slag from copper and lead-smelting furnaces contains iron silicate and oxides of other metals in small amounts. Slag is used in cements, concrete, and roofing materials as well as ballast for roads and railways.

## What is **creosote**?

Creosote is a yellowish, poisonous, oily liquid obtained from the distillation of coal or wood tar. Crude creosote oil, also called dead oil or pitchoil, is obtained by distilling coal tar (coal tar constitutes the major part of the liquid condensate obtained from the "dry" distillation or carbonization of coal to coke) and is used as a wood preservative. Railroad ties, poles, fence posts, marine pilings, and lumber for outdoor use are impregnated with creosote in large cylindrical vessels. This treatment can greatly extend the useful life of wood that is exposed to the weather. Creosote that is distilled from wood tar is used in pharmaceuticals. Other uses of creosote include disinfectants and solvents. In 1986, the United States Environmental Protection Agency (EPA) began restricting the use of creosote as a wood preservative because of its poisonous and carcinogenic nature.

## What is **neatsfoot oil**?

Neatsfoot oil, also called bubulum oil or hoof oil, is a pale yellow, inedible oil that is rendered from the feet and shin bones of cattle by boiling them in water. It was once prized as a leather dressing and as a lubricating oil for delicate machinery.

## What is **carbon black**?

Carbon black is finely divided carbon produced by incomplete combustion of methane or other hydrocarbon gases (by letting the flame impinge on a cool surface). This forms a very fine pigment containing up to 95 percent carbon, which gives a very intense black color that is widely used in paints, inks, and protective coatings, and as a colorant for paper and plastics. It is also used in large amounts by the tire industry in the production of vulcanized rubber.

### What is the name of the chemical used in watches to make them glow in the dark?

Generally, radioactive paints are used to make watch surfaces visible in the dark. These paints, which do not require activation by an outside light source, will glow for several years. In the past, radium was often the active substance used in luminescent paints for watch faces. However, the practice was discontinued when it was found to emit dangerous gamma rays. The radioactive materials used today give off much lower emissions, which are easily blocked by the glass or plastic covering the watch face. These substances include tritium, krypton 85, promethium 147, and thallium 204.

## How is **parchment paper** made?

Most parchment paper is now vegetable parchment. It is made from a base paper of cotton rags or alpha cellulose, known as waterleaf, which contains no sizing or filling materials. The waterleaf is treated with sulfuric acid converting a part of the cellulose into a gelatin-like amyloid. When the sulfuric acid is washed off, the amyloid film hardens on the paper. The strength of the paper is increased and will not disintegrate even when fully wet. Parchment paper can withstand heat and items will not stick to it.

## How is **sandpaper** made?

Sandpaper is a coated abrasive that consists of a flexible-type backing (paper) upon which a film of adhesive holds and supports a coating of abrasive grains. Various types of resins and hide glues are used as adhesives. The first record of a coated abrasive is in thirteenth century China, when crushed seashells were bound to parchment using natural gums. The first known article on coated abrasives was published in 1808 and described how calcined, ground pumice was mixed with varnish and spread on paper with a brush. Most abrasive papers are now made with aluminum oxide or silicon carbide, although the term sandpapering is still used. Quartz grains are also used for wood polishing. The paper used is heavy, tough, and flexible, and the grains are bonded with a strong glue.

## Why is **titanium dioxide** the most widely used **white pigment**?

Titanium dioxide ($TiO_2$), also known as titania, titanic anhydride, titanium oxide, or titantium white, has become the predominant white pigment in the world because of its high refractive index, lack of absorption of visible light, ability to be produced in the right size range, and its stability. It is the whitest known pigment, unrivalled for color, opacity, stain resistance, and durability; it is also non-toxic. The main consuming industries are paint, printing inks, plastics, and ceramics. Titanium dioxide is also used in the manufacture of floor coverings, paper, rubber, and welding rods.

## When and where was **gunpowder** invented?

The explosive mixture of saltpeter (potassium nitrate), sulfur, and charcoal called gunpowder was known in China at least by 850 C.E., and probably was discovered by Chinese alchemists searching for components to make artificial gold. Early mixtures had too little saltpeter (50 percent) to be truly explosive; 75 percent minimum is needed to get a detonation. The first use of the mixture was in making fireworks. Later, the Chinese used it in incendiary-like weapons. Eventually it is thought that the Chinese found the correct proportions to utilize its explosive effects in rockets and "bamboo bullets." However, some authorities still maintain that the "Chinese gunpowder" really had only pyrotechnic qualities, and "true" gunpowder was a European invention. Roger Bacon (1214–1292) had a formula for it and so might have the German monk Berthold Schwartz (1353). Its first European use depended on the development of firearms in the fourteenth century. Not until the seventeenth century was gunpowder used in peacetime, for mining and civil engineering applications.

## How are **colored fireworks** made?

Fireworks existed in ancient China in the ninth century where saltpeter (potassium nitrate), sulfur, and charcoal were mixed to produce the dazzling effects. Magnesium burns with a brilliant white light and is widely used in making flares and fireworks. Various other colors can be produced by adding certain substances to the flame. Strontium compounds color the flame scarlet, barium compounds produce yellowish-green, copper produces a blue-green, lithium creates purple, and sodium results in yellow. Iron and aluminum granules give gold and white sparks, respectively.

## What is the chemical formula for **TNT**?

TNT is the abbreviation for 2,4,6-trinitrotoluene [$C_7H_5N_3O_6$ or $C_6H_2(CH_3)(NO_2)_3$]. TNT is a powerful, highly explosive compound widely used in conventional bombs. Discovered by J. Wilbrand in 1863, it is made by treating toluene with nitric acid and sulfuric acid. This yellow crystalline solid with a low melting point has low shock sensitivity and even burns without exploding. This makes it safe to handle and cast; but once detonated, it explodes violently.

## Who invented **dynamite**?

Dynamite was not an accidental discovery but the result of a methodical search by the Swedish technologist Alfred Nobel (1833–1896). Nitroglycerine had been discovered in 1849 by the Italian organic chemist Ascanio Sobriero (1812–1888), but it was so

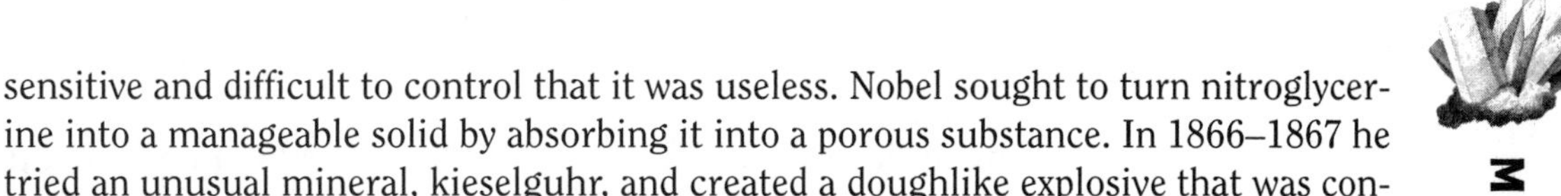



sensitive and difficult to control that it was useless. Nobel sought to turn nitroglycerine into a manageable solid by absorbing it into a porous substance. In 1866–1867 he tried an unusual mineral, kieselguhr, and created a doughlike explosive that was controllable. He also invented a detonating cap incorporating mercury fulminate with which nitroglycerine could be detonated at will. Nobel made a great fortune and bequeathed it to a foundation for awarding prizes for contributions to science, literature, and the promotion of peace.

## When was **plastic** first invented?

In the mid-1850s, Alexander Parkes (1813–1890) experimented with nitrocellulose (or guncotton). Mixed with camphor, it made a hard but flexible transparent material, which he called "Parkesine." He teamed up with a manufacturer to produce it, but there was no demand for it, and the firm went bankrupt. An American, John Wesley Hyatt (1837–1920), acquired the patent in 1868 with the idea of producing artificial ivory for billiard-balls. Improving the formula and with an efficient manufacturing process, he marketed the material, intended for use in making a few household articles, under the name "celluloid." It soon found use in the manufacture of novelty and fancy goods—buttons, letter openers, boxes, hatpins, combs and the like were products often made of celluloid. The material also became the medium for cinematography: celluloid strips coated with a light-sensitive "film" were ideal for shooting and showing moving pictures.

Celluloid was the only plastic material until 1904, when a Belgian scientist, Leo Hendrik Baekeland (1863–1944), succeeded in producing a synthetic shellac from formaldehyde and phenol. Called "bakelite," it was the first of the thermosetting plastics (i.e., synthetic materials that, having once been subjected to heat and pressure, became extremely hard and resistant to high temperatures). Bakelite and other, more versatile plastics, eventually eclipsed celluloid, and by the 1940s, celluloid's markets had shrunk so that it was no longer of commercial importance.

## Why can't **metal objects** be placed in a **microwave**?

Metal objects should not be used in a microwave (except as directed by the manufacturer) because the microwaves are reflected off these materials and do not allow the heat to penetrate the food. The oven could be damaged by an arc between the metal utensils and the cavity interior or door assembly if the cooking load is not large enough to absorb the microwave energy.

## What is **Kevlar**?

The registered trademark Kevlar refers to synthetic fiber called liquid crystalline polymers. Discovered by Stephanie Kwolek (b. 1923), Kevlar is a thin, very strong fiber. It is best known for its use in bulletproof garments.

## Who invented **Teflon**?

In 1938, the American engineer Roy J. Plunkett (1910–1994) at DuPont de Nemours discovered the polymer of tetraluorethylene (PTFE) by accident. This fluorocarbon is marketed under the name of Fluon in Great Britain and Teflon in the United States. Patented in 1939 and first exploited commercially in 1954, PTFE is resistant to all acids and has exceptional stability and excellent electrical insulating properties. It is used in making piping for corrosive materials, in insulating devices for radio transmitters, in pump gaskets, and in computer microchips. In addition, its nonstick properties make PTFE an ideal material for surface coatings. In 1956, French engineer Marc Gregoire discovered a process whereby he could fix a thin layer of Teflon on an aluminum surface. He then patented the process of applying it to cookware, and the no-stick frying pan was created.

## Who invented **Velcro**?

The inspiration for Velcro came about when Swiss engineer George deMaestral took a closer look at those annoying cockleburs that stick so well to things like dog fur and your favorite pair of socks. He recreated what he discovered into a new type of fastener.

## What type of **high-technology material** is used in **space vehicles** for its heat-resistant properties?

Carbon-carbon composites (CCCs) are baseline materials in high temperature applications such as re-entry vehicle nose tips, rocket motor nozzles, and the Space Shuttle Orbiter leading edges. Carbon-carbon is a non-metallic composite material made of carbon fibers in a carbon matrix having operational capabilities ranging from cryogenics to 5,000°F (2,760°C). Numerous fiber types and weave patterns having a wide range of physical properties are available, providing flexibility in the design of carbon-carbon components. Carbon-carbon has a unique combination of light weight, high strength, and temperature resistance.

## Who made the first successful **synthetic gemstone**?

In 1902, Auguste Victor Louis Verneuil (1856–1913) synthesized the first man-made gemstone—a ruby. Verneuil perfected a "flame-fusion" method of producing crystals of ruby and other corundums within a short time period.

# ENERGY

## NON-NUCLEAR FUELS

### What are the three types of **primary energy** that flow continuously on or to the surface of the Earth?

*Geothermal energy* is heat contained beneath the Earth's crust and brought to the surface in the form of steam or hot water. The five main sources of this geothermal reservoir are dry, super-heated steam from steam fields below the Earth's surface, mixed hot water, wet steam, etc., from geysers, etc., dry rocks (into which cold water is pumped to create steam), pressurized water fields of hot water and natural gas beneath ocean beds, and magma (molten rock in or near volcanoes and five to 30 miles [eight to 48 kilometers] below the Earth's crust). Most buildings in Iceland are heated by geothermal energy, and a few communities in the United States, such as Boise, Idaho, use geothermal home heating. Electric power production, industrial processing, space heating, etc., are fed from geothermal sources. The California Geysers project is the world's largest geothermal electric generating complex, with 200 steam wells that provide some 1,300 megawatts of power. The first geothermal power station was built in 1904 at Larderello, Italy.

*Solar radiation* utilization depends on the weather, number of cloudy days, and the ability to store energy for night use. The process of collecting and storing is difficult and expensive. A solar thermal facility (LUZ International Solar Thermal Plant) in the Mojave Desert currently produces 274 megawatts and is used to supplement the power needs of Los Angeles utilities companies. Japan has four million solar panels on roofs, and two-thirds of the houses in Israel have them; 90 percent of Cyprus homes do as well. Solar photo voltaic cells can generate electric current when exposed to the sun. Virtually every spacecraft and satellite since 1958 utilizes this kind of resource, too.

*Tidal and wave energy* contain enormous amounts of energy to be harnessed. The first tidal-powered mill was built in England in 1100; another in Woodbridge, England, built in 1170, has functioned for over 800 years. The Rance River Power Station in France, in operation since 1966, was the first large tidal electric generator plant, producing 160 megawatts. A tidal station works like a hydropower dam, with its turbines spinning as the tide flows through them. Unfortunately, the tidal period of 13.5 hours causes problems in integrating the peak use with the peak generation ability. Ocean wave energy can also be made to drive electrical generators.

## What is the difference between passive **solar energy systems** and active solar energy systems?

Passive solar energy systems use the architectural design, the natural materials, or absorptive structures of the building as an energy saving system. The building itself serves as a solar collector and storage device. An example would be thick-walled stone and adobe dwellings that slowly collect heat during the day and gradually release it at night. Passive systems require little or no investment of external equipment.

Active solar energy systems require a separate collector, a storage device, and controls linked to pumps or fans that draw heat from storage when it is available. Active solar systems generally pump a heat-absorbing fluid medium (air, water, or an antifreeze solution) through a collector. Collectors, such as insulated water tanks, vary in size, depending on the number of sunless days in a locale. Another heat storage system uses eutectic (phase-changing) chemicals to store a large amount of energy in a small volume.

## How does a **solar cell** generate electricity?

A solar cell consists of several layers of silicon-based material. The top, p-type layer, absorbs light energy. This energy frees electrons at the junction layer. The freed electrons collect at the bottom, n-type layer. The loss of electrons from the top layer produces "holes" in the layer that are then filled by other electrons. When a connection, or circuit, is completed between the p-type and n-type layers the flow of electrons creates an electric current.

## What is **biomass energy**?

The catch-all term biomass includes all the living organisms in an area. Wood, crops and crop waste, and wastes of plant, mineral, and animal matter are part of the biomass.



Much of it is in garbage, which can be burned for heat energy or allowed to decay and produce methane gas. However, some crops are grown specifically for energy, including sugar cane, sorghum, ocean kelp, water hyacinth, and various species of trees. It has been estimated that 90 percent of United States waste products could be burned to provide as much energy as 100 million tons of coal (20 percent will not burn, but can be recycled). Biomass energy accounted for less than five percent of total energy consumption in the United States. The use of biomass energy is significantly higher in developing countries where electricity and motor vehicles are scarcer. For example, in India nearly 55 percent of the energy supply is from biomass. In Western Europe, there are over 200 power plants that burn rubbish to produce electricity. France, Denmark, and Switzerland recover 50, 60, and 80 percent of their municipal waste respectively. Biomass can be converted into biofuels such as biogas or methane, methanol, ethanol, etc. However, the process has been more costly than the conventional fossil fuel processes. Rubbish buried in the ground can provide methane gas through an aerobic decomposition. One ton of refuse can produce 8,000 cubic feet (227 cubic meters) of methane.

## Which woods have the best heating quality in a **wood-burning stove**?

Woods that have high heat value, meaning that one cord equals 200 to 250 gallons (757 to 946 liters) of fuel oil or 250 to 300 cubic feet (7 to 8.5 cubic meters) of natural gas, are hickory, beech, oak, yellow birch, ash, hornbeam, sugar maple, and apple.

Woods that have medium heat value, meaning that one cord equals 150 to 200 gallons (567 to 757 liters) of fuel oil or 200 to 250 cubic feet (5.5 to 7 cubic meters) of natural gas, are white birch, douglas fir, red maple, eastern larch, big leaf maple, and elm.

Woods that have a low heat value, meaning that one cord equals 100 to 150 gallons (378 to 567 liters) of fuel oil or 150 to 200 cubic feet (4 to 5.5 cubic meters) of natural gas, are aspen, red alder, white pine, redwood, western hemlock, eastern hemlock, sitka spruce, cottonwood, western red cedar, and lodgepole pine.

## Which **plant** has been investigated as a **source of petroleum**?

A number of plant species have been investigated as potential sources of petroleum. The shrub called the gopher plant (*Euphorbia lathyrus*) produces significant quantities of a milk-like sap—called latex—that is an emulsion of hydrocarbons in water. Another candidate is *Pittosporum resiniferum*, a native of the Philippines. The fruit of this plant, called a petroleum nut, is quite large and the oil harvested from it is frequently used for illumination. Various experiments are under way to use vegetable and seed oils as diesel substitutes, particularly in farm machinery.

## Why are coal, oil, and natural gas called **fossil fuels**?

They are composed of the remains of organisms that lived as long ago as 500 million years. These organisms (such as phytoplankton) became incorporated into the bottom

Researchers measure the output from petroleum producing plants such as the gopher plant.

sediments and then were converted, with time, to oil and gas. Coal is the remains of plants and trees that were buried and subjected to pressure, temperature, and chemical processes (changing into peat and then lignite) for millions of years.

## How and when was **coal formed**?

Coal is formed from the remains of plants that have undergone a series of far-reaching changes, turning into a substance called peat, which subsequently was buried. Through millions of years, the Earth's crust buckled and folded, subjecting the peat deposits to very high pressure and changing the deposits into coal. The Carboniferous, or coal-bearing period occurred about 250 million years ago. Geologists in the United States sometimes divide this period into the Mississippian and the Pennsylvanian periods. Most of the high-grade coal deposits are to be found in the strata of the Pennsylvanian period.

## What **types of coal** are there?

The first stage in the formation of coal converts peat into lignite, a dark-brown type of coal. Lignite is then converted into subbituminous coal as pressure from overlying materials increases. Under still greater pressure, a harder coal called bituminous, or soft, coal is produced. Intense pressure changes bituminous coal into anthracite, the hardest of all coals.

## What is **cannel coal**?

Cannel coal is a type of coal that possesses some of the properties of petroleum. Valued primarily for its quick-firing qualities, it burns with a long, luminous flame. It is made up of coal-like material mixed with clay and shale, and it may also look like black shale, being compact and dull black in color.

## How is underground **coal mined**?

There are two basic types of underground mining methods: room and pillar and longwall. In room and pillar mines, coal is removed by cutting rooms, or large tunnels, in the solid coal, leaving pillars of coal for roof support. Longwall mining takes successive slices over the entire length of a long working face. In the United States, about two-thirds of the coal recovered by underground mining is by room and pillar method; the other third is recovered by longwall mining. Coal seams in the United States range in thickness from a thin film to 50 feet (15 meters) or more. The thickest coal beds are in the western states, ranging from 10 feet (three meters) in Utah and New Mexico to 50 feet (15 meters) in Wyoming. Other places, such as Great Britian, use the longwall method.

## What is a **miner's canary**?

"Miner's canary" refers to the birds used by miners to test the purity of the air in the mines. At least three birds were taken by exploring parties, and the distress of any one bird was taken as an indication of carbon monoxide danger. Some miners used mice rather than birds. This method of safety was used prior to the more sophisticated equipment used today.

## Where are the largest **oil and gas fields** in the world and in the United States?

The Ghawar field, discovered in 1948 in Saudi Arabia, is the largest in the world; it measures 150 × 22 miles (241 × 35 kilometers). The largest oil field in the United States is the Permian Basin, which covers approximately 100,000 square miles in southeast New Mexico and western and northwestern Texas.

## When was the first **oil well** in the United States drilled?

The Drake well at Titusville, Pennsylvania, was completed on August 28, 1859 (some sources list the date as August 27). The driller, William "Uncle Billy" Smith, went down 69.5 feet (21 meters) to find oil for Edwin L. Drake (1819–1880), the well's operator. Within 15 years, Pennsylvania oil field production reached over 10 million 360-pound (163-kilogram) barrels a year.

## When was **offshore drilling** for oil first done?

The first successful offshore oil well was built off the coast at Summerland, Santa Barbara County, California, in 1896.

## Why is **Pennsylvania crude oil** so highly valued?

The waxy, sweet paraffinic oils found in Pennsylvania first became prominent because high quality lubricating oils and greases could be made from them. Similar grade crude oil is also found in West Virginia, eastern Ohio, and southern New York. Different types of crude oil vary in thickness and color, ranging from a thin, clear oil to a thick, tar-like substance.

## What is the process known as **hydrocarbon cracking**?

Cracking is a process that uses heat to decompose complex substances. Hydrocarbon cracking is the decomposition by heat, with or without catalysts, of petroleum or heavy petroleum fractions (groupings) to give materials of lower boiling points. Thermal cracking, developed by William Burton in 1913, uses heat and pressure to break some of the large heavy hydrocarbon molecules into smaller gasoline-grade ones. The cracked hydrocarbons are then sent to a flash chamber where the various fractions (groupings) are separated. Thermal cracking not only doubles the gasoline yield, but has improved gasoline quality, producing gasoline components with good anti-knock characteristics.

## Why was **lead added** to **gasoline** and why is lead-free gasoline used in cars?

Tetraethyl lead was used for more than 40 years to improve the combustion characteristics of gasoline. It reduces or eliminates "knocking" (pinging caused by premature ignition) in large high-performance engines and in smaller high-compression engines. It provides lubrication to the extremely close-fitting engine parts, where oil has a tendency to wash away or burn off. However, lead will ruin and effectively destroy the catalyst presently used in emission control devices installed in new cars. Therefore, only lead-free gasoline must be used.

## When did the use of **lead-free fuel** become mandatory in the United States?

The sale of leaded gasoline for motor vehicles ended in 1996. All vehicles manufactured after July 1974 for sale in the United States were required to use unleaded gasoline.

## What is a **reformulated gasoline**?

Oil companies are being required to offer new gasolines that burn more cleanly and have less impact on the environment. Typically, reformulated gasolines contain lower concentrations of benzene, aromatics, and olefins; less sulfur; a lower Reid vapor pressure (RVP); and some percentage of an oxygenate (non-aromatic component) such as methyl tertiary butyl ether (MTBE). MTBE is a high-octane gasoline blending component produced by the reaction of isobutylene and methanol. It was developed to meet the ozone ambient air quality standards, but its unique characteristics as a water pollutant pose a challenge to the Environmental Protection Agency (EPA) in meeting the requirements of the Clean Air Act, the Safe Drinking Water Act, and the Underground Storage Tank Program. The Clean Air Act called for reformulated gasoline to be sold in the nine worst ozone nonattainment areas beginning January 1, 1995.

## What kinds of **additives** are in **gasoline** and why?

| Additive | Function |
|---|---|
| Antiknock compounds | Increase octane number |
| Scavengers | Remove combustion products of antiknock compounds |
| Combustion chamber | Suppress surface ignition and spark plug deposit modifiers fouling |
| Antioxidants | Provide storage stability |
| Metal deactivators | Supplement storage stability |
| Antirust agents | Prevent rusting in gasoline-handling systems |
| Anti-icing agents | Suppress carburetor and fuel system freezing |
| Detergents | Control carburetor and induction system cleanliness |
| Upper-cylinder lubricants | Lubricate upper cylinder areas and control intake system deposits |
| Dyes | Indicate presence of antiknock compounds and identify makes and grades of gasoline |

## What do the **octane numbers** of gasoline mean?

The octane number is a measure of the gasoline's ability to resist engine knock (pinging caused by premature ignition). Two test fuels, normal heptane and isooctane, are blended for test results to determine octane number. Normal heptane has an octane number of zero and isooctane a value of 100. Gasolines are then compared with these test blends to find one that makes the same knock as the test fuel. The octane rating of the gasoline under testing is the percentage by volume of isooctane required to produce the same knock. For example, if the test blend has 85 percent isooctane, the gasoline has an octane rating of 85. The octane rating that appears on gasoline pumps is an average of research octane determined in laboratory tests with engines running at low speeds, and motor octane, determined at higher speeds.

## When did **gasoline stations** open?

The first service station (or garage) was opened in Bordeaux, France, in December 1895 by A. Barol. It provided overnight parking, repair service, and refills of oil and "motor spirit." In April 1897 a parking and refueling establishment—Brighton Cycle and Motor Co.—opened in Brighton, England.

The pump that would be used to eventually dispense gasoline was devised by Sylanus Bowser of Fort Wayne, Indiana, but in September 1885 it dispensed kerosene. Twenty years later Bowser manufactured the first self-regulating gasoline pump. In 1912, a Standard Oil of Louisiana superstation opened in Memphis, Tennessee, featuring 13 pumps, a ladies' rest room, and a maid who served ice water to waiting customers. On December 1, 1913, in Pittsburgh, Pennsylvania, the Gulf Refining Company opened the first drive-in station as a 24-hour-a-day operation. Only 30 gallons (114 liters) of gasoline were sold the first day.

## What are the advantages and disadvantages of the **alternatives to gasoline** to power automobiles?

Because the emissions of gasoline is a major air pollution problem in most U.S. urban areas, researchers are seeking alternatives. Currently, none of the alternatives deliver as much energy content as gasoline, so more of each of these fuels must be consumed to equal the distance that the energy of gasoline propels the automobile. The most viable alternative is flexible fuel, a combination of methanol and gasoline, which would add at least $300 to car prices for an expensive fuel sensor and a longer fuel tank.

| Alternative | Advantages | Disadvantages |
|---|---|---|
| Electricity from batteries | No vehicle emissions, good for stop-and-go driving | Short-lived bulky batteries; limited trip range |
| Ethanol from corn, biomass, etc. | Relatively clean fuel | Costs, corrosive damage |
| Hydrogen from electrolysis; etc. | Plentiful supply; non-toxic emissions | High cost; highly flammable |
| Methanol from methanol gas, coal, biomass, wood | Cleaner combustion; less volatile | Corrosive; some irritant emissions |
| Natural gas from hydrocarbons and petroleum deposits | Cheaper on energy basis; relatively clean | Cost to adapt vehicle bulky storage; sluggish performance |

## How is **gasohol** made?

Gasohol, a mixture of 90 percent unleaded gasoline and 10 percent ethyl alcohol (ethanol), has gained some acceptance as a fuel for motor vehicles. It is comparable in performance to 100 percent unleaded gasoline with the added benefit of superior antiknock properties (no premature fuel ignition). No engine modifications are needed for the use of gasohol.

Since corn is the most abundant United States grain crop, it is predominantly used in producing ethanol. However, the fuel can be made from other organic raw materials, such as oats, barley, wheat, milo, sugar beets, or sugar cane. Potatoes, cassava (a starchy plant), and cellulose (if broken up into fermentable sugars) are possible other sources. The corn starch is processed through grinding and cooking. The process requires the conversion of a starch into a sugar, which in turn is converted into alcohol by reaction with yeast. The alcohol is distilled and any water is removed until it is 200 proof (100 percent alcohol).

One acre of corn yields 250 gallons (946 liters) of ethanol; an acre of sugar beets yields 350 gallons (1,325 liters), while an acre of sugar can produce 630 gallons (2,385 liters). In the future, motor fuel could conceivably be produced almost exclusively from garbage, but currently its conversion remains an expensive process.

## What are the main components found in motor vehicle **exhaust**?

The main components of exhaust gas are nitrogen, carbon dioxide, and water. Smaller amounts of nitrogen oxides, carbon monoxide, hydrocarbons, aldehydes, and other products of incomplete combustion are also present. The most important air pollutants, in order of amount produced, are carbon monoxide, nitrogen oxides, and hydrocarbons.

## What is **cogeneration**?

Cogeneration is an energy production process involving the simultaneous generation of thermal (steam or hot water) and electric energy by using a single primary heat source. By producing two kinds of useful fuels in the same facility the net energy yield from the primary fuel increases from 30–35 percent to 80–90 percent. Cogeneration can result in significant cost savings and can reduce any possible environmental effects conventional energy production may produce. Cogeneration facilities have been installed at a variety of sites, including oil refineries, chemical plants, paper mills, utility complexes, and mining operations.

## Where are the largest **wind farms**?

Wind farms are clusters of wind turbines that generate electricity. The three largest wind farms in the world are in California: one is at Altamont Pass, east of San Francisco; one is in the Tehachapi Mountains in Kern County; and the third is at San Gorgonio Pass north of Palm Springs. California has the largest wind farms for a number of reasons. The California wind farms are in places where very favorable winds occur. They are also near electric power transmission lines and large cities. Peak winds in these areas occur approximately at the same times as the peak electricity demand in the cities. This increases the value of the electricity to consumers. Another important factor is that the state of California required electric utilities to purchase the electricity from the wind farms at favorable prices to the turbine owners.

### Which states have the greatest **wind energy potential**?

The windiest areas in the United States are in Montana, North Dakota, Wyoming, and the other Great Plains states. North Dakota, South Dakota, and Texas have sufficient wind resources to meet the electricity needs for the entire United States.

### Who invented the **fuel cell**?

The earliest fuel cell, known as a "gas battery," was invented by Sir William Grove (1811–1896) in 1839. Grove's fuel cell incorporated separate test tubes of hydrogen and oxygen, which he placed over strips of platinum. It was later modified by Francis Thomas Bacon (1904–1992) with nickel replacing the platinum. A fuel cell is equivalent to a generator—it converts a fuel's chemical energy directly into electricity.

### Who was **Nikola Tesla**?

Nikola Tesla (1856–1943) was a leading innovator in the field of electricity. Tesla held over 100 patents, among which are patents for alternating current and the seminal patents for radio. Tesla's work for Westinghouse in the late 1880's led to the commercial production of electricity, including the Niagara Falls Power Project in 1895. After a bitter and prolonged public feud, Tesla's alternating current system was proven superior to Edison's direct current system. Tesla was responsible for many other innovations, including the Tesla coil, radio controlled boats, and neon and fluorescent lighting.

### Where was the first **hydroelectric power** plant in the United States built?

Appleton, Wisconsin, is the site of the first hydroelectric generating plant in the United States. It was built in 1882.

## NUCLEAR POWER

### What is the **life of a nuclear power plant**?

While there is some controversy about the subject, the working life of a nuclear power plant as defined by the industry and its regulators is approximately 40 years, which is about the same as that of other types of power stations.

## Where is the **oldest operational nuclear power plant** in the United States?

The oldest operational plant in the United States is the Oyster Creek, New Jersey, plant. It became operational in December 1969. The Yankee Plant at Rowe, Massachusetts, which was constructed in 1960 and closed in 1991, was the first commercially used nuclear power plant to be built in the United States.

## Which **nuclear reactors** have had major, publicly reported accidents?

**Incidents with core damage in nuclear reactors**

| Description of incident | Site | Date | Adult thyroid dose (in rems) |
|---|---|---|---|
| Minor core damage (no release of radiologic material) | Chalk River, Ontario, Canada | 1952 | not applicable |
| | Breeder Reactor Idaho | 1955 | not applicable |
| | Westinghouse Test Reactor | 1960 | not applicable |
| | Detroit Edison Fermi, Michigan | 1966 | not applicable |
| Major core damage (radioiodine released) | | | |
| Noncommercial | Windscale, England | 1957 | 16 |
| | Idaho Falls SL-1, Idaho | 1961 | 0.035 |
| Commercial | Three Mile Island, Pennsylvania | 1979 | 0.005 |
| | Chernobyl, Soviet Union | 1986 | 100 (estimated) |

## What actually happened at **Three Mile Island**?

The Three Mile Island nuclear power plant in Pennsylvania experienced a partial meltdown of its reactor core and radiation leakage. On March 28, 1979, just after 4:00 a.m., a water pump in the secondary cooling system of the Unit 2 pressurized water reactor failed. A relief valve jammed open, flooding the containment vessel with radioactive water. A backup system for pumping water was down for maintenance. Temperatures inside the reactor core rose, fuel rods ruptured, and a partial (52 percent) meltdown occurred, because the radioactive uranium core was almost entirely uncovered by coolant for 40 minutes. The thick steel-reinforced containment building prevented nearly all the radiation from escaping—the amount of radiation released into the atmosphere was one-millionth of that at Chernobyl. However, if the coolant had not been replaced, the molten fuel would have penetrated the reactor containment vessel, where it would have come into contact with the water, causing a steam explosion,

**The estimated working life of a nuclear power plant is approximately 40 years.**

breaching the reactor dome, and leading to radioactive contamination of the area similar to the Chernobyl accident.

## How many **nuclear power plants** are there **worldwide**?

As of 2000, 437 reactors were operational with 34 more under construction.

| Country | Number of Units |
|---|---|
| Argentina | 2 |
| Armenia | 1 |
| Belgium | 7 |
| Brazil | 2 |
| Bulgaria | 6 |
| Canada | 14 |
| China | 3 |
| Czech Republic | 5 |
| Finland | 4 |
| France | 59 |
| Germany | 19 |
| Hungary | 4 |
| India | 14 |
| Japan | 53 |

| | |
|---|---|
| Kazakhstan | 1 |
| Korea, South | 16 |
| Lithuania | 2 |
| Mexico | 2 |
| Netherlands | 2 |
| Pakistan | 2 |
| Romania | 1 |
| Russia | 29 |
| South Africa | 2 |
| Slovakia | 6 |
| Slovenia | 1 |
| Spain | 9 |
| Sweden | 11 |
| Switzerland | 5 |
| Taiwan | 6 |
| Ukraine | 13 |
| United Kingdom | 33 |
| United States | 104 |

As of 2000, the United States had 104 reactors in operation, with none under construction. The plants generated 719,400 million net kilowatt hours or 19.54 percent of domestic electricity in 1999.

## What is the **Rasmussen report**?

Dr. Norman Rasmussen of the Massachusetts Institute of Technology (MIT) conducted a study of nuclear reactor safety for the United States Atomic Energy Commission. The 1975 study cost four million dollars and took three years to complete. It concluded that the odds against a worst-case accident occurring were astronomically large—10 million to one. The worst-case accident projected about three thousand early deaths and 14 billion dollars in property damage due to contamination. Cancers occurring later due to the event might number 1,500 per year. The study concluded that the safety features engineered into a plant are very likely to prevent serious consequences from a meltdown. Other groups criticized the Rasmussen report and declared that the estimates of risk were too low. After the Chernobyl disaster in 1986, some scientists estimated that a major nuclear accident might in fact happen every decade.

## What caused the **Chernobyl** accident?

The worst nuclear power accident in history, which occurred at the Chernobyl nuclear power plant in the Ukraine, will affect, in one form or another, 20 percent of the republic's population (2.2 million people). On April 26, 1986, at 1:23:40 a.m., during unauthorized experiments by the operators in which safety systems were deliberately

### What is a meltdown, and what does it have to do with the "China Syndrome"?

A meltdown is a type of accident in a nuclear reactor in which the fuel core melts, resulting in the release of dangerous amounts of radiation. In most cases the large containment structure that houses a reactor would prevent the radioactivity from escaping. However, there is a small possibility that the molten core could become hot enough to burn through the floor of the containment structure and go deep into the Earth. Nuclear engineers call this type of situation the "China Syndrome." The phrase derives from a discussion on the theoretical problems that could result from a meltdown, when a scientist commented that the molten core could bore a hole through the Earth, coming out—if one happened to be standing in North America—in China. Although the scientist was grossly exaggerating, some took him seriously. In fact, the core would only bore a hole about 30 feet (10 meters) into the Earth, but even this distance would have grave repercussions. All reactors are equipped with emergency systems to prevent such an accident from occurring.

circumvented in order to learn more about the plant's operation, one of the four reactors rapidly overheated and its water coolant "flashed" into steam. The hydrogen formed from the steam reacted with the graphite moderator to cause two major explosions and a fire. The explosions blew apart the 1,000 ton (907 metric ton) lid of the reactor, and released radioactive debris high into the atmosphere. It is estimated that 3.5 percent of the reactor's fuel and 10 percent of the graphite reactor itself was emitted into the atmosphere. Human error and design features (such as a positive void coefficient type of reactor, use of graphite in construction, and lack of a containment building) are generally cited as the causes of the accident. Thirty-one people died from trying to stop the fires. More than 240 others sustained severe radiation sickness. Eventually 150,000 people living near the reactor were relocated; some of whom may never be allowed to return home. Fallout from the explosions, containing radioactive isotope cesium-137, was carried by the winds westward across Europe.

The problems created by the Chernobyl disaster are overwhelming and continue today. Particularly troubling is the fact that by 1990–1991, a five-fold increase had occurred in the rate of thyroid cancers in children in Belarus. A significant rise in general morbidity has also taken place among children in the heaviest-hit areas of Gomel and Mogilev.

# MEASURES AND MEASUREMENT

## What is the **weight** per gallon of common **fuels**?

| One gallon of fuel | Weight in pounds/kilograms |
|---|---|
| Butane | 4.86/2.2 |
| Propane | 4.23/1.9 |
| Kerosene | 6.75/3.1 |
| Gasoline | 6.00/2.7 |
| Aviation gasoline | 6.46–6.99/2.9–3.2 |

## How much does a **barrel of oil weigh**?

A barrel of oil weighs about 306 pounds (139 kilograms).

## How many **gallons** are in a **barrel of oil**?

The barrel, a common measure of crude oil, contains 42 U.S. gallons or 34.97 imperial gallons.

## How do **various energy sources** compare?

Below are listed some comparisons (approximate equivalents) for the energy sources:

| Energy unit | Equivalent |
|---|---|
| 1 BTU of energy | 1 match tip<br>250 calories (International Steam Table)<br>0.25 kilocalories (food calories) |
| 1,000 BTU of energy | 2 5-ounce glasses of wine<br>250 kilocalories (food calories)<br>0.8 peanut butter and jelly sandwiches |
| 1 million BTU of energy | 90 pounds of coal<br>120 pounds of oven-dried hardwood<br>8 gallons of motor gasoline<br>10 therms of dry natural gas<br>11 gallons of propane<br>2 months of the dietary intake of a laborer |
| 1 quadrillion BTU of energy | 45 million short tons of coal<br>60 million short tons of oven-dried hardwood<br>1 trillion cubic feet of dry natural gas<br>170 million barrels of crude oil |

| | |
|---|---|
| | 470 thousand barrels of crude oil per day for 1 year<br>28 days of U.S. petroleum imports<br>26 days of U.S. motor gasoline<br>26 hours of world energy use (1989) |
| 1 barrel of crude oil | 5.6 thousand cubic feet of dry natural gas<br>0.26 short tons (520 pounds) of coal<br>1,700 kilowatt-hours of electricity |
| 1 short ton of coal | 3.8 barrels of crude oil<br>21 thousand cubic feet of dry natural gas<br>6,500 kilowatt-hours of electricity |
| 1,000 cubic feet of natural gas | 0.18 barrels (7.4 gallons) of crude oil<br>0.05 short tons (93 pounds) of coal<br>300 kilowatt-hours of electricity |
| 1,000 kilowatt-hours of electricity | 0.59 barrels of crude oil<br>0.15 short tons (310 pounds) of coal<br>3,300 cubic feet of dry natural gas |

*Note:* One quadrillion equals 1,000,000,000,000,000.

Because of energy losses associated with the generation of electricity, about three times as much fossil fuel is required to generate 1,000 kilowatt-hours: 1.8 barrels of oil, 0.47 short tons of coal, or 10,000 cubic feet (283 cubic meters) of dry natural gas.

## What are the approximate **heating values** of fuels?

| Fuel | BTU | Unit of measure |
|---|---|---|
| Oil | 141,000 | gallon |
| Coal | 31,000 | pound |
| Natural gas | 1,000 | cubic feet |
| Steam | 1,000 | cubic feet |
| Electricity | 3,413 | kilowatt hour |
| Gasoline | 124,000 | gallon |

A BTU (British thermal unit), a common energy measurement, is defined as the amount of energy required to raise the temperature of one pound of water by 1°F.

## What are the **fuel equivalents** to produce one quad of energy?

One quad (meaning one quadrillion) is equivalent to:

$1 \times 10^{15}$ BTU

$252 \times 10^{15}$ calories or $252 \times 10^{12}$ K calories

In fossil fuels, one quad is equivalent to:

180 million gallons (681 million liters) of crude oil

0.98 trillion cubic feet (0.028 trillion cubic meters) of natural gas

37.88 million tons of anthracite coal

38.46 million tons of bituminous coal

In nuclear fuels, one quad is equivalent to 2,500 tons of $U_3O_8$ if only $U_{235}$ is used.

In electrical output, one quad is equivalent to $2.93 \times 10^{11}$ kilowatt-hours electric.

## How much heat will 100 cubic feet of **natural gas** provide?

One hundred cubic feet of natural gas can provide about 100,000 BTUs (British thermal units) of heat. A British thermal unit, a common energy measurement, is defined as the amount of energy required to raise one pound of water by 1°F.

## How is a **heating degree day** defined?

Early this century engineers developed the concept of heating degree days as a useful index of heating fuel requirements. They found that when the daily mean temperature is lower than 65°F (18°C), most buildings require heat to maintain a 70°F (21°C) temperature. Each degree of mean temperature below 65°F (18°C) is counted as "one heating degree day." For every additional heating degree day, more fuel is needed to maintain a 70°F (21°C) indoor temperature. For example, a day with a mean temperature of 35°F (1.5°C) would be rated as 30 heating degree days and would require twice as much fuel as a day with a mean temperature of 50°F (10°C; 15 heating degree days). The heating degree concept has become a valuable tool for fuel companies for evaluation of fuel use rates and efficient scheduling of deliveries. Detailed daily, monthly, and seasonal totals are routinely computed for the stations of the National Weather Service.

## What does the term **cooling degree day** mean?

It is a unit for estimating the energy needed for cooling a building. One unit is given for each degree Fahrenheit above the daily mean temperature when the mean temperature exceeds 75°F (24°C).

## How are **utility meters** read?

Older electric and gas meters have a series of four or five dials, which indicate the amount of energy being consumed. The dials are read from left to right, and if the pointer falls between two numbers, the lower number is recorded. Gas meters are set to read hundreds of cubic feet. New meter models have digital displays.

### How many BTUs are equivalent to one ton of **cooling capacity**?

1 ton = 288,000 BTUs/24 hours or 12,000 BTUs/hour.

### How much **wood** is in a **cord**?

A cord of wood is a pile of logs four feet (1.2 meters) wide and four feet (1.2 meters) high and eight feet (2.4 meters) long. It may contain from 77 to 96 cubic feet of wood. The larger the unsplit logs the larger the gaps, with fewer cubic feet of wood actually in the cord.

## CONSUMPTION AND CONSERVATION

### Does the United States currently produce **enough energy** to meet its consumption needs?

No. From 1958 forward, the United States has consumed more energy than it produced, and the difference has been met by energy imports. In 1999, 71.98 quadrillion BTUs was the total United States energy production; in that same year 96.87 quadrillion BTUs were consumed. In that same time period China produced 30.87 quadrillion BTUs and consumed 31.88 quadrillion BTUs.

### Which countries **consume the most energy**?

**Top Energy-Consuming Countries (1999, in quadrillion BTUs)**

| Country | Energy Consumption |
|---|---|
| United States | 96.87 |
| China | 31.88 |
| U.S.S.R. | 26.01 |
| Japan | 21.71 |
| Germany | 13.98 |
| Canada | 12.52 |
| India | 12.18 |
| France | 10.26 |
| United Kingdom | 9.92 |
| Brazil | 8.51 |

## What is the current **per capita energy consumption** in the United States?

In 1999 in the United States per capita consumption was about 350.9 million BTUs (British Thermal Units). A BTU, a common energy measurement, is defined as the amount of energy required to raise one pound of water by 1°F. Below is listed per capita consumption for representative years:

| Year | Quantity (End-Use) (Million BTU) |
|---|---|
| 1950 | 194 |
| 1960 | 212 |
| 1970 | 270 |
| 1980 | 259 |
| 1990 | 256 |
| 1994 | 260 |
| 1999 | 351 |

(End use energy consumption is total energy consumption less losses incurred in the generation, transmission, distribution, etc. of electricity)

## How long, at the present rate of consumption, will **current major energy reserves** last?

The best estimates indicate there will be enough oil to provide energy to the world for another 50 years. Recoverable reserves of coal and natural gas have expanded, too. The current supply of natural gas, at gradually increased rates of consumption, will last almost 60 years. Known coal supplies should last until about the year 2225. All in all, taking into account the likely discovery of new deposits of fossil fuels and the development of new technologies to get it out of the ground or from under the sea, the total energy supply may soon reach 600 times current world consumption levels.

## How does the **Energy Star** program promote energy efficiency?

Energy Star is a dynamic government/industry partnership that offers businesses and consumers energy-efficient solutions, making it easy to save money while protecting the environment for future generations. In 1992 the U.S. Environmental Protection Agency (EPA) introduced Energy Star as a voluntary labeling program designed to identify and promote energy-efficient products to reduce greenhouse gas emissions. Computers and monitors were the first labeled products. The Energy Star label is now on major appliances, office equipment, lighting, consumer electronics, and more. The EPA has also extended the label to cover new homes and commercial and industrial buildings. Through its partnerships with more than 7,000 private and public sector organizations, Energy Star delivers the technical information and tools that organiza-

## What is a Cargo Quilt?

The Cargo Quilt was developed as an alternative to transporting temperature-sensitive cargo under refrigeration. Once the cargo is loaded on a trailer or container, the Cargo Quilt is draped over the cargo and acts like a thermos bottle. The Cargo Quilt can maintain the temperature of the freight—whether hot or cold—from five to 30 days.

tions and consumers need to choose energy-efficient solutions and best management practices. Energy Star has successfully delivered energy and cost savings across the country, saving businesses, organizations, and consumers more than $5 billion a year. Over the past decade, Energy Star has been a driving force behind the more widespread use of such technological innovations as LED traffic lights, efficient fluorescent lighting, power management systems for office equipment, and low standby energy use.

## How much money can be saved by lowering the setting on a **home furnace thermostat**?

Tests have shown that a 5°F reduction in the home thermostat setting for approximately eight hours will save up to 10 percent in fuel costs.

## How much energy is saved by raising the setting for a **house air conditioner**?

For every 1°F the inside temperature is increased, the energy needed for air conditioning is reduced by 3 percent. If all consumers raised the settings on their air conditioners by 6°F, for example, 190,000 barrels of oil could be saved each day.

## How much fuel can be saved when a home is **properly insulated**?

Insulation of a single-family house with EPS or XPS (extruded polystyrene) over a 50-year period has the potential to save 80 metric tons of heating oil. This in turn corresponds to the fuel consumption of a fully loaded jumbo jet during a flight from Frankfurt to New York.

## How much energy is required to use various **electrical appliances**?

The formula to estimate the amount of energy a specific appliance consumes is:

$$(\text{wattage} \times \text{hours used per day})/1{,}000 = \text{daily kilowatt-hour (kWh) consumption.}$$

Multiply this by the number of days you use the appliance during the year for annual consumption.

The table below indicates the annual estimated energy consumption for various household electrical products.

| Appliance | Time in Use | Est. kilowatt-hours per year |
|---|---|---|
| Clock radio | 24 hours per day | 44 |
| Clothes washer | 2 hours per week | 31 |
| Coffee maker | 30 minutes per day | 128 |
| Dehumidifier | 12 hours per day | 700 |
| Dishwasher (does not include hot water) | 1 hour per day | 432 |
| Electric blanket | 8 hours per day, 120 days per year | 175 |
| Furnace fan | 12 hours per day | 432 |
| Hair dryer | 15 minutes per day | 100 |
| Iron | 1 hour per week | 52 |
| Microwave oven | 2 hours per week | 89 |
| Portable heater | 3 hours per day, 120 days per year | 540 |
| Radio (stereo) | 2 hours per day | 73 |
| Refrigerator (frost free, 16 cubic feet) | 24 hours per day | 642 |
| Refrigerator (frost free, 18 cubic feet) | 24 hours per day | 683 |
| Television (color) | 4 hours per day | 292 |
| Toaster oven | 1 hour per day | 73 |
| Vacuum cleaner | 1 hour per week | 38 |
| VCR | 4 hours per day | 30 |
| Water heater (40 gallon) | 2 hours per day | 2190 |
| Water pump (deep well) | 2 hours per day | 730 |
| Whole house fan | 4 hours per day, 120 days per year | 270 |
| Window fan | 4 hours per day, 180 days per year | 144 |

## Is it possible to **compare** the **energy efficiency** of different brands of appliances?

In 1980, the Federal Trade Commission's Appliance Labeling Rule became effective. It requires EnergyGuide labels be placed on all new refrigerators, freezers, water heaters, dishwashers, clothes washers, room air conditioners, heat pumps, furnaces, and boilers. The bright yellow labels with black lettering identify energy consumption and operating cost for each of the various household appliances. EnergyGuide labels show the estimated yearly electricity consumption to operate the product along with a scale of comparison among similar products. The comparison scale shows the least and most energy used by comparable models.

### When was **gas lighting** invented?

In 1799 Philippe Lebon patented a method of distilling gas from wood for use in a "Thermolamp," a type of lamp. By 1802 William Murdock installed gas lighting in a factory in Birmingham, England. The introduction of widespread, reliable interior illumination enabled dramatic changes in commerce and manufacturing.

### What are the advantages of **compact fluorescent light bulbs** over incandescent light bulbs?

Compact fluorescent light bulbs (CFLs) have the following advantages over incandescent light bulbs when used properly: they last up to 10 times longer, use about a quarter of the energy, and produce 90 percent less heat, while producing more light per watt. For example, a 27 watt compact fluorescent lamp provides about 1800 lumens, compared to 1750 lumens from a 100 watt incandescent lamp. CFLs are most efficient in areas where lights are on for long periods of time.

### When should a **fluorescent light** be turned off to save energy?

Fluorescent lights use a lot of electric current getting started, and frequently switching the light on and off will shorten the lamp's life and efficiency. It is energy-efficient to turn off a fluorescent light only if it will not be used again within an hour or more.

### What changes have been made to make **windows** more efficient?

Until recently, clear glass was the primary glazing material used in windows. Although glass is durable and allows a high percentage of sunlight to enter buildings, it has very little resistance to heat flow. Glazing technology has changed greatly during the past two decades. There are now several types of advanced glazing systems available to help control heat loss or gain. The advanced glazings include double- and triple-pane windows with such coatings as low-emissivity (low-e), spectrally selective, heat-absorbing (tinted), reflective, or a combination of these; windows can also be filled with a gas (typically xenon, argon, or krypton) that helps in insulation.

### How does **driving speed** affect **gas mileage** for most automobiles?

Most automobiles get about 28 percent more miles per gallon of fuel at 50 miles (80.5 kilometers) per hour than at 70 miles (112 kilometers) per hour, and about 21

### How much energy is saved by recycling one aluminum can?

Some sources indicate that one recycled aluminum can save as much energy as it takes to run a TV set for four hours or the energy equivalent of half a gallon (1.9 liters) of gasoline. To manufacture one ton of aluminum, nearly 9,000 pounds (4,086 kilograms) of bauxite and 1,020 pounds (463 kilograms) of petroleum coke are needed. Recycling aluminum cans reduces the need for raw material by 95 percent and reduces the energy needed to produce aluminum by 90 percent.

percent more at 55 miles (88.5 kilometers) per hour than at 70 miles (112 kilometers) per hour.

## Is it more economical to run an automobile with its **windows open** rather than using its **air conditioner**?

At speeds greater than 40 miles (64 kilometers) per hour, less fuel is used in driving an automobile with the air conditioner on and the windows up than with the windows rolled down. This is due to the air drag effect—the resistance that a vehicle encounters as it moves through a fluid medium, such as air. In automobiles, the amount of engine power required to overcome this drag force increases with the cube of the vehicle's speed—twice the speed requires eight times the power. For example, it takes five horsepower for the engine to overcome the air resistance at 40 miles (64 kilometers) per hour; but at 60 miles (97 kilometers) per hour, it takes 18 horsepower; at 80 miles (128 kilometers) per hour, it takes 42 horsepower. Improved aerodynamics, in which the drag coefficient (measure of air drag effect) is reduced, significantly increases fuel efficiency. The average automobile in 1990 had a drag coefficient of about 0.4. In the early 1960s it was 0.5 on average, and it was 0.47 in the 1970s. The lowest maximum level possible for wheeled vehicles is 0.15.

## When is it more economical to **restart an automobile** rather than let it idle?

Tests by the Environmental Protection Agency have shown that it is more economical to turn an engine off rather than let it idle if the idle time exceeds 60 seconds.

## How much gasoline do **underinflated tires** waste?

Underinflated tires waste as much as one gallon (4.5 liters) out of every 20 gallons (91 liters) of gasoline. To save fuel, follow the automaker's guidelines regarding recommended air pressure levels for the tires. However, greater fuel economy can be achieved by inflating tires to the maximum air pressure listed on the sidewall of the tire, resulting in less rolling resistance.

## What is the **energy use** of various modes of **transportation**?

| Mode of transportation | Number of passengers | Energy use (BTUs) per passenger mile |
|---|---|---|
| Bicycle | 1 | 80 |
| Automobile—high economy | 4 | 600 |
| Motorcycle | 1 | 2,100 |
| Bus—inter-city | 45 | 600 |
| Subway train | 1,000 | 900 |
| 747 jet plane | 360 | 3,440 |

## What is the **fuel consumption** of various **aircraft**?

| Aircraft | Number of passengers | Fuel consumption (gallons) per 1,000-mile flight |
|---|---|---|
| 737 | 128 | 1,600 |
| 737 Stretch | 188 | 1,713 |
| 747-400 | 413 | 6,584 |
| SST Concorde | 126 | 6,400 |
| Turboprop DHC-8 | 37 | 985 |
| Fighter plane F-15 | 1 | 750 |
| Military cargo plane C-17 | 126,000 pound of cargo | 5,310 |

# ENVIRONMENT

## ECOLOGY, RESOURCES, ETC.

### What is **biodiversity**?

Biodiversity refers to genetic variability within a species, diversity of populations of a species, diversity of species within a natural community, or the wide array of natural communities and ecosystems throughout the world. Some scientists estimate that there may be between 15 and 100 million species throughout the world. Biodiversity is threatened at the present time more than at any other time in history. In the time since the North American continent was settled, as many as 500 plant and animal species have disappeared. Some recent examples of threats to biodiversity in the United States include: 50 percent of the United States no longer supports its original vegetation; in the Great Plains, 99 percent of the original prairies are gone; and across the United States, we destroy 100,000 acres of wetlands each year.

### What is a **biome**?

It is a plant and animal community that covers a large geographical area. Complex interactions of climate, geology, soil types, water resources, and latitude all determine the kinds of plants and animals that thrive in different places. Fourteen major ecological zones, called "biomes," exist over five major climatic regions and eight zoogeographical regions. Important land biomes include tundra, coniferous forests, deciduous forests, grasslands, savannas, deserts, chaparral, and tropical rainforests.

### What is **limnology**?

Limnology is the study of freshwater ecosystems—especially lakes, ponds, and streams. These ecosystems are more fragile than marine environments since they are

A saguaro cactus forest thrives within this Sonoran Desert biome near Tucson, Arizona.

subject to great extremes in temperature. The chemistry, physics, and biology of these bodies of water are explored. F. A. Forel (1848–1931), a Swiss professor, has been called the father of limnology.

## How does the process work in a **food chain**?

A food chain is the transfer of food energy from the source in plants through a series of organisms with repeated eating and being eaten. The number of steps or "links" in a sequence is usually four to five. The first trophic level (group of organisms that get their energy the same way) is plants; the animals that eat plants (called herbivores) form the second trophic level. The third level consists of primary carnivores (animal-eating animals like wolves) who eat herbivores, and the fourth level are animals (like killer whales) that eat primary carnivores. Food chains overlap because many organisms eat more than one type of food, so that these chains can look more like food webs. In 1891 German zoologist Karl Semper introduced the food chain concept.

## What is a **food web**?

A food web consists of interconnecting food chains. Many animals feed on different foods rather than exclusively on one single species of prey or one type of plant. Animals that use a variety of food sources have a greater chance of survival than those with a single food source. Complex food webs provide greater stability to a living community.

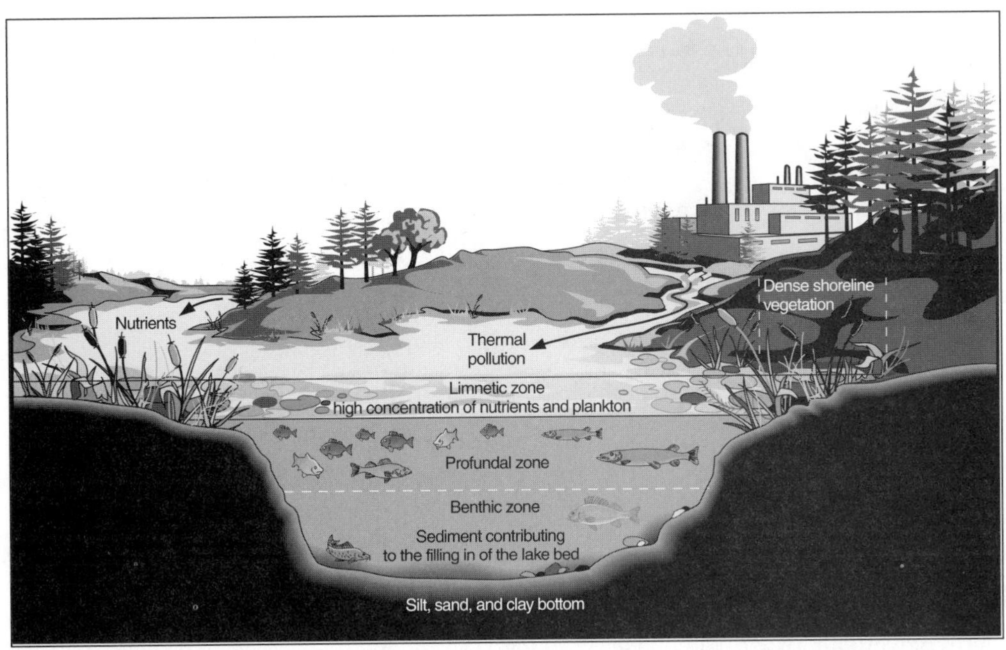


**The structure of a eutrophic lake.**

## What is **killer algae**?

*Caulerpa taxifolia*, "killer algae," was introduced to the Mediterranean Sea in the mid-1980s when the Oceanographic Museum in Monaco dumped the bright green seaweed into the sea while cleaning the aquarium tanks. *Caulerpa taxifolia* now covers 32,000 acres of the coasts of France, Spain, Italy, and Croatia, devastating the Mediterranean ecosystem. The species continues to invade the Mediterranean today and appears to be unstoppable.

## What is **eutrophication**?

Eutrophication is a process in which the supply of plant nutrients in a lake or pond is increased. In time, the result of natural eutrophication may be dry land where water once flowed, caused by plant overgrowth.

Natural fertilizers, washed from the soil, result in an accelerated growth of plants, producing overcrowding. As the plants die off, the dead and decaying vegetation depletes the lake's oxygen supply, causing fish to die. The accumulated dead plant and animal material eventually changes a deep lake to a shallow one, then to a swamp, and finally it becomes dry land.

While the process of eutrophication is a natural one, it has been accelerated enormously by human activities. Fertilizers from farms, sewage, industrial wastes, and some detergents all contribute to the problem.

## How large is the Antarctic ozone hole?

The term "hole" is widely used in popular media when reporting on ozone. However, the concept is more correctly described as a low concentration of ozone. In September 2000, scientists at the National Aeronautics and Space Administration announced that the "hole" over Antarctica, first discovered in 1985, now covered 11 million square miles, making it the largest hole ever recorded.

## How many acres of **wetlands have been lost** in the United States?

Wetlands are the lands between aquatic and terrestrial areas, such as bogs, marshes, swamps, and coastal waters. At one time considered wastelands, scientists now recognize the importance of wetlands to improve water quality, stabilize water levels, prevent flooding, regulate erosion, and sustain a variety of organisms. The United States has lost approximately 100 million acres of wetland areas between colonial times and the 1970s. The 1993 Wetlands Plan established a goal of reversing the trend of 100,000 acres of wetland loss to 100,000 acres of wetland recovery.

## How does **ozone** benefit life on Earth?

Ozone, a form of oxygen with three atoms instead of the normal two, is highly toxic; less than one part per million of this blue-tinged gas is poisonous to humans. In the Earth's upper atmosphere (stratosphere), it is a major factor in making life on Earth possible. About 90 percent of the planet's ozone is in the ozone layer. The ozone belt shields and filters the Earth from excessive ultraviolet (UV) radiation generated by the sun. Scientists predict that a diminished or depleted ozone layer could lead to increased health problems for humans, such as skin cancer, cataracts, and weakened immune systems. Increased UV can also lead to reduced crop yields and disruption of aquatic ecosystems, including the marine food chain. While beneficial in the stratosphere, near ground level it is a pollutant that helps form photochemical smog and acid rain.

## What is the **greenhouse effect**?

The greenhouse effect is a warming near the Earth's surface that results when the Earth's atmosphere traps the sun's heat. The atmosphere acts much like the glass walls and roof of a greenhouse. The effect was described by John Tyndall (1820–1893) in 1861. It was given the greenhouse analogy much later in 1896 by the Swedish chemist Svante Arrhenius (1859–1927). The greenhouse effect is what makes the Earth habitable. Without the presence of water vapor, carbon dioxide, and other gases in the atmosphere, too much heat would escape and the Earth would be too cold to sustain life. Carbon dioxide, methane, nitrous oxide, and other greenhouse gases

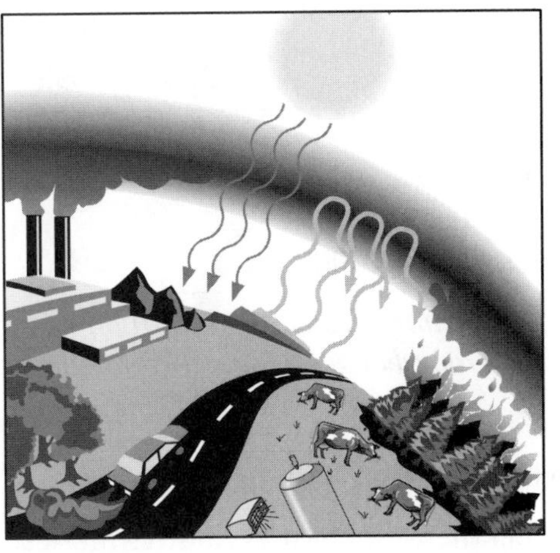

An atmosphere with natural levels of greenhouse gases (left) compared with an atmosphere of increased greenhouse effect (right).

absorb the infrared radiation rising from the Earth and hold this heat in the atmosphere instead of reflecting it back into space.

In the 20th century, the increased build-up of carbon dioxide, caused by the burning of fossil fuels, has been a matter of concern. There is some controversy concerning whether the increase noted in the Earth's average temperature is due to the increased amount of carbon dioxide and other gases, or is due to other causes. Volcanic activity, destruction of the rainforests, use of aerosols, and increased agricultural activity may also be contributing factors.

## What compounds cause **ozone depletion** in the **stratosphere**?

Research in the 1970s linked chlorofluorocarbons (CFCs), such as freon, to the depletion of the ozone layer. In 1978, the use of CFC propellants in spray cans was banned in the United States. In 1987, the Montreal Protocol was signed and the signatory nations committed themselves to a reduction in the use of CFCs and other ozone-depleting substances.

## What are the **greenhouse gases**?

Scientists recognize carbon dioxide ($CO_2$), methane ($CH_4$), chlorofluorocarbons (CFCs), nitrous oxide ($N_20$), and water vapor as significant greenhouse gases. Greenhouse gases account for less that one percent of the Earth's atmosphere. These gases trap heat in the Earth's atmosphere, preventing the heat from escaping back into space. Human activities, such as using gasoline for automobiles for fuel, account for the release of carbon dioxide and nitrogen oxides.

**Emissions of Greenhouse Gases in the United States, 1990–2000 (Million Metric Tons of Gas)**

| Gas | 1990 | 1995 | 1996 | 1997 | 1998 | 1999 | P2000 |
|---|---|---|---|---|---|---|---|
| Carbon Dioxide | 4,969.4 | 5,273.5 | 5,454.8 | 5,533.0 | 5,540.0 | 5,630.7 | 5,805.5 |
| Methane | 31.7 | 31.1 | 29.9 | 29.6 | 28.9 | 28.7 | 28.2 |
| Nitrous Oxide | 1.2 | 1.3 | 1.2 | 1.2 | 1.2 | 1.2 | 1.2 |
| HFCs, PFCs, and $SF_6$ | * | * | * | * | * | * | * |

*P=Preliminary data, *Less than 0.05 million metric tons of gas, HFCs=hydrofluorocarbons, PFCs=perfluorocarbons,* $SF_6$*=sulfur hexafluoride*

## Why is **El Niño** harmful?

Along the west coast of South America, near the end of each calendar year, a warm current of nutrient-poor tropical water moves southward, replacing the cold, nutrient-rich surface water. Because this condition frequently occurs around Christmas, local residents call it El Niño (Spanish for child), referring to the Christ child. In most years the warming lasts for only a few weeks. However, when El Niño conditions last for many months, the economic results can be catastrophic. It is this extended episode of extremely warm water that scientists now refer to as El Niño. During a severe El Niño, large numbers of fish and marine plants may die. Decomposition of the dead material depletes the water's oxygen supply, which leads to the bacterial production of huge amounts of smelly hydrogen sulfide. A greatly reduced fish (especially anchovy) harvest affects the world's fishmeal supply, leading to higher prices for poultry and other animals that normally are fed fishmeal. Anchovies and sardines are also major food sources for marine mammals, such as sea lions and seals. When the food source is in short supply, these animals travel further from their homes in search of food. Not only do many sea lions and seals starve, but also a large proportion of the infant animals die. The major El Niño event of 1997–1998 indirectly caused 2,100 deaths and $33 billion in damage globally.

## What **percentage** of the Earth's surface is **tropical rain forest**?

Rain forests account for approximately 7 percent of the Earth's surface, or about three million square miles (7.7 million square kilometers).

## What is the rate of **species extinction** in the **tropical rain forests**?

Biologists estimate that tropical rain forests contain approximately one-half of the Earth's animal and plant species. These forests contain 155,000 of the 250,000 known plant species and innumerable insect and animal species. Nearly 100 species become extinct each day. This is equivalent to four species per hour. At the current rates, 5–10 percent of the tropical rain forest will become extinct every decade.

### What is red tide and what causes it?

Red tide is a term used for a brownish or reddish discoloration occurring in ocean, river, or lake water. It is caused by the rapid reproduction of a variety of toxic organisms, especially the toxic red dinoflagellates that are members of the genera *Gymnodidium* and *Gonyaulax*. Some red tides are harmless, but millions of fish may be killed during a "bloom," as the build-up is called. Other red tides can poison shellfish and the birds or humans who eat the contaminated food. Scientists do not fully understand why the "bloom" occurs.

## What is the **largest rain forest**?

The Amazon Basin is the world's largest continuous tropical rain forest. It covers about 2.7 million square miles (6.9 million square kilometers).

## How **rapidly** is **deforestation** occurring?

Agriculture, excessive logging, and fires are the major causes of deforestation.

**Change in Forest Area, 1990–2000**

| Area | Total forest, 1990 ('000 ha*) | Total forest, 2000 ('000 ha*) | Forest cover annual change ('000 ha*) | Forest cover annual rate of change (%) |
|---|---|---|---|---|
| Africa | 702,502 | 649,866 | -5,262 | -0.8 |
| Asia | 551,448 | 547,793 | -364 | -0.1 |
| Europe | 1,030,475 | 1,039,251 | 881 | 0.1 |
| North and Central America | 555,002 | 549,304 | -570 | -0.1 |
| Oceania | 201,271 | 197,623 | -365 | -0.2 |
| South America | 922,731 | 885,618 | -3,711 | -0.4 |
| TOTAL WORLD | 3,963,429 | 3,869,455 | -9,391 | -0.2 |

* ha = hectare (1 hectare = 2.47 acres)

## How rapidly is the **tropical rain forest** being **destroyed**?

Tropical rain forests once covered more than four billion acres (1.6 billion hectares) of the Earth. Today, nearly half of the tropical rain forests are gone. Approximately 33.8 million acres of tropical forest are lost per year. This translates to 2.8 million acres lost per month, which can be broken down to 93,000 acres/day, 3,800 acres/hour, or 64 acres/minute. It equals more than the total area of New Hampshire, Vermont, Massachusetts, Rhode Island, Connecticut, New Jersey, and Delaware combined. At the current rate, scientists predict that no rain forests will remain by the middle of the twenty-first century.

## What is the **importance** of the **rain forest**?

Half of all medicines prescribed worldwide are originally derived from wild products, and the United States National Cancer Institute has identified more than two thousand tropical rainforest plants with the potential to fight cancer. Rubber, timber, gums, resins and waxes, pesticides, lubricants, nuts and fruits, flavorings and dyestuffs, steroids, latexes, essential and edible oils, and bamboo are among the forest's products that would be drastically affected by the depletion of the tropical forests.

## What causes the most **forest fires** in the western United States?

Lightning is the single largest cause of forest fires in the western states.

## Which countries have the largest amount of **protected areas**?

Protected areas include national parks, nature reserves, national monuments, and other sites. There are at least 44,3000 protected areas around the world, covering more than 10 percent of the total land area.

| Country | Protected Area (miles$^2$) | Protected Area (km$^2$) | Total % |
|---|---|---|---|
| Venezuela | 217,397 | 563,056 | 61.7 |
| Greenland | 379,345 | 1,025,405 | 45.2 |
| Saudi Arabia | 318,811 | 825,717 | 34.4 |
| United States | 902,091 | 2,336,406 | 24.9 |
| Indonesia | 138,002 | 357,425 | 18.6 |
| Australia | 395,911 | 1,025,405 | 13.4 |
| Canada | 357,231 | 925,226 | 9.3 |
| China | 263,480 | 682,410 | 7.1 |
| Brazil | 215,312 | 557,656 | 6.6 |
| Russia | 204,273 | 529,067 | 3.1 |

## When was the symbol of **Smokey the Bear** first used to encourage forest fire prevention?

The origin of Smokey the Bear can be traced to World War II when the U.S. Forest Service, concerned about maintaining a steady lumber supply for the war effort, wished to educate the public about the dangers of forest fires. They sought volunteer advertising

support from the War Advertising Council, and on August 9, 1944, Albert Staehle, a noted illustrator of animals, created Smokey the Bear. Since 1944, Smokey the Bear has been a national symbol of forest fire prevention not only in America, but also in Canada and Mexico, where he is known as Simon. This public service advertising (PSA) campaign is the longest running PSA campaign in U.S. history. In 1947, a Los Angeles advertising agency coined the slogan "Only you can prevent forest fires." On April 23, 2001, after more than 50 years, the famous ad slogan was revised to "Only you can prevent wildfires" in response to the wildfire outbreaks during 2000. The campaign gained a living mascot in 1950 when a firefighting crew rescued a male bear cub from a forest fire in the Capital Mountains of New Mexico. Sent to the National Zoo in Washington, D.C., to become Smokey the Bear, the animal was a living symbol of forest fire protection until his death in 1976. His remains are buried at the Smokey the Bear State Historical Park in Capitan, New Mexico.

Smokey the Bear has been a recognizable icon for forest fire prevention since 1944.

## What was the United States' **first national park**?

On March 1,1872, an Act of Congress signed by Ulysses S. Grant established Yellowstone National Park as the first national park, inspiring a worldwide national park movement.

## When was the **U.S. National Park Service** established?

The National Park Service was created by an Act signed by President Woodrow Wilson on August 25, 1916.

## What are some of the **largest National Parks**?

The largest National Parks are in Alaska:

| Park | Area (acres) |
|---|---|
| Wrangell-St. Elias | 8,323,148 |
| Gates of the Arctic | 7,523,898 |
| Denali | 4,740,912 |
| Katmai | 3,674,530 |
| Glacier Bay | 3,224,840 |
| Lake Clark | 2,619,733 |
| Kobuk Valley | 1,750,737 |

The five largest parks in the 48 contiguous states are:

| Park | Location | Area (acres) |
|---|---|---|
| Death Valley | California | 3,291,779 |
| Yellowstone | Idaho, Montana, Wyoming | 2,219,791 |
| Everglades | Florida | 1,398,903 |
| Grand Canyon | Arizona | 1,217,403 |
| Glacier | Montana | 1,013,572 |

## Where is **Hawk Mountain Sanctuary**?

Hawk Mountain Sanctuary, founded in 1934 as the first sanctuary in the world to offer protection to migrating hawks and eagles, is near Harrisburg, Pennsylvania, on the Kittatinny Ridge. Each year between the months of August and December, over 15,000 migrating birds pass by. Rare species such as golden eagles may be seen there.

## Who is considered the **founder** of **modern conservation**?

American naturalist John Muir (1838–1914) was the father of conservation and the founder of the Sierra Club. He fought for the preservation of the Sierra Nevada Mountains in California, and the creation of Yosemite National Park. He directed most of the Sierra Club's conservation efforts and was a lobbyist for the Antiquities Act.

Another prominent influence was George Perkins Marsh (1801–1882), a Vermont lawyer and scholar. His outstanding book *Man and Nature* emphasizes the mistakes of past civilizations that resulted in destruction of natural resources. As the conservation movement swept through the country in the last three decades of the 19th century, a number of prominent citizens joined the efforts to conserve natural resources and to preserve wilderness areas. Writer John Burroughs, forester Gifford Pinchot, botanist Charles Sprague Sargent, and editor Robert Underwood Johnson were early advocates of conservation.

John Muir (left), founder of the Sierra Club and modern conservation movement, pictured here with fellow advocate John Burroughs.

## Who coined the term **"Spaceship Earth"**?

American inventor and environmentalist Buckminster Fuller (1895–1983) coined the term "Spaceship Earth" as an analogy of the need for technology to be self-contained and to avoid waste.

## When was the **Environmental Protection Agency** created?

In 1970, President Richard M. Nixon created the Environmental Protection Agency (EPA) as an independent agency of the U.S. government by executive order. The creation of a federal agency by executive order rather than by an act of the legislative branch is somewhat of an exception to the rule. The EPA was established in response to public concern about unhealthy air, polluted rivers and groundwater, unsafe drinking water, endangered species, and hazardous waste disposal. Responsibilities of the EPA include environmental research, monitoring, and enforcement of legislation regulating environmental activities.

## What is a **"green product"**?

Green products are environmentally safe products that contain no chlorofluorocarbons, are degradable (can decompose), and are made from recycled materials. "Deep-green" products are those from small suppliers who build their identities around their

### Who started Earth Day?

The first Earth Day, April 22, 1970, was coordinated by Denis Hayes at the request of Gaylord Nelson, a U.S. senator from Wisconsin. Nelson is sometimes called the father of Earth Day. His main objective was to organize a nationwide public demonstration so large it would get the attention of politicians and force the environmental issue into the political dialogue of the nation. Important official actions that began soon after the celebration of the first Earth Day were: the establishment of the Environmental Protection Agency (EPA); the creation of the President's Council on Environmental Quality; and the passage of the Clean Air Act, establishing national air quality standards. In 1995 Gaylord Nelson received the Presidential Medal of Freedom for his contributions to the environmental protection movement.

claimed environmental virtues. "Greened-up" products come from the industry giants and are environmentally improved versions of established brands.

# EXTINCT AND ENDANGERED PLANTS AND ANIMALS

## When was the term **"dinosaur"** first used?

The term *dinosaur* was first used by Richard Owen in 1841 in his report on British fossil reptiles. The term, meaning "fearful lizard" was used to describe the group of large extinct reptiles whose fossil remains had been found by many collectors.

## What is the name of the **early Jurassic mammal** that is now extinct?

The fossil site of the mammal *hadrocodium wui* was in Yunnan Province, China. This newly described mammal is at least 195 million years old. The estimated weight of the whole mammal is about 0.07 ounces (2 grams). Its tiny skull was smaller than a human thumbnail.

## Did **dinosaurs and humans** ever coexist?

No. Dinosaurs first appeared in the Triassic Period (about 220 million years ago) and disappeared at the end of the Cretaceous Period (about 65 million years ago). Modern

humans (*Homo sapiens*) appeared only about 25,000 years ago. Movies that show humans and dinosaurs existing together are only Hollywood fantasies.

## What were the **smallest and largest dinosaurs**?

*Compsognathus*, a carnivore from the late Jurassic period (131 million years ago), was about the size of a chicken and measured, at most, 35 inches (89 centimeters) from the tip of its snout to the tip of its tail. The average weight was about 6 pounds 8 ounces (3 kg), but they could be as much as 15 pounds (6.8 kg).

The largest species for which a whole skeleton is known is *Brachiosaurus*. A specimen in the Humboldt Museum in Berlin measures 72.75 feet (22.2 meters) long and 46 feet (14 meters) high. It weighed an estimated 34.7 tons (31,480 kilograms). *Brachiosaurus* was a four-footed, plant-eating dinosaur with a long neck and a long tail and lived from about 155 to 121 million years ago.

## How **long** did **dinosaurs live**?

The life span has been estimated at 75 to 300 years. Such estimates are educated guesses. From examination of the microstructure of dinosaur bones, scientists have inferred that they matured slowly and probably had proportionately long life spans.

## How does a **mastodon** differ from a **mammoth**?

Although the words are sometimes used interchangeably, the mammoth and the mastodon were two different animals. The mastodon seems to have appeared first, while a side branch may have led to the mammoth.

The *mastodon* lived in Africa, Europe, Asia, and North and South America. It appeared in the Oligocene epoch (25 to 38 million years ago) and survived until less than one million years ago. It stood a maximum of 10 feet (three meters) tall and was covered with dense woolly hair. Its tusks were aligned straight forward and were nearly parallel to each other.

The *mammoth* evolved less than two million years ago and died out about 10 thousand years ago. It lived in North America, Europe, and Asia. Like the mastodon, the mammoth was covered with dense, woolly hair, with a long, coarse layer of outer hair to protect it from the cold. It was somewhat larger than the mastodon, standing 9 to 15 feet (2.7 to 4.5 meters). The mammoth's tusks tended to spiral outward, then up.

The gradual warming of the Earth's climate and the change in environment were probably primary factors in the animals' extinction. Early man killed many of them as well, perhaps hastening the process.

## Why did **dinosaurs** become **extinct**?

There are many theories as to why dinosaurs disappeared from the Earth about 65 million years ago. Scientists debate whether dinosaurs became extinct gradually or all

### When did the last passenger pigeon die?

At one time, 200 years ago, the passenger pigeon (*Ectopistes migratorius*) was the world's most abundant bird. Although the species was found only in eastern North America, it had a population of three to five billion birds (25 percent of the North American land bird population). Overhunting caused a chain of events that reduced their numbers below minimum threshold for viability. In the 1890s several states passed laws to protect the pigeon, but it was too late. The last known wild bird was shot in 1900. The last passenger pigeon, named Martha, died on September 1, 1914, in the Cincinnati Zoo.

at once. The gradualists believe that the dinosaur population steadily declined at the end of Cretaceous Period. Numerous reasons have been proposed for this. Some claim the dinosaurs' extinction was caused by biological changes that made them less competitive with other organisms, especially the mammals that were just beginning to appear. Overpopulation has been argued, as has the theory that mammals ate too many dinosaur eggs for the animals to reproduce themselves. Others believe that disease—everything from rickets to constipation—wiped them out. Changes in climate, continental drift, volcanic eruptions, and shifts in the Earth's axis, orbit, and/or magnetic field have also been held responsible.

The catastrophists argue that a single disastrous event caused the extinction not only of the dinosaurs but also of a large number of other species that coexisted with them. In 1980, American physicist Luis Alvarez (1911–1988) and his geologist son, Walter Alvarez (b. 1940), proposed that a large comet or meteoroid struck the Earth 65 million years ago. They pointed out that there is a high concentration of the element iridium in the sediments at the boundary between the Cretaceous and Tertiary Periods. Iridium is rare on Earth, so the only source of such a large amount of it had to be outer space. This iridium anomaly has since been discovered at over 50 sites around the world. In 1990, tiny glass fragments, which could have been caused by the extreme heat of an impact, were identified in Haiti. A 110-mile (177-kilometer) wide crater in the Yucatan Peninsula, long covered by sediments, has been dated to 64.98 million years ago, making it a leading candidate for the site of this impact.

A hit by a large extraterrestrial object, perhaps as much as six miles (9.3 kilometers) wide, would have had a catastrophic effect upon the world's climate. Huge amounts of dust and debris would have been thrown into the atmosphere, reducing the amount of sunlight reaching the surface. Heat from the blast may also have caused large forest fires, which would have added smoke and ash to the air. Lack of sunlight would kill off plants and have a domino-like effect on other organisms in the food chain, including the dinosaurs.

It is possible that the reason for the dinosaurs' extinction may have been a combination of both theories. The dinosaurs may have been gradually declining, for whatever reason. The impact of a large object from space merely delivered the coup de grâce.

The fact that dinosaurs became extinct has been cited as proof of their inferiority and that they were evolutionary failures. However, these animals flourished for 150 million years. By comparison, the earliest ancestors of humanity appeared only about three million years ago. Humans have a long way to go before they can claim the same sort of success as the dinosaurs.

In a span of just 200 years the passenger pigeon passed from the world's most abundant bird species into extinction.

## How did the **dodo** become extinct?

The dodo became extinct around 1800. Thousands were slaughtered for meat, but pigs and monkeys, which destroyed dodo eggs, were probably most responsible for the dodo's extinction. Dodos were native to the Mascarene Islands in the Central Indian Ocean. They became extinct on Mauritius soon after 1680 and on Réunion about 1750. They remained on Rodriguez until 1800.

## What is the difference between an **"endangered" species** and a **"threatened" species**?

An "endangered" species is one that is in danger of extinction throughout all or a significant portion of its range. A "threatened" species is one that is likely to become endangered in the foreseeable future.

## Under what **conditions** is a species considered **"endangered"**?

This determination is a complex process that has no set of fixed criteria that can be applied consistently to all species. The known number of living members in a species is not the sole factor. A species with a million members known to be alive but living in only one small area could be considered endangered, whereas another species having a smaller number of members, but spread out in a broad area, would not be considered so threatened. Reproduction data—the frequency of reproduction, the average number of offspring born, the survival rate, etc.—enter into such determinations. In the United States, the director of the U.S. Fish and Wildlife Service (within the Department of the Interior) determines which species are to be considered endangered, based on research and field data from specialists, biologists, botanists, and naturalists.

According to the Endangered Species Act of 1973, a species can be listed if it is threatened by any of the following:

1. The present or threatened destruction, modification, or curtailment of its habitat or range.
2. Utilization for commercial, sporting, scientific, or educational purposes at levels that detrimentally affect it.
3. Disease or predation.
4. Absence of regulatory mechanisms adequate to prevent the decline of a species or degradation of its habitat.
5. Other natural or man-made factors affecting its continued existence.

If the species is so threatened, the director then determines the "critical habitat," that is the species' inhabitation areas that contain the essential physical or biological features necessary for the species' preservation. The critical habitat can include non-habitation areas, which are deemed necessary for the protection of the species.

## Which species have **become extinct** since the **Endangered Species Act was passed** in 1973?

Seven domestic species have become extinct.

| First Listed | Date Delisted (declared extinct) | Species Name |
|---|---|---|
| 3/11/1967 | 9/2/1983 | Cisco, longjaw (*Coregonus alpenae*) |
| 4 /30/1980 | 12/4/1987 | Gambusia, Amistad (*Gambusia amistadensis*) |
| 6/14/1976 | 1/9/1984 | Pearlymussel, Sampson's (*Epioblasma sampsoni*) |
| 3/11/1967 | 9/2/1983 | Pike, blue (*Stizostedion vitreum glaucum*) |
| 10/13/1970 | 1/15/1982 | Pupfish, Tecopa (*Cyprinodon nevadensis calidae*) |
| 3/11/1967 | 12/12/1990 | Sparrow, dusky seaside (*Ammodramus maritimus nigrescens*) |
| 6/4/1973 | 10/12/1983 | Sparrow, Santa Barbara song (*Melospiza melodia graminea*) |

## Which species have been **removed** from the Endangered Species List because they **have recovered**?

Thirteen species have been removed from the Endangered Species List because they have recovered.

| First Listed | Date Delisted | Species Name |
|---|---|---|
| 3/11/1967 | 6/04/1987 | Alligator, American (*Alligator mississippiensis*) |
| 6/2/1970 | 9/12/1985 | Dove, Palau ground (*Gallicolumba canifrons*) |
| 6/2/1970 | 8/25/1999 | Falcon, American peregrine (*Falco peregrinus anatum*) |
| 6/2/1970 | 10/5/1994 | Falcon, Arctic peregrine (*Falco peregrinus tundrius*) |
| 6/2/1970 | 9/12/1985 | Flycatcher, Palau fantail (*Rhipidura lepida*) |
| 3/11/1967 | 3/20/2001 | Goose, Aleutian Canada (*Branta canadensis leucopareia*) |
| 12/30/1974 | 3/9/1995 | Kangaroo, eastern gray (*Macropus giganteus*) |
| 12/30/1974 | 3/9/1995 | Kangaroo, red (*Macropus rufus*) |
| 12/30/1974 | 3/9/1995 | Kangaroo, western gray (*Macropus fuliginosus*) |
| 4/26/1978 | 9/14/1989 | Milk-vetch, Rydberg (*Astragalus perianus*) |
| 6/2/1970 | 9/12/1985 | Owl, Palau (*Pyroglaux podargina*) |
| 6/2/1970 | 2/4/1985 | Pelican, brown (U.S. Atlantic coast, FL, AL) (*Pelecanus occidentalis*) |
| 6/2/1970 | 06/16/1994 | Whale, gray (except where listed) (*Eschrichtius robustus*) |

The status of five species has been changed due to taxonomic revision. New information has been discovered for four other species.

## How many species of **plants and animals** are **threatened or endangered** in the world?

### Summary of Listed Species
### Species and Recovery Plans as of 4/30/2002

| Group | Endangered | | Threatened | | Total Species | U.S. Species with Recovery Plans |
|---|---|---|---|---|---|---|
| | U.S. | Foreign | U.S. | Foreign | | |
| Mammals | 65 | 251 | 9 | 17 | 342 | 53 |
| Birds | 78 | 175 | 14 | 6 | 273 | 75 |
| Reptiles | 14 | 64 | 22 | 15 | 115 | 32 |
| Amphibians | 11 | 8 | 8 | 1 | 28 | 12 |
| Fishes | 71 | 11 | 44 | 0 | 126 | 95 |
| Clams | 62 | 2 | 8 | 0 | 72 | 56 |
| Snails | 21 | 1 | 11 | 0 | 33 | 27 |
| Insects | 35 | 4 | 9 | 0 | 48 | 29 |
| Arachnids | 12 | 0 | 0 | 0 | 12 | 5 |

| | | | | | | |
|---|---|---|---|---|---|---|
| Crustaceans | 18 | 0 | 3 | 0 | 21 | 12 |
| **Animal Subtotal** | **387** | **516** | **128** | **39** | **1070** | **396** |
| Flowering Plants | 568 | 1 | 144 | 0 | 713 | 556 |
| Conifers and Cycads | 2 | 0 | 1 | 2 | 5 | 2 |
| Ferns and Allies | 24 | 0 | 2 | 0 | 26 | 26 |
| Lichens | 2 | 0 | 0 | 0 | 2 | 2 |
| **Plant Subtotal** | **596** | **1** | **147** | **2** | **746** | **586** |
| **GRAND TOTAL** | **983** | **517** | **275** | **41** | **1816** | **982** |

*Total U.S. Endangered—983 (387 animals, 596 plants)*
*Total U.S. Threatened—275 (128 animals, 147 plants)*
*Total U.S. Species—1258 (515 animals, 743 plants)*

## What is the current population and status of the **great whales**?

| Species | Latin Name | Original Population (thousands) | Current Population | Status |
|---|---|---|---|---|
| Sperm | *Physeter macro cephalas* | 2400 | Unknown; 1-2 million maximum | Vulnerable |
| Blue | *Balenoptera musculus* | 226 | Under 5,000 | Endangered |
| Finback | *Balenoptera physalus* | 543 | 50,000–90,00 | Vulnerable |
| Humpback | *Megaptera novaeangliae* | 146 | Around 28,000 | Vulnerable |
| Northern Atlantic right | *Eubalaena glacialis* | 120 | Western North Atlantic 300–350; Eastern North Atlantic near extinction | Endangered |
| Southern right | *Eubalaena australis* | | About 7,000 | Conservation Dependent |
| Sei | *Balaenoptera borealis* | 254 | About 50,000 | Endangered |
| Gray | *Eschrichtius robustus* | 20 | About 27,000 | Conservation Dependent; Critically endangered in Western North Pacific |

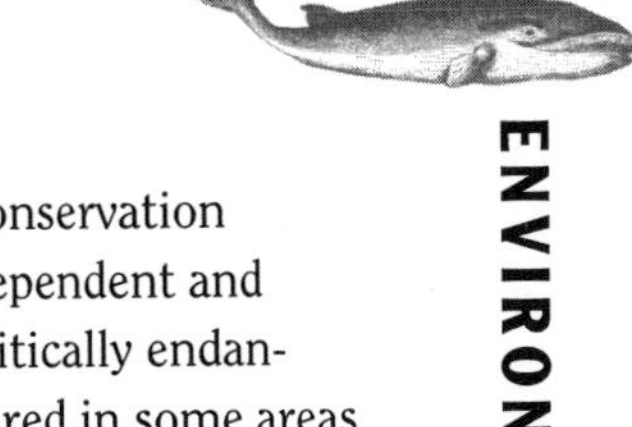



| | | | | |
|---|---|---|---|---|
| Bowhead | *Balaena mysticetus* | 20 | Under 8,500 | Conservation Dependent and critically endangered in some areas |
| Bryde's | *Balaenoptera edeni* | 92 | About 40,000–80,000 | Insufficient data |
| Minke | *Balaenoptera acutorostrata* | 295 | 610,000–1,284,000 | Low risk–near threatened |

## What is the status of the **elephant** in Africa?

From 1979 to 1989, Africa lost half of its elephants from poaching and illegal ivory trade, with the population decreasing from an estimated 1.3 million to 600,000. This led to the transfer of the African elephant from threatened to endangered status in October 1989 by CITES (the Convention on International Trade in Endangered Species). An ivory ban took effect on January 18, 1990. Botswana, Namibia, and Zimbabwe have agreed to restrict the sale of ivory to a single, government-controlled center in each country. All countries have further pledged to allow independent monitoring of the sale, packing, and shipping processes to ensure compliance with all conditions. Finally, all three countries have promised that all net revenues from the sale of ivory will be directed back into elephant conservation for use in monitoring, research, law enforcement, other management expenses or community-based conservation programs within elephant range.

## Are **turtles** endangered?

Worldwide turtle populations have declined due to several reasons, including habitat destruction; exploitation of species by humans for their eggs, leather, and meat; and their becoming accidentally caught in the nets of fishermen. In particular danger are sea turtles, such as Kemp's ridley sea turtle (*Lepidochelys kempii*), which is believed to have a population of only a few hundred. Other species include the Central American river turtle (*Dermatemys mawii*), the green sea turtle (*Chelonia mydas*), and the leatherback sea turtle (*Dermochelys coriacea*). Endangered tortoises include the angulated tortoise (*Geochelone yniphora*), the desert tortoise (*Gopherus agassizii*), and the Galapagos tortoise (*Geochelone elephantopus*).

## What is a **dolphin-safe tuna**?

The order *Cetacea*, composed of whales, dolphins, and porpoises, were spared from the extinction of large mammals at the end of the Pleistocene about 10,000 years ago. But from 1000 B.C.E. on they, especially the whale, have been relentlessly hunted by man for their valuable products. The twentieth century, with its many technological improvements, has become the most destructive period for the *Cetacea*. In 1972, the United States Congress passed the Marine Mammal Protection Act; one of its goals was to reduce the number of small cetaceans (notably Stenalla and Delphinus) killed and injured during commercial fishing operations, such as the incidental catch of dolphins in tuna purse-seines (nets that close up to form a huge ball to be hoisted aboard ship). Dolphins are often found swimming with schools of yellowfin tuna and are caught along with the tuna by fisherman who use purse-seine nets. The dolphins are drowned by this fishing method because they must be able to breathe air to survive. The number of incidental deaths and injuries in 1972 was estimated at 368,000 for United States fishing vessels and 55,078 for non-United States vessels. In 1979 the figures were reduced to 17,938 and 6,837 respectively. But in the 1980s, the dolphins killed by foreign vessels rose dramatically to over 100,000 a year. Most of the slaughter occurs in the eastern Pacific Ocean from Chile to Southern California. In 1999, the United States attempted to weaken the definition of dolphin-safe tuna by allowing tuna fleets to chase and net dolphins. The ruling was not upheld by the United States District Court.

To further reduce the numbers of dolphins killed during tuna catches, the three largest sellers of canned tuna in the United States, spearheaded by the Starkist company, decided that they would not sell tuna that has been caught by these methods harmful to dolphins.

# POLLUTION

*See also: Health and Medicine—Health Hazards, Risks, Etc.*

## How are **hazardous waste** materials classified?

There are four types of hazardous waste materials—corrosive, ignitable, reactive, and toxic.

*Corrosive* materials can wear away or destroy a substance. Most acids are corrosive and can destroy metal, burn skin, and give off vapors that burn the eyes.

*Ignitable* materials can burst into flames easily. These materials pose a fire hazard and can irritate the skin, eyes, and lungs. Gasoline, paint, and furniture polish are ignitable.

*Reactive* materials can explode or create poisonous gas when combined with other chemicals. Combining chlorine bleach and ammonia, for example, creates a poisonous gas.

*Toxic* materials or substances can poison humans and other life. They can cause illness or death if swallowed or absorbed through the skin. Pesticides and household cleaning agents are toxic.

## What is **bioremediation**?

Bioremediation is the degradation, decomposition, or stabilization of pollutants by microorganisms such as bacteria, fungi, and cyanobacteria. Oxygen and organisms are injected into contaminated soil and/or water (e.g. oil spills). The microorganisms feed on and eliminate the pollutants. When the pollutants are gone, the organisms die.

## What is the **Pollutant Standard Index**?

The U.S. Environmental Protection Agency and the South Coast Air Quality Management District of El Monte, California, devised the Pollutant Standard Index to monitor concentrations of pollutants in the air and inform the public concerning related health effects. The scale measures the amount of pollution in parts per million, and has been in use nationwide since 1978.

| PSI Index | Health Effects | Cautionary Status |
|---|---|---|
| 0 | Good | |
| 50 | Moderate | |
| 100 | Unhealthful | |
| 200 | Very unhealthful | Alert: elderly or ill should stay indoors and reduce physical activity. |
| 300 | Hazardous | Warning: General population should stay indoors and reduce physical activity. |
| 400 | Extremely hazardous | Emergency: all people remain indoors windows shut, no physical exertion. |
| 500 | Toxic | Significant harm; same as above. |

## What is the **Toxic Release Inventory** (TRI)?

TRI is a government mandated, publicly available compilation of information on the release of over 650 individual toxic chemicals and toxic chemical categories by manufacturing facilities in the United States. The law requires manufacturers to state the amounts of chemicals they release directly to air, land, or water, or that they transfer to off-site facilities that treat or dispose of wastes. The U.S. Environmental Protection Agency compiles these reports into an annual inventory and makes the information available in a computerized database. In 2000, 23,484 facilities released 7.1 billion pounds (3.2 billion kilograms) of toxic chemicals into the environment. Over 260 million pounds (118 million kilograms) of this total were released into surface water; 1.9

billion pounds (0.86 million kilograms) were emitted into the air; over 4.13 billion pounds (1.87 billion kilograms) were released to land; and over 278 million pounds (126 million kilograms) were injected into underground wells. The total amount of toxic chemicals released in 2000 was 6.7 percent lower than the amount released in 1999.

## Which countries emit the most **carbon dioxide** into the air?

Carbon dioxide is one of the greenhouse gases. In 2000 the ten countries that emitted the most carbon dioxide from the consumption and flaring of fossil fuels were:

| Country | $CO_2$ emissions (in millions of metric tons) |
|---|---|
| United States | 1,571.14 |
| China | 775.01 |
| Russia | 450.70 |
| Japan | 313.69 |
| India | 253.28 |
| Germany | 219.72 |
| Canada | 157.95 |
| United Kingdom | 147.77 |
| Italy | 116.66 |
| S. Korea | 115.33 |
| World | 6,443.38 |

## What **chemicals,** which were once hailed as wonder chemicals that would benefit society, are now **banned or strictly controlled** ?

Dichlorodiphenyl-trichloro-ethene (DDT), polychlorinated biphenyls (PCBs) and chlorofluorocarbons (CFCs) were once widely used. Increased awareness of the detrimental effects on the environment from these chemicals led to all of them being banned or strictly controlled.

## How did **DDT** affect the environment?

Although DDT was synthesized as early as 1874 by Othmar Zeidler, it was the Swiss chemist Paul Müller (1899–1965) who recognized its insecticidal properties in 1939. He was awarded the 1948 Nobel Prize in medicine for his development of dichlorodiphenyl-trichloro-ethene, or DDT. Unlike the arsenic-based compounds then in use, DDT was effective in killing insects and seemed not to harm plants and animals. In the following 20 years it proved to be effective in controlling disease-carrying insects (mosquitoes that carry malaria and yellow fever, and lice that carry typhus) and in killing many plant crop destroyers. Publication of Rachel Carson's *Silent Spring* in

1962 alerted scientists to the detrimental effects of DDT. Increasingly, DDT-resistant insect species and the accumulative hazardous effects of DDT on plant and animal life cycles led to its disuse in many countries during the 1970s.

## Which **industries** release the most **toxic chemicals**?

The metal mining industry released the most toxic chemicals for the year 2000, accounting for 47.3 percent of total chemical releases.

| Industry | Total Releases (pounds) | Percent of Total |
|---|---|---|
| Metal mining | 3,357,765,313 | 47.3 |
| Manufacturing industries | 2,284,399,698 | 32.2 |
| Electric utilities | 1,152,242,786 | 16.2 |
| Hazardous waste/Solvent recovery | 284,950,589 | 4.0 |
| Coal mining | 15,968,001 | 0.2 |
| Petroleum terminals/bulk storage | 3,878,087 | 0.1 |
| Chemical wholesale distributors | 1,611,790 | 0.02 |

## What are **PCBs**?

Polychlorinated biphenyls (PCBs) are a group of chemicals that were widely used before 1970 in the electrical industry as coolants for transformers and in capacitors and other electrical devices. They caused environmental problems because they do not break down and can spread through the water, soil, and air. They have been linked by some scientists to cancer and reproductive disorders and have been shown to cause liver function abnormalities. Government action has resulted in the control of the use, disposal, and production of PCBs in nearly all areas of the world, including the United States.

## How do **chlorofluorocarbons** affect the Earth's ozone layer?

Chlorofluorocarbons (CFCs) are hydrocarbons, such as freon, in which part or all of the hydrogen atoms have been replaced by fluorine atoms. These can be liquids or gases, are non-flammable and heat-stable, and are used as refrigerants, aerosol propellants, and solvents. When released into the air, they slowly rise into the Earth's upper atmosphere, where they are broken apart by ultraviolet rays from the sun. Some of the resultant molecular fragments react with the ozone in the atmosphere, reducing the amount of ozone. The CFC molecules' chlorine atoms act as catalysts in a complex set of reactions that convert two molecules of ozone into three molecules of ordinary hydrogen. This is depleting the beneficial ozone layer faster than it can be recharged by natural processes. The resultant "hole" lets through more ultraviolet light to the Earth's surface and creates health problems for humans, such as cataracts and skin cancer, and disturbs delicate ecosystems (for example, making plants produce fewer seeds). In 1978 the United States government banned the use of fluorocarbon aerosols, and currently aerosol propellants have been changed from fluorocarbons to hydrocarbons, such as butane. The Montreal Protocol of 1987 initiated worldwide cooperation to reduce the use of CFCs.

Since the late 1980s a substantial slowdown has taken place in the atmospheric buildup of two prime ozone-destroying compounds, CFC 11 and CFC 12. Based upon these measurements, experts suggest that concentrations of these CFCs will peak before the turn of the century, allowing the ozone layer to begin the slow process of repairing itself. It is believed that it will take 50 to 100 years for reactions in the atmosphere to reduce the concentrations of ozone-destroying chlorine and bromine back to natural levels. Until then, these chemicals will continue to erode the global ozone layer.

## What are the components of **smog**?

Smog, the most widespread pollutant in the United States, is a photochemical reaction resulting in ground-level ozone. Ozone, an odorless, tasteless gas in the presence of light can initiate a chain of chemical reactions. Ozone is a desirable gas in the stratospheric layer of the atmosphere, but it can be hazardous to health when found near the Earth's surface in the troposphere. The hydrocarbons, hydrocarbon derivations, and nitric oxides emitted from such sources as automobiles are the raw materials for photochemical reactions. In the presence of oxygen and sunlight, the nitric oxides combine with organic compounds, such as the hydrocarbons from unburned gasoline, to produce a whitish haze, sometimes tinged with a yellow-brown color. In this process, a large number of new hydrocarbons and oxyhydrocarbons are produced. These secondary hydrocarbon products may comprise as much as 95 percent of the total organics in a severe smog episode.

## What is an example of how we can **alleviate air pollution**?

The scrubbing of flue gases refers to the removal of sulfur dioxide ($SO_2$) and nitric oxide (NO), which are major components of air pollution. Wet scrubbers use a chemical solvent of lime, limestone, sodium alkali, or diluted sulfuric acid to remove the $SO_2$ formed during combustion. Dry scrubbing uses either a lime/limestone slurry or ammonia sprayed into the flue gases.

## What was the distribution of **radioactive fallout** after the 1986 **Chernobyl accident**?

Radioactive fallout, containing the isotope cesium 137, and nuclear contamination covered an enormous area, including Byelorussia, Latvia, Lithuania, the central por-

tion of the then Soviet Union, the Scandinavian countries, the Ukraine, Poland, Austria, Czechoslovakia, Germany, Switzerland, northern Italy, eastern France, Romania, Bulgaria, Greece, Yugoslavia, the Netherlands, and the United Kingdom. The fallout, extremely uneven because of the shifting wind patterns, extended 1,200 to 1,300 miles (1,930 to 2,090 kilometers) from the point of the accident. Roughly 5 percent of the reactor fuel—or seven tons of fuel containing 50 to 100 million curies—was released. Estimates of the effects of this fallout range from 28,000 to 100,000 deaths from cancer and genetic defects within the next 50 years. In particular, livestock in high rainfall areas received unacceptable dosages of radiation.

## What is **acid rain**?

The term "acid rain" was coined by British chemist Robert Angus Smith (1817–1884) who, in 1872, published *Air & Rain: The Beginnings of a Chemical Climatology*. Since then, acid rain has unfortunately become an increasingly used term for rain, snow, sleet, or other precipitation that has been polluted by acids such as sulfuric and nitric acids.

When gasoline, coal, or oil are burned, their waste products of sulfur dioxide and nitrogen dioxide combine in complex chemical reactions with water vapor in clouds to form acids. The United States alone discharges 40 million metric tons of sulfur and nitrogen oxides into the atmosphere. This, combined with natural emissions of sulfur and nitrogen compounds, has resulted in severe ecological damage. Hundreds of lakes in North America (especially northeastern Canada and United States) and in Scandinavia are so acidic that they cannot support fish life. Crops, forests, and building materials, such as marble, limestone, sandstone, and bronze, have been affected as well, but the extent is not as well documented. However, in Europe, where so many living trees are stunted or killed, the new word "Waldsterben" (forest death) has been coined to describe this new phenomenon.

In 1990, amendments to the U.S. Clean Air Act contained provisions to control emissions that cause acid rain. It included the reductions of sulfur dioxide emissions from 19 million tons to 9.1 million tons annually and the reduction of industrial nitrogen oxide emissions from six to four million tons annually.

| Year | Sulfur Dioxide Emissions (million tons) | Nitrogen Oxide Emissions (million tons) |
|---|---|---|
| 1990 | 15.73 | 6.66 |
| 1995 | 11.87 | 6.09 |
| 1996 | 12.51 | 5.91 |
| 1997 | 12.96 | 6.04 |
| 1998 | 13.13 | 5.97 |
| 1999 | 12.45 | 5.49 |
| 2000 | 11.28 | 5.11 |

### How harmful are balloon releases?

Both latex and metallic balloons can be harmful. A latex balloon can land in water, lose its color, and resemble a jellyfish, which if eaten by sea animals can cause their death because they cannot digest it. A metal balloon can get caught in electric wires and cause power outages.

## How **acidic** is acid rain?

Acidity or alkalinity is measured by a scale known as the pH (potential for Hydrogen) scale. It runs from zero to 14. Since it is logarithmic, a change in one unit equals a tenfold increase or decrease. So a solution at pH 2 is 10 times more acidic than one at pH 3 and 100 times as acidic as a solution at pH 4. Zero is extremely acid, 7 is neutral, and 14 is very alkaline. Any rain below 5.0 is considered acid rain; some scientists use the value of 5.6 or less. Normal rain and snow containing dissolved carbon dioxide (a weak acid) measure about pH 5.6. Actual values vary according to geographical area. Eastern Europe and parts of Scandinavia have 4.3 to 4.5; the rest of Europe is 4.5 to 5.1; eastern United States and Canada ranges from 4.2 to 4.6, and Mississippi Valley has a range of 4.6 to 4.8. The worst North American area, having 4.2, is centered around Lake Erie and Lake Ontario. For comparison, some common items and their pH values are listed below:

| | |
|---|---|
| Concentrated sulfuric acid | 1.0 |
| Lemon juice | 2.3 |
| Vinegar | 3.3 |
| Acid rain | 4.3 |
| Normal rain | 5.0 to 5.6 |
| Normal lakes and rivers | 5.6 to 8.0 |
| Distilled water | 7.0 |
| Human blood | 7.35 to 7.45 |
| Seawater | 7.6 to 8.4 |

## Where did the first **major oil spill** occur?

The first major commercial oil spill occurred on March 18, 1967, when the tanker *Torrey Canyon* grounded on the Seven Stones Shoal off the coast of Cornwall, England, spilling 830,000 barrels (119,000 tons) of Kuwaiti oil into the sea. This was the first major tanker accident. However, during World War II, German U-boat attacks on tankers, between January and June of 1942, off the United States East Coast, spilled 590,000 tons of oil. Although the Exxon *Valdez* was widely publicized as a major spill of 35,000 tons in 1989, it is dwarfed by the deliberate dumping of oil from Sea Island into the Persian Gulf on January 25, 1991. It is estimated that the spill equaled almost

1.5 million tons of oil. A major spill also occurred in Russia in October 1994 in the Komi region of the Arctic. The size of the spill was reported to be as much as two million barrels (286,000 tons).

In addition to the large disasters, day-to-day pollution occurs from drilling platforms where waste generated from platform life, including human waste, and oils, chemicals, mud, and rock from drilling are discharged into the water.

| Date | Cause | Thousands of tons spilled |
|---|---|---|
| 1/42–6/42 | German U-boat attacks on tankers off the East Coast of U.S. during World War II | 590 |
| 3/18/67 | Tanker *Torrey Canyon* grounds off Land's End in the English Channel | 119 |
| 3/20/70 | Tanker *Othello* collides with another ship in Tralhavet Bay, Sweden | 60–100 |
| 12/19/72 | Tanker *Sea Star* collides with another ship in Gulf of Oman | 115 |
| 5/12/76 | *Urquiola* grounds at La Coruna, Spain | 100 |
| 3/16/78 | Tanker *Amoco Cadiz* grounds off Northwest France | 223 |
| 6/3/79 | Itox I oil well blows in Southern Gulf of Mexico | 600 |
| 7/79 | Tankers *Atlantic Express* and *Aegean Captain* collide off Trinidad and Tobago | 300 |
| 2/19/83 | Blowout in Norwuz oil field in the Persian Gulf | 600 |
| 8/6/83 | Fire aboard *Castillo de Beliver* off Cape Town, South Africa | 250 |
| 1/25/91 | Iraq begins deliberately dumping oil into Persian Gulf from Sea Island, Kuwait | 1,450 |
| 1994 | A structure to prevent pipeline leaks fails, spilling oil in the Komi Republic in northern Russia | almost 102,000 |
| 1999 | Tanker *New Carissa* spills some of its oil in Coos Bay, Oregon | 70,000 gallons (238 tons) |

## What are the **sources of oil pollution** in the oceans?

Most oil pollution results from accidental discharges of oil when oil tankers are loaded and unloaded and minor spillage from tankers. Other sources of oil pollution include improper disposal of used motor oil, oil leaks from motor vehicles, routine ship maintenance, leaks in pipelines that transport oil, and accidents at storage facilities and refineries.

| Source | Percent of total oil spillage |
|---|---|
| Runoff from rivers and surface water | 31% |
| Tanker activity (loading, unloading, etc.) | 20% |
| Sewage plants and refineries | 13% |

| | |
|---|---|
| Natural seepage from ocean floor | 9% |
| Smaller craft (fishing vessels, ferries, etc.) | 9% |
| Oil tanker accidents | 3–5% |

## What are the most commonly collected **debris** items found along **ocean coasts**?

| Debris Item | Total number reported (2000) | Percentage of total collected (2000) |
|---|---|---|
| Cigarette butts | 1,027,303 | 20.25% |
| Plastic pieces | 337,384 | 6.65% |
| Food bags/wrappers (plastic) | 284,287 | 5.6% |
| Foamed plastic pieces | 268,945 | 5.3% |
| Caps, lids (plastic) | 255,253 | 5.03% |
| Paper pieces | 219,256 | 4.32% |
| Beverage cans | 184,294 | 3.63% |
| Beverage bottles (glass) | 177,039 | 3.49% |
| Straws | 161,639 | 3.19% |

## What are **Operation Ranch Hand** and **Agent Orange**?

Operation Ranch Hand was the tactical military project for the aerial spraying of herbicides in South Vietnam during the Vietnam Conflict (1961–1975). In these operations Agent Orange, the collective name for the herbicides 2,4-D and 2,4,5-T, was used for defoliation. The name derives from the color-coded drums in which the herbicides were stored. In all, U.S. troops sprayed approximately 19 million gallons (72 million liters) of herbicides over four million acres (1.6 million hectare).

Concerns about the health effects of Agent Orange were initially voiced in 1970, and since then the issue has been complicated by scientific and political debate. In 1993, a 16-member panel of experts reviewed the existing scientific evidence and found strong evidence of a statistical association between herbicides and soft-tissue sarcoma, non-Hodgkin's lymphoma, Hodgkin's disease, and chloracne. On the other hand, they concluded that no connection appeared to exist between exposure to Agent Orange and skin cancer, bladder cancer, brain tumors, or stomach cancer.

## What causes **formaldehyde** contamination in homes?

Formaldehyde contamination is related to the widespread construction use of wood products bonded with urea-formaldehyde resins and products containing formaldehyde. Major formaldehyde sources include subflooring of particle board; wall paneling made from hardwood plywood or particle board; and cabinets and furniture made from particle board, medium density fiberboard, hardwood plywood, or solid wood. Urea-formaldehyde foam insulation (UFFI) has received the most media notoriety and

regulatory attention. Formaldehyde is also used in drapes, upholstery, carpeting, wallpaper adhesives, milk cartons, car bodies, household disinfectants, permanent-press clothing, and paper towels. In particular, mobile homes seem to have higher formaldehyde levels than houses do. Six billion pounds (2.7 billion kilograms) of formaldehyde are used in the United States each year.

The release of formaldehyde into the air by these products (called outgassing) can develop poisoning symptoms in humans. The EPA classifies formaldehyde as a potential human carcinogen (cancer-causing agent).

## Which pollutants lead to **indoor air pollution**?

Indoor air pollution, also known as "tight building syndrome," results from conditions in modern, high energy efficiency buildings, which have reduced outside air exchange, or have inadequate ventilation, chemical contamination, and microbial contamination. Indoor air pollution can produce various symptoms, such as headache, nausea, and eye, nose, and throat irritation. In addition, houses are affected by indoor air pollution emanating from consumer and building products and from tobacco smoke. Below are listed some pollutants found in houses:

| Pollutant | Sources | Effects |
|---|---|---|
| Asbestos | Old or damaged insulation, fireproofing, or acoustical tiles | Many years later, chest and abdominal cancers and lung diseases |
| Biological pollutants | Bacteria, mold and mildew, viruses, animal dander and cat saliva, mites, cockroaches, and pollen | Eye, nose, and throat irritation; shortness of breath; dizziness; lethargy; fever; digestive problems; asthma; influenza and other infectious diseases |
| Carbon monoxide | Unvented kerosene and gas heaters; leaking chimneys and furnaces; wood stoves and fireplaces; gas stoves; automobile exhaust from attached garages; tobacco smoke | At low levels, fatigue; at higher levels, impaired vision and coordination; headaches; dizziness; confusion; nausea. Fatal at very high concentrations |
| Formaldehyde | Plywood, wall paneling, particle board, fiber-board; foam insulation; fire and tobacco smoke; textiles, and glues | Eye, nose, and throat irritations; wheezing and coughing; fatigue; skin rash; severe allergic reactions; may cause cancer |

| Pollutant | Sources | Effects |
|---|---|---|
| Lead | Automobile exhaust; sanding or burning of lead paint; soldering | Impaired mental and physical development in children; decreased coordination and mental abilities; kidneys, nervous system, and red blood cells damage |
| Mercury | Some latex paints | Vapors can cause kidney damage; long-term exposure can cause brain damage |
| Nitrogen dioxide | Kerosene heaters, unvented gas stoves and heaters; tobacco smoke | Eye, nose, and throat irritation; may impair lung function and increase respiratory infections in young children |
| Organic Gases | Paints, paint strippers, solvents, wood preservatives; aerosol sprays; cleansers and disinfectants; moth repellents; air fresheners; stored fuels; hobby supplies; dry-cleaned clothing. | Eye, nose and throat irritation; headaches; loss of coordination; nausea; damage to liver, kidney, and nervous system; some organics cause cancer in animals and are suspected of causing cancer in humans. |
| Pesticides | Products used to kill household pests and products used on lawns or gardens that drift or are tracked inside the house. | Irritation to eye, nose, and throat; damage to nervous system and kidneys; cancer. |
| Radon | Earth and rock beneath the home; well water, building materials. | No immediate symptoms; estimated to cause about 10% of lung cancer deaths; smokers at higher risk. |

# RECYCLING, CONSERVATION, AND WASTE

*See also: Energy—Consumption and Conservation*

## What is the **Resource Conservation and Recovery Act**?

In 1976, the United States Congress passed the Resource Conservation and Recovery Act (RCRA), which was amended in 1984 and 1986. This law requires the EPA to identify hazardous wastes and set standards for their management including generation, transportation, treatment, storage, and disposal of hazardous waste. The law requires all firms that store, treat, or dispose of more than 220 pounds (100 kilograms) of hazardous wastes per month to have a permit stating how such wastes are managed.

## What is the **Toxic Substances Control Act** (TOSCA)?

In 1976, the United States Congress passed the Toxic Substances Control Act (TOSCA). This act requires the premarket testing of toxic substances. When a chemical substance is planned to be manufactured, the producer must notify the Environmental Protection Agency (EPA), and, if the data presented is determined to be inadequate to approve its use, the EPA will require the manufacturer to conduct further tests. Or, if it is later determined that a chemical is present at a level that presents an unreasonable public or environmental risk, or if there is insufficient data to know the chemical's effects, manufacturers have the burden of evaluating the chemical's characteristics and risks. If testing does not convince the EPA of the chemical's safety, the chemical's manufacturing, sale, or use can be limited or prohibited.

## What is the **Superfund Act**?

In 1980, the United States Congress passed the Comprehensive Environmental Response, Compensation, and Liability Act, commonly known as the Superfund program. This law (along with amendments in 1986 and 1990) established a $16.3-billion Superfund financed jointly by federal and state governments and by special taxes on chemical and petrochemical industries (which provide 86 percent of the funding). The purpose of the Superfund is to identify and clean up abandoned hazardous-waste dump sites and leaking underground tanks that threaten human health and the environment. To keep taxpayers from footing most of the bill, cleanups are based on the polluter-pays principle. The EPA is charged with locating dangerous dump sites, finding the potentially liable culprits, ordering them to pay for the entire cleanup and suing them if they don't. When the EPA can find no responsible party, it draws money out of the Superfund for cleanup.

## How many **hazardous waste sites** are there?

As of 2000, there were 1,279 sites.

## Which **states** have the most **hazardous waste sites** (Superfund) sites?

There are hazardous waste sites in all of the 50 states except North Dakota. The states with the most hazardous waste sites are:

| State | Number of hazardous waste sites |
|---|---|
| New Jersey | 113 |
| California | 99 |
| Pennsylvania | 97 |
| New York | 88 |
| Michigan | 69 |
| Florida | 53 |
| Washington | 48 |
| Illinois | 44 |
| Wisconsin | 41 |
| Texas | 38 |

## What is the **NIMBY syndrome**?

NIMBY is the acronym for "Not In My Back Yard," referring to major community resistance to new incinerator sitings, landfills, prisons, roads, etc. NIMFY is "Not In My Front Yard."

## What are **Brownfields**?

The Environmental Protection Agency defines Brownfields as abandoned, idled, or underused industrial or commercial sites where expansion or redevelopment is complicated by real or perceived environmental contamination. Real estate developers perceive Brownfields as inappropriate sites for redevelopment. There are approximately 450,000 Brownfields in the United States, with the heaviest concentrations being in the Northeast and Midwest.

## Which **government agencies** regulate the storage of **nuclear waste**?

The Department of Energy, Environmental Protection Agency, and Nuclear Regulatory Commission are responsible for the disposal of spent nuclear fuel and other radioactive waste. The Department of Energy (DOE) has the responsibility for developing permanent disposal capacity for spent fuel and other high-level radioactive waste. The Environmental Protection Agency (EPA) has responsibility for developing envi-

ronmental standards to evaluate the safety of a geologic repository. The Nuclear Regulatory Commission (NRC) has responsibility for developing regulations to implement the EPA safety standards and for licensing the repository.

## How is **nuclear waste** stored?

Nuclear wastes consist either of fission products formed from the splitting of uranium, cesium, strontium, or krypton, or from transuranic elements formed when uranium atoms absorb free neutrons. Wastes from transuranic elements are less radioactive than fission products; however, these elements remain radioactive far longer—hundreds of thousands of years. The types of waste are irradiated fuel (spent fuel) in the form of 12-foot (4-meter) long rods, high-level radioactive waste in the form of liquid or sludge, and low-level waste (non-transuranic or legally high-level) in the form of reactor hardware, piping, toxic resins, water from fuel pool and other items that have become contaminated with radioactivity.

Currently, most spent nuclear fuel in the United States is safely stored in specially designed pools at individual reactor sites around the country. If pool capacity is reached, licensees may move toward use of above-ground dry storage casks. The three low-level radioactive waste disposal sites are Barnwell located in South Carolina, Hanford located in Washington, and Envirocare located in Utah. Each site accepts low-level radioactive waste from specific regions of the country.

Most high-level nuclear waste has been stored in double-walled, stainless-steel tanks surrounded by three feet (one meter) of concrete. The current best storage method, developed by the French in 1978, is to incorporate the waste into a special molten glass mixture, then enclose it in a steel container and bury it in a special pit. The Nuclear Waste Policy Act of 1982 specified that high-level radioactive waste would be disposed of underground in a deep geologic repository. Yucca Mountain, Nevada, was chosen as the single site to be developed for disposal of high-level radioactive waste. However, as of the late 1990s, selection of the Yucca Mountain site was controversial due to dormant volcanoes in the vicinity and known earthquake fault lines.

## What examples can be given that suggest the average United States citizen lives in a **"throw-away society"**?

Consumers in the United States dispose of astounding amounts of solid waste, including:

Enough aluminum to rebuild the country's entire commercial airline fleet every three months

About 18 billion disposable diapers per year, which if linked end-to-end would reach to the moon and back seven times

About two billion disposable razors, 10 million computers, and eight million television sets each year

### When was metal recycling started?

The first metal recycling in the United States occurred in 1776 when patriots in New York City toppled a statue of King George III, which was melted down and made into 42,088 bullets.

About 2.5 million nonreturnable plastic bottles each hour.

Some 14 billion catalogs (an average of 54 per American) and 38 billion pieces of junk mail each year

## When was the first **garbage incinerator** built?

The first garbage incinerator in the United States was built in 1885 on Governor's Island in New York Harbor.

## How much **garbage** does the **average American** generate?

According to the Environmental Protection Agency, nearly 230 million tons of municipal waste was generated in 1999. This is equivalent to 4.6 pounds (2.1 kilograms) per person per day, or approximately 1,700 pounds (770 kilograms) per year. The total amount of waste is distributed as follows:

| Waste Product | Percent of Total |
|---|---|
| Paper and paperboard | 38.1 |
| Glass | 5.5 |
| Metals | 7.8 |
| Plastics | 10.5 |
| Rubber and leather | 2.7 |
| Textiles | 3.9 |
| Wood | 5.3 |
| Food wastes | 10.9 |
| Yard wastes | 12.1 |
| Other wastes | 3.2 |

## Why did the ship ***Mobro*** gain international fame?

The *Mobro* was the Long Island, New York, garbage barge that was loaded with garbage and could not dispose of it. It was rejected by six states and three countries, drawing attention to the landfill capacity shortage in the Northeast. The garbage was finally burned in Brooklyn and the ash brought to a landfill near Islip.

The United States is currently home to more than 9,000 landfill sites.

## How critical is the problem of **landfilling** in the United States?

Landfilling has been an essential component of waste management for several decades. In 1960, 62 percent of all garbage was sent to landfills, and by 1980 the figure had risen to 81 percent. By 1990, 84 percent of the 269 million tons of municipal solid waste that was generated was sent to landfills. An increased awareness of the benefits of recycling has brought a decline in the actual number of landfills from 4,482 in 1995 to 2,142 in 2000 as well as a decrease in the amount of municipal solid waste that is sent to landfills. Figures for 2000 indicate that only 60 percent of the municipal solid waste generated was sent to landfills. The total amount of recycled waste increased from 8 percent to 33 percent between 1990 and 2000.

## How much **methane fuel** does a ton of garbage make?

Over a period of 10 to 15 years, a ton of garbage will make 14,126 cubic feet (400 cubic meters) of fuel, although a landfill site will generate smaller amounts for 50 to 100 years. One ton of garbage can produce more than 100 times its volume in methane over a decade. Landfill operators tend not to maximize methane production.

## How much space does a **recycled ton of paper** save in a landfill?

Each ton (907 kilograms) saves more than three cubic yards of landfill space.

## When was paper recycling started?

Paper recycling was actually born in 1690 in the United States when the first paper mill was established by the Rittenhouse family on the banks of Wissahickon Creek, near Philadelphia. The paper at this mill was made from recycled rags.

## What **natural resources** are saved by **recycling paper**?

One ton (907 kilograms) of recycled waste paper would save an average of 7,000 gallons (26,460 liters) of water, 3.3 cubic yards (2.5 cubic meters) of landfill space, three barrels of oil, 17 trees, and 4,000 kilowatt-hours of electricity—enough energy to power the average home for six months. It would also reduce air pollution by 74 percent.

## How much **newspaper** must be recycled to **save one tree**?

One 35 to 40 foot (10.6 to 12 meter) tree produces a stack of newspapers four feet (1.2 meters) thick; this much newspaper must be recycled to save a tree.

## How much **waste paper** does a **newspaper** generate?

An average yearly newspaper subscription (for example, the *San Francisco Chronicle*) received every day produces 550 pounds (250 kilograms) of waste paper per subscription per year. The average *New York Times* Sunday edition produces eight million pounds (3.6 million kilograms) of waste paper.

## How many **paper mills** in the United States use waste paper to produce new products?

Nearly 80 percent of the paper mills in the United States use some recovered fiber in manufacture of new products. An additional 200 mills used only recovered fiber for manufacture.

## How much **paper** is **recycled** in the United States?

In 1999, 47 million tons, or 45 percent, of all paper used was recovered for recycling. Of the paper that is recovered, 81 percent is recycled into new products by American paper mills, 16 percent is exported to the recycling market, and the rest is used in the United States for other products such as insulation, molded packaging hydromulch, compost, and kitty litter.

### What state was first to create mandatory deposits on beverage containers?

In 1971, the state of Oregon was the first to create legislation mandating deposits on beverage containers. The deposit was five cents per container.

## What problems may be encountered when **polyvinyl chloride (PVC) plastics** are burned?

Chlorinated plastics, such as PVC, contribute to the formation of hydrochloric acid gases. They also may be a part of a mix of substances containing chlorine that form a precursor to dioxin in the burning process. Polystyrene, polyethylene, and polyethylene terephthalate (PET) do not produce these pollutants.

## How can **plastics** be made **biodegradable**?

Plastic does not rust or rot. This is an advantage in its usage, but when it comes to disposal of plastic the advantage turns into a liability. Degradable plastic has starch in it so that it can be attacked by starch-eating bacteria to eventually disintegrate the plastic into bits. Chemically degradable plastic can be broken up with a chemical solution that dissolves it. Used in surgery, biodegradable plastic stitches slowly dissolve in the body fluids. Photodegradable plastic contains chemicals that disintegrate over a period of one to three years when exposed to light. One-quarter of the plastic yokes used to package beverages are made from a plastic called Ecolyte, which is photodegradable.

## How many **curbside recycling programs** are there in the United States?

The first municipal-wide curbside recycling program was started in 1974 in University City, Missouri, where city officials designed and distributed a container for collecting newspapers. Curbside recycling programs continue to increase in the United States. In 1997, there were 7,375 curbside recycling programs. By 1999 the number of curbside recycling programs had increased to 9,349.

## How much has **recycling** of municipal solid waste grown since 1960?

In 1960 the recycling rate was 6.4 percent. It grew to 28 percent in 1999. The United States composted 5.6 percent of the municipal waste, incinerated 15 percent, and landfilled the remaining 57 percent.

## What do the **numbers** inside the **recycling symbol** on plastic containers mean?

The Society of the Plastics Industry developed a voluntary coding system for plastic containers to assist recyclers in sorting plastic containers. The symbol is designed to be imprinted on the bottom of the plastic containers. The numerical code appears inside three arrows that form a triangle. A guide to what the numbers mean is listed below. The most commonly recycled plastics are polyethylene terephthalate (PET) and high-density polyethylene (HDPE).

| Code | Material | Examples |
|---|---|---|
| 1 | Polyethylene terephthalate (PET/PETE) | 2-liter soft drink bottle |
| 2 | High-density polyethylene (HDPE) | Milk and water jugs |
| 3 | Vinyl (PVC) | Plastic pipes, shampoo bottles |
| 4 | Low-density polyethylene (LDPE) | Produce bags, food storage containers |
| 5 | Polypropylene (PP) | Squeeze bottles, drinking straws |
| 6 | Polystyrene (PS) | Fast-food packaging, packaging |
| 7 | Other | Food containers |

## What **products** are made from **recycled plastic**?

| Resin | Common Uses | Products Made from Recycled Resin |
|---|---|---|
| HDPE | Beverage bottles, milk jugs, milk and soft drink crates, pipe, cable, film | Motor oil bottles, detergent bottles, pipes and pails |
| LDPE | Film bags such as trash bags, coatings, and plastic bottles | New trash bags, pallets, carpets, fiberfill, non-food |
| PET | Soft drink, detergent, and juice bottles | Bottles/containers |
| PP | Auto battery cases, screw-on caps and lids; some yogurt and margarine tubs, plastic film | Auto parts, batteries, carpets |
| PS | Housewares, electronics, fast food carry-out packaging, plastic utensils | Insulation board, office equipment, reusable cafeteria trays |
| PVC | Sporting goods, luggage, pipes, auto parts. In packaging for shampoo bottles, blister packaging, and films | Drainage pipes, fencing, house siding |

A new clothing fiber called Fortrel EcoSpun is made from recycled plastic soda bottles. The fiber is knit or woven into garments such as fleece for outerwear or long underwear. The processor estimates that every pound of Fortrel EcoSpun fiber results in 10 plastic bottles being kept out of landfills.

## When offered a choice between **plastic or paper bags** for your groceries, wich should you choose?

The answer is neither. Both are environmentally harmful and the question of which is the more damaging has no clear-cut answer. On one hand, plastic bags degrade slowly in landfills and can harm wildlife if swallowed, and producing them pollutes the environment. On the other hand, producing the brown paper bags used in most supermarket uses trees and pollutes the air and water. Overall, white or clear polyethylene bags require less energy for manufacture and cause less damage to the environment than do paper bags not made from recycled paper. Instead of having to choose between paper and plastic bags, you can bring your own reusable canvas or string containers to the store, and save and reuse any paper or plastic bags you get.

## Is **washing dishes by hand** better for the environment than using an **automatic dishwasher**?

Dishwashers often save energy and water compared to hand washing. Depending on the brand, dishwashers typically consume 7.5 to 12 gallons (28 to 45 liters) of water per normal wash. Hand-washing a day's worth of dishes may use up to 15 gallons (57 liters) of water. One university study found that dishwashers consume about 37 percent less water than washing by hand.

Several steps can be taken for additional energy savings when using a dishwasher. The setting on a home's water heater can be turned down to 120°F (49°C) if the dishwasher has a booster heater. While some machines feature a no-heat, air-dry setting, simply opening the door after the final rinse to let the dishes air dry will save energy. Prewashing the dishes before loading generally wastes water since most machines can handle even heavily soiled plates.

## How many **automobile tires** are scrapped each year and what can be done with them?

In 2001, approximately 281 million tires were discarded in the United States. Nearly 75 percent of the discarded tires were sent to one of the three major markets for scrap tires—tire-derived fuel (TDF), civil engineering and ground rubber applications. An additional 20–25 million tires were sent to landfills. From 1979–1992 the only market for scrap tires was fuel. At the end of 2001, 83 facilities used TDF on a regular basis. TDF is used in a variety of combustion technologies, including cement kilns, pulp and paper mill boilers, utility and industrial boilers, and dedicated scrap-tire-to-energy facilities. Discarded tires are used for construction materials in lieu of sand, stone, and clean fill. They can also be recycled into new products such as floor mats, blasting mats, and muffler hangers. Tires that have been ground into crumb can be used in a variety of molded or die cut products such as traffic cone bases, mud flaps, and moisture barriers. Whole tires can be used in artificial reefs, for erosion control, and to stabilize mine tailing ponds.

## What is a **WOBO**?

A WOBO (world bottle) is the first mass-produced container designed for secondary use as a building product. It was conceived by Albert Heineken of the Heineken beer family. The beer bottles were designed in a special shape to be used, when empty, as glass bricks for building houses. The actual building carried out with WOBOs was only a small shed and a double garage built on the Heineken estate at Noordwijk, near Amsterdam. Although not implemented, WOBO was a sophisticated and intelligent design solution to what has emerged as a major environmental issue in recent years.

# BIOLOGY

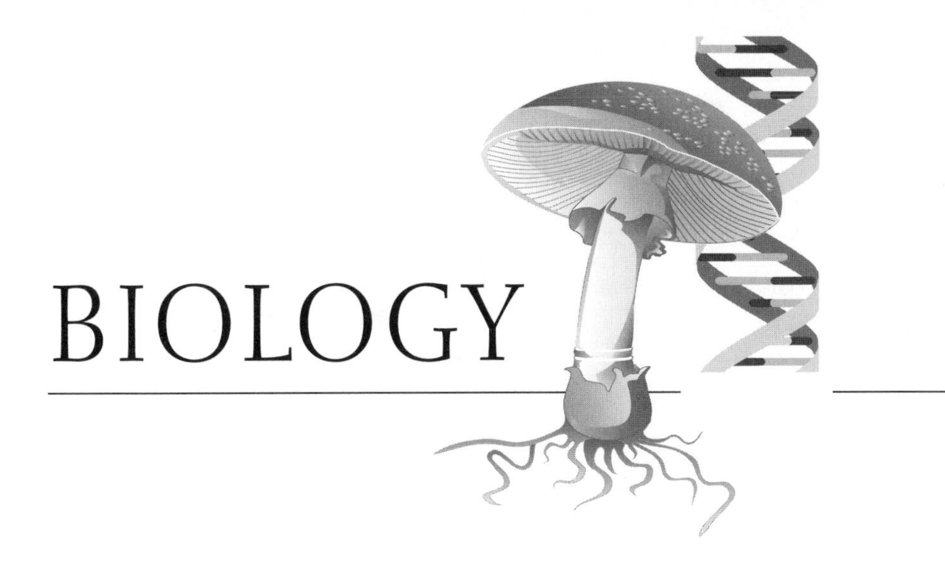

## CELLS

### What is the **cell theory**?

The cell theory is the concept that all living things are made up of essential units called "cells." The cell is the simplest collection of matter that can live. There are diverse forms of life existing as single-celled organisms. More complex organisms, including plants and animals, are multicellular—cooperatives of many kinds of specialized cells that could not survive for long on their own. All cells are related by division from earlier cells and have been modified in various ways during the long evolutionary history of life on Earth. Everything an organism does occurs fundamentally at the cellular level.

### What **scientists** made **important discoveries** associated with the cell?

In the late 1600s, Robert Hooke (1635–1703) was the first to see a cell, initially in a section of cork, and then in bones and plants. In 1824, Henri Dutrochet (1776–1847) proposed that animals and plants had similar cell structures. Robert Brown (1773–1858) discovered the cell nucleus in 1831, and Matthias Schleiden (1804–1881) named the nucleolus (the structure within the nucleus now known to be involved in the production of ribosomes) around that time. Schleiden and Theodor Schwann (1810–1882) described a general cell theory in 1839, the former stating that cells were the basic unit of plants and Schwann extending the idea to animals. Robert Remak (1815–1865) was the first to describe cell division in 1855. Chromosomes were named and observed in the nucleus of a cell in 1888 by Wilhelm von Waldeyen-Hartz (1836–1921). Walther Flemming (1843–1905) was the first individual to follow chromosomes through the entire process of cell division.

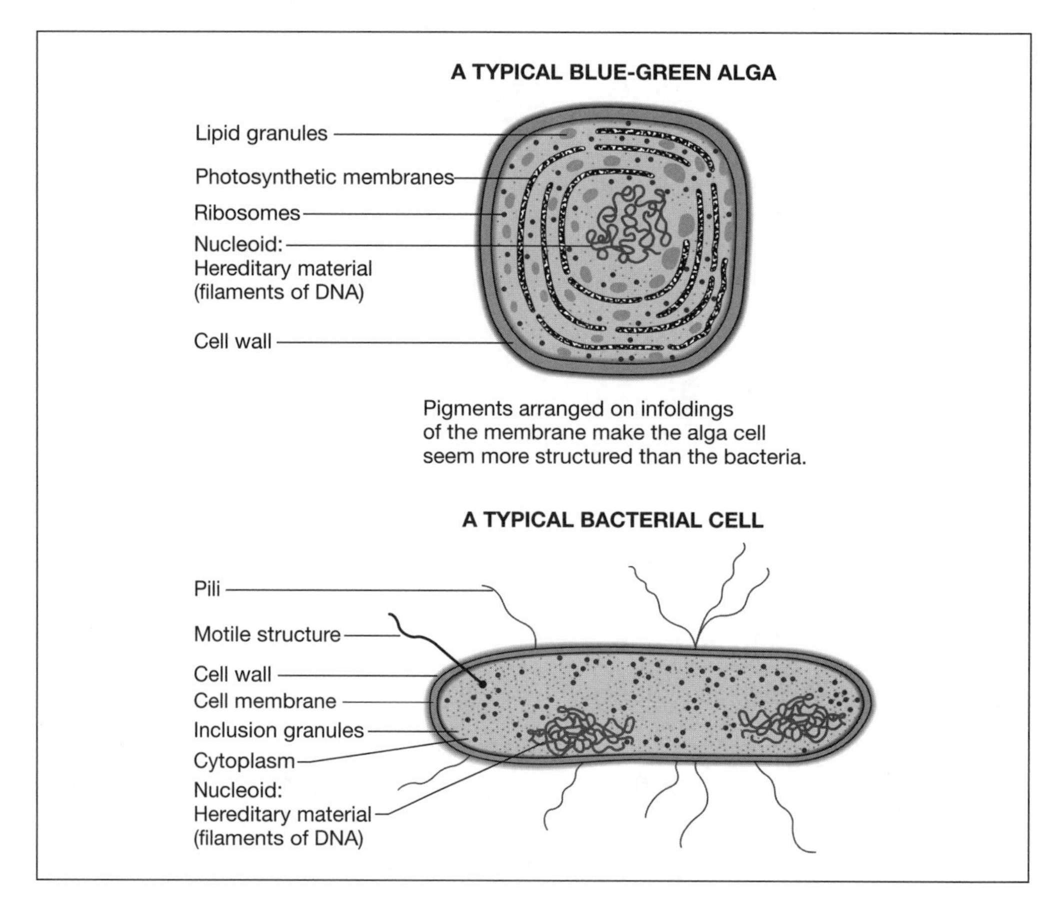


**Two typical prokaryotic cells: A blue-green algae and a bacteria.**

## What is the difference between **prokaryotic** and **eukaryotic cells**?

Every organism is composed of one of two structurally different types of cells: prokaryotic cells or eukaryotic cells. Only monera have prokaryotic cells. Protists, plants, fungi, and animals all have eukaryotic cells.

| **Prokaryotic** | **Eukaryotic** |
|---|---|
| No nucleus | Nucleus |
| No membrane-bound organelles | Many organelles |
| 1-10µm in size | 2–1,000µm in size |
| Evolved 3.5 billion years ago | Evolved 1.5 billion years ago |

## How many **mitochondria** are there in a cell?

The number of mitochondria varies according to the type of cell, but each cell in the human liver has over 1,000 mitochondria. A mitochondrion (singular form) is a self-replicating, double-membraned body found in the cytoplasm of all eukaryotic (having a nucleus) cells. The number of mitochondria per cell varies between one and 10,000, and averages about 200. The mitochondria are the sites for much of the metabolism necessary for the production of ATP, lipids, and protein synthesis.

## What are the **functions** of **organelles** of **prokaryotic** and **eukaryotic** cells in plants and animals?

| Structure | Function | Prokaryotes | Eukaryotes (Animals) | Eukaryotes (Plants) |
|---|---|---|---|---|
| Cell wall | Protects, supports cell | Yes | No | Yes |
| Cytoskeleton | Structural support; cell movement | No | Yes | Yes |
| Flagella and cilia | Motility or moving fluids over surfaces | Yes | Often present | Mostly absent except for sperm, respiratory and digestive epithelia |
| Plasma membrane | Regulates what passes into and out of cell; cell-to-cell recognition | Yes | Yes | Yes |
| Nucleus | Control center of cell; directs protein synthesis and cell reproduction | No | Yes | Yes |
| Chromosomes | Contain hereditary information | No | Yes | Yes |
| Nucleolus | Assembles ribosomes | No | Yes | Yes |
| Ribosomes | Sites of protein synthesis | Yes | Yes | Yes |
| Mitochondria | Power plant of the cell; site of oxidative metabolism | No | Yes | Yes |
| Chloroplast | Site of photosynthesis | No | No | Yes |
| Lysosomes | Digest worn-out mitochondria and cell debris; play role in cell death | No | Yes | Yes |
| Endoplasmic reticulum | Forms compartments and vesicles | No | Yes | Yes |
| Golgi complex | Modifies and packages proteins for export from cell; forms secretory vesicles | No | Yes | Yes |

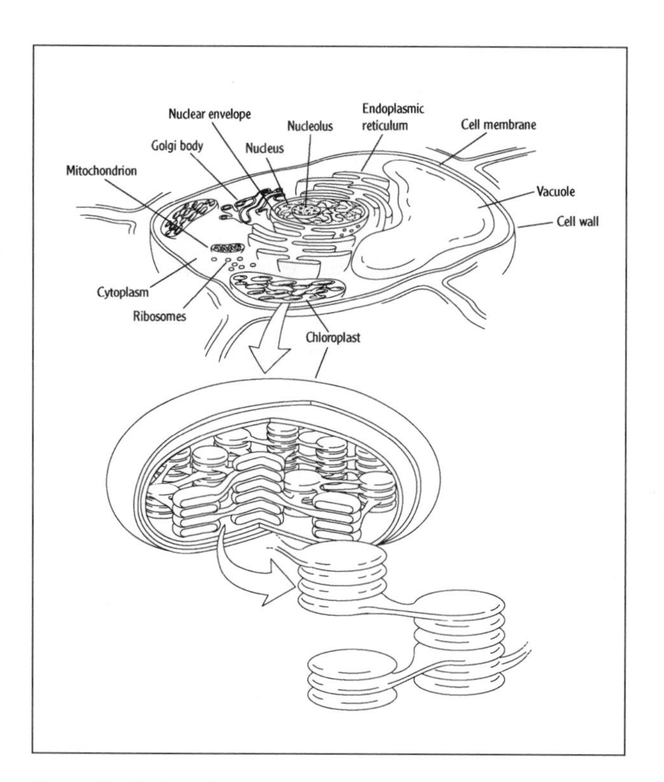


Plant cell with chloroplast.

## How many **chloroplasts** are in plant cells?

Chloroplasts are the functional units where photosynthesis takes place—the process whereby green plants use light energy for the synthesis of sugar from carbon dioxide and water, with oxygen released as a by-product. They contain the green pigments chlorophyll a and b, which trap light energy for photosynthesis. A unicellular plant may have only a single large chloroplast, whereas a plant leaf cell may have as many as 20 to 100.

## Do all cells have a **nucleus**?

Red blood cells are the only cells in the human body that do not have a nucleus. As a result, they cannot divide. They are produced in bone marrow at the rate of 140,000 per minute. They exist in the body's circulatory system for about 120 days before being destroyed in the liver.

## What are the stages in the type of cell division called **mitosis**?

Cell division in eukaryotes (higher organisms) consists of two stages: mitosis, the division of the nucleus, and cytokinesis, the division of the whole cell. The first process in the actual division of the cell is mitosis. In mitosis, the replicated chromosomes are maneuvered so that each new cell gets a full complement of chromosomes—one of each. The process is divided into four phases: prophase, metaphase, anaphase, and telophase. Nuclear division of sex cells is called meiosis. Sexual reproduction generally requires two parents and it always involves two events (meiosis and fertilization).

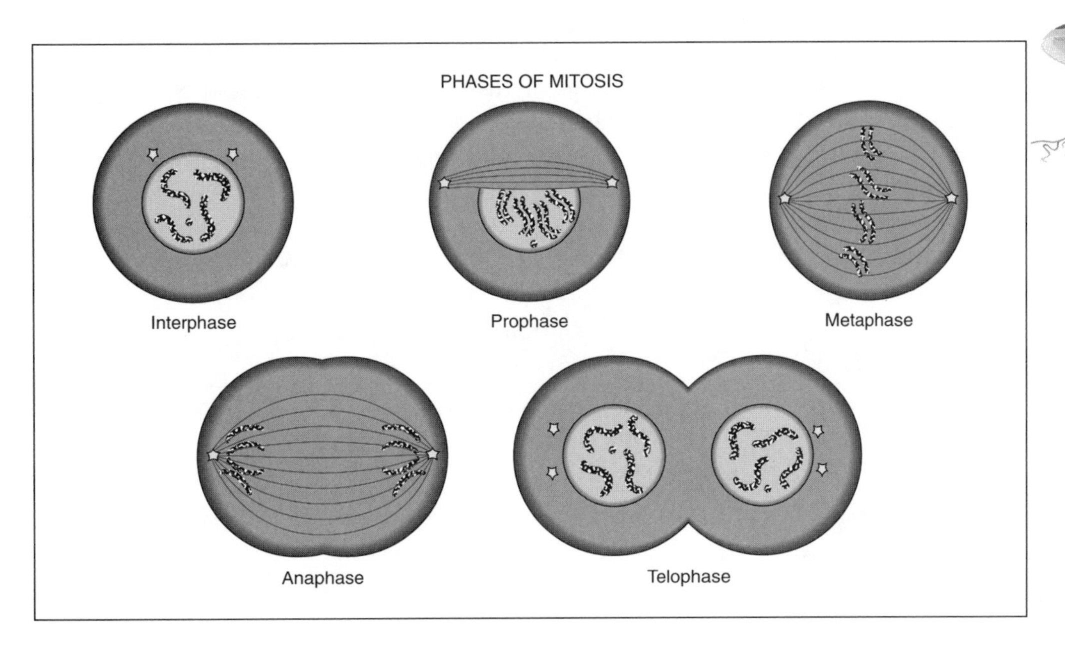


The phases of mitosis.

# EVOLUTION AND GENETICS

## What is the significance of **Darwin's finches**?

In his studies on the Galapagos Islands, Charles Darwin observed patterns in animals and plants that suggested to him that species changed over time to produce new species. Darwin collected several species of finches. The species were all similar, but each had developed beaks and bills specialized to catch food in a different way. Some species had heavy bills for cracking open tough seeds. Others had slender bills for catching insects. One species used twigs to probe for insects in tree cavities. All the species resembled one species of South American finch. In fact, all the plants and animals of the Galapagos Islands were similar to those of the nearby (600 miles/1,000 km away) coast of South America. Darwin felt that the simplest explanation for this similarity was that a few species of plants and animals from South America must have migrated to the Galapagos Islands. These few plants and animals then changed during the years they lived in their new home, giving rise to many new species. Evolutionary theory proposes that species change over time in response to environmental challenges.

## What is **punctuated equilibrium**?

Punctuated equilibrium is a model of macroevolution first detailed in 1972 by Niles Eldredge and Stephen J. Gould. It can be considered either a rival or supplementary model to the more gradual-moving model of evolution posited by neo-Darwinism. The punctuated equilibrium model essentially asserts that most of geological history shows periods of little evolutionary change, followed by short (geologically-speaking, a few million years) periods of time of rapid evolutionary change.

## Which **biological events** occurred during the **geologic time divisions**?

**Cenozoic Era (Age of Mammals)**

| Period | Epoch | Beginning date in est. millions of years | Plants and microorganisms | Animals |
|---|---|---|---|---|
| Quaternary | Holocene (Recent) | 10,000 years ago | Decline of woody plants and rise of herbaceous plants | Age of *Homo sapiens*; humans dominate |
| | Pleistocene | 1.9 | Extinction of many species (from 4 ice ages) | Extinction of many large mammals (from 4 ice ages) |
| Tertiary | Pliocene | 6.0 | Development of grasslands; decline of forests; flowering plants | Large carnivores; many grazing mammals; first known human-like primates |
| | Miocene | 25.0 | | Many modern mammals evolve |
| | Oligocene | 38.0 | Spread of forests; flowering plants, rise of monocotyledons | Apes evolve; all present mammal families evolve; saber-toothed cats |
| | Eocene | 55.0 | Gymnosperms and angiosperms dominant | Beginning of age of mammals; modern birds |
| | Paleocene | 65.0 | | Evolution of primate mammals |

**Mesozoic Era (Age of Reptiles)**

| Period | Epoch | Beginning date in est. millions of years | Plants and microorganisms | Animals |
|---|---|---|---|---|
| Cretaceous | | 135.0 | Rise of angiosperms; gymnosperms decline | Dinosaurs reach peak and then become extinct; toothed birds |

| Period | Epoch | Beginning date in est. millions of years | Plants and microorganisms | Animals |
|---|---|---|---|---|
| | | | | become extinct; first modern birds; primitive mammals |
| Jurassic | | 200.0 | Ferns and gymnosperms common | Large, specialized dinosaurs; insectivorous marsupials |
| Triassic | | 250.0 | Gymnosperms and ferns dominate | First dinosaurs; egg-laying mammals. |
| | | **Paleozic Era (Age of Ancient Life)** | | |
| Permian | | 285.0 | Conifers evolve. | Modern insects like reptiles; extinction of many Paleozic invertebrates |
| Carboniferous (divided into Mississippian and Pennsylvanian periods by some in the U.S.) | | 350.0 | Forests of ferns and gymnosperms; swamps; club mosses and horsetails | Ancient sharks abundant; many echinoderms, mollusks and insect forms; first reptiles; spread of ancient amphibians |
| Devonian | | 410.0 | Terrestrial plants established; first forests; gymnosperms appear | Age of fish; amphibians; wingless insects and millipedes appear |
| Silurian | | 425.0 | Vascular plants appear; algae dominant | Fish evolve; marine arachnids dominant; first insects; crustaceans |
| Ordovician | | 500.0 | Marine algae dominant; terrestrial plants first appear | Invertebrates dominant; first fish appear |
| Cambrian | | 570.0 | Algae dominant | Age of marine invertebrates |

**Precambrian Era**

| Period | Epoch | Beginning date in est. millions of years | Plants and microorganisms | Animals |
|---|---|---|---|---|
| Archeozoic and Proterozoic Eras | | 3800.0 | Bacterial cells; then primitive algae and fungi; marine protozoans | Marine invertebrates at end of period |
| Azoic | | 4600.0 | Origin of the Earth. | |

Austrian monk and scientist Gregor Mendel experimented with garden peas to establish the principle of heredity.

## How did **humans evolve**?

It has been proposed that the Homo lineage of modern humans (*Homo sapiens*) originated from a hunter of nearly five feet tall, *Homo habilis,* who is widely presumed to have evolved from an australopithecine ancestor. Near the beginning of the Pleistocene epoch (two million years ago), *Homo habilis* is thought to have transformed into *Homo erectus* (Java Man), who used fire and possessed culture. Middle Pleistocene populations of *Homo erectus* are said to show steady evolution toward the anatomy of *Homo sapiens* (Neanderthals, Cro-Magnons, and modern humans), 120,000 to 40,000 years ago. Pre-modern *Homo sapiens* built huts and made clothing.

## Who is generally known as the **founder of genetics**?

Gregor Mendel (1822–1884), an Austrian monk and biologist, is considered the founder

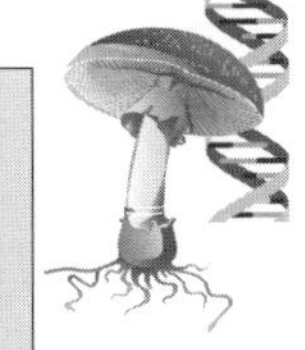



**Did Charles Darwin have any nicknames?**

Darwin had several nicknames. As a young naturalist on board the H.M.S. *Beagle,* he was called "Philos" because of his intellectual pursuits and "Flycatcher" when his shipmates tired of him filling the ship with his collections. Later in his life, when he became a leader in the scientific community, journalists refered to him as "The Sage of Down" or "The Saint of Science," but his friend Thomas Henry Huxley privately called him "The Czar of Down" and the "Pope of Science." His own favorite nickname was "Stultis the Fool," and he often signed letters to scientific friends with "Stultis." This name referred to his habit of trying experiments most people would prejudge to be fruitless or fool's experiments.

of genetics. Using his knowledge of statistics to analyze biological phenomena, Mendel discovered specific and regular ratios that he used to formulate the laws of heredity. It was the English biologist William Bateson (1861–1926), however, who brought Mendel's work to the attention of the scientific world and who coined the term "genetics."

## What is meant by **Mendelian inheritance**?

Mendelian inheritance refers to genetic traits carried through heredity; the process was studied and described by Austrian monk Gregor Mendel (1822–1884). Mendel was the first to deduce correctly the basic principles of heredity. Mendelian traits are also called single gene or monogenic traits, because they are controlled by the action of a single gene or gene pair. More than 4,300 human disorders are known or suspected to be inherited as Mendelian traits, encompassing autosomal dominant (e.g., neurofibromatosis), autosomal recessive (e.g., cystic fibrosis), sex-linked dominant and recessive conditions (e.g., color-blindness and hemophilia).

Overall, incidence of Mendelian disorders in the human population is about one percent. Many non-anomalous characteristics that make up human variation are also inherited in Mendelian fashion.

## What is the significance of ***On the Origin of Species***?

Charles Darwin (1809–1882) first proposed a theory of evolution based on natural selection in his treatise *On the Origin of Species.* The publication of *On the Origin of Species* ushered in a new era in our thinking about the nature of man. The intellectual revolution it caused and the impact it had on man's concept of himself and the world were greater than those caused by the works of Newton and others. The effect was immediate, the first edition being sold out on the day of publication (November 24,

**The viceroy butterfly (left) has naturally evolved to resemble the monarch butterfly—a species more unpalatable to predators.**

1859). *Origin* has been referred to as "the book that shook the world." Every modern discussion of man's future, the population explosion, the struggle for existence, the purpose of man and the universe, and man's place in nature rests on Darwin.

The work was a product of his analyses and interpretations of his findings from his voyages on the H.M.S. *Beagle* as a naturalist. In Darwin's day, the prevailing explanation for organic diversity was the story of creation in the book of Genesis in the Bible. *Origin* was the first publication to present scientifically sound, well-organized evidence for evolution. Darwin's theory of evolution was based on natural selection in which the best—the fittest—survive, and if there is a difference in genetic endowment among individuals, the race will, by necessity, steadily improve. It is a two-step process: the first consists of the production of variation, and the second of the sorting of this variability by natural selection in which the favorable variations tend to be preserved.

## Who coined the phrase **"survival of the fittest"**?

Although frequently associated with Darwinism, this phrase was coined by Herbert Spencer (1820–1903), an English sociologist. It is the process by which organisms that are less well-adapted to their environment tend to perish and better-adapted organisms tend to survive.

## What is **Batesian mimicry**?

In 1861, Henry Walter Bates (1825–1892), a British naturalist, proposed that a nontoxic species can evolve (especially in color and color pattern) to look like a toxic or unpalatable species, or to act like a toxic species, to avoid being eaten by a predator. The classic example is the viceroy butterfly, which resembles the unpalatable monarch butterfly. This is called Batesian mimicry. In another example, the larva of the hawk moth puffs up its head and thorax when disturbed, looking like the head of a small poisonous snake, complete with eyes. The mimicry even involves behavior; the larva weaves its head back and forth and hisses like a snake. Subsequently, Fritz Müller (1821–1897), a German-born zoologist, discovered that all the species of similar appearance become distasteful to predators. This phenomenon is called Müllerian mimicry.

## When was the **Scopes (monkey) trial**?

John T. Scopes (1900–1970), a high-school biology teacher, was brought to trial by the State of Tennessee in 1925 for teaching the theory of evolution. He challenged a recent law passed by the Tennessee legislature that made it unlawful to teach in any public school any theory that denied the divine creation of man. He was convicted and sentenced, but the decision was reversed later and the law repealed in 1967.

At the present time, pressure against school boards still affects the teaching of evolution. Recent drives by anti-evolutionists either have tried to ban the teaching of evolution or have demanded "equal time" for "special creation" as described in the biblical book of Genesis. This has raised many questions about the separation of church and state, the teaching of controversial subjects in public schools, and the ability of scientists to communicate with the public. The gradual improvement of the fossil record, the result of comparative anatomy, and many other developments in biological science have contributed toward making evolutionary thinking more generally accepted.

## What is the **Red Queen hypothesis**?

This hypothesis, also called the law of constant extinction, is named after the Red Queen in Lewis Carroll's *Through the Looking Glass,* who said, "Now here, you see, it takes all the running you can do to keep in the same place." The idea is that an evolutionary advance by one species represents a deterioration of the environment for all remaining species. This places pressure on those species to advance just to keep up.

## What is the difference between a **plasmid** and a **prion**?

A plasmid is a small, circular, self-replicating DNA molecule separate from the bacterial chromosome. Plasmids do not normally exist outside the cell and are generally beneficial to the bacterial cell. Plasmids are often used to pick up foreign DNA for use in genetic engineering. A prion is an infectious form of a protein or malformed protein that may increase in number by converting related proteins to more prions. Prions may cause a number of degenerative brain diseases such as "mad cow disease" or Creutzfeldt-Jakob disease in humans.

## What is **polymerase chain reaction**?

Polymerase chain reaction, or PCR, is a laboratory technique that amplifies or copies any piece of DNA very quickly without using cells. The DNA is incubated in a test tube with a special kind of DNA polymerase, a supply of nucleotides, and short pieces of syntethic single-stranded DNA that serve a primers for DNA synthesis. With automation, PCR can make billions of copies of a particular segment of DNA in a few hours. Each cycle of the PCR procedure takes only about five minutes. At the end of the cycle, the DNA segment—even one with hundreds of base pairs—has been doubled. A PCR machine repeats the cycle over and over. PCR is much faster than the days it takes to clone a piece of DNA by making a recombinant plasmid and letting it replicate within bacteria.

PCR was developed by the biochemist Kary Mullis (1994–) in 1983 while working for Cetus Corporation, a California biotechnology firm. In 1993, Mullis won the Nobel Prize in Chemistry for developing PCR.

## What is **genetic engineering**?

Genetic engineering, also known popularly as molecular cloning or gene cloning, is the artificial recombination of nucleic acid molecules in the test tube, their insertion into a virus, bacterial plasmid, or other vector system, and the subsequent incorporation of the chimeric molecules into a host organism in which they are capable of continued propagation. The construction of such molecules has also been termed gene manipulation because it usually involved the production of novel genetic combinations by biochemical means.

Genetic engineering techniques include cell fusion and the use of recombinant DNA (RNA) or gene-splicing. In cell fusion, the tough outer membranes of sperm and egg cells are stripped off by enzymes, and then the fragile cells are mixed and combined with the aid of chemicals or viruses. The result may be the creation of a new life form from two species. Recombinant DNA techniques transfer a specific genetic activity from one organism to the next through the use of bacterial plasmids (small circular pieces of DNA lying outside the main bacterial chromosome) and enzymes, such as restriction endonucleases (which cut the DNA strands); reverse transcriptase (which makes a DNA strand from an RNA strand); DNA ligase (which joins DNA strands together); and tag polymerase (which can make a double-stranded DNA molecule from a single stranded "primer" molecule). The process begins with the isolation of suitable DNA strands and fragmenting them. After these fragments are combined with vectors, they are carried into bacterial cells where the DNA fragments are "spliced" on to plasmid DNA that has been opened up. These hybrid plasmids are now mixed with host cells to form transformed cells. Since only some of the transformed cells will exhibit the desired characteristic or gene activity, the transformed cells are separated and grown individually in cultures. This methodology has been successful in producing large quantities of hormones (such as insulin) for the biotechnology industry. However, it is more difficult to transform animal and plant cells. Yet the technique exists to make plants resistant to diseases and to make animals grow larger. Because genetic engineering interferes with the processes of heredity and can alter the genetic structure of our own species, there is much concern over the ethical ramifications of such power, as well as the possible health and ecological consequences of the creation of these bacterial forms. Some applications of genetic engineering in the various fields are listed below:

*Agriculture:* Crops having larger yields, disease- and drought-resistancy; bacterial sprays to prevent crop damage from freezing temperatures; and livestock improvement through changes in animal traits.

*Industry:* Use of bacteria to convert old newspaper and wood chips into sugar; oil- and toxin-absorbing bacteria for oil spill or toxic waste clean-ups; and yeasts to accelerate wine fermentation.

*Medicine:* Alteration of human genes to eliminate disease (experimental stage); faster and more economical production of vital human substances to alleviate deficiency and disease symptoms (but not to cure them) such as insulin, interferon (cancer therapy), vitamins, human growth hormone ADA, antibodies, vaccines, and antibiotics.

*Research:* Modification of gene structure in medical research, especially cancer research. Food processing: Rennin (enzyme) in cheese aging.

## What was the **first commercial use** of **genetic engineering**?

Recombinant DNA technology was first used commercially to produce human insulin from bacteria. In 1982, genetically engineered insulin was approved for use by diabetics. Insulin is normally produced by the pancreas, and the pancreas of slaughtered animals such as swine or sheep was used as a source of insulin. To provide a reliable source of human insulin, researchers obtained DNA from human cells carrying the gene with the information for making human insulin. Researchers made a copy of DNA carrying this insulin gene and moved it into a bacterium. When the bacterium was grown in the lab, the microbe split from one cell into two cells, and both cells got a copy of the insulin gene. Those two microbes grew, then divided into four, those four into eight, the eight into sixteen, and so forth. With each cell division, the two new cells each had a copy of the gene for human insulin. And because the cells had a copy of the genetic "recipe card" for insulin, they could make the insulin protein.

## What is the **largest protein** produced by genetic engineering?

Blood-clotting factor VIII, produced by Bayer, is the largest protein so far prepared by genetic engineering and is the first genetically engineered drug. Assembled from

2,332 amino acids, it has a molecular weight of 300,000. (By comparison, human insulin consists of only 51 amino acids.) As a coagulating agent, factor VIII plays a critical life-sustaining role. In hemophiliacs, this protein is missing or does not function properly. As a result, even a minor incident of untreated bleeding can lead to death.

## What is the **Human Genome Project**?

Begun in 1990, the U.S. Human Genome Project is an effort coordinated by the Department of Energy and the National Institutes of Health. The project originally was planned to last 15 years, but effective resource and technological advances have accelerated the expected completion date to 2003. Project goals are to identify all the approximately 30,000 genes in human DNA, determine the sequences of the three billion chemical base pairs that make up human DNA, store this information in databases, improve tools for data analysis, transfer related technologies to the private sector, and address the ethical, legal, and social issues (ELSI) that may arise from the project. The goal is not only to pinpoint these genes, but also to decode the biochemical information down to the so called "letters" of inheritance, the four basic constituents of all genes called nucleotides: A (adenine), C (cytosine), G (guanine), and T (thymine). Since these letters are linked in pairs of sequences in the double helix of DNA, this means that three billion pairs are involved in this process. The sequencing of the human genome has been hailed as the most ground-breaking scientific event of our time. It will, in equal measure, offer insight into our collective history and our individual identities and open up untold possibilities for the diagnosis, treatment, and prevention of disease. Not since Watson and Crick's discovery of the structure of DNA has a scientific investigation been greeted with such fanfare.

## What is **cloning**?

A clone is a group of cells derived from the original cell by fission (one cell dividing into two cells) or by mitosis (cell nucleus division with each chromosome splitting into two). It perpetuates an existing organism's genetic make-up. Gardeners have been making clones (copies) of plants for centuries by taking cuttings of plants to make genetically identical copies. For plants that refuse to grow from cuttings, or for the animal world, modern scientific techniques have greatly extended the range of cloning. The technique for plants starts with taking a cutting of a plant, usually the "best" one in terms of reproductivity or decorativeness or other standard. Since all the plant's cells contain the genetic information from which the entire plant can be reconstructed, the cutting can be taken from any part of the plant. Placed in a culture medium having nutritious chemicals and a growth hormone, the cells in the cutting divide, doubling in size every six weeks until the mass of cells produces small, white, globular points called embryoids. These embryoids develop roots, or shoots, and begin to look like tiny plants. Transplanted into compost, these plants grow into exact copies of the parent plant. The whole process takes 18 months. This process, called tissue culture, has been used to make clones of oil palm, asparagus, pineapples, strawberries, brussels sprouts, cauliflower, bananas, carnations, ferns, etc. Besides making high productive copies of the best plant available, this method controls viral diseases that are passed through seed generations.



## Can **human beings** be **cloned**?

In theory, yes. There are, however, many technical obstacles to human cloning, as well as moral, ethical, philosophical, religious, and economic issues to be resolved before a human being could be cloned. Most scientists would agree that cloning a human is unsafe under current conditions.

Nuclear transplantation or somatic cell nuclear transfer is the concept of moving the cell nucleus and its genetic material from one cell to another. Somatic cell nuclear transfer may be used to make tissue that is genetically compatible with that of the recipient, and it can be used in the treatment of specific disease.

| | **Nuclear Transplantation** | **Human Reproductive Cloning** |
|---|---|---|
| End product | Cells growing in a petri dish | Human being |
| Purpose | To treat a specific disease of tissue generation | Replace or duplicate a human |
| Time frame | A few weeks (growth in culture) | 9 months |
| Surrogate mother needed | No | Yes |
| Human created | No | Yes |
| Ethical implications | Similar to all embryonic cell research | Highly complex issues |
| Medical implications | Similar to any cell-based therapy | Safety and long-term efficacy concerns |

## What was the **first animal** to be successfully **cloned**?

In 1970, British molecular biologist John B. Gurdon (1933–) cloned a frog. He transplanted the nucleus of an intestinal cell from a tadpole into a frog's egg that had had its nucleus removed. The egg developed into an adult frog that had the tadpole's genome in all of its cells and was therefore a clone of the tadpole.

## What was the **first mammal** to be successfully **cloned**?

The first mammal cloned from adult cells was Dolly, a ewe, born in July 1996. Dolly was born in a research facility in Scotland. Ian Wilmut led the team of biologists that removed a nucleus from a mammary cell of an adult ewe and transplanted it into an

**Dolly the sheep, the world's first adult animal to be successfully cloned.**

enucleated egg extracted from a second ewe. Electrical pulses were administered to fuse the nucleus with its new host. When the egg began to divide and develop into an embryo, it was transplanted into a surrogate mother ewe. Dolly was the genetic twin of the ewe that donated the mammary cell nucleus. On April 13, 1998, Dolly gave birth to Bonnie.

## What are **stem cells**?

Stem cells are derived from totipotent cells of the early embryo. As totipotent cells they have the ability to differentiate into all of the cell types ultimately present in the adult, including muscle, blood, nerves or any other tissue.

## Who originated the idea called **panspermia**?

Panspermia is the idea that microorganisms, spores, or bacteria attached to tiny particles of matter have traveled through space, eventually landing on a suitable planet and initiating the rise of life there. The word itself means "all-seeding." The British scientist Lord Kelvin (1824–1907) suggested in the 19th century that life may have arrived here from outer space, perhaps carried by meteorites. In 1903, the Swedish chemist Svante Arrhenius (1859–1927) put forward the more complex panspermia idea that life on Earth was "seeded" by means of extraterrestrial spores, bacteria, and microorganisms coming here on tiny bits of cosmic matter.

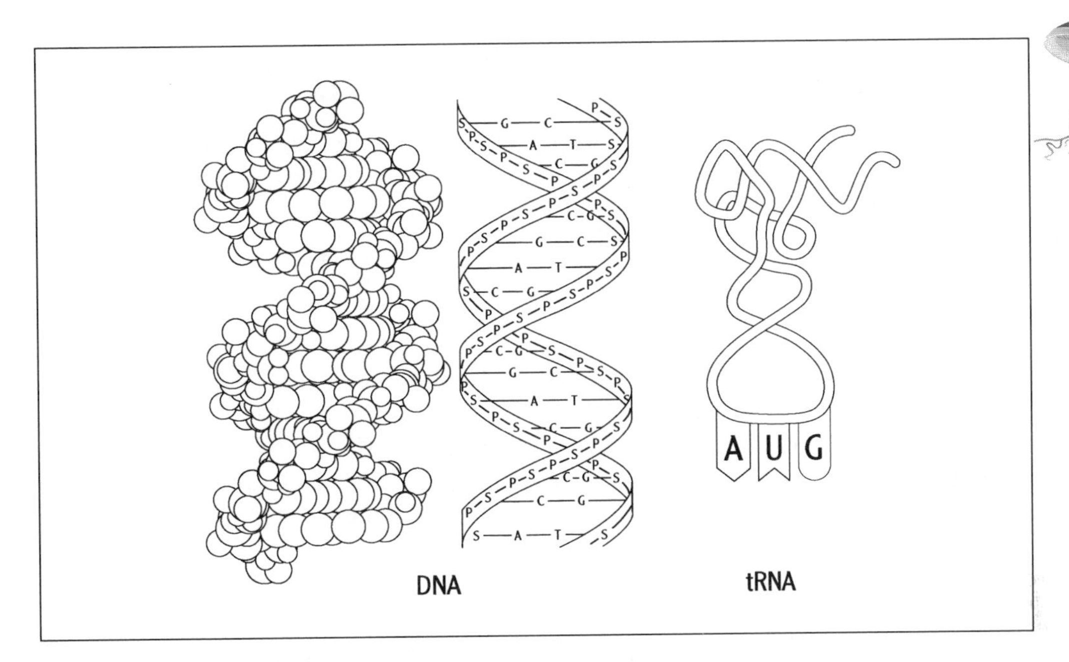


Double helix of DNA and RNA.

## What is the **difference** between **DNA** and **RNA**?

DNA (deoxyribonucleic acid) is a nucleic acid formed from a repetition of simple building blocks called nucleotides. The nucleotides consist of phosphate ($PO_4$), sugar (deoxyribose), and a base that is either adenine (A), thymine (T), guanine (G), or cytosine (C). In a DNA molecule, this basic unit is repeated in a double helix structure made from two chains of nucleotides linked between the bases. The links are either between A and T or between G and C. The structure of the bases does not allow other kinds of links. The famous double helix structure resembles a twisted ladder. The 1962 Nobel Prize in physiology or medicine was awarded to James Watson (b. 1928), Francis Crick (b. 1916), and Maurice Wilkins (b. 1916) for determining the molecular structure of DNA.

RNA (ribonucleic acid) is also a nucleic acid, but it consists of a single chain and the sugar is ribose rather than deoxyribose. The bases are the same except that the thymine (T), which appears in DNA is replaced by another base called uracil (U), which links only to adenine (A).

## What is **p53**?

Discovered in 1979, p53—sometimes referred to as "The Guardian Angel of the Genome"—is a gene that, when a cell's DNA is damaged, acts as an "emergency brake" to halt the resulting cycle of cell division that can lead to tumor growth and cancer. It also acts as an executioner, programming damaged cells to self-destruct before their

## How much DNA is in a typical human cell?

If the DNA in a single human cell were stretched out and laid end-to-end, it would measure approximately 6.5 feet (two meters). The average human body contains 10 to 20 billion miles (16 to 32 billion kilometers) of DNA distributed among trillions of cells.

altered DNA can be replicated. However, when it mutates, p53 can lose its suppressive powers or have the devastating effect of actually promoting abnormal cell growth. Indeed, p53 is the most commonly mutated gene found in human tumors. Scientists have discovered a compound that could restore function to a mutant p53. Such a discovery may lead to the development of anti-cancer drugs targeting the mutant p53 gene.

## Which **organism** has the **largest number** of **chromosomes**?

*Ophioglossum reticulatum,* a species of fern, has the largest number of chromosomes with more than 1,260 (630 pairs).

## Who is considered the **founder of embryology**?

Kaspar Friedrich Wolff (1733–1794), a German surgeon, is regarded as the founder of embryology. Wolff produced his revolutionary work *Theoria generationis* in 1759. Until that time it was generally believed that each living organism developed from an exact miniature of the adult within the seed or sperm. Wolff introduced the idea that cells in a plant or animal embryo are initially unspecified (undefined) but later differentiate to produce the separate organs and systems with distinct types of tissues.

## What does **"ontogeny recapitulates phylogeny"** mean?

Ontogeny is the course of development of an organism from fertilized egg to adult; phylogeny is the evolutionary history of a group of organisms. So the phrase, originating in 19th-century biology, means that as an embryo of an advanced organism grows, it will pass through stages that look very much like the adult phase of less advanced organisms. For example, at one point the human embryo has gills and resembles a tadpole.

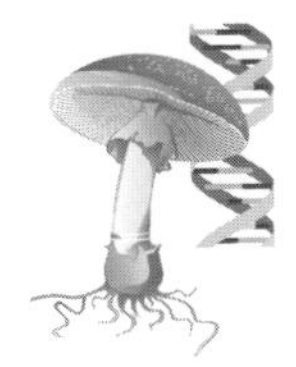



## What is a **biological clock**?

First recognized by the Chinese in the third century B.C.E., the biological clock is an intrinsic mechanism that controls the rhythm of various metabolic activities of plants and animals. Some, such as mating, hibernation, and migration, have a yearly cycle; others, such as ovulation and menstrual cycles of women, follow a lunar month. The majority, however, have a 24-hour, day-night cycle called the circadian rhythm. This day-night cycle, first recognized in plants over 250 years ago and existing in virtually all species of plants and animals, regulates these organisms' metabolic functions: plants opening and closing their petals or leaves, germination and flowering functions, changes in human body temperature, hormone secretion, blood sugar and blood pressure levels, and sleep cycles.

Research in chronobiology, the study of these daily rhythms, reveals that many accidents occur between 1 and 6 a.m., that most babies are born in the morning hours, that heart attacks tend to occur between 6 and 9 a.m., and that most Olympic records are broken in the late afternoon. The clock regulator may be the pineal gland located in the heads of animals (including humans).

## Is there any scientific basis for **biorhythms**?

There is little, if any, scientific support for this theory, which claims there are three precise cycles that control human behavior. These are: a physical cycle of 23 days; an emotional cycle of 28 days; and an intellectual cycle of 33 days. Hazardous critical days are proposed to occur when two or more cycles intersect.

In contrast, biological rhythms such as activity cycles, feeding cycles, and sleeping cycles are well known. They vary from individual to individual and most are tied to the 24-hour rotation period of the Earth. Biological rhythms are real; biorhythms are a hoax.

## What is the **"Spiegelman monster"**?

Sol Spiegelman, an American microbiologist, conducted an experiment to identify the smallest molecule capable of replicating itself. He began with a virus called QB, which consisted of a single molecule of ribonucleic acid (RNA) composed of 4,500 nucleotides (units of nucleic acid).

Usually this virus is able to make copies of itself only by invading a living cell because it requires a cellular enzyme called replicase. When Spiegelman added the replicase along with a supply of free nucleotides in a test tube to the virus, the virus replicated itself for several generations when a mutant appeared having fewer than 4,500 nucleotides. Being smaller, this mutant replicated faster than the original virus.

Then another mutant appeared and displaced the first, and so it went. Finally, the virus degenerated into a little piece of ribonucleic acid with only 220 nucleotides, the smallest bit necessary for recognizing the replicase. This little test tube monster could continue to replicate at high speed, provided it had a source of building blocks.

### Who was regarded as the **founder** of **biochemistry**?

Jan Baptista van Helmont (1577–1644) is called the father of biochemistry because he studied and expressed vital phenomena in chemical terms. The term "biochemistry," coined by F. Hoppe-Seyler in 1877, is the science dealing with the dynamics of living chemical processes or metabolism. It was formed from both the chemists' animal and vegetable chemistry, and the biologists' and doctors' physiological, zoological, or biological chemistry.

Helmont devoted his life to the study of chemistry as the true key to medicine and is considered to be one of the founders of modern pathology because he studied the external agents of diseases as well as the anatomical changes caused by diseases.

### What was the first **amino acid** to be discovered?

Asparagine, isolated from the asparagus plant, was discovered in 1806 by the French chemist Nicolas-Louis Vauquelin.

## CLASSIFICATION, MEASUREMENTS, AND TERMS

### What is **bioinformatics**?

Bioinformatics is the field of science in which biology, computer science, and information technology merge into a single discipline. The ultimate goal of the field is to enable the discovery of new biological insights as well as to create a global perspective from which unifying principles in biology can be discerned. There are three important sub-disciplines within bioinformatics: 1) the development of new algorithms and statistics with which to assess relationships among members of the large data sets; 2) the analysis and interpretation of various types of data including nucleotide and amino acid sequences, protein domains, and protein structures; and 3) the development and implementation of tools that enable efficient access and management of different types of information.

## What is **radiocarbon dating**?

Radiocarbon dating is a process for determining the age of a prehistoric object by measuring its radiocarbon content. The technique was developed by an American chemist, Dr. Willard F. Libby (1908–1980), in the late 1940s. All living things contain radiocarbon (carbon 14), an isotope that occurs in a small percentage of atmospheric carbon dioxide as a result of cosmic ray bombardment. After an animal or plant dies, it no longer absorbs radiocarbon and the radiocarbon present begins to decay (break down by releasing particles) at an exact and uniform rate. Its half-life of 5,730 years made it useful for measuring prehistory and events occurring within the past 35,000 to 50,000 years. A recent development, called the Accelerated Mass Spectrometer, which separates and detects atomic particles of different mass, can establish more accurate dates with a smaller sample. The remaining radiocarbon can be measured and compared to that of a living sample. In this way, the age of the 50,000-year-old or less animal or plant (or more precisely the elapsed time since its death) can be determined.

Since Libby's work, other isotopes having longer half-lives have been used as "geologic clocks" to date very old rocks. The isotope uranium-238 (decaying to lead-206) has a half-life of 4.5 billion years, uranium-235 (decaying to lead-207) has a value of 704 million years, thorium-232 (decaying to lead-278) has a half-life of 14 billion years, rubidium-87 (decaying to strontium-87) has a half-life value of 48.8 billion years, potassium-40 (decaying to argon-40) has a value of 1.25 billion years, and samarium-147 (decaying to neodymium-143) has a value of 106 billion years. These isotopes are used in dating techniques of gas formation light emission (called thermoluminescence). Other ways to date the past include dating by tree rings (counting its annual growth rings), and dating by thermoremanent magnetism (the magnetic field of the rock is compared to a date chart of changes in the Earth's magnetic field).

## Who coined the term **biology**?

Biology was first used by Karl Burdach (1776–1847) to denote the study of man. Jean Baptiste Pierre Antoine de Monet Lamarck (1744–1829) gave the term a broader meaning in 1812. He believed in the integral character of science. For the special sciences, chemistry, meteorology, geology, and botany-zoology, he coined the term "biology." Lamarckism epitomizes the belief that changes acquired during an individual's lifetime as the result of active, quasi-purposive, functional adaptations can somehow be imprinted upon the genes, thereby becom-

Jean Baptiste Lamarck's theories on evolution preceded and influenced those of Charles Darwin.

ing part of the heritage of succeeding generations. Today, very few professional biologists believe that anything of the kind occurs—or can occur.

Biology is the science that deals with living things (Greek *bios,* "life"). Once broadly divided into two areas—zoology (Greek *zoon,* "animal"), the study of animals, and botany (Greek *botanes,* "plant"), the study of plants—biology is now divided and sub-divided into hundreds of special fields involving the structure, function, and classification of forms of life. These include anatomy, ecology, embryology, evolution, genetics, paleontology, and physiology.

## Where did the term **molecular biology** originate?

Warren Weaver (1898–1978), the director of the Rockefeller Foundation's Division of Natural Sciences, originated the term molecular biology. Weaver used X-ray diffraction to investigate the molecular basis of inheritance and the structure of biological macromolecules. He called this relatively new field molecular biology in a 1938 report.

## What is **gnotobiotics**?

Gnotobiotics is the scientific study of animals or other organisms that are raised in germ-free environments or ones that contain only specifically known germs. These animals are first removed from the womb and then are placed in sterilized cages called isolators. Scientists are able to use these animals to determine how specific agents, such as viruses, bacteria, and fungi, affect the body.

Taxonomist Carolus Linnaeus devised a classification system for animal and plant species still used to this day.

## What are the **five kingdoms** presently used to categorize living things?

Carolus Linnaeus (1707–1778) in 1735 divided all living things into two kingdoms in his classification system that was based on similarities and differences of organisms. However, since then, fungi seemed not to fit nicely into either kingdom. Although fungi were generally considered plants, they had no chlorophyll, roots, stems, or leaves, and hardly resemble any true plant. They also have several features found in the animal kingdom, as well as unique features characteristic of themselves alone. So fungi were placed into a third kingdom. In 1959, R. H. Whittaker proposed the current five kingdom system, based on new evidence from biochemical techniques and electron microscope observations that revealed fundamental differences among organisms. Each kingdom is listed below.

*Monera*—One-celled organisms lacking a membrane around the cell's genetic matter. Prokaryote (or procaryote) is the term used for this condition where the genetic material lies free in the cytoplasm of the cell without a membrane to form the nucleus of the cell. The kingdom consists of bacteria and blue-green algae (also called blue-green bacteria or cyanobacteria). Bacteria do not produce their own food but blue-green algae do. Blue-green algae, the primary form of life 3.5 to 1.5 billion years ago, produce most of the world's oxygen through photosynthesis.

*Protista*—Mostly single-celled organisms with a membrane around the cell's genetic material called eukaryotes (or eucaryotes) because of the nuclear membrane and other organelles found in the cytoplasm of the cell. Protista consists of true algae, diatoms, slime molds, protozoa, and euglena. Protistans are diverse in their modes of nutrition, etc. They may be living examples of the kinds of ancient single cells that gave rise to the kingdoms of multicelled eukaryotes (fungi, plants, and animals).

*Fungi*—One-celled or multicelled eukaryotes (having a nuclear membrane or a membrane around the genetic material). The nuclei stream between cells, giving the appearance that cells have multiple nuclei. This unique cellular structure, along with the unique sexual reproduction pattern, distinguishes the fungi from all other organisms. Consisting of mushrooms, yeasts, molds, etc., the fungi do not produce their own food.

*Plantae*—Multicellular organisms having cell nuclei and cell walls, which directly or indirectly nourish all other forms of life. Most use photosynthesis (a process by which green plants containing chlorophyll utilize sunlight as an energy source to synthesize complex organic material, especially carbohydrates from carbon dioxide, water, and inorganic salts) and most are autotrophs (produce their own food from inorganic matter).

*Animalia*—Multicellular organisms whose eukaryotic cells (without cell walls) form tissues (and from tissues form organs). Most get their food by ingestion of other organisms; they are heterotrophs (cannot produce their food from inorganic elements). Most are able to move from place to place (mobile) at least during part of the life cycle.

## Who devised the current animal and plant **classification system**?

The naming and organizing of the millions of species of plants and animals is frequently called taxonomy; such classifications provide a basis for comparisons and gen-

### How many different organisms have been identified by biologists?

Approximately 1.5 million different species of plants, animals and microorganisms have been described and formally named. Some biologists believe another 10 million species are still to be discovered, classified, and named.

eralizations. A common classification format is a hierarchical arrangement in which a group is classified within a group, and the level of the group is denoted in a ranking.

Carolus Linnaeus (1707–1778) composed a hierarchical classification system for plants (1753) and animals (1758) using a system of nomenclature (naming) that continues to be used today. Every plant and animal was given two scientific names (binomial method) in Latin, one for the species and the other for the group or genus within the species. He categorized the organisms by perceived physical differences and similarities. Although Linnaeus started with only two kingdoms, contemporary classifiers have expanded them into five kingdoms. Each kingdom is divided into two or more phyla (major groupings; phylum is the singular form). Members within one phylum are more closely related to one another than they are to members of another phylum. These phyla are subdivided into parts again and again; members of each descending level have a closer relationship to each other than those of the level above. Generally, the ranking of the system, going from general to specific, are Kingdom, Phylum (plant world uses the term division), Class, Order, Family, Genus, and Species. In addition, intermediate taxonomic levels can be created by adding the prefixes "sub" or "super" to the name of the level, for example, "subphylum" or "superfamily." Zoologists working on parts of the animal classification many times are not uniform in their groupings. The system is still evolving and changing as new information emerges and new interpretations develop. Below is listed a comparison of hierarchy for four of the five kingdoms.

| Taxonomic level | Human | Grasshopper | White Pine | Typhoid Bacterium |
|---|---|---|---|---|
| Kingdom | Animalia | Animalia | Plantae | Protista |
| Phylum | Chordata | Anthropoda | Tracheophyta | Schizomycophyta |
| Class | Mammalia | Insecta | Gymnospermae | Schizomycetes |
| Order | Primates | Orthoptera | Coniferales | Eubacteriales |
| Family | Hominidae | Arcridiidae | Pinaceae | Bacteriaceae |
| Genus | *Homo* | *Schistocerca* | *Pinus* | *Eberthella* |
| Species | *sapiens* | *americana* | *strobus* | *typhosa* |

## Are any **bacteria visible** to the **naked eye**?

*Epulopiscium fishelsoni*, which lives in the gut of surgeonfish, was first identified in 1985 and mistakenly classified as a protozoan. Later studies analyzed the organism's genetic material and proved it to be a bacterium of unprecedented size—0.015 inches (0.38 millimeters) in diameter, or about the size of a period in a small-print book.

## How **fast** do **bacteria reproduce**?

Bacteria can reproduce very rapidly in a favorable environment, whether in a laboratory culture or a natural habitat. For example, *Escherichia coli* growing under optimal conditions can divide every 20 minutes. A laboratory culture started with a single cell can produce a colony of $10^7$ to $10^8$ bacteria in about 12 hours.

## What is **anthrax**?

*Bacillus anthracis,* the etiologic agent of anthrax, is a large, gram-positive, non-motile, spore-forming bacterial rod. The three virulence factor of *B. anthracis* are edema toxin, lethal toxin, and a capsular antigen. Human anthrax has three major clinical forms: cutaneous, inhalation and gastrointestinal. If left untreated, anthrax in all forms can lead to septicemia (blood poisoning) and death.

## What are **diatoms**?

Diatoms are microscopic algae in the phylum bacillarrophyte of the protista kingdom. Yellow or brown in color, almost all diatoms are single-celled algae, dwelling in fresh and salt water, especially in the cold waters of the North Pacific Ocean and the Antarctic. Diatoms are an important food source for marine plankton (floating animal and plant life) and many small animals.

Diatoms have hard cell walls; these "shells" are made from silica that they extract from the water. It is unclear how they accomplish this. When they die, their glassy shells, called frustules, sink to the bottom of the sea, which hardens into rock called diatomite. One of the most famous and accessible diatomites is the Monterrey Formation along the coast of central and southern California.

## How is a **fairy ring** formed?

A fairy ring, or fungus ring, is found frequently in a grassy area. There are three types: those that do not affect the surrounding vegetation, those that cause increased growth, and those that damage the surrounding environment. The ring is started from a mycelium (the underground, food-absorbing part of a fungus). The fungus growth is on the outer edge of the grassy area because the inner band of decaying mycelium "use-up" the resources in the soil at the center. This creates a ring effect. Each succeeding generation is farther out from the center.

## What is the scientific study of **fungi** called?

Mycology is the science concerning fungi. In the past, fungi have been classified in other kingdoms, but currently they are recognized as a separate kingdom based on their unique cellular structure and their unique pattern of sexual reproduction.

Fungi are heterotrophs (cannot produce their own food from inorganic matter). They secrete enzymes that digest food outside their bodies and their fungal cells absorb the products. Their activities are essential in the decomposition of organic material and cycling of nutrients in nature.

Some fungi, called saprobes, obtain nutrients from non-living organic matter. Other fungi are parasites; they obtain nutrients from the tissues of living host organisms. The great majority of fungi are multicelled and filamentous. A mushroom is a modified reproductive structure in or upon which spores develop. Each spore dispersed from it may grow into a new mushroom.

## What is a **lichen**?

Lichens are organisms that grow on rocks, tree branches, or bare ground. They are composed of a green algae and a colorless fungus living together symbiotically. They do not have roots, stems, flowers, or leaves. The fungus, having no chlorophyll, cannot manufacture its own food, but can absorb food from the algae that it enwraps completely, providing protection from the sun and moisture.

This relationship between the fungus and algae is called symbiosis (a close association of two organisms not necessarily to both their benefits). Lichens were the first recognized and are still the best examples of this phenomenon. A unique feature of lichen symbiosis is that it is so perfectly developed and balanced as to behave as a single organism.

## Who was first to coin the word **virus**?

The English physician Edward Jenner (1749–1823), founder of virology and a pioneer in vaccination, first coined the word "virus." Using one virus to immunize against another one was precisely the strategy Jenner used when he inoculated someone with cowpox (a disease that attacks cows) to make them immune to smallpox. This procedure is called vaccination, from the word vaccine (the Latin name for cowpox). Vaccines are usually a very mild dose of the disease-causing bacteria or virus (weakened or dead). These vaccines stimulate the creation of antibodies in the body that recognize and attack a particular infection. A virus is a minute parasitic organism that reproduces only inside the cell of its host. Viruses replicate by invading host cells and taking over the cell's "machinery" for DNA replication. Viral particles then can break out of the cells, causing disease.

## Who were the founders of **modern bacteriology**?

French chemist Louis Pasteur is considered one of the founders of modern bacteriology.

The German bacteriologist Robert Koch (1843–1910) and the French chemist Louis Pasteur (1822–1895) are considered the founders. Pasteur devised a way to heat food or beverages at a temperature low enough not to ruin them, but high enough to kill most of the microorganisms that would cause spoilage and disease. This process is called pasteurization in his honor. By demonstrating that tuberculosis was an infectious disease caused by a specific bacillus and not by bad heredity, Koch laid the groundwork for public health measures that would significantly reduce such diseases. His working methodologies for isolating microorganisms, his laboratory procedures, and his four postulates for determination of disease agents gave medical investigators valuable insights into the control of bacterial infections.

## Why are **Koch's postulates** significant?

The German bacteriologist Robert Koch (1843–1910) developed four rules in his study of disease-producing organisms, which later investigators found useful. The following conditions must be met to prove that a bacterium causes a particular disease:

1. The microorganism must be found in large numbers in all diseased animals but not in healthy ones.
2. The organism must be isolated from a diseased animal and grown outside the body in a pure culture.
3. When the isolated microorganism is injected into other healthy animals, it must produce the same disease.
4. The suspected microorganism must be recovered from the experimental hosts, isolated, compared to the first microorganism, and found to be identical.

# PLANT WORLD

## PHYSICAL CHARACTERISTICS, FUNCTIONS, ETC.

### If you were to compress the history of the earth into a single year, what would be some major "dates" in **plant evolution**?

| Time (million years) | Event | Date |
|---|---|---|
| 3600 | first algae | Mar. 21 |
| 433 | land plants appear | Nov. 27 |
| 400 | ferns and gymnosperms | Nov. 30 |
| 300 | major coal deposits formed | Dec. 8 |
| 65 | flowering plants appear | Dec. 26 |

### What is the **best type** of **pollination**?

Effective pollination occurs when viable pollen is transferred to plant's stigmas, ovule-bearing organs, or ovules (seed precursors). Without pollination, there would be no fertilization. Since plants are immobile organisms, they usually need external agents to transport their pollen from where it is produced in the plant to where fertilization can occur. This situation produces cross-pollination, wherein one plant's pollen is moved by an agent to another plant's stigma. Some plants are able to self-pollinate—transfer their own pollen to their own stigmas. But of the two methods, cross-pollination seems the better, for it allows new genetic material to be introduced.

Cross-pollination agents include insects, wind, birds, mammals, and water. Many times flowers offer one or more "rewards" to attract these agents—sugary nectar, oil, solid food bodies, perfume, a place to sleep, or sometimes the pollen itself. Other times the plant can "trap" the agent into transporting the pollen. Generally, plants use color and fragrances as attractants to lure these agents. For example, a few orchids use a combination of smell and color to mimic the female of certain species of bees and wasps so successfully that the corresponding males will attempt to mate with them. Through this process (pseudocopulation) the orchids achieve pollination. While some plants cater to a variety of agents, other plants are very selective and are pollinated by a single species of insect only. This extreme pollinator specificity tends to maintain the purity of a plant species.

Plant structure can accommodate the type of agent used. For example, plants such as grasses and conifers, whose pollen is carried by the wind, tend to have a simple structure lacking petals, with freely exposed and branched stigmas to catch airborne pollen and dangling anthers (pollen-producing parts) on long filaments. This type of anther allows the light, round pollen to be easily caught by the wind. These plants are found in areas such as prairies and mountains, where insect agents are rare. In contrast, semi-enclosed, nonsymmetrical, long-lived flowers such as iris, rose, and snapdragon have a "landing platform" and nectar in the flower base to accommodate insect agents such as the bee. The sticky, abundant pollen can easily become attached to the insect to be borne away to another flower.

## What is **tropism**?

Tropism is the movement of a plant in response to a stimulus. The categories include:

*Chemotropism*—a response to chemicals by plants in which incurling of leaves may occur.

*Geotropism or gravitropism*—a response to gravity in which the plant moves in relation to gravity. Shoots of a plant are negatively geotropic (growing upward), while roots are positively geotropic (growing downward).

*Hydrotropism*—a response to water or moisture in which roots grow toward the water source.

*Paraheliotropism*—a response by the plant leaves to avoid exposure to the sun.

*Phototropism*—a response to light in which the plant may be positively phototropic (moving toward the light source) or negatively phototropic (moving away from the light source). Main axes of shoots are usually positively phototropic, whereas roots are generally insensitive to light.

*Thermotropism*—a response to temperature by plants.

*Thigmotropism or haptotropism*—a response to touch by the climbing organs of a plant. For example, the plant's tendrils may curl around a support in a spring-like manner.

# TREES AND SHRUBS

*See also: Environment—Ecology, Resources, etc.; Minerals and Other Materials*

## Is **hemlock** poisonous?

There are two species known commonly as hemlock: *Conium maculatum* and *Tsuga canadensis*. *Conium maculatum* is a weedy plant. All parts of the plant are poisonous. In ancient times minimal doses of the plant were used to relieve pain with a great risk of poisoning. *Conium maculatum* was used to carry out the death sentence in ancient times. The Greek philosopher Socrates was condemned to death by drinking a potion made from hemlock. It should not be confused with *Tsuga canadensis*, a member of the evergreen family. The leaves of the *Tsuga canadensis* may be used to make tea.

## What are the **longest-lived tree species** in the **United States**?

Of the 850 different species of tree in the United States, the oldest species is the bristlecone pine (*Pinus longaeva*), which grows in the deserts of Nevada and southern California (especially in the White Mountains). Some of these trees are believed to be over 4,600 years old. The potential life span of these pines is estimated to be 5,500 years. But these ages are very young when compared to the oldest surviving species in the world, which is the maiden-hair tree (*Ginkgo biloba*) of Zhexiang, China. This tree first appeared during the Jurassic era some 160 million years ago. Also called icho, or the ginkyo ("silver apricot"), this species has been cultivated in Japan since 1100 B.C.E.

**Longest-Lived Tree Species in the United States**

| Name of tree | Number of years |
|---|---|
| Bristlecone pine (*Pinus longaeva*) | 3,000–4,700 |
| Giant sequoia (*Sequoiadendron giganteum*) | 2,500 |
| Redwood (*Sequoia sempervirens*) | 1,000–3,500 |
| Douglas fir (*Pseudotsuga menziesii*) | 750 |
| Bald cypress (*Taxodium distichum*) | 600 |

## How are **tree rings** used to date historical events?

Tree fragments of unknown age and the rings of living trees can be compared in order to establish the date when the fragment was part of a living tree. Thus, tree rings can be used to establish the year in which an event took place as long as the event involved the maiming or killing of a tree. Precise dates can be established for the building of a medieval cathedral or an American Indian pueblo; the occurrence of an earthquake, landslide, volcanic eruption, or a fire; and even the date when a panel of wood was cut for a Dutch painting. Every year, the tree produces an annular ring composed of one wide, light ring and one narrow, dark ring. During spring and early summer, tree stem cells grow rapidly and are larger; this produces the wide, light ring. In winter, growth is greatly reduced and cells are much smaller; this produces the narrow, dark ring. In the coldest part of winter or the dry heat of summer, no cells are produced.

**Tree rings reflect the age of a tree and shed light on climatic conditions such as seasonal rainfall.**

## Who first recorded that the **number of rings** in the cross-section of a tree trunk tell its **age**?

The painter and engineer Leonardo da Vinci (1452– 1519) noticed this phenomenon. He also saw that the year's dampness can be determined by the space between the tree's rings. The farther apart the rings, the more moisture there was in the ground around the tree.

## Why are fall leaves **bright red in some years** and **dull in others**?

Two factors are necessary in the production of red autumn leaves. There must be warm, bright, sunny days during which the leaves manufacture sugar. Warm days must be followed by cool nights with temperatures below 45°F (7.2°C). This weather combination traps the sugar and other materials in the leaves. This results in the manufacture of red anthocyanin. A warm cloudy day restricts the formation of bright colors. With decreased sunlight, sugar production is decreased and this small amount of sugar is transported back to the trunk and roots, where it has no color effect.

## How many **leaves** are on a **mature tree**?

An old, still healthy, oak tree is estimated to have approximately 250,000 leaves.

## Prior to the **chestnut blight** devastation of the American chestnut, what was the proportion of chestnut trees in the eastern forest?

In 1900 the American chestnut (*Castenea dentata Marsh*) was an important forest tree widespread across eastern North America until the chestnut blight fungus (*Cryphonectria parasitica*) devastated it, reducing it to small trees and stumps. In central and southern Pennsylvania, New Jersey, and southern New England, the chestnut formed almost half of the hardwood forest. In its entire range, the species dominated the deciduous forests making up almost one-quarter of the trees.

## Which U.S. city has the **greatest number of trees**?

According to a survey of 20 cities, Houston, Texas, has the most trees, with 956,700.

## Why do tree **leaves turn color** in the fall?

The carotenoids (pigments in the photosynthesizing cells), which are responsible for the fall colors, are present in the leaves during the growing season. However, the colors are eclipsed by the green chlorophyll. Toward the end of summer, when the chlorophyll production ceases, the other colors of the carotenoids (such as yellow, orange, red, or purple) become visible. Listed below are the autumn leaf colors of some common trees.

| Tree | Color |
|---|---|
| Sugar maple and sumac | Flame red and orange |
| Red maple, dogwood, sassafras, and scarlet oak | Dark red |
| Poplar, birch, tulip tree, willow | Yellow |
| Ash | Plum purple |
| Oak, beech, larch, elm, hickory, and sycamore | Tan or brown |
| Locust | Stays green until leaves drop |
| Black walnut and butternut | Drops leaves before they turn color |

## What part of a **chestnut tree** is not killed by chestnut blight?

The roots survive the blight and sprout again to attempt a new tree.

## What is the **tallest tree**?

The tallest tree ever measured was the Australian eucalyptus (*Eucalyptus regnans*) at Watts River, Victoria, Australia. In 1872 it was reported to measure 435 feet (132 meters) tall and it was most likely over 500 feet (152 meters) originally. According to the National Register of Big Trees 2002–03, the tallest living tree is a Coastal redwood (*Sequoia sempervirens*) in Redwood National Park, California at 321 feet (98 meters).

## What is the **Joshua tree**?

The Mormon pioneers named this treelike plant after the prophet Joshua, because its extended branches resembled the outstretched arm of the prophet Joshua as he pointed with his spear to the city of Ai. The plant, *Yucca brevifolia*, is found in the southwestern United States.

## Which tree has the **largest measured trunk** circumference?

The tree in the United States with the largest circumference of 85.3 feet (26 meters) is a Giant sequoia *(Sequoiadendron giganteum)* in Sequoia National Park, California.

## What is a **banyan tree**?

The banyan tree (*Ficus benghalensis*), a native of tropical Asia, is a member of the Ficus or fig genus. It is a magnificent evergreen, sometimes 100 feet (30.48 meters) in height. As the massive limbs spread horizontally, the tree sends down roots that develop into secondary, pillar-like supporting trunks. Over a period of years a single tree may spread to occupy a tremendous area, as much as 2,000 feet (610 meters) around the periphery.

## What are the distinguishing characteristics of **pine**, **spruce,** and **fir** trees?

The best way to tell the difference between the three trees is by their cones and leaves:

**Pines**

| | |
|---|---|
| White Pine | Five needles in each bundle; needles soft and 3-5 inches long. Cones can be 4-8 inches long |
| Scotch Pine | Two needles in each bundle. Needles are stiff, yellow green, 1.5-3 inches long. Cones are 2-5 inches long. |

**Spruce**

| | |
|---|---|
| White Spruce | Dark green needles are rigid, but not prickly, grow from all sides of the twig and are less than an inch long. Cones are 1-2.5 inches long and hang downward. |
| Blue Spruce | Needles are about an inch long, silvery blue, very stiff and prickly; needles grow from all sides of the branch. Cones are 3.5 inches long. |

**Fir**

| | |
|---|---|
| Balsam Fir | Needles are flat, 1-1.5 inches long and arranged in pairs opposite each other. Cones are upright, cylindrical and 2-4 inches long. |
| Fraser Fir | Looks like a balsam but needles are smaller and more rounded. |
| Douglas Fir | Single needles, 1-1.5 inches long and very soft. Cone scales have bristles that stick out. |

## Which **conifers** in North America **lose their leaves** in winter?

Dawn redwood trees (*Metasequoia*) are deciduous. Their leaves are bright green in summer and turn coppery red in the fall before they drop. Previously known only as a fossil, the tree was found in China in 1941 and has been growing in the United States

since the 1940s. The U.S. Department of Agriculture distributed seeds to experimental growers in the United States, and the dawn redwood tree now grows all over the country.

The only native conifers that shed all of their leaves in the fall are the bald cypress (*Taxodium distichum*) and the Larch (*Larix larcina*).

## Does the **rose family** produce any trees?

The apple, pear, peach, cherry, plum, mountain ash, and hawthorn trees are members of the rose family (*Rosaceae*).

## What is a **monkey ball tree**?

The osage orange (*Maclura pomifera*) produces large, green orange-like fruits. These are roughly spherical, 3.5 to 5 inches (8.8 to 12.7 centimeters) in diameter, and have a coarse, pebbly surface.

## Why do the **leaves of mimosa** ("sensitive plant") close in **response to touch**?

When the plant perceives touch, the stimulus is relayed to rest of the plant. Electrical signals then turn on motor cells that in turn control the leaf movement.

## What is the **fastest growing** land plant?

Bamboo is the quickest to gain height. Bamboo can grow one meter, or almost three feet, in twenty-four hours. This growth is produced partly by cell division and partly by cell enlargement.

## How is **glycerine** used to preserve leaves?

Glycerine and water are generally used to preserve coarser leaves, such as magnolia, rhododendron, beech, holly, heather, or Japanese maple. The leaves should be fresh. For autumn colors, the leaves should be picked just as they are turning color. The preserving solution is made by adding two parts boiling water to one part glycerine. Place the leaf stems, split if necessary, into the warm solution so they are covered about three to four inches (seven to 10 centimeters). When drops of glycerine form on the leaves, enough has been absorbed. Wipe off any excess oil. Entire branches may also be treated by using a container large enough to allow the solution to completely cover the leaves. The solution for this method is made with equal quantities of glycerine and water. The leaves are drained on thick sections of newspapers for a few days, then washed with a little soap and water and pegged on a line to dry.

# FLOWERS AND OTHER PLANTS

*See also: Environment—Extinct and Endangered Plants and Animals*

## Who is known as the **founder** of **botany**?

The ancient Greek, Theophrastus (ca. 372–ca. 287 B.C.E.), is known as the father of botany. His two large botanical works, *On the History of Plants* and *On the Causes of Plants,* were so comprehensive that 1,800 years went by before any new discovery in botany was made. He integrated the practice of agriculture into botany and established a theory of plant growth and the analysis of plant structure. He related plants to their natural environment and identified, classified, and described 550 different plants.

## What are the **parts of a flower**?

*Sepal*—found on the outside of the bud or on the underside of the open flower. It serves to protect the flower bud from drying out. Some sepals ward off predators by their spines or chemicals.

*Petals*—serve to attract pollinators and are usually dropped shortly after pollination occurs.

*Nectar*—contains varying amounts of sugar and proteins that can be secreted by any of the floral organs. It usually collects inside the flower cup near the base of the cup formed by the flower parts.

*Stamen*—is the male part of a flower and consists of a filament and anther where pollen is produced.

*Pistil*—is the female part, which consists of the stigma, style, and ovary containing ovules. After fertilization the ovules mature into seeds.

## What is meant by an **"imperfect" flower**?

An imperfect flower is one that is unisexual, having either stamens (male parts) or pistils (female parts) but not both.

## How does a **bulb** differ from a **corm**, a **tuber**, and a **rhizome**?

Many times the term "bulb" is applied to any underground storage organ in which a plant stores energy for its dormant period. Dormancy is one of nature's devices a plant utilizes to get through difficult weather conditions (winter cold or summer drought).

*A true bulb*—consists of fleshy scales containing a small basal plate (a modified stem from which the roots emerge) and a shoot. The scales that surround the embryo are modified leaves that contain the nutrients for the bulb during dormancy and early growth. Some bulbs have a tunic (a paper-thin covering)

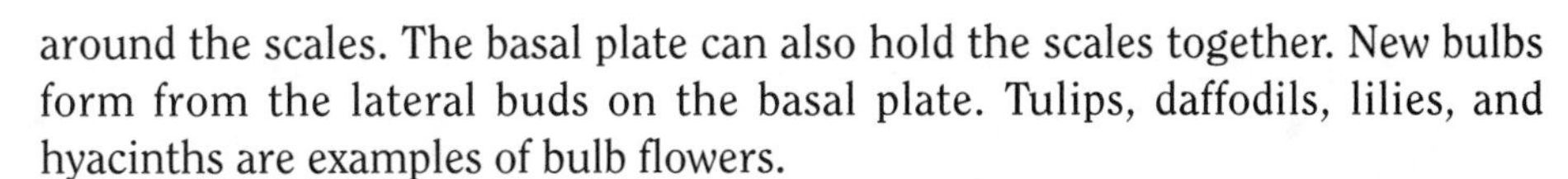


around the scales. The basal plate can also hold the scales together. New bulbs form from the lateral buds on the basal plate. Tulips, daffodils, lilies, and hyacinths are examples of bulb flowers.

*A corm*—is actually a stem that has been modified into a mass of storage tissue. The eye(s) at the top of the corm is a growing point. The corm is covered by dry leaf bases similar to the tunic-covering of the bulb. Roots grow from the basal plate on the underside of the corm. New corms form on top of or beside the old one. Corm-type flowers include gladiolus, freesia, and crocus.

*A tuber*—is a solid underground mass of stem like a corm but it lacks both a basal plate and a tunic. Roots and shoots grow from "eyes" (growth buds) out of its sides, bottom, and sometimes its top. Some tubers are roundish. Others are flattened and lumpy. Some examples of tubers are gloxinia, caladium, ranunculuses, and anemone.

*A tuberous root*—is a swollen root that has taken in moisture and nutrients. It resembles a tuber. New growth occurs on the base of the old stem, where it joins the root. A tuberous root can be divided by cutting off a section with an eye-bearing portion from where the old stem was attached. Dahlias have tuberous roots.

*A rhizome or a rootstock*—is a thickened, branching storage stem that usually grows laterally along or slightly below the soil surface. Roots develop downward on the bottom surface, while buds and leaves sprout upwards from the top of the rhizome. It is propagated by cutting the parent plant into sections. Japanese, Siberian, and bearded irises; cannas; calla lilies; and trillium are rhizomes.

## How are **carnivorous plants** categorized?

Carnivorous plants are plants that attract and catch animal prey digesting and absorbing the body juices for the nutrients. There are between 450 and 500 species and 12 genera of carnivorous plants classified according to the nature of their trapping mechanisms. Active traps display rapid motion in their capture of prey. The Venus fly trap (*Dionaea muscipula*) and the bladderwort (*Utricularia vulgaris*) are active traps. Semi-active traps employ a two-stage trap in which the prey is caught in the trap's adhesive fluid. As the prey struggles, the plant is triggered to slowly tighten its grip. The sundew (species *Drosera*) and butterwort (*Pinguicula vulgaris*) are semi-active traps. Passive traps entice insects by nectar. The insects fall into a reservoir of water and drown. An example of the passive trap is the pitcher plant (five genera). The Green Swamp Nature Preserve in southeastern North Carolina has the most numerous types of carnivorous plants.

## What plants are considered to be **child safe**?

These are some plants that are considered to be child safe even if they are swallowed by a child:

**The slender-leaved sundew, a carnivorous plant that attracts and feeds on insects.**

African violet
Aster
Begonia
Boston fern
California poppy
Coleus
Dandelion
Easter lily
Gardenia
Impatiens
Jade plant
Marigold
Norfolk Island pine
Petunia
Purple passion
Rose
Spider Plant
Swedish ivy
Tiger lily
Violet
Wandering Jew
Zebra plant

## What are the **symbolic meanings** of **herbs** and **plants**?

*Aloe*—Healing, protection, affection

*Angelica*—Inspiration

*Arbor vitae*—Unchanging friendship

*Bachelor's buttons*—Single blessedness

*Basil*—Good wishes, love

*Bay*—Glory



*Carnation*—Alas for my poor heart

### Are poinsettias poisonous to children and pets?

It is *not true* that poinsettias are poisonous to children and pets. Even if a child ate an entire plant, an upset stomach is probably all he or she would experience.

*Chamomile*—Patience
*Chives*—Usefulness
*Clover, White*—Think of me
*Coriander*—Hidden worth
*Cumin*—Fidelity
*Fennel*—Flattery
*Fern*—Sincerity
*Geranium, Oak-leaved*—True friendship
*Goldenrod*—Encouragement
*Heliotrope*—Eternal love
*Holly*—Hope
*Hollyhock*—Ambition
*Honeysuckle*—Bonds of love
*Horehound*—Health
*Hyssop*—Sacrifice, cleanliness
*Ivy*—Friendship, continuity
*Lady's mantle*—Comforting
*Lavender*—Devotion, virtue
*Lemon balm*—Sympathy
*Marjoram*—Joy, Happiness
*Mints*—Eternal refreshment
*Morning glory*—Affectation
*Nasturtium*—Patriotism
*Oak*—Strength
*Oregano*—Substance
*Pansy*—Thoughts
*Parsley*—Festivity
*Pine*—Humility
*Poppy, Red*—Consolation

*Rose*—Love
*Rosemary*—Remembrance
*Rudbeckia*—Justice
*Rue*—Grace, clear vision
*Sage*—Wisdom, immortality
*Salvia, Blue*—I think of you
*Salvia, Red*—Forever mine
*Savory*—Spice, interest
*Sorrel*—Affection
*Southernwood*—Constancy, jest
*Sweet-pea*—Pleasures
*Sweet woodruff*—Humility
*Tansy*—Hostile thoughts
*Tarragon*—Lasting interest
*Thyme*—Courage, strength
*Valerian*—Readiness
*Violet*—Loyalty, devotion
*Violet, Blue*—Faithfulness
*Violet, Yellow*—Rural happiness
*Willow*—Sadness
*Zinnia*—Thoughts of absent friends

What flowers are designated as **symbolic of each month** of the year?

| Month | Flower |
|---|---|
| January | Carnation |
| February | Violet |
| March | Jonquil |
| April | Sweet Pea |
| May | Lily of the Valley |
| June | Rose |
| July | Larkspur |
| August | Gladiola |
| September | Aster |
| October | Calendula |
| November | Chrysanthemum |
| December | Narcissus |

## What are the **most common colors** found in garden flowers offered in nurseries?

Several surveys have come up with the following estimates:

| Color | Estimate of popularity |
|---|---|
| white | 28% |
| yellow | 19% |
| red or purplish | 17% |
| blue | 16% |
| pink | 13% |
| orange | 4% |
| lavender or violet | 3% |

## What do the **different colors** and varieties of **roses symbolize**?

| Rose | Meaning |
|---|---|
| Yellow rose | Jealousy; unfaithfulness |
| Red rosebud | Youth; beauty |
| White rose | Silence |
| Lancaster rose | Union |
| Burgundy rose | Unconscious beauty |
| Musk rose | Capricious beauty |
| Dog rose | Pleasure and pain |
| Cabbage rose | Ambassador of love |
| Bridal rose | Happy love |
| Carolina rose | Dangerous love |
| May rose | Precocity |
| Moss rose | Voluptuousness |
| Christmas rose | Tranquility |
| Pompon rose | Gentility |

## Which variety of **orchid** is commonly used for corsages?

The pale purple Cattelya orchid, named for the English botanist William Cattley, is often used in corsages.

## Is there a **national flower** for the United States?

The United States adopted the rose as the national flower on October 7, 1986, after many years of deliberation. Other flowers suggested were the dogwood, the mountain laurel, and the columbine.

## What is the **national flower** of **different countries**?

| Country | National flower |
|---|---|
| Argentina | Ceibo |
| Australia | Wattle |
| Belgium | Poppy |
| Bolivia | Cantua buxifolia |
| Brazil | Cattleya |
| Canada | Sugar Maple |
| Chile | Chilean bellflower |
| China | Narcissus |
| Costa Rica | Cattleya |
| Denmark | Clover |
| Ecuador | Cinchona |
| Egypt | Lotus (water lily) |
| England | Rose |
| France | Fleur-de-lis |
| Germany | Cornflower |
| Greece | Violet |
| Holland | Tulip |
| Honduras | Rose |
| India | Lotus (Zizyphus) |
| Ireland | Shamrock |
| Italy | Lily |
| Japan | Chrysanthemum |
| Mexico | Prickly Pear |
| Newfoundland | Pitcher Plant |
| New Zealand | Silver fern |
| Norway | Heather |
| Persia | Rose |
| Poland | Poppy |
| Russia | Sunflower |
| Scotland | Thistle |
| South Africa | Protea cynaroides |
| Spain | Pomegranate |
| Sweden | Twinflower |
| Switzerland | Edelweiss |
| Wales | Leek |

## What special significance does the **passionflower** have?

Spanish friars of the sixteenth century first gave the name to this flower. They saw in the form of the passionflower (*Passiflora*) a representation of the passion of Christ: the

### Does the familiar phrase "open sesame" have anything to do with sesame seeds?

Sesame seeds burst open when they ripen, so the phrase is probably inspired by that fact. Middle Easterners were familiar wth the plant, and sesame seeds and flour are still used in Near Eastern cooking.

flowers have five petals and five sepals, which was thought to symbolize the 10 faithful apostles present at the crucifixion; the corona of five filaments was believed to resemble Christ's crown of thorns; the five stamens represented the five wounds in Christ's body, and the three stigmas stood for the nails driven into his hands and feet. Most species of passionflower are native to the tropical areas of the Western Hemisphere.

## How was **anise** used in ancient times?

The Romans brought the licorice-flavored herb from Egypt to Europe, where they used it for payment of taxes. It became popular flavoring for cakes, cookies, bread, and candy.

## What **wildflower** was used by Native Americans to **make red dye**?

Native Americans painted their faces and dyed their clothes red with the root of the wildflower bloodroot *(Sanguinaria),* also called redroot, Indian paint, and sometimes tetterwort. Bloodroot, found in shady, damp woodsy soils, blooms in May with two-inch-wide white flowers.

## What is the **first wildflower to bloom** in the spring in the northern United States?

The first flower of the northern spring is unusual and interesting, but rarely seen, because it blooms in the swamp. Commonly called skunk cabbage (*Spathyema foetidus*), it appears in February. The first spring flower that is generally known in New England and the Middle West is the *Hepatica,* or liverleaf, which blooms in March or early April.

## What is **wormwood**?

*Artemisia absinthium,* known as wormwood, is a hardy, spreading, fragrant perennial, two to four feet (61 to 122 centimeters) tall. It is native to Europe but widely naturalized in North America. The liqueur absinthe is flavored from this plant.

## What is the greatest number of leaves a **clover** can have?

A fourteen-leafed white clover (*Trifolium repens*) and a fourteen-leafed red clover (*Trifolium pratense*) have been found in the United States.

## What is a **weed**?

The dictionary definition is any plant considered undesirable, unattractive, or troublesome, especially a plant growing where it is not wanted in cultivated ground. Some celebrated authors have had their own definition of a weed. Ralph Waldo Emerson wrote, "What is a weed? A plant whose virtues have not yet been discovered." James Russell Lowell wrote in 1848, "A weed is no more than a flower in disguise." Ella Wheeler Wilcox, a Wisconsin poet (more famous for her expression "Laugh, and the world laughs with you; Weep, and you weep alone") wrote in her poem "The Weed" that "a weed is but an unloved flower." Finally, Shakespeare wrote in *Richard III,* "great weeds do grow apace."

## What is the origin of the name **Jimson weed**?

Jimson weed (*Datura stramonium*) is a corruption of the name Jamestown weed, given to a highly poisonous plant. The Jamestown, Virginia, colonists were familiar with this weed. It is also known as thorn apple, mad apple, stinkwort, angel's trumpet, devil's trumpet, stinkweed, dewtry, and white man's weed. Every part of this plant is potentially deadly if consumed in even moderate amounts. Even so, some of the alkaloids found in this plant are used by doctors as a preanesthetic.

## What are **luffa sponges**?

Luffas are nonwoody vines of the cucumber family. The interior fibrous skeletons of the fruit are used as sponges. The common name is sometimes spelled loofah. Dishcloth gourd, rag gourd, and vegetable sponge are other popular names for this sponge.

## Does **ivy cause damage** growing on a brick wall?

Experts at the New York Botanical Garden say it is possible for ivy to damage walls that are already in bad condition. Mortar that is in good shape is not normally subject to any damage.

A thick growth of ivy traps moisture against the walls. The adhesive disks that attach the ivy to the wall are subject to decay. This and other organic matter forms humic acid, which is capable of dissolving carbonate rock, like marble and lime mortar.

## Why do cats like **catnip**?

Catnip (*Nepeta cataria*) is a hardy perennial herb attractive to cats. Also known as catmint or catnep, it belongs to the mint family. The whole cat family (*Felidae*) reacts to catnip. Mountain lions, lynx, tigers, and lions roll over, rub their face, extend their claws, and do a body twist when they smell catnip's pungent odor. The oil from the leaves of catnip probably excites cats because it contains a chemical called trans-neptalactone, which closely resembles an excretion in a female cat's urine.

## Why are some mushrooms called **toadstools**?

The term toadstool goes back to the Middle Ages when it was popularly associated with the toad that was thought to be poisonous. Toadstools, with their stool-like shape, are mushrooms that are inedible or poisonous. It is said that when a toad is alarmed, the warts on its back secrete a poisonous substance called bufonenin.

## What are **living stones**?

Various succulent plants from the stony deserts of South Africa that mimic their surroundings are given this name. Each shoot or plant is made up of two grossly swollen leaves virtually fused together and colored to resemble a pebble. Large daisy-like flowers are borne from between the leaf pair.

## What is unique about the **water-lily** *Victoria amazonica?*

It is very big! Found only on the Amazon River, this water-lily has leaves that are up to six feet (1.8 meters) in diameter. The 12-inch (30-centimeter) flowers open at dusk on two successive nights.

# GARDENING, FARMING, ETC.

## What is considered the **most fertile, productive** type of **soil**?

There are three broad categories of garden soils: clay, sandy, and loam. Clay soils are heavy with the particles sticking close together. Working clay soil with your fingers you should get a shiny ball in your hand. Most plants have a hard time getting at the nutrients in clay soil, and the soil tends to get waterlogged. Clay soils can be good for a few deep rooted plants, such as mint, peas, and broad beans.

Sandy soils are light and have particles that do not stick together. Rolling sandy soil in your hand produces crumbs that split apart. A sandy soil is good for many alpine and arid plants, some herbs such as tarragon and thyme, and vegetables such as onions, carrots and tomatoes.

Loam soils are considered the best garden soils, because they are a well balanced mix of smaller and larger materials. They give up nutrients to plant roots easily and they drain well, but loam also retains water very well. Loam soil forms a ball in your fingers, but it feels gritty and shouldn't shine like clay soil.

## What do the numbers on a bag of **fertilizer** indicate?

The three numbers, such as 15–20–15, refer to the percentages by weight of macronutrients found in the fertilizer. The first number stands for nitrogen, the second for phosphorus, and the third for potassium. In order to determine the actual amount of each element in the fertilizer, multiply the percentage by the fertilizer's total weight in pounds. For example, in a 50-pound bag of 15–20–15, there are 7.5 pounds of nitrogen, 10 pounds of phosphorus, and 7.5 pounds of potassium. The remaining pounds are filler.

## What does **"pH"** mean when applied to **garden soil**?

Literally, pH stands for "potential of hydrogen" and is the term used by soil scientists to represent the hydrogen ion concentration in a soil sample. The relative alkalinity and acidity is commonly expressed in terms of the symbol pH. The neutral point in the scale is seven. Soil testing below seven is said to be acid; soil testing above pH seven is alkaline. The pH values are based on logarithms with a base of ten. Thus, a soil testing pH 5 is ten times as acidic as soil testing pH 6, while a soil testing pH 4 is one hundred times as acidic as soil testing pH 6.

## What is the **best soil pH** for growing plants?

Nutrients such as phosphorous, calcium, potassium, and magnesium are most available to plants when the soil pH is between 6.0 and 7.5. Under highly acid (low pH) conditions, these nutrients become insoluble and relatively unavailable for uptake by plants. High soil pH can also decrease the availability of nutrients. If the soil is more alkaline than pH 8, phosphorous, iron, and many trace elements become insoluble and unavailable for plant uptake.

## Is there a quick and easy way to test for **soil acidity** or **alkalinity**?

Some gardeners make a simple taste and smell test to check the soil. Acid soil smells and tastes sour. Some put a soil sample in a jar of vinegar. If the vinegar starts to bubble, your soil has plenty of lime. If there are no bubbles, lime the soil with four ounces (113 grams) of lime for every square yard (0.84 square meter).

### What are the guidelines for gardening by the moon?

Guidelines for lunar gardening are simple. The waxing moon occurs between the new moon and full moon, and includes the first and second quarter phases. The waning moon occurs between the full moon and new moon, and includes the third and fourth quarters. Generally all activities for growth and increase, especially plants that produce above the ground, should take place during the waxing moon. Vegetables and fruits intended to be eaten immediately should be gathered at the waxing moon. Cutting, controlling, and harvesting for food to be conserved or preserved, as well as planting crops that yield below the ground, should take place during the waning moon.

## When is the **best time to work the soil**?

Although it is possible to prepare the soil at any time of year, fall digging is the best time. Dig in the fall and leave the ground rough. Freezing and thawing during winter breaks up clods and aerates the soil. Insects that otherwise survive the winter are mostly turned out. The soil settling during winter will lessen the likelihood of air pockets in the soil when planting the following spring. Fall preparation provides time for soil additives such as manure and compost to break down before planting time.

## How can **garden soil** be used as **potting soil**?

The garden soil must be pasteurized and then mixed with coarse sand and peat moss. Soil may be pasteurized by putting the soil in a covered baking dish in the oven. When a meat thermometer stuck in the soil has registered 180°F (82°C) for 30 minutes, the soil is done.

## What is the composition of **synthetic soil**?

Synthetic soil is composed of a variety of organic and inorganic materials. Inorganic substances used include pumice, calcinated clay, cinders, vermiculite, perlite, and sand. Vermiculite and perlite are used for water retention and drainage. Organic materials used include wood residues, manure, sphagnum moss, plant residues, and peat. Sphagnum peat moss is also helpful for moisture retention and lowers the pH of the mixture. Lime may be added to offset the acidity of peat. Synthetic soil may also be referred to as growing medium, soil mixes, potting mixture, plant substrate, greenhouse soil, potting soil, and amended soil. Most synthetic soils are deficient in important mineral nutrients, which can be added during the mixing process or with water.

## What is meant by the term **double-digging**?

Double-digging produces an excellent deep planting bed for perennials, especially if the area is composed of heavy clay. It involves removing the top 10 inches (25 centimeters) of soil and moving it to a holding area, then spading the next 10 inches (25 centimeters) and amending this layer with organic matter and/or fertilizer. Then the soil from the "first" digging is amended as well, then replaced.

## What is meant by **xeriscaping**?

A xeriscape, a landscape of low water-use plants, is the modern approach to gardening in areas that experience water shortages. Taken from the Greek word xeros, meaning dry, this type of gardening uses drought-resistant plants and low maintenance grasses, which require water only every two to three weeks. Drip irrigation, heavy mulching of plant beds, and organic soil improvements are other xeriscape techniques that allow better water absorption and retention, which in turn decrease garden watering time.

## When a plant is said to be **"double dormant,"** what does that mean?

Plants that are double dormant require a unique sort of layering or stratification in order for the seeds to germinate. The seeds of these plants must have a period of warmth and moisture followed by a cold spell. Both the seed coat and the seed embryo require this double dormancy if they are to germinate. In nature this process usually takes two years. Some well-known plants that live the life of double dormancy include some lilies, dogwood, junipers, lilac, tree peony, and viburnum.

## What is meant by the phrase **rain shadow** in gardening?

The ground in the lee of a wall or solid fence receives less rainfall than the ground on the windward side. The wall or fence creates an area of rain shadow.

## What does the term **hydroponics** mean?

This term refers to growing plants in some medium other than soil; the inorganic plant nutrients (such as potassium, sulphur, magnesium, and nitrogen) are continuously supplied to the plants in solution. Hydroponics is mostly used in areas where there is little soil or unsuitable soil. Since it allows precise control of nutrient levels and oxygenation of the roots, it is often used to grow plants used for research purposes. Julius von Sachs (1832–1897), a researcher in plant nutrition, pioneered modern hydroponics. Research plants have been grown in solution culture since the mid-1800s. William Gericke, a scientist at the University of California, defined the word hydroponics in 1937. In the 50 years that hydroponics has been used on a commercial basis, it has been adapted to many situations. NASA will be using hydroponics in the space station for crop production and to recycle carbon dioxide into oxygen. Although successful for research, hydroponics has many limitations and may prove frustrating for the amateur gardener.

## How many years can **seeds** be kept?

Seeds stored in an airtight container and kept in a cool, dry place are usable for a long time. The following table indicates how long commonly used seed can be kept for planting:

| Vegetable | Years |
|---|---|
| Beans | 3 |
| Beets | 3 |
| Cabbage | 4 |
| Carrots | 1 |
| Cauliflower | 4 |
| Corn, sweet | 2 |
| Cucumbers | 5 |
| Eggplant | 4 |
| Kale | 3 |
| Lettuce | 4 |
| Melons | 4 |
| Onions | 1 |
| Peas | 1 |
| Peppers | 2 |
| Pumpkin | 4 |
| Radishes | 3 |
| Spinach | 3 |
| Squash | 4 |
| Swiss chard | 3 |
| Tomatoes | 3 |
| Turnips | 5 |

## How are **seedlings hardened off** before planting?

Hardening off is a gardening term for gradually acclimatizing seedlings raised indoors to the outdoor environment. Place the tray of seedlings outdoors for a few hours each day in a semi-protected spot. Lengthen the amount of time they stay out by an hour or so each day; at the end of the week, they will be ready for planting outdoors.

## What is the difference between **container-grown**, **balled-and-burlapped**, and **bare-rooted** plants?

Container-grown plants have been grown in some kind of pot—usually peat, plastic, or clay—for most or all of their lives. Balled-and-burlapped plants have been dug up with the soil carefully maintained around their roots in burlap. Bare-rooted plants have also been dug from their growing place but without retaining the root ball. Typically, plants from a mail-order nursery come bare-rooted with their roots protected with damp sphagnum moss. Bare-rooted plants are the most susceptible to damage.

## What is meant by **companion planting**?

There is not a lot of scientific documentation on this subject, but gardeners and farmers have noticed for years certain affinities that some plants have when planted near other plants. Nasturtiums, for example, lure aphids away from apple trees and attract blackfly away from vegetables. Onions and garlic act as both a fungicide and an insecticide, possibly because they accumulate sulfur very efficiently, and many pests avoid the odor. Some plants do not make good neighbors. Below is a sample list of good and bad companion plants:

| | Close Neighbors | Distant Neighbors |
|---|---|---|
| Bush Beans | Potatoes, lettuce, tomatoes | Onions |
| Carrots | Leaf lettuce, onions, tomatoes | — |
| Corn | Potatoes, beans, cucumbers | — |
| Cucumbers | Beans, corn | Potatoes |
| Lettuce | Carrots, cucumbers | — |
| Onions | Tomatoes, lettuce | Beans |
| Potatoes | Beans, corn | Cucumbers |
| Tomatoes | Onions, carrots | Potatoes |

## Which plants are best for **container** gardening?

Most vegetables can be grown in a container—even large ones like pumpkins. Miniature varieties of vegetables are better because they require less space and develop earlier. Fluorescent lights help leaf crops to grow indoors even in winter. Most root crops are best grown outdoors. Fruit crops such as tomatoes can be grown indoors, but need warm temperatures and at least six hours of summer sunshine. Some of the plants that may be grown are: bush beans, pole beans, beets, broccoli, cabbage, carrots, cucumbers, kale, lettuce, onions, peppers, summer squash, and tomatoes.

## What is the difference between an **arboretum** and a **botanical garden**?

An arboretum is technically a garden or collection of trees, often rare ones grown for study, research, or ornamentation. In practice, most arboretums also display shrubs and other plants. A botanical garden is primarily an institution for research in the field of botany and horticulture. The modern botanic garden has large collections of growing plants in a greenhouse and outdoors usually in elaborate gardens, in addition to research laboratories, a library and herbarium.



### Who established the first **botanical garden** in the United States?

John Bartram planned and laid out a botanic garden of five to six acres (two to 2.5 hectare) in 1728. It is located in Philadelphia, Pennsylvania.

### What is a **Shakespeare garden**?

A Shakespeare garden contains plants referred to by William Shakespeare in his plays and poems. Not all of the 200 flowers and herbs Shakespeare mentions will grow in the United States. Here is a list of gardens you may visit:

Golden Gate Park, San Francisco, California
Huntington Botanical Gardens, San Marino, California
Northwestern University, Evanston, Illinois
Ellis Park, Cedar Rapids, Iowa
Vassar College, Poughkeepsie, New York
Anne Hathaway Cottage, Wessington Springs, South Dakota

### What is **Ikebana**?

Ikebana is the Japanese expression for "the arrangement of living material in water." It is the ancient Japanese art of flower arrangement. Ikebana follows certain ancient rules that aim at achieving perfect harmony, beauty, and balance. Some describe Ikebana as sculpture with flowers. In Japan it has been practiced for fourteen hundred years. Buddhist monks in the sixth century practiced the art using pebbles, rock, and wood with plants and flowers. In Japan Ikebana was evolved and practiced exclusively by men—priests first, then warriors and noblemen. Today, of course, Ikebana is practiced by millions of women as well as men, although the great flower schools in Japan are mostly headed by men.

### What was a **victory garden**?

During World War I, patriots grew "liberty gardens." In World War II, U.S. Secretary of Agriculture Claude R. Wickard encouraged householders to plant vegetable gardens wherever they could find space. By 1945 there were said to be 20 million victory gardens producing about 40 percent of all American vegetables in many unused scraps of land. Such sites as the strip between a sidewalk and the street, town squares, and the land around Chicago's Cook County jail were used. The term "victory garden" derives from an English book by that title written by Richard Gardner in 1603.

## What is a recommended size for a beginner's **vegetable garden**?

It really depends on the amount of space available, how much produce is desired from a garden, and how much work a person is willing to put into it. A modest-sized, 10 × 20 foot (3 × 6 meter) plot, laid out in traditional rows, is quite manageable in terms of weeding, cultivating, planting, and harvesting. Even a 10 × 10 foot (3 × 3 meter) plot will suffice for a salad or "kitchen" garden, with plenty of greens and herbs for salads and seasonings on a daily basis. "Intensive" gardening methods, where plants are arranged in blocks rather than rows, allow for increased yield in an even smaller space. One 4 × 4 foot (1 × 1 meter) block, with a vertical frame at one end, can provide salad vegetables for one person throughout the growing season, though two blocks would provide a wider variety of vegetables.

## Should **tomato plants** be staked?

There are only a few advocates who continue to recommend non-staking. They argue for the natural sprawling growing method because it tends to give a greater yield. Staking keeps the plants off the ground, where they are susceptible to disease and attacks by snails. Staked tomatoes are much easier to harvest and make better use of garden space. They ripen faster and more evenly when staked. The method of staking has different advocates as well. Some say a sturdy five-foot stick is best, while others argue for a wire cage, and still others claim a wooden teepee arrangement is best.

## Is there a **"best" time to weed** in the vegetable garden?

Weeding is usually the most unpopular and the most time-consuming garden chore. Some studies (using weeding with peas and beans as examples) have shown that weeding done during the first three to four weeks of vegetable growth produced the best crops and that unabated weed growth after that time did not significantly reduce the vegetable yields.

## What are the best annual and perennial plants to grow to attract **butterflies**?

Ageratum, cosmos, globe candytuft, heliotrope, lantana, marigold, mexican sunflower, torch flower, nasturtium, sweet alyssum, buddleias (known as butterfly bush), dianthus, violas, and zinnia attract butterflies.

## Which flowers should be planted in the garden to attract **hummingbirds**?

Scarlet trumpet honeysuckle, weigela, butterfly bush, beardtongue, coralbells, red-hot-poker, foxglove, beebalm, nicotiana, petunia, summer phlox, and scarlet sage provide brightly colored (in shades of reds and orange), nectar-bearing attractants for hummingbirds.

## Why should **lawn clippings** be left on the grass after mowing?

The clippings are a valuable source of nutrients for the lawn. They provide nitrogen, potassium, and phosphorous to feed the new grass and reduce the need for fertilizer. Young, tender, short clippings decompose fast. Furthermore, when clippings are left on the lawn instead of being added to the trash collection, the amount of waste added to landfills is decreased.

## What is **snowmold** and how do you treat it?

Snowmold is a lawn disease common in the northern United States, characterized by a white, cottony growth. The fungus *Fusarium nivale* often grows beneath the snow as it melts in early spring. Avoiding late fall fertilizing in wet areas can prevent the spread of this disease. The lawn may be treated at the first sign of the disease with a fungicide and again in 10 to 14 days.

## How can **daffodils** be encouraged to bloom the year after they are planted?

Try fertilizing daffodils (*Narcissus pseudonarcissus*) as soon as the new shoots appear. This helps the roots renew themselves and will also aid in leaf and flower development. If they don't bloom, the problem could be overcrowding, which hinders flower production. Try digging the bulbs up every third or fifth year, separating them, and re-spacing them.

## How can **geraniums** be kept alive during the winter?

While they must be kept from freezing, geraniums can survive the winter happily in a cool sunny spot, such as a cool greenhouse, a bay window, or a sunny unheated basement. They need only occasional watering while in this semi-dormant state. Cuttings from these plants can be rooted in late winter or early spring for a new crop of geraniums (some sources suggest rooting in the fall). In homes without a suitable cool and sunny spot, the plants can be forced into dormancy by allowing the soil to dry completely, then gently knocking the soil off of the roots. While the plants can simply be hung from the rafters in a cool (45° to 50°F, or 7° to 10°C), slightly humid room, they will do better if put into individual paper bags, with the openings tied shut. The plants should be checked regularly. The leaves will dry and shrivel, but if the stems shrivel, the plants should be lightly misted with water. If any show mold or rot, cut off the affected sections, move the plants to a drier area, and leave the bags open for a day or two. In the early spring, prune the stems back to healthy green tissue and pot in fresh soil.

## What is meant by the **chilling requirement** for fruit trees?

When a fruit tree's fruiting period has ended, a dormant period must follow, during which the plant rests and regains strength for another fruit set the following year. The

John Chapman, the legendary apple tree planter who came to be known as Johnny Appleseed.

length of this set is measured in hours and occurs at temperatures between 32° and 45°F (0° to 7.2°C). A cherry tree requires about 700 hours of chilling time.

## What is a **five-in-one tree**?

These very curious trees consist of a rootstock with five different varieties of the same fruit—usually apples—grafted to it. The blooming period is usually magnificent with various colors of blooms appearing on the same tree.

## What is meant by **espaliering** a fruit tree and why is it done?

To espalier a fruit tree means to train it to grow flat against a surface. It can be grown in small places such as against a wall, and it will thrive even if its roots are underneath sidewalks or driveways. Since many fruit trees must be planted in pairs, espaliered fruit trees can be planted close together, providing pollen for each other, yet taking up little space.

## How did the **navel orange** originate?

Every navel orange is derived from a mutant tree that appeared on a plantation in Brazil in the early nineteenth century. A bud from the mutant tree was grafted onto another tree, whose branches were then grafted onto another, and so forth.

## How are **seedless grapes** grown?

Since seedless grapes cannot reproduce in the conventional way that grapes usually do (i.e., dropping seeds), growers have to take cuttings from other seedless grape plants and root them. Although the exact origin of seedless grapes is unknown, they might have been first cultivated in present-day Iran or Afghanistan thousands of years ago. Initially, the first seedless grape was a genetic mutation in which the hard seed casing failed to develop—the mutation is called stenospermoscarpy. One modern seedless grape commonly bought today is the green Thompson seedless grape, from which 90 percent of all raisins are made.

### Did Johnny Appleseed really plant apple trees?

John Chapman (1774–1845), called Johnny Appleseed, did plant apple orchards in the Midwest. He also encouraged the development of orchards farther west by giving pioneers free seedlings. His depiction as a barefoot tramp roaming the countryside scattering seeds at random from a bag slung over his shoulder, however, is more popular legend than fact. He was a curious figure who often preached from the Bible and from religious philosophy to passers-by. At the time of his death in 1845 he was a successful businessman who owned thousands of acres of orchards and nurseries.

## Is a **seedless watermelon** natural?

Seedless watermelon was first introduced in 1988 after 50 years of research. A seedless watermelon plant requires pollen from a regular, seeded watermelon. Farmers frequently plant seeded and seedless plants close together and depend on bees to pollinate the seedless plants. The white "seeds" found in seedless watermelon serve to hold a fertilized egg and embryo. Because the seedless melon is sterile and no fertilization takes place, the pod doesn't harden and become the familiar black watermelon seed.

## What is **pleaching**?

Pleaching is a method of shearing closely planted trees or shrubs into a high wall of foliage. Many kinds of trees have been used, including maples, sycamores, and lindens. Because of the time needed in caring for pleached allees, as the walls of foliage are called, they are infrequently seen in American gardens, but they are frequently observed in European ones.

## What is a **dwarf conifer**?

Conifers are evergreen shrubs and trees with needle-shaped leaves, cones, and resinous wood, such as the pines, spruces, firs, and junipers. After 20 years, dwarf or slow-growing forms of these otherwise tall trees are typically about three feet (91 centimeters) tall.

## What is the secret of **bonsai**, the Japanese art of growing dwarf trees?

These miniature trees with tiny leaves and twisted trunks can be centuries old. To inhibit growth of the plants, they have been carefully deprived of nutrients, pruned of their fastest-growing shoots and buds, and kept in small pots to reduce the root systems. Selective pruning, pinching out terminal buds, and wiring techniques are devices used to control the shape of the trees. Bonsai possibly started during the Chou dynasty (900–250 B.C.E.) in China, when emperors made miniature gardens that were dwarf representations of the provincial lands that they ruled.

## Can a branch of the **dogwood tree** be forced into bloom?

Forcing dogwood (genus *Cornus*) into bloom is similar to forcing forsythia. Bring the dogwood branch indoors when the buds begin to swell. Put the branch in water and set it in a sunny window.

## What are **chia seeds**?

The minuscule black chia seeds are gathered from a type of wild sage that is found in the southwest United States and Mexico. Chia seeds have a very high protein content. They cannot be sprouted in the conventional manner because of their highly mucilaginous nature. However, if spread on hollow earthenware vessels (often in the shape of animals) that are made especially for the purpose, the seeds will soon produce a green blanket of protein-rich chia sprouts ready for plucking.

## What is the difference between **poison ivy**, **oak**, and **sumac**?

These North American woody plants grow in almost any habitat and are quite similar in appearance. Each has alternating compound leaves of three leaflets each, berrylike fruits, and rusty brown stems. But poison ivy (*Rhus radicans*) acts more like a vine than a shrub at times and can grow high into trees. Its grey fruit is not hairy, and its leaves are slightly lobed. On the other hand, poison oak (*Rhus toxicodendron*) is often shrubby, but it can climb. Its leaflets are lobed and resemble oak leaves; its fruit is hairy. Poison sumac (*Rhus vernix*) grows only in North American wet acid swamps. This shrub can grow as high as 12 feet (3.6 meters). The fruit it produces hangs in a cluster and is grayish-brown in color. Poison sumac has sharply pointed, dark green, compound, alternating leaves; and greenish-yellow inconspicuous flowers. All parts of poison ivy, poison oak, and poison sumac can cause serious dermatitis.

## What is a natural way to get rid of **poison ivy**?

Poison ivy can be killed by spraying or treating the plants with a saltwater solution. Large plants can be killed by cutting the vines at or below ground level and soaking the base with brine. A second application after two weeks may be needed. Do not burn the plants; smoke and ash may cause the rash on exposed parts of the body, eyes, nasal passages, and lungs.

## What is the **railroad worm**?

The apple maggot (*Rhagoletis pomonella*), which becomes the apple fruit fly, is frequently called the railroad worm. Inhabiting orchards in the eastern United States and

Canada, the larvae feed on the fruit pulp of apples, plums, cherries, etc., and cause damage to fruit crops.

### Which **insects** are common problems in **strawberry** cultivation?

Earwigs, slugs, and snails are big problems in some areas. Strawberries (*Fragaria*) are also bothered by Japanese beetles, aphids, thrips, weevils, nematodes, and mites.

### How can **fruit trees** be protected from being eaten by **field mice**?

Valuable trees, especially newly planted fruit trees, can be protected by wrappings or guards of wire, wood veneer, or plastic. Other controls, such as pieces of lava rocks soaked in garlic, are effective as repellents, and garlic sprays will repel most rodents.

### Which type of **fence** protects a garden from **deer**?

A post and wire-mesh fence with a sharp-angled, narrow gate, which people and small animals can navigate, but deer cannot, plus the installation of a motion-sensing security light with a beeper helps keep deer away. Electric fences are also a good deterrent, but for smaller gardens snow fencing works well.

### How can **squirrels** be kept away from vegetable and flower gardens?

Squirrels like to take a bite out of tomatoes, cucumbers, and melons, dig up bulbs, and ruin anything colorful in the flower garden. The traditional recommendation of spreading mothballs around is apparently not too successful. A better method is laying down one to two inch (2.5 to five centimeter) mesh sheets of chicken wire. Squirrels will avoid the mesh, apparently because they fear getting their toes stuck in it. Another method to try is sprinkling hot pepper around the plants, renewing it after it rains.

### How do you keep **cats** away from catnip growing in the garden?

Instead of transplanting plants, which can bruise the leaves and release the oil that attracts cats, grow catnip directly from seeds. Try not to disturb the leaves after the plant grows. Once the scent is released, it is difficult to keep cats away.

### Before chemical sprays came along to control plant disease and pests, what were some of the **traditional spray formulas** used on plants?

Gardeners have used kitchen cupboard and organic plant materials for a very long time. Baking soda spray is a good fungicide. Mix two tablespoons of baking soda in four pints of water. Garlic spray is made by crushing a large garlic bulb into two pints

Agricultural chemist George Washington Carver developed important methods of crop management and hundreds of new uses for crops.

of water. Boil for five minutes and allow to cool. Garlic spray is a good insecticide and fungicide. The weed horsetail, the leaves of elder, and the leaves of the fern bracken have been made into sprays and used to fight mildew, black spot, many fungi and bacteria attacking garden plants.

## What were some of the accomplishments of **Dr. George Washington Carver**?

Because of the work of Dr. George Washington Carver (1864–1943) in plant diseases, soil analysis, and crop management, many southern farmers who adopted his methods increased their crop yields and profits. Carver developed recipes using cowpeas, sweet potatoes, and peanuts. He eventually made 118 products from sweet potatoes, 325 from peanuts, and 75 from pecans. He promoted soil diversification and the adoption of peanuts, soybeans, and other soil-enriching crops. His other work included developing plastic material from soy beans, which Henry Ford later used in part of his automobile. He extracted dyes and paints from the Alabama red clay and worked with hybrid cotton. Carver was a widely talented man who became an almost mythical American folk hero.

## When was the first practical **greenhouse** built?

French botanist Jules Charles constructed one in 1599 in Leiden, Holland, which housed tropical plants grown for medicinal purposes. The most popular plant there was an Indian date called the tamarind, whose fruit was made into a curative drink.

### When was the **first plant patent** issued?

Henry F. Bosenberg, a landscape gardener, received U.S. Plant Patent no. 1 on August 18, 1931, for a climbing or trailing rose.

# ANIMAL WORLD

## PHYSICAL CHARACTERISTICS, ETC.

*See Also: Environment—Extinct and Endangered Plants and Animals*

### When was the **first zoo** in the United States established?

The Philadelphia Zoological Garden, chartered in 1859, was the first zoo in the United States. The zoo was delayed by the Civil War, financial difficulties, and restrictions on transporting wild animals. It opened in 1874 on 33 acres, and 282 animals were exhibited.

### Which animal has the **longest gestation** period?

The animal with the longest gestation period is not a mammal; it is the viviparous amphibian the Alpine black salamander, which can have a gestation period of up to 38 months at altitudes above 4,600 feet (1,402 meters) in the Swiss alps. It bears two fully metamorphosed young.

### How **long** do animals, in particular mammals, **live**?

Of the mammals, humans and fin whales live the longest. Below is the maximum life span for several animal species.

| Animal | Latin name | Maximum life span in years |
|---|---|---|
| Marion's tortoise | *Testudo sumeirii* | 152+ |
| Quahog | *Venus mercenaria* | ca. 150 |
| Common box tortoise | *Terrapene carolina* | 138 |

| Animal | Latin name | Maximum life span in years |
|---|---|---|
| European pond tortoise | *Emys orbicularis* | 120+ |
| Spur-thighed tortoise | *Testudo graeca* | 116+ |
| Fin whale | *Balaenoptera physalus* | 116 |
| Human | *Homo sapiens* | 116 |
| Deep-sea clam | *Tindaria callistiformis* | ca. 100 |
| Killer whale | *Orcinus orca* | ca. 90 |
| European eel | *Anguilla anguilla* | 88 |
| Lake sturgeon | *Acipenser fulvescens* | 82 |
| Freshwater mussel | *Margaritana margaritifera* | 80 to 70 |
| Asiatic elephant | *Elephas maximus* | 78 |
| Andean condor | *Vultur gryphus* | 72+ |
| Whale shark | *Rhiniodon typus* | ca. 70 |
| African elephant | *Loxodonta africana* | ca. 70 |
| Great eagle-owl | *Bubo bubo* | 68+ |
| American alligator | *Alligator mississipiensis* | 66 |
| Blue macaw | *Ara macao* | 64 |
| Ostrich | *Struthio camelus* | 62.5 |
| Horse | *Equus caballus* | 62 |
| Orangutan | *Pongo pygmaeus* | ca. 59 |
| Bataleur eagle | *Terathopius ecaudatus* | 55 |
| Hippopotamus | *Hippopotamus amphibius* | 54.5 |
| Chimpanzee | *Pan troglodytes* | 51 |
| White pelican | *Pelecanus onocrotalus* | 51 |
| Gorilla | *Gorilla gorilla* | 50+ |
| Domestic goose | *Anser a. domesticus* | 49.75 |
| Grey parrot | *Psittacus erythacus* | 49 |
| Indian rhinoceros | *Rhinoceros unicornis* | 49 |
| European brown bear | *Ursus arctos arctos* | 47 |
| Grey seal | *Halichoerus gryphus* | 46+ |
| Blue whale | *Balaenoptera musculus* | ca. 45 |
| Goldfish | *Carassius auratus* | 41 |
| Common toad | *Bufo bufo* | 40 |
| Roundworm | *Tylenchus polyhyprus* | 39 |
| Giraffe | *Giraffa camelopardalis* | 36.25 |
| Bactrian camel | *Camelus ferus* | 35+ |
| Brazilian tapir | *Tapirus terrestris* | 35 |
| Domestic cat | *Felis catus* | 34 |
| Canary | *Serinus caneria* | 34 |
| American bison | *Bison bison* | 33 |
| Bobcat | *Felis rufus* | 32.3 |
| Sperm whale | *Physeter macrocephalus* | 32+ |

| Animal | Latin name | Maximum life span in years |
|---|---|---|
| American manatee | *Trichechus manatus* | 30 |
| Red kangaroo | *Macropus rufus* | ca. 30 |
| African buffalo | *Syncerus caffer* | 29.5 |
| Domestic dog | *Canis familiaris* | 29.5 |
| Lion | *Panthera leo* | ca. 29 |
| African civet | *Viverra civetta* | 28 |
| Theraphosid spider | *Mygalomorphae* | ca. 28 |
| Red deer | *Cervus elaphus* | 26.75 |
| Tiger | *Panthera tigris* | 26.25 |
| Giant panda | *Ailuropoda melanoleuca* | 26 |
| American badger | *Taxidea taxus* | 26 |
| Common wombat | *Vombatus ursinus* | 26 |
| Bottle-nosed dolphin | *Tursiops truncatus* | 25 |
| Domestic chicken | *Gallus g. domesticus* | 25 |
| Grey squirrel | *Sciurus carolinensis* | 23.5 |
| Aardvark | *Orycteropus afer* | 23 |
| Domestic duck | *Anas platyrhynchos domesticus* | 23 |
| Coyote | *Canis latrans* | 21+ |
| Canadian otter | *Lutra canadensis* | 21 |
| Domestic goat | *Capra hircus domesticus* | 20.75 |
| Queen ant | *Myrmecina graminicola* | 18+ |
| Common rabbit | *Oryctolagus cuniculus* | 18+ |
| White or beluga whale | *Delphinapterus leucuas* | 17.25 |
| Platypus | *Ornithorhynchus anatinus* | 17 |
| Walrus | *Odobenus rosmarus* | 16.75 |
| Domestic turkey | *Melagris gallapave domesticus* | 16 |
| American beaver | *Castor canadensis* | 15+ |
| Land snail | *Helix spiriplana* | 15 |
| Guinea pig | *Cavia porcellus* | 14.8 |
| Hedgehog | *Erinaceus europaeus* | 14 |
| Burmeister's armadillo | *Calyptophractus retusus* | 12 |
| Capybara | *Hydrochoerus hydrochaeris* | 12 |
| Chinchilla | *Chinchilla laniger* | 11.3 |
| Giant centipede | *Scolopendra gigantea* | 10 |
| Golden hamster | *Mesocricetus auratus* | 10 |
| Segmented worm | *Allolobophora longa* | 10 |
| Purse-web spider | *Atypus affinis* | 9+ |
| Greater Egyptian gerbil | *Gerbillus pyramidum* | 8+ |
| Spiny starfish | *Marthasterias glacialis* | 7+ |
| Millipede | *Cylindroiulus landinensis* | 7 |
| Coypu | *Myocastor coypus* | 6+ |

| Animal | Latin name | Maximum life span in years |
|---|---|---|
| House mouse | *Mus musculus* | 6 |
| Malagasy brown-tailed mongoose | *Salanoia concolor* | 4.75 |
| Cane rat | *Thryonomys swinderianus* | 4.3 |
| Siberian flying squirrel | *Pteromys volans* | 3.75 |
| Common octopus | *Octopus vulgaris* | 2 to 3 |
| Pygmy white-toothed shrew | *Suncus etruscus* | 2 |
| Pocket gopher | *Thomomys talpoides* | 1.6 |
| Monarch butterfly | *Danaus plexippus* | 1.13 |
| Bedbug | *Cimex lectularius* | 0.5 (182 days) |
| Black widow spider | *Latrodectus mactans* | 0.27 (100 days) |
| Common housefly | *Musca domesticus* | 0.04 (17 days) |

## What are the **largest** and **smallest** living animals?

| Largest animals | Name | Length and weight |
|---|---|---|
| Sea mammal | Blue or sulphur-bottom whale (*Balaenoptera musculus*) | 100–110 feet (30.5–33.5 meters) long; weighs 135–209 tons (122.4–189.6 tonnes) |
| Land mammal | African bush elephant (*Loxodonta africana*) | Bull is 10.5 feet (3.2 meters) tall at shoulder; weighs 5.25–6.2 tons (4.8–5.6 tonnes) |
| Living bird | North African ostrich (*Struthio c. camelus*) | 8 to 9 feet (2.4–2.7 meters) tall; weighs 345 pounds (156.5 kilograms) |
| Fish | Whale shark (*Rhincodon typus*) | 41 feet (12.5 meters) long; weighs 16.5 tons (15 tonnes) |
| Reptile | Saltwater crocodile (*Crocodylus porosus*) | 14–16 feet (4.3–4.9 meters) long; weighs 900–1,500 pounds (408–680 kilograms) |
| Rodent | Capybara (*Hydrochoerus hydrochaeris*) | 3.25–4.5 feet (1–1.4 meters) long; weighs 250 pounds (113.4 kilograms) |
| **Smallest animals** | **Name** | **Length and weight** |
| Sea mammal | Commerson's dolphin (*Cephalorhynchus* commersonii) | 4–5.6 feet (1.25–1.7 meters); weighs 50–70 pounds (22.7–31.8 kilograms) |
| Land mammal | Bumblebee or Kitti's hog-nosed bat (*Craseonycteris thong longyai*) or the pygmy shrew (*Suncus erruscus*) | 1 inch (2.54 centimeters) long; weighs .062 to .07 ounces (1.6–2 grams)<br>1.5–2 inches (3.8–5 centimeters) long; weighs 0.052–.09 ounces (1.5–2.6 grams) |
| Bird | Bee hummingbird (*Mellisuga helenea*) | 2.25 inches (5.7 centimeters) long; weighs 0.056 ounces (1.6 grams) |

Prior to hibernation, bears in the wild gorge on food—sometimes from human sources.

| Smallest animals | Name | Length and weight |
|---|---|---|
| Fish | Dwarf goby (*Trimmatam nanus*) | 0.35 inches (8.9 millimeters) long |
| Reptile | Gecko (*Spaerodactylus ariasae*) | 0.63 inches (1.6 centimeters) long |
| Rodent | Pygmy mouse (*Baiomys taylori*) | 4.3 inches (10.9 centimeters) long; weighs 0.24–0.28 ounces (6.8–7.9 grams) |

## Do bears in zoos **hibernate**?

Bears do not hibernate in zoos because temperatures in cages and enclosures remain warm throughout the year and the bears are constantly fed by keepers. Hibernation occurs only with lack of food and temperatures below the freezing point.

## How do animals and people **identify smells**?

The sense of smell allows animals and humans as well as other organisms to identify food, mates, and predators, and provides sensory pleasure (e.g. flowers) and warnings of danger (e.g. chemical dangers). There are specialized receptor cells in the nose that have proteins that bind chemical odorants and cause the receptor cells to send electrical signals to the olfactory bulb of the brain. Cells in the olfactory bulb relay this information to olfactory areas of the forebrain to generate the perception of smells.

## Besides humans, which animals are the **most intelligent**?

According to Edward O. Wilson, a behavioral biologist, the ten most intelligent animals are the following:

1. Chimpanzee (two species)
2. Gorilla
3. Orangutan
4. Baboon (seven species, including drill and mandrill)
5. Gibbon (seven species)
6. Monkey (many species, especially macaques, the patas, and the Celebes black ape)
7. Smaller toothed whale (several species, especially killer whale)
8. Dolphin (many of the approximately 80 species)
9. Elephant (two species)
10. Pig

## Do animals other than humans have **fingerprints**?

It is known that gorillas and other primates have fingerprints. Of special interest, however, is that our closest relative, the chimpanzee, does not. Koala bears also have fingerprints. Researchers in Australia have determined that the fingerprints of koala bears closely resemble those of human fingerprints in size, shape, and pattern.

## Do animals have **color vision**?

Most reptiles and birds appear to have a well-developed color sense. Most mammals, however, are color-blind. Apes and monkeys have the ability to tell colors apart; dogs and cats seem to be color-blind and see only shades of black, white, and gray.

## Can animals **regenerate** parts of their bodies?

Regeneration does occur in some animals; however, it progressively declines the more complex the animal species becomes. Among primitive invertebrates (lacking a backbone), regeneration frequently occurs. For example, a planarium (flatworm) can split symmetrically, each part becoming a clone of the other. In higher invertebrates regeneration occurs in echinoderms (such as starfish) and arthropods (such as insects and crustaceans). Regeneration of appendages (limbs, wings, and antennae) occurs in insects (such as cockroaches, fruit flies, and locusts) and in crustaceans (such as lobsters, crabs, and crayfish). For example, regeneration of the crayfish's missing claw occurs at its next molt (shedding of its hard cuticle exterior shell/skin for the growing and the subsequent hardening of a new cuticle exterior). However, sometimes the regenerated claw does not achieve the same size of the missing claw. But after every molt (occurring two to three times a year) it grows and will eventually become nearly

as large as the original claw. On a very limited basis, some amphibians and reptiles can replace a lost leg or tail.

## What is the frequency that separates animal **hearing** from human?

The frequency of a sound is the pitch. Frequency is expressed in Hertz or Hz. Sounds are classified as infrasounds (below the human range of hearing), sonic range (within the range of human hearing) and ultrasound (above the range of human hearing).

| Animal | Frequency range heard (Hz) |
|---|---|
| Dog | 15 to 50,000 |
| Human | 20 to 20,000 |
| Cat | 60 to 65,000 |
| Dolphin | 150 to 150,000 |
| Bat | 1,000 to 120,000 |

## Do animals have **blood types**?

The number of recognized blood groups varies from species to species:

| Species | Number of blood groups |
|---|---|
| Pig | 16 |
| Cow | 12 |
| Chicken | 11 |
| Horse | 9 |
| Sheep | 7 |
| Dog | 7 |
| Rhesus monkey | 6 |
| Mink | 5 |
| Rabbit | 5 |
| Mouse | 4 |
| Rat | 4 |
| Cat | 2 |

## Do all animals have **red blood**?

The color of blood is related to the compounds that transport oxygen. Hemoglobin, containing iron, is red and is found in all vertebrates (animals having a backbone) and a few invertebrates (animals lacking a backbone). Annelids (segmented worms) have either a green pigment, chlorocruorin, or a red pigment, hemerythrin. Some crustaceans (arthropods having divided bodies and generally having gills) have a blue pigment, hemocyanin, in their blood.

## Do any animals snore?

Many animals have been observed snoring occasionally, including dogs, cats, cows, oxen, sheep, buffaloes, elephants, camels, lions, leopards, tigers, gorillas, chimpanzees, horses, mules, zebras, and elands.

## Which animals can **run faster than a human**?

The cheetah, the fastest mammal, can accelerate from zero to 45 miles (64 kilometers) per hour in two seconds; it has been timed at speeds of 70 miles (112 kilometers) per hour over short distances. In most chases, cheetahs average around 40 miles (63 kilometers) per hour. Humans can run very short distances at almost 28 miles (45 kilometers) per hour maximum. Most of the speeds given in the table below are for distances of one-quarter mile (0.4 kilometer).

| Animal | Maximum speed (mph) | Maximum speed (kph) |
|---|---|---|
| Cheetah | 70 | 112.6 |
| Pronghorn antelope | 61 | 98.1 |
| Wildebeest | 50 | 80.5 |
| Lion | 50 | 80.5 |
| Thomson's gazelle | 50 | 80.5 |
| Quarter horse | 47.5 | 76.4 |
| Elk | 45 | 72.4 |
| Cape hunting dog | 45 | 72.4 |
| Coyote | 43 | 69.2 |
| Gray fox | 42 | 67.6 |
| Hyena | 40 | 64.4 |
| Zebra | 40 | 64.4 |
| Mongolian wild ass | 40 | 64.4 |
| Greyhound | 39.4 | 63.3 |
| Whippet | 35.5 | 57.1 |
| Rabbit (domestic) | 35 | 56.3 |
| Mule deer | 35 | 56.3 |
| Jackal | 35 | 56.3 |
| Reindeer | 32 | 51.3 |
| Giraffe | 32 | 51.3 |
| White-tailed deer | 30 | 48.3 |
| Wart hog | 30 | 48.3 |
| Grizzly bear | 30 | 48.3 |
| Cat (domestic) | 30 | 48.3 |
| Human | 27.9 | 44.9 |

# NAMES

## What names are used for **male** and **female animals**?

| Animal | Male name | Female name |
|---|---|---|
| Alligator | Bull | |
| Ant | | Queen |
| Ass | Jack, jackass | Jenny |
| Bear | Boar or he-bear | Sow or she-bear |
| Bee | Drone | Queen or queen bee |
| Camel | Bull | Cow |
| Caribou | Bull, stag, or hart | Cow or doe |
| Cat | Tom, tomcat, gib, gibeat, boarcat, or ramcat | Tabby, grimalkin, malkin, pussy, or queen |
| Chicken | Rooster, cock, stag, or chanticleer | Hen, partlet, or biddy |
| Cougar | Tom or lion | Lioness, she-lion, or pantheress |
| Coyote | Dog | Bitch |
| Deer | Buck or stag | Doe |
| Dog | Dog | Bitch |
| Duck | Drake or stag | Duck |
| Fox | Fox, dog-fox, stag, reynard, or renard | Vixen, bitch, or she-fox |
| Giraffe | Bull | Cow |
| Goat | Buck, billy, billie, billie-goat or he-goat | She-goat, nanny, nannie, or nannie-goat |
| Goose | Gander or stag | Goose or dame |
| Guinea pig | Boar | |
| Horse | Stallion, stag, horse, stud, slot, stable horse, sire, or rig | Mare or dam |
| Impala | Ram | Ewe |
| Kangaroo | Buck | Doe |
| Leopard | Leopard | Leopardess |
| Lion | Lion or tom | Lioness or she-lion |
| Lobster | Cock | Hen |
| Manatee | Bull | Cow |
| Mink | Boar | Sow |
| Moose | Bull | Cow |
| Mule | Stallion or jackass | She-ass or mare |
| Ostrich | Cock | Hen |
| Otter | Dog | Bitch |

| Animal | Male name | Female name |
|---|---|---|
| Owl | | Jenny or howlet |
| Ox | Ox, beef, steer, or bullock | Cow or beef |
| Partridge | Cock | Hen |
| Peafowl | Peacock | Peahen |
| Pigeon | Cock | Hen |
| Quail | Cock | Hen |
| Rabbit | Buck | Doe |
| Reindeer | Buck | Doe |
| Robin | Cock | |
| Seal | Bull | Cow |
| Sheep | Buck, ram, male-sheep, or mutton | Ewe or dam |
| Skunk | Boar | |
| Swan | Cob | Pen |
| Termite | King | Queen |
| Tiger | Tiger | Tigress |
| Turkey | Gobbler or tom | Hen |
| Walrus | Bull | Cow |
| Whale | Bull | Cow |
| Woodchuck | He-chuck | She-chuck |
| Wren | | Jenny or jennywren |
| Zebra | Stallion | Mare |

## What names are used for **juvenile animals**?

| Animal | Name for young |
|---|---|
| Ant | Antling |
| Antelope | Calf, fawn, kid, or yearling |
| Bear | Cub |
| Beaver | Kit or kitten |
| Bird | Nestling |
| Bobcat | Kitten or cub |
| Buffalo | Calf, yearling, or spike-bull |
| Camel | Calf or colt |
| Canary | Chick |
| Caribou | Calf or fawn |
| Cat | Kit, kitten, kitling, kitty, or pussy |
| Cattle | Calf, stot, or yearling (m. bullcalf or f. heifer) |
| Chicken | Chick, chicken, poult, cockerel, or pullet |
| Chimpanzee | Infant |

| | |
|---|---|
| Cicada | Nymph |
| Clam | Littleneck |
| Cod | Codling, scrod, or sprag |
| Condor | Chick |
| Cougar | Kitten or cub |
| Cow | Calf (m. bullcalf; f. heifer) |
| Coyote | Cub, pup, or puppy |
| Deer | Fawn |
| Dog | Whelp or puppy |
| Dove | Pigeon or squab |
| Duck | Duckling or flapper |
| Eagle | Eaglet |
| Eel | Fry or elver |
| Elephant | Calf |
| Elk | Calf |
| Fish | Fry, fingerling, minnow, or spawn |
| Fly | Grub or maggot |
| Frog | Polliwog or tadpole |
| Giraffe | Calf |
| Goat | Kid |
| Goose | Gosling |
| Grouse | Chick, poult, squealer, or cheeper |
| Horse | Colt (m.), foal, stot, stag, filly (f.), hog-colt, youngster, yearling, or hogget |
| Kangaroo | Joey |
| Leopard | Cub |
| Lion | Shelp, cub, or lionet |
| Louse | Nit |
| Mink | Kit or cub |
| Monkey | Suckling, yearling, or infant |
| Mosquito | Larva, flapper, wriggler, or wiggler |
| Muskrat | Kit |
| Ostrich | Chick |
| Otter | Pup, kitten, whelp, or cub |
| Owl | Owlet or howlet |
| Oyster | Set seed, spat, or brood |
| Partridge | Cheeper |
| Pelican | Chick or nestling |
| Penguin | Fledgling or chick |
| Pheasant | Chick or poult |
| Pigeon | Squab, nestling, or squealer |
| Quail | Cheeper, chick, or squealer |
| Rabbit | Kitten or bunny |
| Raccoon | Kit or cub |

| Animal | Name for young |
|---|---|
| Reindeer | Fawn |
| Rhinoceros | Calf |
| Sea Lion | Pup |
| Seal | Whelp, pup, cub, or bachelor |
| Shark | Cub |
| Sheep | Lamb, lambkin, shearling, or yearling |
| Skunk | Kitten |
| Squirrel | Dray |
| Swan | Cygnet |
| Swine | Shoat, trotter, pig, or piglet |
| Termite | Nymph |
| Tiger | Whelp or cub |
| Toad | Tadpole |
| Turkey | Chick or poult |
| Turtle | Chicken |
| Walrus | Cub |
| Weasel | Kit |
| Whale | Calf |
| Wolf | Cub or pup |
| Woodchuck | Kit or cub |
| Zebra | Colt or foal |

## What names are used for **groups of animals**?

| Animal | Group name |
|---|---|
| Ants | Nest, army, colony, state, or swarm |
| Bees | Swarm, cluster, nest, hive, or erst |
| Caterpillars | Army |
| Eels | Swarm or bed |
| Fish | School, shoal, haul, draught, run, or catch |
| Flies | Business, hatch, grist, swarm, or cloud |
| Frogs | Arm |
| Gnats | Swarm, cloud, or horde |
| Goldfish | Troubling |
| Grasshoppers | Cloud |
| Hornets | Nest |
| Jellyfish | Smuck or brood |
| Lice | Flock |
| Locusts | Swarm, cloud, or plague |
| Minnows | Shoal, steam, or swarm |
| Oysters | Bed |

| Animal | Group name |
|---|---|
| Sardines | Family |
| Sharks | School or shoal |
| Snakes | Bed, knot, den, or pit |
| Termites | Colony, nest, swarm, or brood |
| Toads | Nest, knot, or knab |
| Trout | Hover |
| Turtles | Bale or dole |
| Wasps | Nest, herd, or pladge |



# INSECTS, SPIDERS, ETC.

## How many **species of insects** are there?

Estimates of the number of recognized insect species range from about 750,000 to upward of one million—but some experts think that this represents less than half of the number that exists in the world. About 7,000 new insect species are described each year, but unknown numbers are lost annually from the destruction of their habitats, mainly tropical forests.

## Why are insects often **found in amber**?

People have long been infatuated with amber, the fossilized form of ancient tree resin, a semiprecious stone used for jewelry and mosaics. Amber from the Dominican Republic contains an average of one insect in every hundred pieces. Some pieces of amber contain thousands of insects—both whole insects and insect fragments. These insects were probably crawling or lodged on the outside of a tree about 30 million years ago and became trapped by a glob of sticky tree resin that continued to ooze around the animal matter and eventually fossilized. Scientists are able to study these insects, many of which are extinct, and may be found to be the missing link to modern-day species.

## What is the **most destructive insect** in the world?

The most destructive insect is the desert locust (*Schistocera gregaria*), the locust of the Bible, whose habitat ranges from the dry and semi-arid regions of Africa and the Middle East, through Pakistan and northern India. This short-horn grasshopper can eat its own weight in food a day, and during long migratory flights a large swarm can consume 20,000 tons (18,144,000 kilograms) of grain and vegetation a day.

The desert locust—the world's most destructive insect.

## Who introduced the **gypsy moth** into the United States?

In 1869, Professor Leopold Trouvelot brought gypsy moth egg masses from France to Medford, Massachusetts. His intention was to breed the gypsy moth with the silkworm to overcome a wilt disease of the silkworm. He placed the egg masses on a window ledge, and evidently the wind blew them away. About 10 years later these caterpillars were numerous on trees in that vicinity, and in 20 years, trees in eastern Massachusetts were being defoliated. In 1911, a contaminated plant shipment from Holland also introduced the gypsy moth to that area. These pests have now spread to 25 states, especially in the northeastern United States. Scattered locations in Michigan and Oregon have also reported occurrences of gypsy moth infestations.

## Are there any **natural predators** of **gypsy moth caterpillars**?

About 45 kinds of birds, squirrels, chipmunks, and white-footed mice eat this serious insect pest. Among the 13 imported natural enemies of the moth, two flies, *Compislura concinnata* (a tachnid fly) and *Sturnia scutellata*, parasitize the caterpillar. Other parasites and various wasps have also been tried as controls, as well as spraying and male sterilization. Originally from Europe, this large moth (*Porthetria dispar*) lays its eggs on the leaves of oaks, birches, maples, and other hardwood trees. When the yellow, hairy caterpillars hatch from the eggs, they devour the leaves in such quantities that the tree becomes temporarily defoliated. Sometimes this causes the tree to die. The caterpillars grow from half an inch (three millimeters) to about two

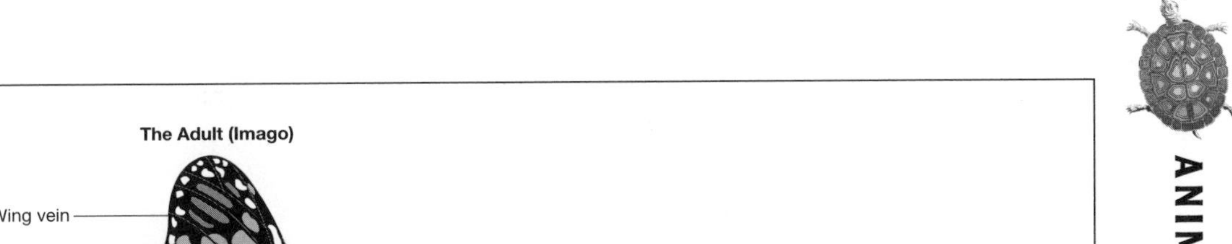

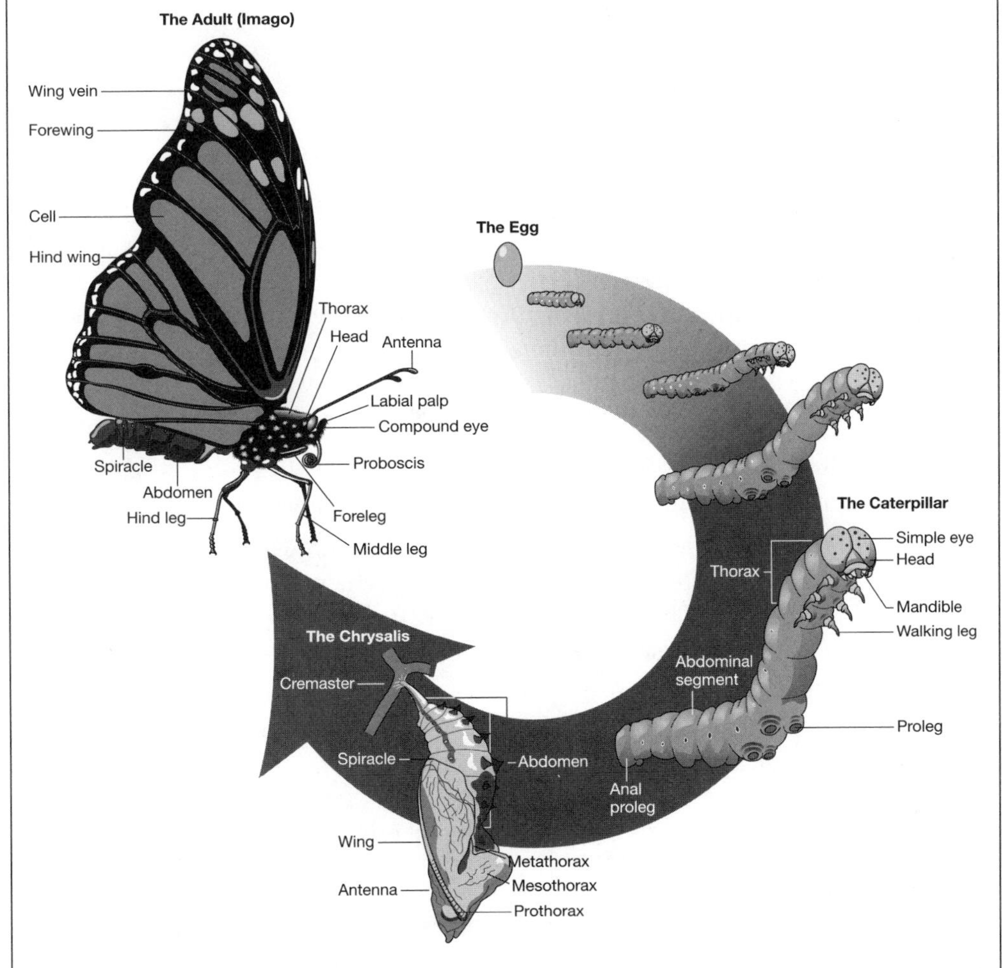


**The life cycle of a butterfly represents complete metamorphoses.**

inches (5.1 centimeters) before they spin a pupa, in which they will metamorphose into adult moths.

## What are some **beneficial insects**?

Beneficial insects include bees, wasps, flies, butterflies, moths, and others that pollinate plants. Many fruits and vegetables depend on insect pollinators for the production of seeds. Insects are an important source of food for birds, fish, and many animals. In some countries such insects as termites, caterpillars, ants, and bees are eaten as food by people. Products derived from insects include honey and beeswax, shellac, and silk. Some predators such as mantises, ladybugs or lady beetles, and lacewings feed on other harmful insects. Other helpful insects are parasites that live on or in the body of harmful insects. For example, some wasps lay their eggs in caterpillars that damage tomato plants.

## What are the stages of **insect metamorphosis**?

There are two types of metamorphoses (marked structural changes in the growth processes): complete and incomplete. In complete metamorphosis, the insect (such as the ant, moth, butterfly, termite, wasp, or beetle) goes through all the distinct stages of growth to reach adulthood. In incomplete metamorphosis, the insect (such as the grasshopper, cricket, or louse) does not go through all the stages of complete metamorphoses.

**Complete metamorphosis**

*Egg*—One egg is laid at a time or many (as much as 10,000).

*Larva*—What hatches from the eggs is called "larva." A larva can look like a worm.

*Pupa*—After reaching its full growth, the larva hibernates, developing a shell or "pupal case" for protection. A few insects (e.g., the moth) spin a hard covering called a "cocoon." The resting insect is called a pupa (except the butterfly, which is called a chrysalis), and remains in the hibernation state for several weeks or months.

*Adult*—During hibernation, the insect develops its adult body parts. When it has matured physically, the fully grown insect emerges from its case or cocoon.

**Incomplete metamorphosis**

*Egg*—One egg or many eggs are laid.

*Early-stage nymph*—Hatched insect resembles an adult, but smaller in size. However, those insects that would normally have wings have not yet developed them.

*Late-stage nymph*—At this time, the skin begins to molt (shed), and the wings begin to bud.

*Adult*—The insect is now fully grown.

## Which **butterfly gardens** can the public visit?

Butterfly gardens are gardens that cultivate special plants to attract various species of butterfly. There are many butterfly gardens open to the public throughout the United States. Butterfly gardens are located in zoos, botanic gardens, arboretums, nature preserves, universities, and commercial gardens. The Cecil B. Day Butterfly Center at Callaway Gardens in Pine Mountain, Georgia, is perhaps the largest. Over 100 species visit the outdoor garden. There is also Butterfly World at Tradewinds Park South in Coconut

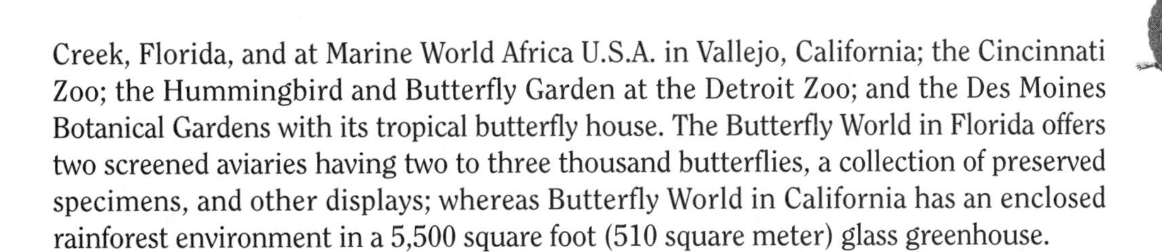

Creek, Florida, and at Marine World Africa U.S.A. in Vallejo, California; the Cincinnati Zoo; the Hummingbird and Butterfly Garden at the Detroit Zoo; and the Des Moines Botanical Gardens with its tropical butterfly house. The Butterfly World in Florida offers two screened aviaries having two to three thousand butterflies, a collection of preserved specimens, and other displays; whereas Butterfly World in California has an enclosed rainforest environment in a 5,500 square foot (510 square meter) glass greenhouse.



## Has the U.S. selected a **national insect**?

A group of citizens has petitioned the United States Congress to name the monarch butterfly as the national insect, but, to date, they have not been successful.

## How does a **butterfly** differ from a **moth**?

| Characteristic | Butterflies | Moths |
|---|---|---|
| Antennae | Knobbed | Unknobbed |
| Active-time of day | Day | Night |
| Coloration | Bright | Dull |
| Resting wing position | Vertically above body | Horizontally beside body |

*Note:* While these guidelines generally hold true, there are exceptions. Moths have hairy bodies, and most have tiny hooks or bristles linking the fore-wing to the hind-wing; butterflies do not have either characteristic.

## What is the most popular **state insect**?

The honeybee is by far the most popular state insect, having been selected by sixteen states: Arkansas, California (nicknamed the Beehive State), Georgia, Kansas, Louisiana, Maine, Mississippi, Missouri, Nebraska, New Jersey, North Carolina, Oklahoma, South Dakota, Tennessee, Vermont, and Wisconsin.

## Do butterflies see **color**?

Butterflies have highly developed sensory capabilities. They have the widest visual spectrum of any animal and are able to see from the red end of the spectrum all the way to the near ultraviolet. They are therefore able to distinguish colors that humans are unable to see.

## What are **"killer bees"**?

Africanized honeybees—the term entomologists prefer rather than killer bees—are a hybrid originating in Brazil, where African honeybees were imported in 1956. The

### Who discovered the "dance of the bees"?

In 1943, Karl von Frisch (1886–1982) published his study on the dance of the bees. It is a precise pattern of movements performed by returning forager (worker) honeybees in order to communicate the direction and distance of a food source to the other workers in the hive. The dance is performed on the vertical surface of the hive and two kinds of dances have been recognized: the round dance (performed when food is nearby) and the waggle dance (done when food is farther away).

breeders, hoping to produce a bee better suited to producing more honey in the tropics, instead found that African bees soon hybridized with and mostly displaced the familiar European honeybees. Although they produce more honey, Africanized honeybees are more dangerous than European bees because they attack intruders in greater numbers. Since their introduction, they have been responsible for approximately 1,000 human deaths. In addition to such safety issues, concern is growing regarding the effect of possible hybridization on the U.S. beekeeping industry.

In October 1990, the bees crossed the Mexican border into the United States; they reached Arizona in 1993. In 1996, six years after their arrival in the United States, Africanized honey bees could be found in parts of Texas, Arizona, New Mexico, and California. They are also found in Nevada. Their migration northward has slowed partially because they are a tropical insect and can't live in colder climates. Experts have suggested two possible ways of limiting the spread of the Africanized honeybees. The first is drone-flooding, a process by which large numbers of European drones are kept in areas where commercially reared European queen bees mate, thereby ensuring that only limited mating occurs between Africanized drones and European queens. The second method is frequent re-queening, in which a beekeeper replaces a colony's queen with one of his or her own choosing. The beekeeper can then be assured that the queens are European and that they have already mated with European drones.

## The sting of which kind of bee is the **most dangerous** to humans?

Africanized honey bees are fiercely defensive, attacking en masse with little or no provocation.

## What are **migratory beekeepers**?

A migratory beekeeper is a person who transports his or her bee colonies to different areas to produce better honey or to collect fees for pollinating such crops as fruit trees, almonds, and alfalfa. They frequently travel north in the spring and summer to

A formicine ant overpowers a termite soldier.

pollinate crops and then back south in the fall and winter to maintain the colonies in the warmer southern weather. Approximately 1,000 migratory beekeepers operate in the United States, transporting approximately two million bee colonies a year.

## How many bees are in a **bee colony**?

On the average, a bee colony contains from 50,000 to 70,000 bees, which produce a harvest of from 60 to 100 pounds of honey per year (five to eight gallons). A little more than a third of the honey produced by the bees is retained in the hive to sustain the population.

## How many **flowers** need to be tapped for bees to gather enough nectar to produce **one pound of honey**?

Bees must gather four pounds of nectar, which requires the bees to tap about two million flowers, to produce one pound of honey. The honey is gathered by worker bees, whose life span is three to six weeks, long enough to collect about a teaspoon of nectar.

## Do **termites** have any natural predators?

Birds, ants, spiders, lizards, and dragonflies have been seen preying on young, winged termites when they emerge and fly from a home colony to establish new colonies. This time is generally when termites are most vulnerable to predators.

## How much weight can an ant carry?

Ants are the "superweight lifters" of the animal kingdom. They are strong in relation to their size and can carry objects ten to twenty times their own weight—some species can carry objects up to fifty times their own weight. Ants are able to carry these objects great distances and even climb trees while carrying them. This is comparable to a 100-pound person picking up a small car, carrying it seven to eight miles on his back and then climbing the tallest mountain while still carrying the car!

## How are **ants** distinguished from **termites**?

Both insect orders—ants (order *Hymenoptera*) and termites (order *Isoptera*)—have segmented bodies with multi-jointed legs. Listed below are some differences.

| Characteristic | Ant | Termite |
|---|---|---|
| Wings | Two pairs with the front pair being much longer than the back pair | Two pairs of equal length |
| Antenna | Bends at right angle | Straight |
| Abdomen | Wasp-waist (pinched in) | No wasp-waist |

## Which insect has the **best sense of smell**?

Giant male silk moths may have the best sense of smell in the world. They can smell a female's perfume nearly seven miles (11 kilometers) away.

## Do male **mosquitoes** bite humans?

No. Male mosquitoes live on plant juices, sugary saps, and liquids arising from decomposition. They do not have a biting mouth that can penetrate human skin as female mosquitoes do. In some species, the females, who lay as many as 200 eggs, need blood to lay their eggs.

## What is a **"daddy longlegs"**?

The name applies to two different kinds of invertebrates. The first is a harmless, non-biting long-legged arachnid. Also called a harvestman, it is often mistaken for a spider, but it lacks the segmented body shape that a spider has. Although it has the same number of legs (eight) as a spider, the harvestman's legs are far longer and thinner. These very long legs enable it to raise its body high enough to avoid ants or other

small enemies. Harvestmen are largely carnivorous, feeding on a variety of small invertebrates, such as insects, spiders, and mites. They never spin webs as spiders do. They also eat some juicy plants and in captivity can be fed almost anything edible, from bread and milk to meat. Harvestmen also need to drink frequently. The term "daddy longlegs" also is used for a cranefly—a thin-bodied insect with long, thin legs that has a snoutlike proboscis with which it sucks water and nectar.

## How many **eggs** does a **spider** lay?

The number of eggs varies according to the species. Some larger spiders lay over 2,000 eggs, but many tiny spiders lay one or two and perhaps no more than a dozen during their lifetime. Spiders of average size probably lay a hundred or so. Most spiders lay all their eggs at one time and enclose them in a single egg sac; others lay eggs over a period of time and enclose them in a number of egg sacs.

## Which is stronger—steel or the silk from a **spider's web**?

Spider silk. Well known for its strength and elasticity, the strongest spider silk has tensile strength second only to fuse quartz fibers and five times greater than that of steel of equivalent weight. It is also five times as impact-resistant as bulletproof Kevlar. Tensile strength is the longitudinal stress that a substance can bear without tearing apart. The tension necessary to bring a compound thread 0.004 inches (0.01 centimeters) in diameter to the breaking point was found to be 2.8 ounces (80 grams).

In terms of tensile strength, spider web silk is many times stronger than steel.

## How long does it take the average spider to **weave a complete web**?

The average orb-weaver spider takes 30 to 60 minutes to completely spin its web. The order Araneae (the spiders) constitutes the largest division in the class Arachnida, containing about 32,000 species. These species of spiders use silk to capture their food in a variety of ways, ranging from the simple trip wires used by the large bird-eating spiders to the complicated and beautiful webs spun by the orb spiders. Some species produce funnel-shaped webs, and other communities of spiders build communal webs.

A completed web features several spokes leading from the initial structure. The number and nature of the spokes depend on the species. The spider replaces any damaged threads by gathering up the thread in front of it and producing a new one behind it. The orb web must be replaced every few days because it loses its stickiness (and its ability to entrap food).

The largest aerial webs are spun by the tropical orb weavers of the genus Nephila, measuring up to 18 feet, 9 inches (six meters) in circumference. The smallest webs are made by the *Glyphesis cottonae* and cover about 0.75 square inches (4.84 square centimeters).

## How long have **cockroaches** been on the earth?

The earliest cockroach fossils are about 280 million years old. Some cockroaches measured three to four inches (7.5 to 10 centimeters) long. Cockroaches (order Dictyoptera) are nocturnal scavenging insects that eat not only human food but book-bindings, ink, and whitewash as well.

## How does an **earthworm** "recycle" the earth?

It is estimated that there are more than 4,000 earthworm species. These animals have long been known to provide many useful ecological services, including making the ground suitable for plants by aerating and draining the soil with their burrows. In addition, worms can consume 30 percent of their body weight each day in plant, animal, and other matter. Much of this is in turn excreted on the earth's surface in the form of "casts." The worms then re-bury these casts with their borrowing process, thereby essentially recycling the earth.

## How many pairs of legs does a **centipede** have?

Centipedes, or members of the class Chilopoda, always have an uneven number of pairs of walking legs, varying from 15 to more than 171. The true centipedes (order Scolopendromorpha) have 21 or 23 pairs of legs. Common house centipedes (*Scutigera coleoptrato*) have 15 pairs of legs.

## How do **fleas** jump so far?

The jumping power of fleas comes both from strong leg muscles and from pads of a rubber-like protein called resilin. The resilin is located above the flea's hind legs. To jump, the flea crouches, squeezing the resilin, and then it relaxes certain muscles. Stored energy from the resilin works like a spring, launching the flea. A flea can jump well both vertically and horizontally. Some species can jump 150 times their own length. To match that record, a human would have to spring over the length of two

## How is the light in fireflies produced?

The light produced by fireflies (*Photinus pyroles*), or lightning bugs, is a kind of heatless light called bioluminescence caused by a chemical reaction in which the substance luciferin undergoes oxidation when the enzyme luciferase is present. The flash is a photon of visible light that radiates when the oxidating chemicals produce a high-energy state and revert back to their normal state. The flashing is controlled by the nervous system and takes place in special cells called photocytes. The nervous system, photocytes, and the tracheal end organs control the flashing rate. The air temperature also seems to be correlated with the flashing rate. The higher the temperature, the shorter the interval between flashes—eight seconds at 65°F (18.3°C) and four seconds at 82°F (27.7°C). Scientists are uncertain as to why this flashing occurs. The rhythmic flashes could be a means of attracting prey or enabling mating fireflies to signal in heliographic codes (that differ from one species to another), or they could serve as a warning signal.

and a quarter football fields—or the height of a 100-story building—in a single bound. The common flea (*Pulex irritans*) has been known to jump 13 inches (33 centimeters) in length and 7.25 inches (18.4 centimeters) in height.

## What is the life span of a **fruit fly**?

The length of adult life can vary considerably. Under ideal conditions, an adult *Drosophila melanogaster* can live as long as 40 days. In crowded conditions, its life span may drop to 12 days. Under normal laboratory conditions, however, adults generally die after only six or seven days.

## What causes the **Mexican jumping bean** to move?

The bean moth (*Carpocapa saltitans*) lays its eggs in the flower or in the seed pod of the spurge, a bush known as *Euphorbia sebastiana*. The egg hatches inside the seed pod, producing a larva or caterpillar. The jumping of the bean is caused by the active shifting of weight inside the shell as the caterpillar moves. The jumps of the bean are stimulated by sunshine or by heat from the palm of the hand.

# AQUATIC LIFE

*See also: Biology—Fungi, Bacteria, Algae, etc.*

## What has been the impact of **zebra mussels**?

Zebra mussels (*Dreissena polymorpha* [*Pallas*]) are black and white striped bi-valve mollusks. They were probably introduced to North America in 1985 or 1986 via a discharge of ballast water into Lake St. Clair. Zebra mussels are hard-shelled species that adhere to hard surfaces with byssal threads and colonize. High densities of zebra mussels have been found in the intakes, pipes, and heat exchangers of water users throughout the world. They can clog power plant, industrial and public drinking water intakes, foul boat hulls and engine cooling water systems, and disrupt aquatic ecosystems. Water-processing facilities must be cleaned manually to rid the systems of the mussels. Zebra mussels are a threat to surface water resources because they reproduce quickly, have free-swimming larvae, grow rapidly, lack competitors for space or food, and have no predators.

## How is the **age of fish** determined?

One way to determine the age of a fish is by its scales, which have growth rings just as trees do. Scales have concentric bony ridges or "circuli," which reflect the growth patterns of the individual fish. The portion of the scale that is embedded in the skin contains clusters of these ridges (called "annuli"); each cluster marks one year's growth cycle.

## How do fish swimming in a school **change their direction** simultaneously?

The movement, which confuses predators, happens because fish detect pressure changes in the water. The detection system, called the lateral line, is found along each side of the fish's body. Along the line are clusters of tiny hairs inside cups filled with a jellylike substance. If a fish becomes alarmed and turns sharply, it causes a pressure wave in the water around it. This wave pressure deforms the "jelly" in the lateral line of nearby fish. This moves the hairs which trigger nerves, and a signal is sent to the brain telling the fish to turn.

## At what **speeds** do fish **swim**?

The maximum swimming speed of a fish is somewhat determined by the shape of its body and tail and by its internal temperature. The cosmopolitan sailfish (*Istiophorus platypterus*) is considered to be the fastest fish species, at least for short distances, swimming at greater than 60 miles (95 kilometers) per hour. Some American fishermen believe, however, that the bluefin tuna (*Thunnus thynnus*) is the fastest, but the fastest speed recorded for them so far is 43.4 miles (69.8 kilometers) per hour. Data is extremely difficult to secure because of the practical difficulties in measuring the speeds. The yellowfin tuna (*Thunnus albacares*) and the wahoe (*Acanthocybium solandri*) are also fast, timed at 46.35 miles (74.5 kilometers) per hour and 47.88 miles (77 kilometers) per hour during 10- to 20-second sprints. Flying fish swim at 40+ miles (64+ kilometers) per hour, dolphins at 37 miles (60 kilometers) per hour, trout at 15 miles (24 kilometers) per hour, and blenny at five miles (eight kilometers) per hour. Humans can swim 5.19 miles (8.3 kilometers) per hour.

## How can you tell male and female **lobsters** apart?

The differences between male and female lobsters can be seen only when they are turned on their backs. In the male lobster, the two swimmerets (forked appendages used for swimming) nearest the carapace (the solid shell) are hard, sharp, and bony; in the female the same swimmerets are soft and feathery. The female also has a receptacle that appears as a shield wedged between the third pair of walking legs. During mating, the male deposits sperm into this receptacle where it remains for as long as several months until the female uses it to fertilize her eggs as they are laid.

## How are **pearls** created?

Pearls are formed in saltwater oysters and freshwater clams. There is a curtain-like tissue called the mantle within the body of these mollusks. Certain cells on the side of the mantle toward the shell secrete nacre, also known as mother-of-pearl, during a specific stage of the shell-building process. A pearl is the result of an oyster's reaction to a foreign body, such as a piece of sand or a parasite, within the oyster's shell. The oyster neutralizes the invader by secreting thin layers of nacre around the foreign body, eventually building it into a pearl. The thin layers are alternately composed of calcium carbonate, argonite, and conchiolin. Irritants intentionally placed within an oyster result in the production of what are called cultured pearls.

## How are **coral reefs** formed?

Coral reefs grow only in warm, shallow water. The calcium carbonate skeletons of dead corals serve as a framework upon which layers of successively younger animals attach themselves. Such accumulations, combined with rising water levels, slowly lead to the formation of reefs that can be hundreds of meters deep and long. The coral animal, or polyp, has a columnar form; its lower end is attached to the hard floor of the reef; the upper end is free to extend into the water. A whole colony consists of thousands of individuals. There are two kinds of corals, hard and soft, depending on the type of skeleton secreted. The polyps of hard corals deposit around themselves a solid skeleton of calcium carbonate (chalk), so most swimmers see only the skeleton of the coral; the animal is in a cuplike formation into which it withdraws during the daytime.

## What gives **coral** their **colors**?

Coral have a symbiotic relationship with zooxanthellae. Zooxanthellae are photosynthetic dinoflagellates (one-celled animals) that give coral their characteristic colors of pink, purple and green. Coral that have expelled the zooxanthellae appear white.

## What are **giant tube worms**?

These worms were found in 1977 when the submersible *Alvin* was exploring the ocean floor of the Galapagos Ridge (located 1.5 miles [2.4 kilometers] below the Pacific Ocean surface and 200 miles [322 kilometers] from the Galapagos Islands). *Riftia pachyptila Jones*, named after worm expert Meredith Jones of the Smithsonian Museum of Natural History, were discovered near the hydrothermal (hot water) ocean vents. Growing to lengths of five feet (1.5 meters), the worms lack both mouth and gut, and are topped with feathery plumes composed of over 200,000 tiny tentacles. The phenomenal growth of these worms is due to their internal food source—symbiotic bacteria, over 100 billion per ounce of tissue—that live within the worms' troposome tissues. To these troposome tissues, the tube worms transport absorbed oxygen from the water, together with carbon dioxide and hydrogen sulfide. Utilizing this supply, the bacteria living there in turn produce carbohydrates and proteins that the worms need to thrive.

This was only one of *Alvin*'s discoveries during its historic voyage. Scientists expected to find a "desert" at these ocean depths where no light penetrated. Most of the world's organisms rely on photosynthesis (the use of light to make organic compounds) at the base of their food chains. But in these depths, giant tube worms, vent crabs, and mollusks thrive because these vent communities depend on chemoautotropic (chemically self-feeding) bacteria, which derive their life-sustaining energy from the oxidation of substances spewing from the vents, or in symbiotic relationships, such as that with the giant tube worms.

## What is a **mermaid's purse**?

Mermaid's purses are the protective cases in which the eggs of dogfish, skates, and rays are released into the environment. The rectangular purse is leathery and has long tendrils streaming from each corner. The tendrils anchor the case to seaweed or rocks and protects the embryos during the six to nine months it takes for them to hatch. Empty cases often wash up on beaches.

## How much electricity does an **electric eel** generate?

An electric eel (*Electrophorus electricus*) has current-producing organs made up of electric plates on both sides of its vertebral column running almost its entire body length. The charge—on the average of 350 volts, but as great as 550 volts—is released

### How do dolphins sleep?

Dolphins apparently sleep by shutting down half of their brain and resting quietly in the water.

by the central nervous system. The shock consists of four to eight separate charges, which last only two- to three-thousandths of a second each. These shocks, used as a defense mechanism, can be repeated up to 150 times per hour without any visible fatigue to the eel. The most powerful electric eel, found in the rivers of Brazil, Colombia, Venezuela, and Peru, produces a shock of 400 to 650 volts.

## How do **salmon** find the way to their **spawning grounds**?

Scientists do not know exactly how a salmon "remembers" the way back to its native stream after an ocean journey possibly lasting several years and covering several thousand miles. They agree, however, that salmon, like homing pigeons, appear to have an innate compass or "search recognition" mechanism that operates independently of astronomical or physical signs. Some scientists theorize that this internal compass uses the infinitely small electrical voltages generated by the ocean currents as they travel through the earth's magnetic field. Others believe that the salmon's homing mechanism may take its cues from the varying salinities of the water or specific smells encountered along the journey.

## What are the main classes of **mollusk shells**?

Scientists recognize five main classes of mollusks: gastropods, bivalves, tooth shells, chitons, and cephalopods. A sixth class, monoplacophora, was once thought to be extinct, but scientists have discovered them in very deep ocean waters. They are now considered very rare.

Most shells belong in one of two main classes: gastropods or bivalves. Three-quarters of the world's mollusks, or about 60,000 species, are classified as gastropods, which possess one-piece shells that are usually coiled. Limpets, cones, olives, murex, cowries, and whelks belong to this class. The 11,000 species of bivalves have two-piece shells, normally hinged along one side. Clams, oysters, cockles, and mussels are some familiar bivalves.

The three minor classes have far fewer species. Scientists recognize about 500 kinds of tooth shells. These shells are tapered, hollow tubes that curve slightly so as to resemble long needles or elephant tusks, prompting some collectors to refer to them as tusk shells. The shells of chitons, of which there are about 600 species, are made up of eight separate movable plates that are kept in place by a leathery oval band called a girdle or belt. They are also called "coat of mail" shells because of their resemblance to armor. The 650 species of cephalopods are quite different from other mollusks. Some have shells that surround their soft bodies. For example, the well-known chambered nautilus inhabits a shell consisting of a series of gradually larger chambers separated by paper-thin walls. Others, such as cuttlefish and squid, have shells inside their bodies to help support them. Octopuses, another kind of cephalopod, have no shells.

## How many kinds of **sharks** are there and how many are dangerous?

The United Nations' Food and Agricultural Organization lists 354 species of sharks, ranging in length from six inches (15 centimeters) to 49 feet (15 meters). While 35 species are known to have attacked humans at least once, only a dozen do so on a regular basis. The relatively rare Great White shark (*Carcharodan carcharias*) is the largest predatory fish. The largest specimen accurately measured was 20 feet, 4 inches (6.2 meters) long and weighed 5,000 pounds (2,270 kilograms).

## How far from shore do **shark attacks** occur?

In a study of 570 shark attacks, it was found that most shark attacks occur near shore. These data are not surprising since most people who enter the water stay close to the shore.

| Distance from shore | Percentage of shark attacks | Percentage of people who swim at this distance |
|---|---|---|
| 50 feet (15 m) | 31 | 39 |
| 100 feet (30 m) | 11 | 15 |
| 200 feet (60 m) | 9 | 12 |
| 300 feet (90 m) | 8 | 11 |
| 400 feet (120 m) | 2 | 2 |
| 500 feet (150 m) | 3 | 5 |
| 1000 feet (300 m) | 6 | 9 |
| 1 mile (1.6 km) | 8 | 6 |
| >1 mile (>1.6 km) | 22 | 1 |

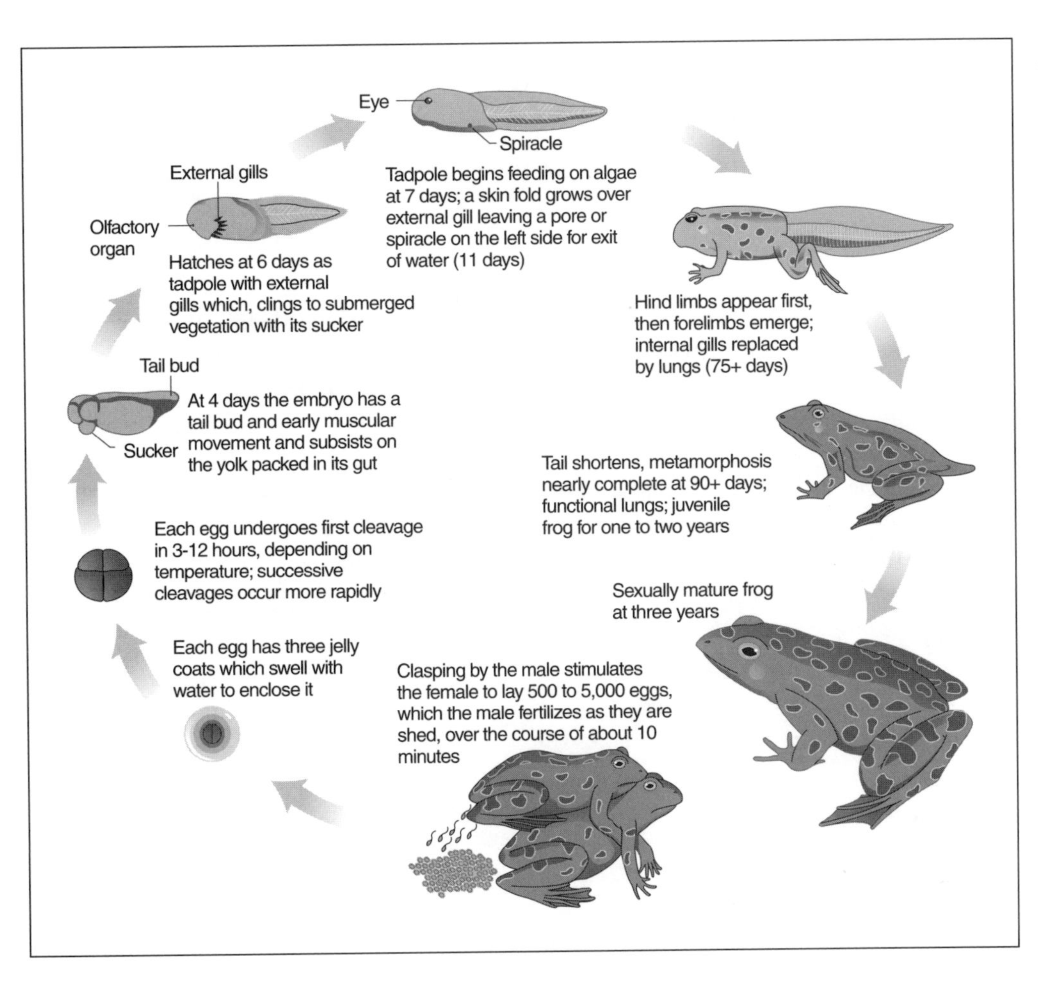


**The life cycle of a frog.**

# REPTILES AND AMPHIBIANS

## What is the difference between a **reptile** and an **amphibian**?

Reptiles are clad in scales, shields, or plates, and their toes have claws; amphibians have moist, glandular skins, and their toes lack claws. Reptile eggs have a thick, hard or parchment-like shell that protects the developing embryo from moisture loss, even on dry land; the eggs of amphibians lack this protective outer covering and are always laid in water or in damp places. Young reptiles are miniature replicas of their parents in general appearance if not always in coloration and pattern; juvenile amphibians pass through a larval, usually aquatic, stage before they metamorphose (change in form and structure) into the adult form. Reptiles include alligators, crocodiles, turtles, and snakes; amphibians include salamanders, toads, and frogs.

## Which **poisonous snakes** are native to the United States?

| Snake | Average Length |
|---|---|
| Rattlesnakes | |
| Eastern diamondback (*Crotalus adamateus*) | 33–65 in (84–165 cm) |
| Western diamondback (*Crotalus atrox*) | 30–65 in (76–419 cm) |
| Timber rattlesnake (*Crotalus horridus horridus*) | 32–54 in (81–137 cm) |
| Prairie rattlesnake (*Crotalus viridis viridis*) | 32–46 in (81–117 cm) |
| Great Basin rattlesnake (*Crotalus viridis lutosus*) | 32–46 in (81–117 cm) |
| Southern Pacific rattlesnake (*Crotalus viridis helleri*) | 30–48 in (76–122 cm) |
| Red diamond rattlesnake (*Crotalus ruber ruber*) | 30–52 in (76–132 cm) |
| Mojave rattlesnake (*Crotalus scutulatus*) | 22–40 in (56–102 cm) |
| Sidewinder (*Crotalus cerastes*) | 18–30 in (46–76 cm) |
| Moccasins | |
| Cottonmouth (*Agkistrodon piscivorus*) | 30–50 in (76–127 cm) |
| Copperhead (*Agkistrodon contortrix*) | 24–36 in (61–91 cm) |
| Cantil (*Agkistrodon bilineatus*) | 30–42 in (76–107 cm) |
| Coral snakes | |
| Eastern coral snake (*Micrurus fulvius*) | 16–28 in (41–71 cm) |

## What is the **fastest snake** on land?

The black mamba *(Dendroaspis polylepis),* a deadly poisonous African snake that can grow up to 13 feet (four meters) in length, has been recorded reaching a speed of seven miles (11 kilometers) per hour. A particularly aggressive snake, it chases animals at high speeds holding the front of its body above the ground.

## How fast can a **crocodile** run on land?

In smaller crocodiles, the running gait can change into a bounding gallop that can achieve speeds of two to 10 miles (three to 17 kilometers) per hour.

## How is the gender of baby **alligators** determined?

The gender of an alligator is determined by the temperature at which the eggs are incubated. High temperatures of 90°–93°F (32°–34°C) result in males; low temperatures of 82°–86°F (28°–30°C) yield females. This determination takes place during the second and third week of the two-month incubation. Further temperature fluctuations before or after this time do not alter the gender of the young. The heat from the decaying matter on top of the nest incubates the eggs.

### How does the **Surinam toad** nurture its young?

Unlike most toads and frogs, the female Surinam toad carries her eggs in special pockets in the skin on her back. Each egg develops in its own pocket in the female's skin. The tadpoles' tails are "plugged in" to the mother's system, similar to the placenta of mammals, exchanging nutrients and gases. The tadpoles develop quickly, undergoing metamorphosis while still in the pockets. Upon transformation into miniature frogs they break free of their pocket walls to begin independent lives.

### What are a **turtle's** upper and lower **shell** called?

The turtle (order Testudines) uses its shell as a protective device. The upper shell is called the dorsal carapace and the lower shell is called the ventral plastron. The shell's sections are referred to as the scutes. The carapace and the plastron are joined at the sides.

## BIRDS

### What names are used for **groups of birds**?

A group of birds in general is called a congregation, flight, flock, volery, or volley.

| Bird | Group name |
|---|---|
| Bitterns | Siege or sedge |
| Budgerigars | Chatter |
| Chickens | Flock, run, brood, or clutch |
| Coots | Fleet or pod |
| Cormorants | Flight |
| Cranes | Herd or siege |
| Crows | Murder, clan, or hover |
| Curlews | Herd |
| Doves | Flight, flock, or dole |
| Ducks | Paddling, bed, brace, flock, flight, or raft |
| Eagles | Convocation |
| Geese | Gaggle or plump (on water), flock (on land), skein (in flight), or covert |
| Goldfinches | Charm, chattering, chirp, or drum |
| Grouses | Pack or brood |
| Gulls | Colony |
| Hawks | Cast |
| Hens | Brood or flock |

| Bird | Group name |
|---|---|
| Herons | Siege, sege, scattering, or sedge |
| Jays | Band |
| Larks | Exaltation, flight, or ascension |
| Magpies | Tiding or tittering |
| Mallards | Flush, sord, or sute |
| Nightingales | Watch |
| Partridges | Covey |
| Peacocks | Muster, ostentation, or pride |
| Penguins | Colony |
| Pheasants | Nye, brood, or nide |
| Pigeons | Flock or flight |
| Plovers | Stand, congregation, flock, or flight |
| Quails | Covey or bevy |
| Sparrows | Host |
| Starlings | Chattering or murmuration |
| Storks | Mustering |
| Swallows | Flight |
| Swans | Herd, team, bank, wedge, or bevy |
| Teals | Spring |
| Turkeys | Rafter |
| Turtle doves | Dule |
| Woodpeckers | Descent |
| Wrens | Herd |

## How sensitive is the **hearing** of birds?

In most species of birds, the most important sense after sight is hearing. Birds' ears are close to their bodies and are covered by feathers. The feathers, however, do not have barbules, which would obstruct sound. Nocturnal raptors, such as the great horned owl, have a very well-developed sense of hearing in order to be able to capture their prey in total darkness.

## How do birds learn to sing the **distinctive melody** of their respective species?

The ability to learn the proper song appears to be influenced by both heredity and experience. Scientists have speculated that a bird is genetically programmed with the ability to recognize the song of its own species and with the tendency to learn its own song. As a bird begins to sing, it goes through a stage of practice (which closely resembles the babbling of human infants) through which it perfects the notes and structure of its distinctive song. In order to produce a perfect imitation, the bird must apparently hear the song from an adult during its first months of life.

## Which birds lay the **largest** and **smallest eggs**?

The elephant bird (*Aepyornis maximus*), an extinct flightless bird of Madagascar, also known as the giant bird or roc, laid the largest known bird eggs. Some of these eggs measured as much as 13.5 inches (34 centimeters) in length and 9.5 inches (24 centimeters) in diameter. The largest egg produced by any living bird is that of the North African ostrich (*Struthio camelus*). The average size is six to eight inches (15 to 20.5 centimeters) in length and four to six inches (five to 15 centimeters) in diameter.

The smallest mature egg, measuring less than 0.39 inch (one centimeter) in length, is that of the vervain hummingbird (*Mellisuga minima*) of Jamaica.

Generally speaking, the larger the bird, the larger the egg. However, when compared with the bird's body size, the ostrich egg is one of the smallest eggs, while the hummingbird's egg is one of the largest. The Kiwi bird of New Zealand lays the largest egg, relative to body size, of any living bird. Its egg weighs up to one pound (0.5 kilogram).

## Why do birds **migrate** annually?

Migratory behavior in birds is inherited; however, birds will not migrate without certain physiological and environmental stimuli. In the late summer, the decrease in sunlight stimulates the pituitary gland and the adrenal gland of migrating birds, causing them to produce the hormones prolactin and corticosterone respectively. These hormones in turn cause the birds to accumulate large amounts of fat just under the skin, providing them with enough energy for the long migratory flights. The hormones also cause the birds to become restless just prior to migration. The exact time of departure, however, is dictated by not only by the decreasing sunlight and hormonal changes, but also by such conditions as the availability of food and the onset of cold weather.

The major wintering areas for North American migrating birds are the southern United States and Central America. Migrating ducks follow four major flyways south: the Atlantic flyway, the Mississippi flyway, the central flyway, and the Pacific flyway. Some bird experts propose that the birds return north to breed for several reasons: (1) Birds return to nest because there is a huge insect supply for their young; (2) The higher the Earth's latitude in the summer in the Northern Hemisphere, the longer the daylight available to the parents to find food for their young; (3) Less competition exists for food and nesting sites in the north; (4) In the north, there are fewer mammal predators for nesting birds (which are particularly vulnerable during the nesting stage); (5) Birds migrate south to escape the cold weather, so they return north when the weather improves.

## Which bird migrates the **greatest distance**?

The arctic tern (*Sterna paradisaea*) migrates the longest distance of any bird. They breed from subarctic regions to the very limits of land in the arctic of North America and Eurasia. At the end of the northern summer, the arctic tern leaves the north on a migration of more than 11,000 miles (17,699 kilometers) to its southern home in Antarctica. A tern tagged in July on the arctic coast of Russia was recovered the next May near Fremantle, Australia, a record 14,000 miles (22,526 kilometers) distant.

## When do the **swallows** come back to Capistrano in California?

According to legend, every year the swallows are expected to return to the Mission San Juan Capistrano, California, on St. Joseph's Day, March 19th, and depart on October 23rd, when they migrate to the Southern Hemisphere. The birds can actually arrive anytime during the month of March and leave anytime in October. The number that return each year has been declining, with the growth of the town and the large number of tourists blamed for the decrease. The legend began when a local innkeeper who considered the birds a nuisance destroyed their nests as an attempt to drive the birds away. One of the mission fathers called the swallows to the mission for shelter and the swallows have returned every year since then.

## Do all **birds fly**?

No. Among the flightless birds, the penguins and the ratites are the best known. Ratites include emus, kiwis, ostriches, rheas, and cassowaries. They are called ratite because they lack a keel on the breastbone. All of these birds have wings but lost their power to fly millions of years ago. Many birds that live isolated on oceanic islands (for example, the great auk) apparently became flightless in the absence of predators and the consequent gradual disuse of their wings for escape.

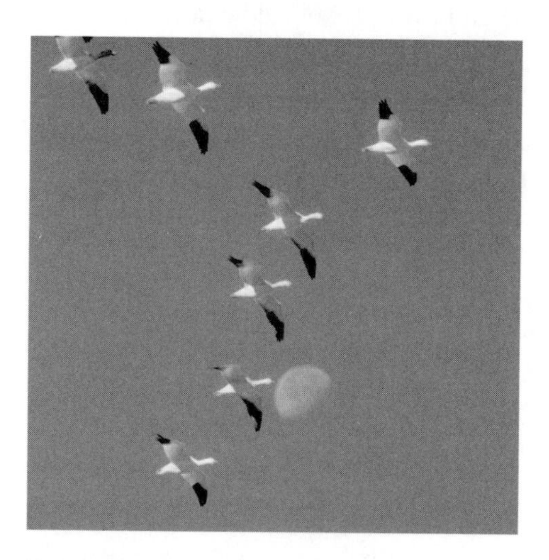

**The common flying "V" formation observed in these migratory snow geese is thought to improve aerodynamics and thus efficiency.**

## What bird has the **biggest wing span**?

Three members of the albatross family, the Wandering Albatross (*Diomedea exulans*), the Royal Albatross (*Diomedea epomophora*), and the Amsterdam Island Albatross (*Diomeda amsterdiamensis*) have the greatest wingspan of any bird species with a spread of 8–11 feet (2.5–3.3 meter).

## Why do **geese** fly in formation?

Aerodynamicists have suspected that long-distance migratory birds, such as geese and swans, adapt the "V" formation

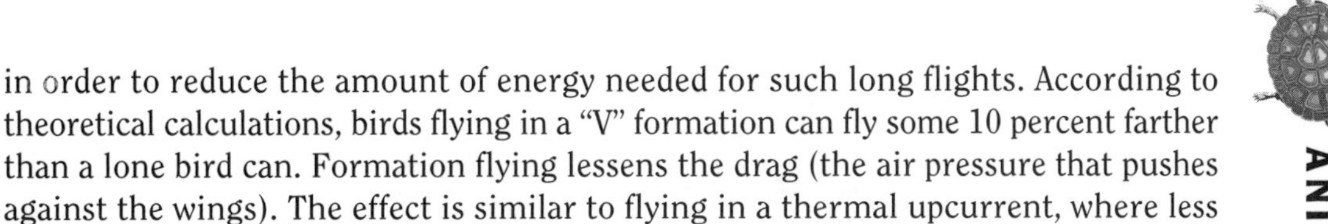



in order to reduce the amount of energy needed for such long flights. According to theoretical calculations, birds flying in a "V" formation can fly some 10 percent farther than a lone bird can. Formation flying lessens the drag (the air pressure that pushes against the wings). The effect is similar to flying in a thermal upcurrent, where less total lift power is needed. In addition, when flying, each bird creates behind it a small area of disturbed air. Any bird flying directly behind it would be caught in this turbulence. In the "V" formation of Canada geese, each bird flies not directly behind the other, but to one side or above the bird in front.

## How fast does a **hummingbird** fly and how far does the hummingbird migrate?

Hummingbirds fly at speeds up to 71 miles (80 kilometers) per hour. Small species beat their wings 50 to 80 times per second, and even faster during courtship displays. For comparison, the following table lists the flight speeds of some other birds:

| Bird | Speed (mph) | Speed (kph) |
|---|---|---|
| Peregrine falcon | 168–217 | 270.3–349.1 |
| Swift | 105.6 | 169.9 |
| Merganser | 65 | 104.6 |
| Golden plover | 50–70 | 80.5–112.6 |
| Mallard | 40.6 | 65.3 |
| Wandering albatross | 33.6 | 54.1 |
| Carrion crow | 31.3 | 50.4 |
| Herring gull | 22.3–24.6 | 35.9–39.6 |
| House sparrow | 17.9–31.3 | 28.8–50.4 |
| Woodcock | 5 | 8 |

The longest migratory flight of a hummingbird documented to date is the flight of a rufous hummingbird from Ramsey Canyon, Arizona, to near Mt. Saint Helens, Washington, a distance of 1,414 miles (2,277 kilometers). Bird-banding studies are now in progress to verify that a few rufous hummingbirds do make a 11,000–11,500 mile (17,699–18,503 kilometer) journey along a super Great Basin High route, a circuit that could take a year to complete. Hummingbird studies, however, are difficult to complete because so few banded birds are recovered.

## Is it best to stop **feeding hummingbirds** after Labor Day?

A difference of opinion exists on this question. Some experts contend that hummingbirds can become over-reliant on feeders and, by not experiencing a decrease in their food supply, will not migrate. Others, however, believe that feeders put out at migrating time have little or no effect on hummingbirds leaving a particular area. Some believe that food supplies do not factor into the mechanism that triggers migration, arguing instead that decreasing daylight activates the necessary biochemical messages. Others believe that a decline in another food staple, insects, is the primary cause for the move south. In addition, some birders keep feeders up during migration to help stragglers. However, until further studies are conducted on the subject, experts generally recommend deactivating feeders in Canada and extreme northern states when it is time for migration and taking them down in the extreme southern portions of the United States by mid-October to prevent birds possibly being caught in unseasonably cold weather.

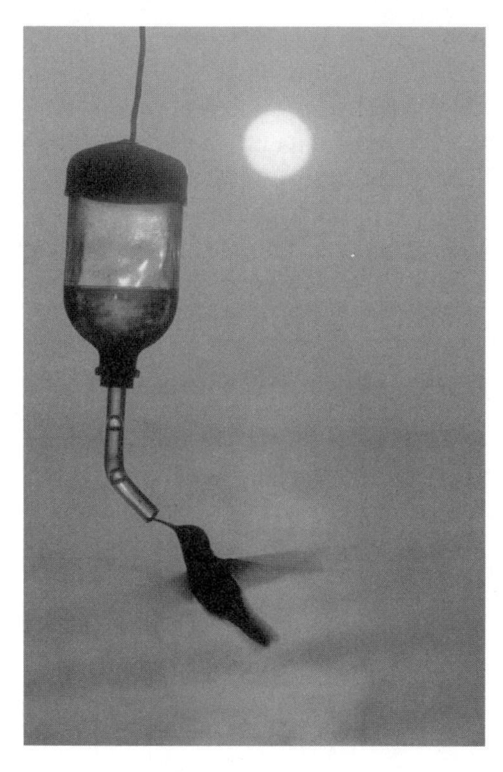

This ruby-throated hummingbird represents the only species native to eastern North America.

## How **fast** does a **hummingbird's wing** move?

Hummingbirds are the only family of birds that can truly hover in still air for any length of time. They need to do so in order to hang in front of a flower while they perform the delicate task of inserting their slim, sharp bills into its depths to drink nectar. Their thin wings are not contoured into the shape of aerofoils and do not generate lift in this way. Their paddle-shaped wings are, in effect, hands that swivel at the shoulder. They beat them in such a way that the tip of each wing follows the line of a figure eight lying on its side. The wing moves forward and downwards into the front loop of the eight, creating lift. As it begins to come up and go back, the wing twists through 180 degrees so that once again it creates a downward thrust. The hummingbird's method of flying does have one major limitation: the smaller a wing, the faster it has to beat in order to produce sufficient downward thrust. An average-sized hummingbird beats its wings 25 times per second. The bee hummingbird, native to Cuba, is only two inches (five centimeters) long and beats its wings at an astonishing 200 times per second.

## What is unusual about the way the **emperor penguin's eggs** are incubated?

Each female emperor penguin (*Aptenodytes forsteri*) lays one large egg. Initially, both sexes share in incubating the egg by carrying it on his or her feet and covering it with a fold of skin. After a few days of passing the egg back and forth, the female leaves to feed in the open water of the Arctic Ocean. Balancing their eggs on their feet, the male

penguins shuffle about the rookery, periodically huddling together for warmth during blizzards and frigid weather. If an egg is inadvertently orphaned, a male with no egg will quickly adopt it. Two months after the female's departure, the chick hatches. The male feeds it with a milky substance he regurgitates until the female returns. Now padded with blubber, the females take over feeding the chicks with fish they have stored in their crops. The females do not return to their mate, however, but wander from male to male until one allows her to take his chick. It is then the males' turn to feed in open water and restore the fat layer they lost while incubating.

## What are the natural **predators** of the **penguin**?

The leopard seal (*Hydrurga leptonyx*) is the principal predator of both the adult and juvenile penguin. The penguin may also be caught by a killer whale while swimming in open water. Eggs and chicks that are not properly guarded by adults are often devoured by skuas and sheathbills.

## What animal is closely related with the **Canary Islands**?

Ancient explorers named the Canary Islands *Canaria* from the Latin word Canis (dog) because of the large, fierce dogs that they found inhabiting them. The Canaries are also the namesake of canary birds, which are native to the islands.

## When was the **bald eagle** adopted as the national bird of the United States?

On June 20, 1782, the citizens of the newly independent United States of America adopted the bald or "American" eagle as their national emblem. At first the heraldic artists depicted a bird that could have been a member of any of the larger species, but by 1902, the bird portrayed on the seal of the United States of America had assumed its proper white plumage on head and tail. The choice of the bald eagle was not unanimous; Benjamin Franklin (1706–1790) preferred the wild turkey. Oftentimes a tongue-in-cheek humorist, Franklin thought the turkey a wily but brave, intelligent, and prudent bird. He viewed the eagle on the other hand as having "a bad moral character" and "not getting his living honestly," preferring instead to steal fish from hardworking fishhawks. He also found the eagle a coward that readily flees from the irritating attacks of the much smaller kingbird.

The bald eagle was heralded as a national emblem of the United States of America in 1782.

## Which state was the first to "officially" name a **state bird**?

In 1926, Kentucky officially named the cardinal as its state bird.

## How does a **homing pigeon** find its way home?

Scientists currently have two hypotheses to explain the homing flight of pigeons. Neither has been proved to the satisfaction of all the experts. The first hypothesis involves an "odor map." This theory proposes that young pigeons learn how to return to their original point of departure by smelling different odors that reach their home in the winds from varying directions. They would, for example, learn that a certain odor is carried on winds blowing from the east. If a pigeon were transported eastward, the odor would tell it to fly westward to return home. The second hypothesis proposes that a bird may be able to extract its home's latitude and longitude from the Earth's magnetic field. It may be proven in the future that neither theory explains the pigeon's navigational abilities or that some synthesis of the two theories is plausible.

## What is the name of the bird that perches on the black **rhinoceros' back**?

The bird, a relative of the starling, is called an oxpecker (a member of the Sturnidae family). Found only in Africa, the yellow-billed oxpecker (*Buphagus africanus*) is widespread over much of western and central Africa, while the red-billed oxpecker (*Buphagus erythrorhynchus*) lives in eastern Africa from the Red Sea to Natal.

Seven to eight inches (17–20 centimeters) long with a coffee-brown body, the oxpecker feeds on more than 20 species of ticks that live in the hide of the black rhinoceros (*Diceros bicornis*), also called the hook-lipped rhino. The bird spends most of its time on the rhinoceros or on other animals, such as the antelope, zebra, giraffe, or buffalo. The bird has even been known to roost on the body of its host.

The relationship between the oxpecker and the rhinoceros is a type of symbiosis (a close association between two organisms in which at least one of them benefits) called mutualism. The rhinoceros' relief of its ticks and the bird's feeding clearly demonstrates mutualism (a condition in which both organisms benefit). In addition, the oxpecker, having much better eyesight than the nearsighted rhinoceros, alerts its host with its shrill cries and flight when danger approaches.

## What year was the **European starling** (*Sturnus vulgaris*) imported into the United States?

Eugene Schieffelin imported the European starling into the United States in 1890. Schieffelin wanted to establish every bird found in Shakespeare's works. He also imported English sparrows to New York City in 1860.

### Why don't **woodpeckers** get headaches?

Woodpeckers' skulls are particularly sturdy to withstand the force of the blows as they hammer with their beaks. They are further aided by strong neck muscles to support their heads.

## Will wild birds reject baby birds that have been **touched by humans**?

No. Contrary to popular belief, birds generally will not reject hatchlings touched by human hands. The best thing to do for newborn birds that have fallen or have been pushed out of the nest is to locate the nest as quickly as possible and gently put them back.

## Why don't birds get electrocuted when they **sit on wires**?

In general, birds do not get electrocuted while just sitting on power transmission wires. Most electrocutions happen when a bird opens its wingspan and completes a circuit by bridging the gap between two live wires or a live wire and a grounded wire, or other parts such as transformers and grounded metal crossarms.

## What can an **orphaned wild bird** eat?

An orphaned songbird needs to be fed every 20 minutes during daylight hours for several weeks. The food should be placed deep in its throat. A soft-billed bird (such as a warbler or catbird) may be given grated carrots, chopped hard-boiled eggs, cottage cheese, fresh fruit, or custard. A young hard-billed bird (such as a sparrow or finch) may be given the same food, but rape, millet, and sunflower seeds should be added to this diet when the bird becomes well-developed. A mixture of dry baby cereal and the yolk of a hard-boiled egg moistened with milk can also be given.

## What species of birds will nest in **bird houses**?

In general, the only birds that will occupy birdhouses are species that normally chisel nesting holes or use such ready-made cavities as hollow trunks or holes already excavated by other species. Different species of birds require birdhouses of different dimensions, especially concerning the diameter of the entrance hole. Some birds that will nest in birdhouses are: bluebirds, chickadees, finches, flycatchers, purple martins, nuthatches, sparrows, starlings, titmice, woodpeckers, and wrens. In addition, larger birds such as some ducks and owls may be attracted.

## How can **bluebirds** be encouraged to nest in a particular location?

Bluebirds may be attracted by providing nesting boxes and perches, having an area of low or sparse vegetation, and planting nearby trees, vines, or shrubs such as blueberries, honeysuckle, and crabapples. Bluebirds prefer open countryside with low undergrowth. Parks, golf courses, and open lawns are their preferred habitats. In the last 40 years eastern bluebird populations have declined 90 percent, coinciding with the disappearing farmland, widespread use of pesticides, and an increase in nest competitors (house sparrows and European starlings). Artificial nesting boxes sometimes provide more secure nesting places than do natural nest sites, because artificial structures can be built to resist predators. Bluebird boxes can be made with entrance holes small enough to exclude starlings (1.5 inches [4 centimeters] in diameter) and can have special raccoon guards on mounting poles. Mounted three to six feet (one to two meters) above ground to discourage predators, the nesting boxes should be no closer together than 100 feet (30 meters). The box should have a tree within 50 feet (15 meters) so that the fledglings can perch. The box should have a 4-inch by 4-inch floor (10.6 by 10.6 centimeter), walls eight to 12 inches (20–30.5 centimeters) in height, with the entrance hole six to 10 inches (15–25.5 centimeters) above the floor.

# MAMMALS

## Which mammal has the shortest **gestation period**? Which one has the longest?

Gestation is the period of time between fertilization and birth in oviparous animals. The shortest gestation period known is 12 to 13 days, shared by three marsupials: the American or Virginian opossum (*Didelphis marsupialis*); the rare water opossum, or yapok (*Chironectes minimus*) of central and northern South America; and the eastern native cat (*Dasyurus viverrinus*) of Australia. The young of each of these marsupials are born while still immature and complete their development in the ventral pouch of their mother. While 12 to 13 days is the average, the gestation period is sometimes as short as eight days. The longest gestation period for a mammal is that of the African elephant (*Loxodonta africana*) with an average of 660 days, and a maximum of 760 days.

## Do any **mammals fly**?

Bats (order Chiroptera with 986 species) are the only truly flying mammals, although several gliding mammals are referred to as "flying" (such as flying squirrel and flying lemur). The "wings" of bats are double membranes of skin stretching from the sides of

the body to the hind legs and tail, and are actually skin extensions of the back and belly. The wing membranes are supported by the elongated fingers of the forelimbs (or arms). Nocturnal (active at night), ranging in length from 1.5 inches (25 millimeters) to 1.3 feet (40.6 centimeters), and living in caves or crevices, bats inhabit most of the temperate and tropical regions of both hemispheres. The majority of species feed on insects and fruit, while some tropical species eat pollen and nectar of flowers, and insects found inside them. Moderate-sized species usually prey on small mammals, birds, lizards, and frogs, and some eat fish. But true vampire bats (three species) eat the blood of animals by making an incision in the animal's skin—from these bats, animals can contract rabies. Most bats do not find their way around by sight but have evolved a sonar system, called "echolocation," for locating solid objects. Bats emit vocal sounds through the nose or mouth while flying. These sounds, usually above the human hearing range, are reflected back as echoes. This method enables bats, when flying in darkness, to avoid solid objects and to locate the position of flying insects. Bats have the most acute sense of hearing of any land animal, hearing frequencies as high as 120 to 210 kilohertz. The highest frequency humans can hear is 20 kilohertz.

## What names are used for **groups** or **companies** of **mammals**?

| Mammal | Group name |
|---|---|
| Antelopes | Herd |
| Apes | Shrewdness |
| Asses | Pace, drove, or herd |
| Baboons | Troop |
| Bears | Sloth |
| Beavers | Family or colony |
| Boars | Sounder |
| Buffaloes | Troop, herd, or gang |
| Camels | Flock, train, or caravan |
| Caribou | Herd |
| Cattle | Drove or herd |
| Deer | Herd or leash |
| Elephants | Herd |
| Elks | Gang or herd |
| Foxes | Cloud, skulk, or troop |
| Giraffes | Herd, corps, or troop |
| Goats | Flock, trip, herd, or tribe |
| Gorillas | Band |
| Horses | Haras, stable, remuda, stud, herd, string, field, set, team, or stable |
| Jackrabbits | Husk |
| Kangaroos | Troop, mob, or herd |
| Leopards | Leap |

| Mammal | Group name |
|---|---|
| Lions | Pride, troop, flock, sawt, or souse |
| Mice | Nest |
| Monkeys | Troop or cartload |
| Moose | Herd |
| Mules | Barren or span |
| Oxen | Team, yoke, drove, or herd |
| Porpoises | School, crowd, herd, shoal, or gam |
| Reindeer | Herd |
| Rhinoceri | Crash |
| Seals | Pod, herd, trip, rookery, or harem |
| Sheep | Flock, hirsel, drove, trip, or pack |
| Squirrels | Dray |
| Swine | Sounder, drift, herd, or trip |
| Walruses | Pod or herd |
| Weasels | Pack, colony, gam, herd, pod, or school |
| Whales | School, gam, mob, pod, or herd |
| Wolves | Rout, route, or pack |
| Zebras | Herd |

## What are the only mammals on Earth that cannot **jump**?

It might not be surprising to learn that neither the rhinoceros nor the elephant can jump, since their enormous weight makes the feat difficult. However, the third mammal that cannot jump is the pronghorn sheep, which was called an "antelope" in the famous song "Home on the Range." The pronghorn sheep's inability to jump has been a particular disadvantage in its North American home, where fences have prevented populations from migrating and hindered the pronghorn's ability to find mates and breed.

## How does the **breath-holding capability** of a human compare with other mammals?

| Mammal | Average time in minutes |
|---|---|
| Human | 1 |
| Polar bear | 1.5 |
| Pearl diver (human) | 2.5 |
| Sea otter | 5 |
| Platypus | 10 |
| Muskrat | 12 |
| Hippopotamus | 15 |
| Sea cow | 16 |

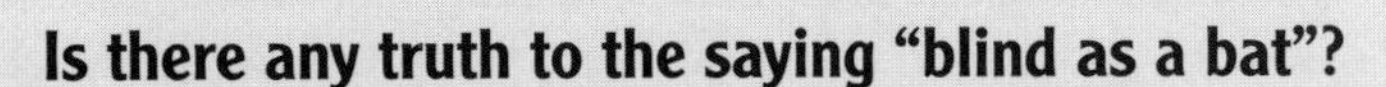

### Is there any truth to the saying "blind as a bat"?

The saying "blind as a bat" is not true. Although bats rely on sound to navigate and find food, they have all the elements found in a normal mammalian eye and they do see.

| Mammal | Average time in minutes |
|---|---|
| Beaver | 20 |
| Porpoise | 6 |
| Seal | 15 to 28 |
| Greenland whale | 60 |
| Sperm whale | 90 |
| Bottle-nosed whale | 120 |

## How does a human's **heartbeat** compare with those of other mammals?

| Mammal | Resting Heart Rate (beats per minute) |
|---|---|
| Human | 75 |
| Horse | 48 |
| Cow | 45–60 |
| Dog | 90–100 |
| Cat | 110–140 |
| Rat | 360 |
| Mouse | 498 |

## How does a **bat** catch flying insects in **total darkness**?

Bats use sound waves for communication and navigation. They emit supersonic radiation ranging from as low as 200 Hz to as high as 30,000 Hz. The sounds are emitted through the bat's nostrils and are aided by a complex flap structure to provide precise directivity to the radiation. Echo returns from the emissions allow a bat to pick out a tiny flying insect some distance ahead. Highly sensitive ears and an ability to maneuver with great agility enables many bats to fly around in a darkened cave, catching insects without fear of collision.

## What are some animals that have **pouches**?

Marsupials (meaning "pouched" animals) differ from all other living mammals in their anatomical and physiological features of reproduction. Most female marsupials—kangaroos, bandicoots, wombats, banded anteaters, koalas, opossums, wallabies, tasmanian devils, etc.—possess an abdominal pouch (called a marsupium), in which their young are carried. In some small terrestrial marsupials, however, the marsupium is not a true pouch but merely a fold of skin around the mammae (milk nipples).

The short gestation period in marsupials (in comparison to other similarly sized mammals) allows their young to be born in an "undeveloped" state. Consequently, these animals have been viewed as "primitive" or second-class mammals. However, some now see that the reproductive process of marsupials has an advantage over that of placental mammals. A female marsupial invests relatively few resources during the brief gestation period, more so during the lactation (nursing period) when the young are in the marsupium. If the female marsupial loses its young, it can conceive again sooner than a placental mammal in a comparable situation.

## How long do **wombats** live and what do they eat?

Native to Australia and Tasmania, the common wombat or coarse-haired wombat (*Vombatus ursinus*) lives between five and 26 years (26 years in zoos). It dines mostly on grasses, roots, mushrooms, fresh shoots, and herbaceous plants. A wombat looks like a small bear in appearance, has a thick heavy body ranging from 2.3 to 4 feet (70 to 120 centimeters) and weighs 33 to 77 pounds (15 to 35 kilograms). Its rough fur ranges from yellowish buff, to gray, to dark brown or black. This marsupial resembles a rodent in its manner of feeding and in its tooth structure—all its teeth are rootless and ever growing to compensate for their wear. Shy, it lives in a burrow and is an active digger.

## What freshwater mammal is **venomous**?

The male duck-billed platypus (*Ornithorhynchus anatinus*) has venomous spurs located on its hind legs. When threatened, the animal will drive them into the skin of a potential enemy, inflicting a painful sting. The venom this action releases is relatively mild and generally not harmful to humans.

## Which **mammals lay eggs** and **suckle** their young?

The duck-billed platypus (*Ornithorhynchus anatinus*), the short-nosed echidna or spiny anteater (*Tachyglossus aculeatus*), and the long-nosed echidna (*Zaglossus bruijni*), indigenous to Australia, Tasmania, and New Guinea, are the only three species of mammals that lay eggs (a non-mammalian feature) but suckle their young (a mammalian feature). These mammals (order Monotremata) resemble reptiles in that they lay rubbery shell-covered eggs that are incubated and hatched outside the mother's body. In addition, they resemble reptiles in their digestive, reproductive, and excretory systems, and in a number of anatomical details (eye structure, presence of certain

### What is the difference between porpoises and dolphins?

Marine dolphins (family Delphinidae) and porpoises (family Phocoenidae) consist of about 40 species. The chief differences between dolphins and porpoises occur in the snout and teeth. True dolphins have a beak-like snout and cone-shaped teeth. True porpoises have a rounded snout and flat or spade-shaped teeth.

skull bones, pectoral [shoulder] girdle and rib and vertebral structures). They are, however, classed as mammals because they have fur and a four-chambered heart, nurse their young from gland milk, are warm-blooded, and have some mammalian skeletal features.

## How deep do **marine mammals dive**?

Below are listed the maximum depths and the longest durations of time underwater by various aquatic mammals:

| Mammal | Maximum depth | | Maximum time underwater |
|---|---|---|---|
| | Feet | Meters | |
| Weddell seal | 1,968 | 600 | 70 minutes |
| Porpoise | 984 | 300 | 15 minutes |
| Bottle-nosed whale | 1,476 | 450 | 120 minutes |
| Fin whale | 1,148 | 350 | 20 minutes |
| Sperm whale | > 6,562 | > 2,000 | 90 minutes |

## How do the **great whales** compare in weight and length?

| Whale | Average weight | | Greatest length | |
|---|---|---|---|---|
| | Tons | Kilograms | Feet | Meters |
| Sperm | 35 | 31,752 | 59 | 18 |
| Blue | 84 | 76,204 | 98.4 | 30 |
| Finback | 50 | 45,360 | 82 | 25 |
| Humpback | 33 | 29,937 | 49.2 | 15 |
| Right | 50 (est.) | 45,360 (est.) | 55.7 | 17 |
| Sei | 17 | 15,422 | 49.2 | 15 |
| Gray | 20 | 18,144 | 39.3 | 12 |
| Bowhead | 50 | 45,360 | 59 | 18 |
| Bryde's | 17 | 15,422 | 49.2 | 15 |
| Minke | 10 | 9,072 | 29.5 | 9 |

Przewalski's horse—ancestor of all domestic horse breeds—is thought to be extinct in the wild.

## What is the name of the **seal-like animal** in Florida?

The West Indian manatee (*Trichechus manatus*), in the winter, moves to more temperate parts of Florida, such as the warm headwaters of the Crystal and Homosassa rivers in central Florida or the tropical waters of southern Florida. When the air temperature rises to 50°F (10°C), it will wander back along the Gulf coast and up the Atlantic coast as far as Virginia. Long-range offshore migrations to the coast of Guyana and South America have been documented. This large, plant-eating, water mammal may have been the inspiration for the mermaid legend. In 1893, when the population of manatees in Florida was reduced to several thousand, the state gave it legal protection from being hunted or commercially exploited. However, many animals continue to be killed or injured by the encroachment of humans. Entrapment in locks and dams, collisions with barges and power boat propellers, etc., cause at least 30 percent of the manatee deaths, which total 125 to 130 annually.

## What is the only **four-horned animal** in the world?

The four-horned antelope (*Tetracerus quadricornis*) is a native of central India. The males have two short horns, usually four inches (10 centimeters) in length, between their ears, and an even shorter pair, one to two inches (2.5 to five centimeters) long, between the brow ridges over their eyes. Not all males have four horns, and in some the second pair eventually falls off. The females have no horns at all.

### How many vertebrae are in the neck of a giraffe?

A giraffe neck has seven vertebrae, the same as other mammals, but the vertebrae are greatly elongated.

## What is **Przewalski's horse**?

This native of Mongolia and northeastern China is the last truly wild horse species. Named for Nikolai Przewalski (1839–1888), the Russian colonel who reported its existence in 1870, Przewalski's horse (*Equus przewalskii*) is a stocky, short-legged animal with a dun-colored coat, a pale muzzle and belly, and dark legs, mane, and tail. Its short mane is bristly and erect. This horse has the unusual feature of possessing 66 chromosomes rather than the customary 64 found in a domestic horse. Przewalski's horse was believed to have become extinct in the wild in 1968. However, 1,000 or so survived in zoos and wildlife parks.

In June 1994, a small herd bred in captivity was returned to the wild in Mongolia. They were kept in large enclosures for two years while they adjusted to the harsh climate. The herd is now flourishing and is expected to survive in the wild.

## Why were **Clydesdale horses** used as war horses?

The Clydesdales were among a group of European horses referred to as the "Great Horses," which were specifically bred to carry the massively armored knights of the Middle Ages. These animals had to be strong enough to carry a man wearing as much as 100 pounds (45 kilograms) of armor as well as up to 80 pounds (36 kilograms) of armor on their own bodies. However, the invention of the musket quickly ended the use of Clydesdales and other Great Horses on the battlefield as speed and maneuverability became more important than strength.

## How many letters are permissible in the name of a **thoroughbred horse**?

Before the name of a thoroughbred can become "official" in the United States, Canada, and Puerto Rico, it must be submitted to the Jockey Club for approval. One of their requirements limits the name to not more than three pronounceable words with a maximum of eighteen letters.

## What is the difference between an **African elephant** and an **Indian elephant**?

The African elephant (*Loxodonta africana*) is the largest living land animal, weighing up to 8.25 tons (7,500 kilograms) and standing 10 to 13 feet (three to four meters) at the shoulder. The Indian elephant (*Elephas maximus*) weighs about six tons (5,500 kilograms) with a shoulder height of 10 feet (three meters). Other differences are:

| African elephant | Indian elephant |
|---|---|
| Larger ears | Smaller ears |
| Gestation period of about 670 days | Gestation period of about 610 days |
| Ear tops turn backwards | Ear tops turn forwards |
| Concave back | Convex back |
| Three toenails on hind feet | Four toenails on hind feet |
| Larger tusks | Smaller tusks |
| Two finger-like lips at tips of trunk | One lip at tip of trunk |

## Why do **cows** have four stomachs?

The stomachs of cows, as well as all ruminants, are divided into four sections—the rumen, reticulum, omasum and abomasums. Ruminants eat rapidly and do not chew much of their food completely before they swallow it. The liquid part of their food enters the reticulum first, while the solid part of their food enters the rumen where it softens. Bacteria in the rumen initially break it down as a first step in digestion. Ruminants later regurgitate it into the mouth where they chew their cud. Cows chew their cud about six to eight times per day, spending a total of five to seven hours in rumination. The chewed cud goes directly into the other chambers of the stomach, where various microorganisms assist in further digestion.

## What is the difference between a **pig** and a **hog**?

In the United States, the term "pig" refers to younger domesticated swine weighing less than 120 pounds (50 kilograms), while the term "hog" refers to older swine weighing more than this. In Great Britain all domesticated swine are referred to as pigs.

## What is the difference between a **rabbit** and a **hare**?

Rabbits are generally smaller than hares and have shorter hind legs and shorter ears. Hares are born with a full coat of fur and their eyes open. Mothers drop their young either on the bare ground or in a slight depression in the ground. Rabbits are born hairless and blind; their nests are lined with grass, bark, and soft stems and covered with a layer of rabbit fur. Rabbits often live in a group while hares live most of the time by themselves.

## Is there a **cat** that lives in the **desert**?

The sand cat (*Felis margarita*) is the only member of the cat family tied directly to desert regions. Found in North Africa, the Arabian peninsula, the deserts of Turk-

### Why are Dalmatians "firehouse dogs"?

Before automobiles, coaches and carriages were often accompanied by dogs that kept horses company and guarded them from theft. Dalmatians were particularly well known for the strong bond they formed with horses, and firemen, who often owned the strongest and speediest horses in the area, kept the dogs at the station to deter horse thieves. Although fire engines have replaced horses, Dalmatians have remained a part of firehouse life, both for the appeal of these beautiful dogs and for their nostalgic tie to the past.

menistan in Uzbekistan, and in western Pakistan, the sand cat has adapted to extremely arid desert areas. The padding on the soles of its feet is well-suited to the loose, sandy soil, and it can live without drinking "free" water. Having sandy or grayish-ochre dense fur, its body length is 17.5 to 22 inches (45 to 57 centimeters). Mainly nocturnal (active at night) the cat feeds on rodents, hares, birds, and reptiles.

The Chinese desert cat (*Felis bieti*) does not live in the desert as its name implies, but inhabits the steppe country and mountains. Likewise, the Asiatic desert cat (*Felis silvestris ornata*) inhabits the open plains of India, Pakistan, Iran, and Asiatic Russia.

## What other names are used for a **cougar**?

A cougar is also known as a puma, painter, screamer, mountain lion, silver ghost, and catamount.

## What is the only **American canine** that can **climb trees**?

The gray fox (*Urocyon cinereoargenteus*) is the only American canine that can climb trees.

## How long does a **gray wolf** live?

The gray wolf (*Canis lupus*), also known as the timber wolf, is the largest and most widespread species of the family Canidae. It can live in the wild for less than 10 years, but under human care, up to 20 years. In many areas, however, it has been hunted and killed because it is believed to pose a threat to humans and domesticated animals (cattle, sheep, and reindeer). It may soon suffer the fate of the red wolf (*Canis rufus*), which once flourished in the southeast and south central United States. The red wolf has been declared biologically extinct in the wild, now existing in captivity and in a small reintroduced population of captive-bred animals in North Carolina. The endangered gray wolf is declining faster in the New World than in the Old. In the United

States, it is limited to Alaska (10,000), Northern Minnesota (1,200), and Isle Royale, Michigan (20), with perhaps a few packs in Wisconsin, northern Michigan, and the Rocky Mountain area; in Canada, they number 15,000. Strongly social, living in packs, and weighing between 75 and 175 pounds (43 to 80 kilograms), in physical appearance the gray wolf resembles a large domestic dog, such as the Alaskan malamute.

## Which **bear** lives in a **tropical rain forest**?

The Malayan sun bear (*Ursus malayanus*) is one of the rarest animals in the tropical forests of Sumatra, Malay Peninsula, Borneo, Burma, Thailand, and southern China. The smallest bear, with a length of 3.3 to 4.6 feet (one to 1.4 meters) and weighing 60 to 143 pounds (27 to 65 kilograms), it has a strong, stocky body and black coloring. With powerful paws having long, curved claws to help it climb trees in the dense forests, it is an expert tree climber. The sun bear tears at tree bark to expose insects, larvae, and the nests of bees and termites. Fruit, coconut palm, and small rodents, too, are part of its diet. Sleeping and sunbathing during the day, it is active at night. Unusually shy and retiring, cautious and intelligent, the sun bear is declining in population as the forests are being destroyed.

## What is the **largest terrestrial mammal** in North America?

The bison (*Bison bison*), which weighs 3,100 pounds (1,400 kilograms) and is six feet (two meters) high.

## Do **camels** store water in their humps?

The hump or humps do not store water, since they are fat reservoirs. The ability to go long periods without drinking water—up to ten months if there is plenty of green vegetation and dew to feed on—results from a number of physiological adaptations. One major factor is that camels can lose up to 40 percent of their body weight with no ill effects. A camel can also withstand a variation of its body temperature by as much as 14°F (–10°C). A camel can drink 30 gallons of water in ten minutes and up to 50 gallons over several hours. A one-humped camel is called a dromedary or Arabian camel; a Bactrian camel has two humps and lives in the wilds of the Gobi desert. Today, the Bactrian is confined to Asia, while most of the Arabian camels are on African soil.

## How many quills does a **porcupine** have?

For its defensive weapon, the average porcupine has about 30,000 quills or specialized hairs, comparable in hardness and flexibility to slivers of celluloid and so sharply pointed they can penetrate any hide. The quills that do the most damage are the short ones that stud the porcupine's muscular tail. With a few lashes, the porcupine can send a rain of quills that have tiny scale-like barbs into the skin of its adversary. The

The nine-banded armadillo generally gives birth to four same-sexed young.

quills work their way inward because of their barbs and the involuntary muscular action of the victim. Sometimes the quills can work themselves out, but other times the quills pierce vital organs, and the victim dies.

Slow-footed and stocky, porcupines spend much of their time in the trees, using their formidable incisors to strip off bark and foliage for their food, and supplement their diets with fruits and grasses. Porcupines have a ravenous appetite for salt; as herbivores (plant-eating animals) their diets have insufficient salt. So natural salt licks, animal bones left by carnivores (meat-eating animals), yellow pond lilies, and other items having a high salt content (including paints, plywood adhesives, and the sweated-on clothing of humans) have a strong appeal to porcupines.

## Why do **nine-banded armadillos** always have four offspring of the same sex?

The one feature that distinguishes the nine-banded armadillo (*Dasypus novemcinctus*) is that the female almost always gives birth to four young of the same sex. This consistency results from the division of the one fertilized egg into four parts to produce quadruplets.

## What is a **capybara**?

The capybara (*Hydrochoerus hydrochoeris*) is the largest of all living rodents. Also called the water hog, water pig, water cary, or carpincho, it looks like a huge guinea pig. Its body length can be 3.25 to 4.5 feet (one to 1.3 meters), and it usually weighs between

120 and 130 pounds (54 to 59 kilograms) or more. A native of northern South America, this rodent leads a semi-aquatic life, feeding on aquatic plants and grasses. A subspecies, native to Panama, is smaller and weighs between 60 and 75 pounds (27 to 34 kilograms).

### What is **chamois**?

The chamois (of the family Bovidae) is a goat-like animal living in the mountainous areas of Spain, central Europe (the Alps and Apennines), south central Europe, the Balkans, Asia Minor, and the Caucasus. Agile and surefooted, with acute senses, it can jump 6.5 feet (two meters) in height and 19.5 feet (six meters) in distance, and run at speeds of 31 miles (50 kilometers) per hour. Its skin has been made into "shammy" leather for cleaning glass and polishing automobiles, although more commonly today the shammy or chamois skins sold are simply specially treated sheepskin.

### How many different kinds of **squirrels** are found in the United States?

North American members of the squirrel family can be divided into six groups: marmots, prairie dogs, ground squirrels, chipmunks, tree squirrels, and flying squirrels. All American squirrels, except the flying squirrels, are diurnal in habits. The marmots are the largest American squirrels.

### What is the chemical composition of a **skunk's spray**?

The chief odorous components of the spray have been identified as crotyl mercaptan, isopentyl mercaptan, and methyl crotyl disulfide in the ratio of 4:4:3. The liquid is an oily, pale-yellow, foul-smelling spray that can cause severe eye inflammation. This defensive weapon is discharged from two tiny nipples located just inside the skunk's anus—either as a fine spray or a short stream of rain-sized drops. Although the liquid's range is 6.5 to 10 feet (two to three meters), its smell can be detected 1.5 miles (2.5 kilometers) downwind.

## PETS

### What is the **oldest breed** of dog?

Dogs are the oldest domestic animal, originating 12,000 to 14,000 years ago. They are believed to be descendants of wild canines, most likely wolves, which began to frequent human settlements where food was more readily available. The more aggressive canines were probably driven off or killed, while the less dangerous ones were kept to

**Wolves, like this red wolf, are believed to be the ancestors of all domestic dog breeds.**

guard, hunt, and later herd other domesticated animals, such as sheep. Attempts at selectively breeding desirable traits likely began soon after.

The oldest purebred dog is believed to be the Saluki. Sumerian rock carvings in Mesopotamia that date to about 7000 B.C.E. depict dogs bearing a striking resemblance to the Saluki. The dogs are 23 to 28 inches (58 to 71 centimeters) tall with a long, narrow head. The coat is smooth and silky and can be white, cream, fawn, gold, red, grizzle (bluish-gray) and tan, black and tan, or tricolor (white, black, and tan). The tail is long and feathered. The Saluki has remarkable sight and tremendous speed, which makes it an excellent hunter.

The oldest American purebred dog is the American Foxhound. It descends from a pack of foxhounds belonging to an Englishman named Robert Brooke who settled in Maryland in 1650. These dogs were crossed with other strains imported from England, Ireland, and France to develop the American Foxhound. This dog stands 22 to 25 inches (56 to 63.5 centimeters) tall. It has a long, slightly domed head, with a straight, squared-out muzzle. The coat is of medium length and can be any color. They are used primarily for hunting.

## Which breeds of dogs are **best for families with young children**?

Research has shown that golden retriever, Labrador retriever, beagle, collie, bichon frise, cairn terrier, pug, coonhound, boxer, basset hound or mixes of these breeds are best for families with young children.

## How do the **bones** of an **adult human** compare to those of an **adult dog**?

| | **Adult Human** | **Adult Dog** |
|---|---|---|
| Number of bones | 206 | 321 |
| Number of vertebrae | 33 | 50 |
| Number of joints | over 200 | over 300 |
| Age of maturity | 18 | 2 |
| Longest bone | femur (thighbone) | ulna (forearm bone) |
| Smallest bone | ossicles (ear bones) | ossicles (ear bones) |
| Number of ribs | 12 on each side | 18 on each side |

## What are the different **classifications of dogs**?

Dogs are divided into groups according to the purpose for which they have been bred.

| **Group** | **Purpose** | **Representative breeds** |
|---|---|---|
| Sporting dogs | Retrieving game birds and water fowl | Cocker spaniel, English setter, English springer spaniel, golden retriever, Irish setter, Labrador retriever, pointer. |
| Hounds | Hunting | Basenji, beagle, dachshund, foxhound, greyhound, saluki, Rhodesian ridgeback. |
| Terriers | Hunting small animals such as rats and foxes | Airedale terrier, Bedlington terrier, bull terrier, fox terrier, miniature schnauzer, Scottish terrier, Skye terrier, West Highland white terrier. |
| Toy dogs | Small companions or lap dogs | Chihuahua, Maltese, Pekingese, Pomeranian, pug, Shih Tzu, Yorkshire terrier. |
| Herding dogs | Protect sheep and other livestock | Australian cattle dog, bouviers des Flandres, collie, German shepherd, Hungarian puli, Old English sheepdog, Welsh corgi. |
| Working dogs | Herding, rescue, and sled dogs | Alaskan malamute, boxer, Doberman pinscher, great Dane, mastiff, St. Bernard, Siberian husky. |
| Non-sporting dogs | No specific purpose, not toys | Boston terrier, bulldog, Dalmatian, Japanese akita, keeshond, Lhasa apso, poodle. |

## Which breeds of dogs are the **most dangerous**?



According to one study, the dog breeds responsible for the most fatal attacks on people are:

| Breed | Known fatal attacks |
|---|---|
| Pit bull | 57 |
| Rottweiler | 19 |
| German shepherd | 17 |
| Husky | 12 |
| Malamute | 12 |
| Doberman pinscher | 8 |
| Chow | 6 |
| Great Dane | 5 |
| St. Bernard | 4 |
| Akita | 4 |

## Which dogs are the **easiest to train**?

In a study of 56 popular dog breeds the top breeds to train were Shetland sheepdogs, Shih Tzus, miniature toy and standard poodles, Bichons Frises, English Springer Spaniels and Welsh Corgis.

## What are the top ten **dog names**?

In one survey, the top ten dog names were:

1. Brandy
2. Lady
3. Max
4. Rocky
5. Sam
6. Heidi
7. Sheba
8. Ginger
9. Muffin
10. Bear

## Why do **dogs hear** more than humans?

A dog's ears are highly mobile, allowing it to scan its environment for sounds. The ears capture the sounds and funnel them down to the eardrum. Dogs can hear sounds from four times farther away than humans.

## Why do dogs **howl** at **sirens**?

The high pitch of a siren is very similar to the pitch of a dog's howl. A dog's howl is a way of communicating with other dogs—either to indicate location or to define territory. When a dog responds to an ambulance or fire engine siren, he is "returning the call of the wild."

## What breeds of dogs **do not shed**?

Poodles, Kerry blue terriers, and schnauzers do not shed.

## Which breed is known as the **wrinkled dog**?

The shar-pei, or Chinese fighting dog, is covered with folds of loose skin. It stands 18 to 20 inches (46 to 51 centimeters) tall and weighs up to 50 pounds (22.5 kilograms). Its solid-colored coat can be black, red, fawn, or cream. The dog originated in Tibet or the northern provinces of China some 2,000 years ago. The People's Republic of China put such a high tax on shar-peis, however, that few people could afford to keep them, and the dog was in danger of extinction. But a few specimens were smuggled out of China, and the breed has made a comeback in the United States, Canada, and the United Kingdom. Although bred as a fighting dog, the shar-pei is generally an amiable companion.

## Which breed is known as the **voiceless dog**?

The basenji dog does not bark. When happy, it will make an appealing sound described as something between a chortle and a yodel. It also snarls and growls on occasion. One of the oldest breeds of dogs, and originating in central Africa, the basenji was often given as a present to the pharaohs of ancient Egypt. Following the decline of the Egyptian civilization, the basenji was still valued in central Africa for its hunting prowess and its silence. The dog was rediscovered by English explorers in the 19th century, although it was not widely bred until the 1940s.

The basenji is a small, lightly built dog with a flat skull and a long, rounded muzzle. It measures 16 to 17 inches (40 to 43 centimeters) in height at the shoulder and weighs 22 to 24 pounds (10 to 11 kilograms). The coat is short and silky in texture. The feet, chest, and tail tip are white; the rest of the coat is chestnut red, black, or black and tan.

## What **food odors** do dogs like best?

In a study of different foods, researchers found that liver and chicken ranked higher than everything else, including hamburgers, fish, vegetables, and fresh fruit.

## Where did the **pug dog** originate?

The pug's true origin is unknown, but it has existed in China, its earliest known source, for 1,800 years. A popular pet in Buddhist monasteries in Tibet, it next appeared in Japan and then in Europe. It probably was introduced into Holland by the

### How is the age of a dog or cat computed in human years?

When a cat is one year old, it is about 20 years old in human years. Each additional year is multiplied by four. Another source counts the age of a cat slightly differently. At age one, a cat's age equals 16 human years. At age two, a cat's age is 24 human years. Each additional year is multiplied by four.

When a dog is one year old, it is about 15 years old in human years. At age two it is about 24; after age two, each additional year is multiplied by four.

traders of the Dutch East India Company. The name "pug dog" may have come from the dog's facial resemblance to a marmoset monkey. This popular pet in the 1700s was called a "pug," so the term "pug dog" distinguished the dog from the "pug" monkey.

A pug has a square, short, compact body, either silver or apricot-fawn in color. Its muzzle is black, short, blunt, and square; its average weight is 14 to 18 pounds (6.4 to 8.2 kilograms). The pug is often described by the motto "Multum in Parvo"—a lot of dog in a small space.

## What is the **rarest breed** of dog?

The Tahltan bear dog, of which only a few remain, is thought to be the rarest dog. In danger of extinction, this breed was once used by the Tahltan Indians of western Canada to hunt bear, lynx, and porcupine.

## What was the **contribution to medical science** of a dog named **Marjorie**?

Marjorie was a diabetic black-and-white mongrel that was the first creature to be kept alive by insulin, a substance that controls the level of sugar in the blood.

## What is the **newest method** of tagging a dog?

There is now a computer-age dog tag. A microchip is implanted painlessly between the dog's shoulder blades. The semiconductor carries a 10-digit code, which can be read by a scanner. When the pet is found, the code can be phoned into a national database to locate the owner. The microchip can store license number, medical condition, and the owner's address and phone number.

## Do **dogs and cats** have **good memories**?

Dogs do have long-term memories, especially for those whom they love. Cats have a memory for things that are important to their lives. Some cats seem to have extraordinary "memories" for finding places. Taken away from their homes, they seem able to remember where they live. The key to this "homing" ability could be a built-in celestial navigation, similar to that used by birds, or the cats' navigational ability could be attributed to the cats' sensitivity to Earth's magnetic fields. When magnets are attached to cats, their normal navigational skills are disrupted.

## What is the **original breed of domestic cat** in the United States?

The American shorthair is believed by some naturalists to be the original domestic cat in America. It is descended from cats brought to the New World from Europe by the early settlers. The cats readily adapted to their new environment. Selective breeding to enhance the best traits began early in the 20th century.

The American shorthair is a very athletic cat with a lithe, powerful body, excellent for stalking and killing prey. Its legs are long, heavy, and muscular, ideal for leaping and for coping with all kinds of terrain. The fur, in a wide variety of color and coat patterns, is thick enough to protect the animal from moisture and cold, but short enough to resist matting and snagging.

Although this cat makes an excellent house pet and companion, it remains very self-sufficient. Its hunting instinct is so strong that it exercises the skill even when well-provided with food. The American shorthair is the only true "working cat" in the United States.

## What is a **tabby cat**?

"Tabby," the basic feline coat pattern, dates back to the time before cats were domesticated. The tabby coat is an excellent form of camouflage. Each hair has two or three dark and light bands, with the tip always dark. There are four variations on the basic tabby pattern.

The mackerel (also called striped or tiger) tabby has a dark line running down the back from the head to the base of the tail, with several stripes branching down the sides. The legs have stripes, and the tail has even rings with a dark tip. There are two rows of dark spots on the stomach. Above the eyes is a mark shaped like an "M" and dark lines run back to the ears. Two dark necklace-like bands appear on the chest.

The blotched, or classic, tabby markings seem to be the closest to those found in the wild. The markings on the head, legs, tail, and stomach are the same as the mackerel tabby. The major difference is that the blotched tabby has dark patches on the shoulder and side, rimmed by one or several lines.

The spotted tabby has uniformly shaped round or oval dark spots all over the body and legs. The forehead has an "M" on it, and a narrow, dark line runs down the back.

The Abyssinian tabby has almost no dark markings on its body; they appear only on the forelegs, the flanks, and the tail. The hairs are banded except on the stomach, where they are light and unicolored.

### How can pets be treated to remove skunk odor?

From a pet store, purchase one of the products specifically designed to counteract skunk odor. Most of these are of the enzyme or bacterial enzyme variety and can be used without washing the pet first. A dog may also be given a bath with tomato juice, diluted vinegar, or neuthroleum-alpha, or you could try mint mouthwash, aftershave, or soap and water.

## What controls the formation of the color **points** in a **Siamese cat**?

The color points are due to the presence of a recessive gene, which operates at cooler temperatures, limiting the color to well-defined areas—the mask, ears, tail, lower legs, and paws—the places at the far reaches of the cardiovascular system of the cat.

There are four classic varieties of Siamese cats. Seal-points have a pale fawn to cream colored coat with seal-brown markings. Blue-points are bluish-white with slate blue markings. Chocolate-points are ivory colored with milk-chocolate brown colored markings. Lilac-points have a white coat and pinkish-gray markings. There are also some newer varieties with red, cream, and tabby points.

The Siamese originated in Thailand (once called Siam) and arrived in England in the 1880s. They are medium-sized and have long, slender, lithe bodies, with long heads and long, tapering tails. Extroverted and affectionate, Siamese are known for their loud, distinctive voices, which are impossible to ignore.

## Why do **cats' eyes shine** in the dark?

A cat's eyes contain a special light-conserving mechanism called the *tapetum lucidum*, which reflects any light not absorbed as it passes through the retina of each eye. The retina gets a second chance (so to speak) to receive the light, aiding the cat's vision even more. In dim light, when the pupils of the cat's eyes are opened the widest, this glowing or shining effect occurs when light hits them at certain angles. The *tapetum lucidum*, located behind the retina, is a membrane composed of 15 layers of special, glittering cells that all together act as a mirror. The color of the glow is usually greenish or golden, but the eyes of the Siamese cat reflect a luminous ruby red.

## Why and how do cats **purr**?

Experts cannot agree on how or why cats purr, or on where the sound originates. Some think that the purr is produced by the vibration of blood in a large vein in the chest cavity. Where the vein passes through the diaphragm, the muscles around the vein contract, nipping the blood flow and setting up oscillations. These sounds are magnified by the air in the bronchial tubes and the windpipe. Others think that purring is the vibrations of membranes, called false vocal cords, located near the vocal cords. No one knows for sure why a cat purrs, but many people interpret the sound as one of contentment.

## Why do cats have **whiskers**?

The function of a cat's whiskers is not fully understood. They are thought to have something to do with the sense of touch. Removing them can disturb a cat for some time. Some people believe that the whiskers act as antennae in the dark, enabling the cat to identify things it cannot see. The whiskers may help the cat to pinpoint the direction from which an odor is coming. In addition, the cat is thought to point some of its whiskers downwards to guide it when jumping or running over uneven terrain at night.

## What are the top ten **cat names**?

In one survey, the top ten cat names were:

1. Kitty
2. Smokey
3. Shadow
4. Tiger
5. Boo (Boo Boo)
6. Boots
7. Molly
8. Tigger
9. Spike
10. Princess

## Which plants are **poisonous to cats**?

Certain common houseplants are poisonous to cats, which should not be allowed to eat the following:

Caladium (Elephant's ears)
Dieffenbachia (Dumb cane)
*Euphorbia pulcherrima* (Poinsettia)
Hedera (True ivy)
Mistletoe
Oleander
Philodendron
*Prunus laurocerasus* (Common or cherry laurel)
Rhododendron (Azalea)
*Solanum capiscastrum* (Winter or False Jerusalem cherry)

## Which types of **birds** make the best pets?

There are several birds that make good house pets and have a reasonable life expectancy:

| Bird | Life expectancy in years | Considerations |
|---|---|---|
| Finch | 2–3 | Easy care |
| Canary | 8–10 | Easy care; males sing |
| Budgerigar (parakeet) | 8–15 | Easy care |
| Cockatiel | 15–20 | Easy care; easy to train |
| Lovebird | 15–20 | Cute, but not easy to care for or train |
| Amazon parrot | 50–60 | Good talkers, but can be screamers |
| African grey parrot | 50–60 | Talkers; never scream |

## What kind of care do **tadpoles** need, and what do they eat?

Keep the frog eggs and the tadpoles that hatch from them in water at all times, changing half of the water volume no more than once a week. The best diet is probably baby cereal having a high protein content, fresh greens, and bits of egg yolk. Provide a rock island when the legs of the tadpoles appear. A five-gallon (19-liter) tank is sufficient for a half dozen tadpoles. When they mature (lose their tails and have grown legs) they should be released into a pond or by the lake shore.

## What do you feed a **hermit crab**?

Hermit crabs are not particular and will eat a variety of foods, including algae, beef heart, brine shrimp, earthworms, fish, flake food, fresh shrimp, scallops, tube food, and almost all other commercially prepared foods. Live, fresh, frozen, dry, or freeze-dried, it makes no difference. Hermit crabs may be fed individually two to three times a week, but be careful not to overfeed or underfeed them. Pieces of meat, such as thawed shrimp, scallops, or beef heart, may be soaked in a liquid vitamin complex and presented on the end of a toothpick to the crab. Algae, which is a part of the crab's diet in the wild, may be grown in a separate container with a few small pieces of coral or ordinary shells on the bottom. Within a few weeks, the shells will be covered with algae and may be placed in the display tank for the crabs to pick clean. Fresh spinach and lettuce may be used as a substitute for algae.

## What are some unusual animals that have been **White House pets**?

Several unusual animals have resided at the White House. In 1825, the Marquis de Lafayette (1757–1834) toured America and was given an alligator by a grateful citizen. While Lafayette was the guest of President John Quincy Adams (1767–1848), the alligator took up residence in the East Room of the White House for several months. When Lafayette departed, he took his alligator with him. Mrs. John Quincy Adams also kept unusual pets: silkworms that feasted on mulberry leaves. Other residents kept a horned-toad, another a green snake, and still another a kangaroo rat. Theodore Roosevelt brought home a badger that was presented to him as he campaigned in Kansas. The Abraham Lincoln household contained rabbits and a pair of goats named Nanny and Nanko. President Calvin Coolidge kept a raccoon as a pet instead of eating it for Thanksgiving dinner, as was intended by the donors from the State of Mississippi. Given the name Rebecca, the raccoon was kept in a large pen near the President's office.

Other unusual White House pets were:

| | |
|---|---|
| Martin Van Buren | Two tiger cubs |
| William Henry Harrison | Billy goat; Durham cow |
| Andrew Johnson | Pet mice |
| Theodore Roosevelt | Lion, hyena, wildcat, coyote, five bears, zebra, barn owl, snakes, lizards, roosters, raccoon |
| William Howard Taft | Cow |
| Calvin Coolidge | Raccoons, donkey, bobcat, lion cubs, wallaby, pigmy hippo, bear |

# HUMAN BODY

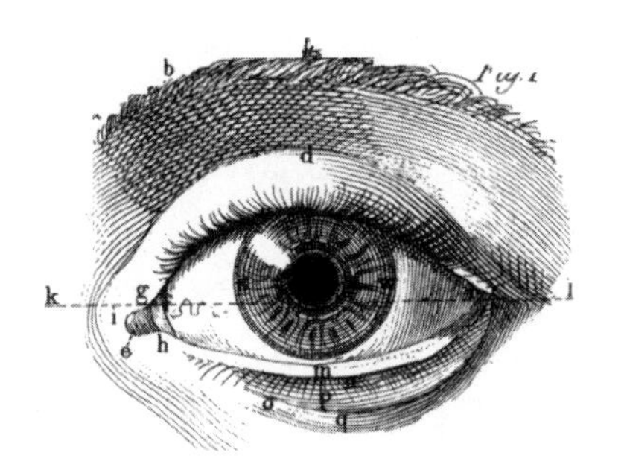

## FUNCTIONS, PROCESSES, AND CHARACTERISTICS

### Which **chemicals** constitute the human body?

About 24 elements are used by the body in its functions and processes.

**Major elements**

| Element | Percentage | Function |
|---|---|---|
| Oxygen | 65.0 | Part of all major nutrients of tissues; vital to energy production |
| Carbon | 18.5 | Essential life element of proteins, carbohydrates, and fats; building blocks of cells |
| Hydrogen | 9.5 | Part of major nutrients; building blocks of cells |
| Nitrogen | 3.3 | Essential part of proteins, DNA, RNA; essential to most body functions |
| Calcium | 1.5 | Forms nonliving bone parts; a messenger between cells |
| Phosphorous | 1.0 | Important to bone building; essential to cell energy |

Potassium, sulfur, sodium, chlorine, and magnesium each occur at 0.35 percent or less. There are also traces of iron, cobalt, copper, manganese, iodine, zinc, fluorine, boron, aluminum, molybdenum, silicon, chromium, and selenium.

### How many **chromosomes** are in a human body cell?

A human being normally has 46 chromosomes (23 pairs) in all but the sex cells. Half of each pair is inherited from the mother's egg; the other, from the father's sperm.

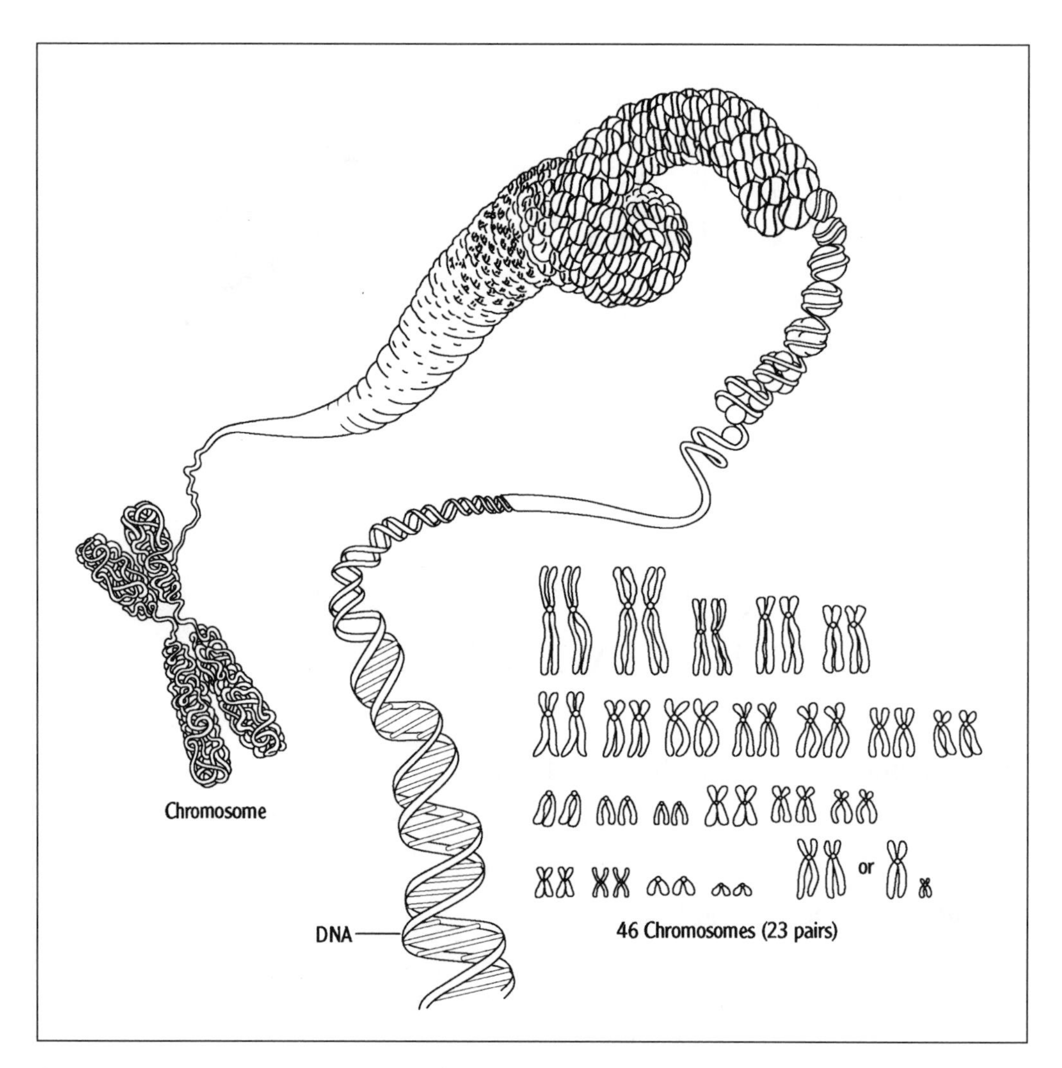


Chromosomes, magnified.

When the sperm and egg unite in fertilization, they create a single cell, or zygote, with 46 chromosomes. When cell division occurs, the 46 chromosomes are duplicated; this process is repeated billions of times over, with each of the cells containing the identical set of chromosomes. Only the gametes, or sex cells, are different. In their cell division, the members of each pair of chromosomes are separated and distributed to different cells. Each gamete has only 23 chromosomes.

Chromosomes contain thousands of genes, each of which has information for a specific trait. That information is in the form of a chemical code, and the chemical compound that codes this genetic information is deoxyribonucleic acid, or DNA. A gene can be seen as a sequence of DNA that is coded for a specific protein. These proteins determine specific physical traits (such as height, body shape, color of hair, eyes,

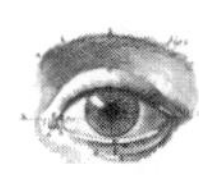



skin, etc.), body chemistry (blood type, metabolic functions, etc.), and some aspects of behavior and intelligence. More than 150 human disorders are inherited, and genes are thought to determine susceptibility to many diseases.

## **How many cells** are in the human body?

Sources give figures that vary from 50 to 75 trillion cells.

## What is the **average life span of cells** in the human body?

The human body is self-repairing and self-replenishing. According to one estimate, almost 200 billion cells die each hour. In a healthy body, dying cells are simultaneously replaced by new cells.

| Cell type | Length of time |
|---|---|
| Blood cells | |
| red blood cells | 120 days |
| lymphocytes | Over 1 year |
| other white cells | 10 hours |
| platelets | 10 days |
| Bone cells | 25–30 years |
| Brain cells* | Lifetime |
| Colon cells | 3–4 days |
| Liver cells | 500 days |
| Skin cells | 19–34 days |
| Spermatozoa | 2–3 days |
| Stomach cells | 2 days |

*Brain cells are the only cells that do not divide further during a person's lifetime. They either last the entire lifetime or, if a cell in the nervous system dies, it is not replaced.

## What is **DNA fingerprinting**?

DNA, or genetic, fingerprinting is a method of determining identity, family relationships, etc. Formulated by Alec Jeffreys (1950– ), a British geneticist, the genetic fingerprint is based on the assumption that every person (except identical twins) has a unique sequence of DNA (deoxyribonucleic acid)—a substance present in the nucleus of every cell that determines individual characteristics.

Within the DNA molecule, the sequence of the genetic information is repeated many times along the DNA structure, which resembles a long twisting ladder. The length of the sequence, the number of repetitions, and its precise location within the DNA chain seems to be unique in all cases but identical twins, with the odds being 30 billion to one of a duplication. A process has been developed that translates these sequences into a visual record that resembles a series of bars on X-ray film. In this process the technician isolates the DNA from blood, saliva, hair follicles, or semen; then the DNA strand is separated into thousands of shorter pieces by placing the DNA into an enzyme solution. Finally, the fragments, which are placed in a gelatin-like material, are subjected to a strong electrical current that separates them according to size and electrical behavior.

In criminal investigations, hair, blood, and skin samples left by the criminal can yield a DNA fingerprint, which can be matched against any suspect's DNA fingerprint. DNA fingerprinting is also used to determine the father in paternity cases, since the infant's DNA strand contains the genetic coding of the parents.

## What does the term **ergonomics** mean?

The study of human capability and psychology in relation to the worker's working environment and equipment is variously known as ergonomics, human engineering, human factors engineering, or engineering psychology. Ergonomics is based on the premise that tools humans use and the environment they work in should be matched with their capabilities and limitations, rather than forcing humans to adapt to the physical environment. Researchers in ergonomics try to determine optimum conditions in communication, cognition, reception of sensory stimuli, physiology, and psychology, and examine the effect of adverse conditions. Specific areas of study include design of work areas (including seats, desks, consoles, and cockpits) in terms of human physical size, comfort, strength, and vision; effects of physiological stresses such as work speed, work load, decision making, fatigue, and demands on memory and perception; and design of visual displays to enhance the quality and speed of interpretation.

## How does the **immune system** work?

The immune system has two main components: white blood cells and antibodies circulating in the blood. The antigen–antibody reaction forms the basis for this immunity. When an antigen (*anti*body *gen*erator)—such as a harmful bacterium, virus, fungus, parasite, or other foreign substance—invades the body, a specific antibody is generated to attack the antigen. The antibody is produced by B lymphocytes (B cells) in the spleen or lymph nodes. An antibody may either destroy the antigen directly or it may "label" it so that a white blood cell (called a macrophage, or scavenger cell) can engulf the foreign intruder. After a human has been exposed to an antigen, a later exposure to the same antigen will produce a faster immune system reaction. The necessary antibodies will be produced more rapidly and in larger amounts. Artificial immunization uses this antigen–antibody reaction to protect the human body from certain diseases by exposing the body to a safe dose of antigen to produce effective antibodies as well as a "readiness" for any future attacks of the harmful antigen.

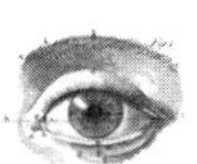

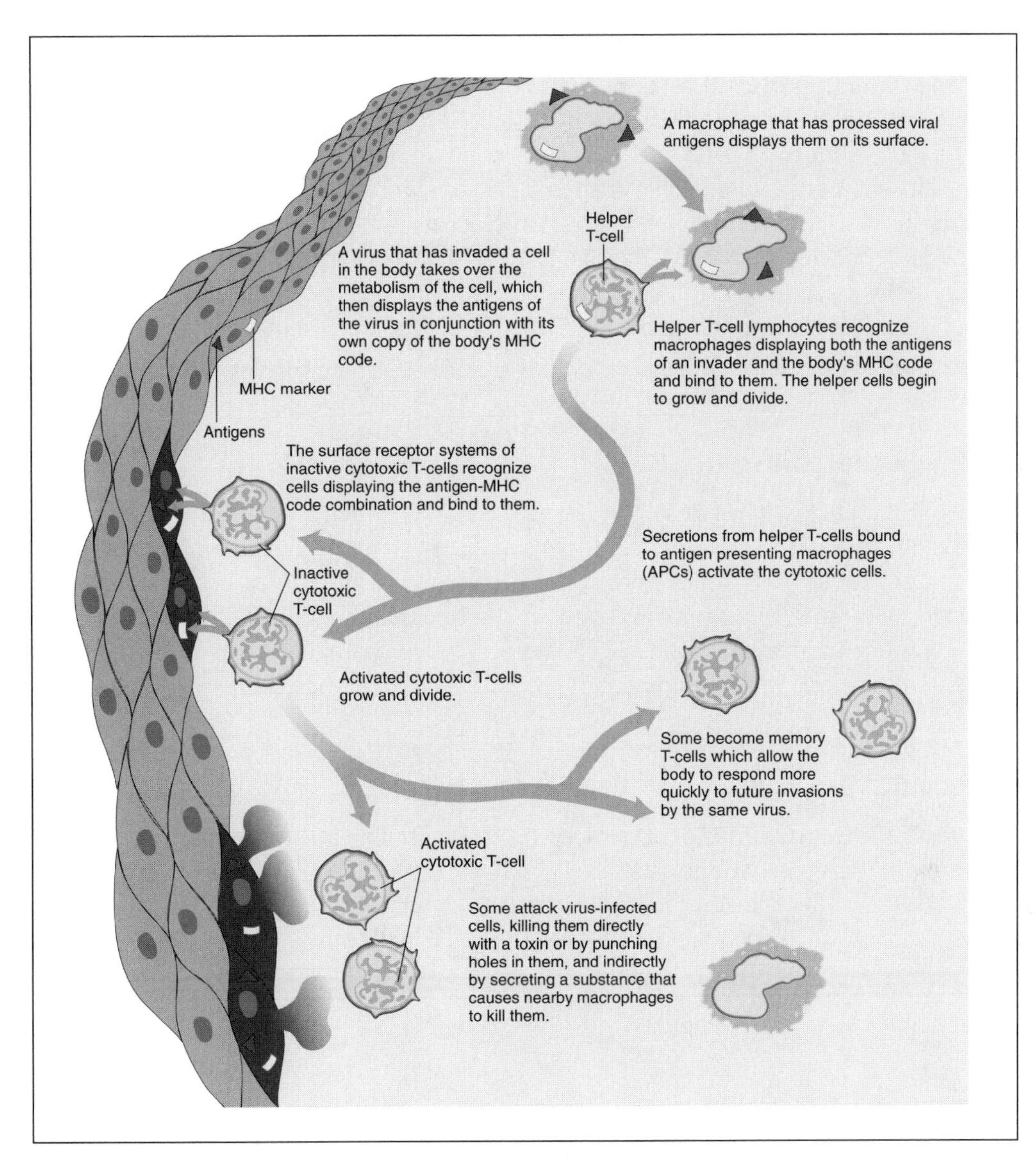


**How the immune system works.**

## How do **T cells** differ from **B lymphocytes**?

Lymphocytes are one variety of white blood cells and are part of the body's immune system. The immune system fights invading organisms that have penetrated the body's general defenses. T cells, responsible for dealing with most viruses, for handling some bacteria and fungi, and for cancer surveillance, are one of the two main classes of lymphocytes. T lymphocytes, or T cells, compose about 60–80 percent of the lymphocytes circulating in the blood. They have been "educated" in the thymus to perform particular functions. Killer T cells are sensitized to multiply when they come into contact with antigens (foreign proteins) on abnormal body cells (cells that have been invaded by viruses, cells in transplanted tissue, or tumor cells). These killer T cells attach themselves to the abnormal cells and release chemicals (lymphokines) to destroy them. Helper T cells assist killer cells in their activities and control other aspects of the immune response. When B lymphocytes, which compose approximately 10–15 percent of total lymphocytes, contact the antigens on abnormal cells, the lymphocytes enlarge and divide to become plasma cells. Then the plasma cells secrete vast numbers of immunoglobulins or antibodies into the blood, which attach themselves to the surfaces of the abnormal cells, to begin a process that will lead to the destruction of the invaders.

## What are **endorphins**?

Endorphins and closely related chemicals called enkephalins are part of a larger group called opiods, which have properties very much like drugs such as heroin or morphine. They can act not only as pain killers but also can induce a sense of well-being or euphoria. Clinical applications of endorphin research include possible treatments for some forms of mental illness; treatment or control of pain for chronic pain sufferers; development of new anesthetics; and the development of non-addictive, safe, and effective pain relievers.

## Is it true that people **need less sleep** as they get older?

As a person ages, the time spent in sleeping changes. The following table shows how long a night's sleep generally lasts.

| Age | Sleep time (in hours) |
|---|---|
| 1–15 days | 16–22 |
| 6–23 months | 13 |
| 3–9 years | 11 |
| 10–13 years | 10 |
| 14–18 years | 9 |
| 19–30 years | 8 |
| 31–45 years | 7.5 |
| 45–50 years | 6 |
| 50+ years | 5.5 |

## What is **REM** sleep?

REM sleep is rapid eye movement sleep. It is characterized by faster breathing and heart rates than NREM (non-rapid eye movement) sleep. The eyes move rapidly, and dreaming, often with elaborate story lines, occurs. The only people who do not have REM sleep are those who have been blind from birth. REM sleep usually occurs in four

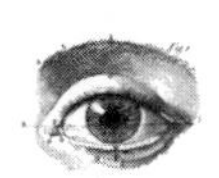



to five periods, varying from five minutes to about an hour, growing progressively longer as sleep continues.

Scientists do not understand why dreaming is important, but they think the brain is either cataloging the information it picked up during the day and throwing out the data it does not want, or is creating scenarios to work through situations causing emotional distress. Regardless of its function, most people who are deprived of sleep or dreams become disoriented, unable to concentrate, and may even have hallucinations.

## Why is it difficult to remember **dreams**?

Almost all dreams occur during rapid eye movement (REM) sleep. It appears that the content of dreams is stored in short-term memory and cannot be transferred into long-term memory unless they are somehow articulated. Sleep studies show that when individuals who believe they never dream are awakened at various intervals during the night they are in the middle of a dream.

## Why do people **snore** and how loud can snoring be?

Snoring is produced by vibrations of the soft palate, usually caused by any condition that hinders breathing through the nose. It is more common while sleeping on the back. Research has indicated that a snore can reach 69 decibels, as compared to 70–90 decibels for a pneumatic drill.

## How many **calories** does a person burn while **sleeping**?

A 150-pound (68-kilogram) person burns one calorie per minute during bed rest. The approximate caloric expenditure of other activities for a person weighing 150 pounds are given below. Actual numbers may vary, depending on the vigor of the exercise, air temperature, clothing, etc.

| Activity | Calories used per hour |
| --- | --- |
| Aerobic dance | 684 |
| Basketball | 500 |
| Bicycling (5.5 mph) | 210 |
| Bicycling (13 mph) | 660 |
| Bowling | 220–270 |
| Calisthenics | 300 |
| Circuit weight training | 756 |
| Digging | 360–420 |
| Gardening | 200 |
| Golfing (using power cart) | 150–220 |
| Golfing (pulling cart) | 240–300 |

| Activity | Calories used per hour |
|---|---|
| Golfing (carrying clubs) | 300–360 |
| Football | 500 |
| Handball(social) | 600–660 |
| Handball (competitive) | >660 |
| Hoeing | 300–360 |
| Housework | 180 |
| In-line skating | 600 |
| Jogging (5–10 mph) | 500–800 |
| Lawn mowing (power) | 250 |
| Lawn mowing (hand) | 420–480 |
| Racquetball | 456 |
| Raking leaves | 300–360 |
| Rowing machine | 415 |
| Sitting | 100 |
| Skiing (cross-country) | 600–660 |
| Skiing (downhill) | 570 |
| Snow shoveling | 420–480 |
| Square dancing | 350 |
| Standing | 140 |
| Swimming moderately | 500–700 |
| Tennis (doubles) | 300–360 |
| Tennis (singles) | 420–480 |
| Vacuuming | 240–300 |
| Volleyball | 350 |
| Walking (2 mph) | 150–240 |
| Walking (3.5 mph) | 240–300 |
| Walking (4 mph) | 300–400 |
| Walking (5 mph) | 420–480 |

## Who were the doctor and patient involved in the first studies on **digestion** performed by direct observation of the patient's stomach?

Alexis St. Martin, a French Canadian, was accidentally wounded by a shotgun blast in 1822. Fortunately, William Beaumont (1785–1853), an army surgeon, was nearby and began treatment of the wound immediately. St. Martin's recuperation lasted nearly three years, and the enormous wound healed except for a small opening leading into his stomach. A fold of flesh covered this opening; when this was pushed aside the interior of the stomach was exposed to view. Through the opening, Beaumont was able to extract and analyze gastric juice and stomach contents at various stages of digestion, observe changes in secretions, and note the stomach's muscular movements. The results of his experiments and observations formed the basis of our modern knowledge of digestion. Today, the use of X-rays and other medical instruments provides the same diagnostic function.

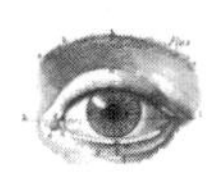



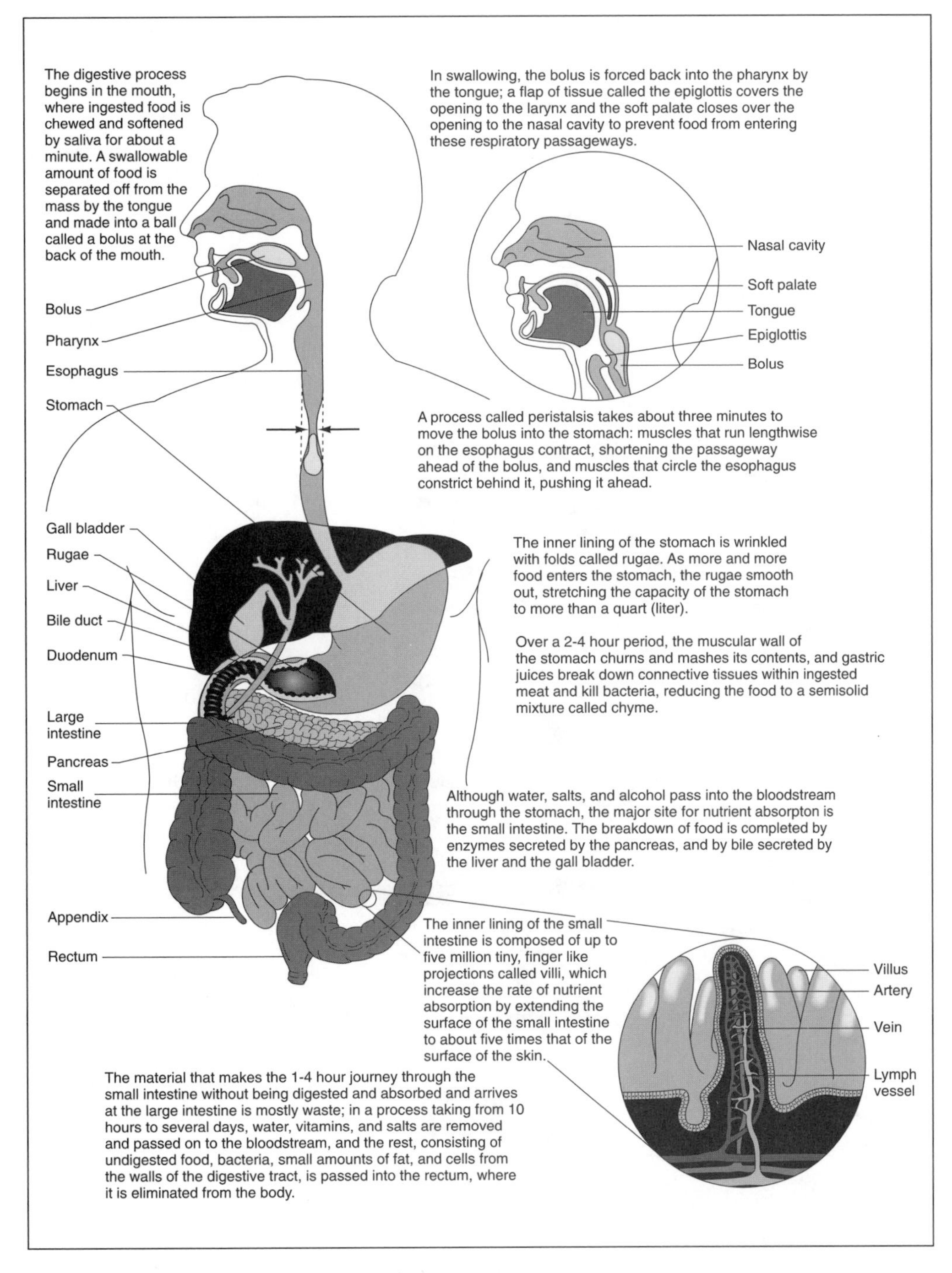


**The digestive process.**

## When a person **swallows** solid or liquid food, what prevents it from going down the windpipe?

Once food is chewed, voluntary muscles move it to the throat. In the pharynx (throat), automatic involuntary reflexes take over. The epiglottis closes over the larynx (voice box), which leads to the windpipe. A sphincter at the top of the esophagus relaxes, allowing the food to enter the digestive tract.

## **How long** does it take food to **digest**?

The stomach holds a little under two quarts (1.9 liters) of semi-digested food that stays in the stomach three to five hours. The stomach slowly releases food to the rest of the digestive tract. Fifteen hours or more after the first bite started down the alimentary canal, the final residue of the food is passed along to the rectum and is excreted through the anus as feces.

Claude Bernard, pioneer in the field of physiology.

## What is the length of the human **intestine**?

The small intestine is about 22 feet (seven meters) long. The large intestine is about five feet (1.5 meters) long.

## Who is considered the founder of **physiology**?

As an experimenter, Claude Bernard (1813–1878) enriched physiology by his introduction of numerous new concepts into the field. The most famous of these concepts is that of the milieu intérieur or internal environment. The complex functions of the various organs are closely interrelated and are all directed to

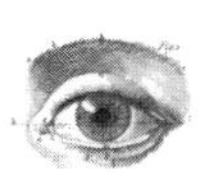



maintaining the constancy of internal conditions despite external changes. All cells exist in this aqueous (blood and lymph) internal environment, which bathes the cells and provides a medium for the elementary exchange of nutrients and waste material.

## Who coined the term **homeostasis**?

Walter Bradford Cannon (1871–1945), who elaborated on Claude Bernard's (1813–1878) concept of the milieu intérieur (interior environment), used the term homeostasis to describe the body's ability to maintain a relative constancy in its internal environment.

## How much **heat** is **lost through the head** when a person is not wearing a hat?

Between 7 percent and 55 percent of total body heat can be lost through the head. The amount of blood going to the head is controlled by cardiac output, and the harder the body works, the more blood is circulated to the head, and the more heat is quickly radiated away.

## In the United States, what is the **average height and weight** for a man and a woman?

The average female is five feet, 3.75 inches (1.62 meters) tall and weighs 152 pounds (69.09 kilograms). The average male is five feet, nine inches (1.75 meters) tall and weighs 180 pounds (81.82 kilograms). Between 1960 and 2000 the average American male became two inches (five centimeters) taller and 45 pounds (20.45 kilograms) heavier, while the average American woman also grew two inches (five centimeters) taller and gained 18 pounds (8.18 kilogram).

## What are **desirable weights** for men and women?

The National Heart, Lung, and Blood Institute (NHLBI), in cooperation with the National Institute of Diabetes and Digestive and Kidney Diseases, released guidelines for weight for adults in 1998. These guidelines define degrees of overweight and obesity in terms of body mass index (BMI). Body mass index is based on weight and height and is strongly correlated with total body fat content. The Body Mass Index (BMI) incorporates both your height and weight to assess your weight-related level of risk for heart disease, diabetes, and high blood pressure. To find your BMI, multiply your weight in pounds by 700, divide by your height in inches, then divide by your height again. For example, if you are 5 feet, 10 inches and weigh 185 pounds, the calculation would be as follows:

185 pounds × 700 = 129,500

129,500/70 inches = 1,850

1,850/70 inches = 26.4

A BMI of 25 or less indicates low risk; 25–29 indicates overweight; and a BMI of over 30 indicates obesity. BMI numbers apply to both men and women. Very muscular individuals, such as athletes, may have a high BMI without health risks.

| **BMI** | **19** | **20** | **21** | **22** | **23** | **24** | **25** | **26** | **27** | **28** | **29** | **30** |
|---|---|---|---|---|---|---|---|---|---|---|---|---|
| **Height (inches)** | | | | | | **Body Weight (pounds)** | | | | | | |
| 58 | 91 | 96 | 100 | 105 | 110 | 115 | 119 | 124 | 129 | 134 | 138 | 143 |
| 59 | 94 | 99 | 104 | 109 | 114 | 119 | 124 | 128 | 133 | 138 | 143 | 148 |
| 60 | 97 | 102 | 107 | 112 | 118 | 123 | 128 | 133 | 138 | 143 | 148 | 153 |
| 61 | 100 | 106 | 111 | 116 | 122 | 127 | 132 | 137 | 143 | 148 | 153 | 158 |
| 62 | 104 | 109 | 115 | 120 | 126 | 131 | 136 | 142 | 147 | 153 | 158 | 164 |
| 63 | 107 | 113 | 118 | 124 | 130 | 135 | 141 | 146 | 152 | 158 | 163 | 169 |
| 64 | 110 | 116 | 122 | 128 | 134 | 140 | 145 | 151 | 157 | 163 | 169 | 174 |
| 65 | 114 | 120 | 126 | 132 | 138 | 144 | 150 | 156 | 162 | 168 | 174 | 180 |
| 66 | 118 | 124 | 130 | 136 | 142 | 148 | 155 | 161 | 167 | 173 | 179 | 186 |
| 67 | 121 | 127 | 134 | 140 | 146 | 153 | 159 | 166 | 172 | 178 | 186 | 191 |
| 68 | 125 | 131 | 138 | 144 | 151 | 158 | 164 | 171 | 177 | 184 | 190 | 197 |
| 69 | 128 | 135 | 142 | 149 | 155 | 162 | 169 | 176 | 182 | 189 | 196 | 203 |
| 70 | 132 | 139 | 146 | 153 | 160 | 167 | 174 | 181 | 188 | 195 | 202 | 209 |
| 71 | 136 | 143 | 150 | 157 | 165 | 172 | 179 | 186 | 193 | 200 | 208 | 215 |
| 72 | 140 | 147 | 154 | 162 | 169 | 177 | 184 | 191 | 199 | 206 | 213 | 221 |
| 73 | 144 | 151 | 159 | 166 | 174 | 182 | 189 | 197 | 204 | 212 | 219 | 227 |
| 74 | 148 | 155 | 163 | 171 | 179 | 186 | 194 | 202 | 210 | 218 | 225 | 233 |
| 75 | 152 | 160 | 168 | 176 | 184 | 192 | 200 | 208 | 216 | 224 | 232 | 240 |
| 76 | 156 | 164 | 172 | 180 | 189 | 197 | 205 | 213 | 221 | 230 | 238 | 246 |

| **BMI** | **31** | **32** | **33** | **34** | **35** | **36** | **37** | **38** | **39** | **40** | **41** | **42** |
|---|---|---|---|---|---|---|---|---|---|---|---|---|
| **Height (inches)** | | | | | | **Body Weight (pounds)** | | | | | | |
| 58 | 148 | 153 | 158 | 162 | 167 | 172 | 177 | 181 | 186 | 191 | 196 | 201 |
| 59 | 153 | 158 | 163 | 168 | 173 | 178 | 183 | 188 | 193 | 198 | 203 | 208 |
| 60 | 158 | 163 | 168 | 174 | 179 | 184 | 189 | 194 | 199 | 204 | 209 | 215 |
| 61 | 164 | 169 | 174 | 180 | 185 | 190 | 195 | 201 | 206 | 211 | 217 | 222 |
| 62 | 169 | 175 | 180 | 186 | 191 | 196 | 202 | 207 | 213 | 218 | 224 | 229 |
| 63 | 175 | 180 | 186 | 191 | 197 | 203 | 208 | 214 | 220 | 225 | 231 | 237 |
| 64 | 180 | 186 | 192 | 197 | 204 | 209 | 215 | 221 | 227 | 232 | 238 | 244 |
| 65 | 186 | 192 | 198 | 204 | 210 | 216 | 222 | 228 | 234 | 240 | 246 | 252 |
| 66 | 192 | 198 | 204 | 210 | 216 | 223 | 229 | 235 | 241 | 247 | 253 | 260 |
| 67 | 198 | 204 | 211 | 217 | 223 | 230 | 236 | 242 | 249 | 255 | 261 | 268 |

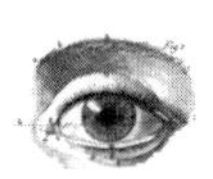



| BMI | 31 | 32 | 33 | 34 | 35 | 36 | 37 | 38 | 39 | 40 | 41 | 42 |
|---|---|---|---|---|---|---|---|---|---|---|---|---|
| **Height (inches)** | | | | | | **Body Weight (pounds)** | | | | | | |
| 68 | 203 | 210 | 216 | 223 | 230 | 236 | 243 | 249 | 256 | 262 | 269 | 276 |
| 69 | 209 | 216 | 223 | 230 | 236 | 243 | 250 | 257 | 263 | 270 | 277 | 284 |
| 70 | 216 | 222 | 229 | 236 | 243 | 250 | 257 | 264 | 271 | 278 | 285 | 292 |
| 71 | 222 | 229 | 236 | 243 | 250 | 257 | 265 | 272 | 279 | 286 | 293 | 301 |
| 72 | 228 | 235 | 242 | 250 | 258 | 265 | 272 | 279 | 287 | 294 | 302 | 309 |
| 73 | 235 | 242 | 250 | 257 | 265 | 272 | 280 | 288 | 295 | 302 | 310 | 318 |
| 74 | 241 | 249 | 256 | 264 | 272 | 280 | 287 | 295 | 303 | 311 | 319 | 326 |
| 75 | 248 | 256 | 264 | 272 | 279 | 287 | 295 | 303 | 311 | 319 | 327 | 335 |
| 76 | 254 | 263 | 271 | 279 | 287 | 295 | 304 | 312 | 320 | 328 | 336 | 344 |

| BMI | 43 | 44 | 45 | 46 | 47 | 48 | 49 | 50 | 51 | 52 | 53 | 54 |
|---|---|---|---|---|---|---|---|---|---|---|---|---|
| **Height (inches)** | | | | | | **Body Weight (pounds)** | | | | | | |
| 58 | 205 | 210 | 215 | 220 | 224 | 229 | 234 | 239 | 244 | 248 | 253 | 258 |
| 59 | 212 | 217 | 222 | 227 | 232 | 237 | 242 | 247 | 252 | 257 | 262 | 267 |
| 60 | 220 | 225 | 230 | 235 | 240 | 245 | 250 | 255 | 261 | 266 | 271 | 276 |
| 61 | 227 | 232 | 238 | 243 | 248 | 254 | 259 | 264 | 269 | 275 | 280 | 285 |
| 62 | 235 | 240 | 246 | 251 | 256 | 262 | 267 | 273 | 278 | 284 | 289 | 295 |
| 63 | 242 | 248 | 254 | 259 | 265 | 270 | 278 | 282 | 287 | 293 | 299 | 304 |
| 64 | 250 | 256 | 262 | 267 | 273 | 279 | 285 | 291 | 296 | 302 | 308 | 314 |
| 65 | 258 | 264 | 270 | 276 | 282 | 288 | 294 | 300 | 306 | 312 | 318 | 324 |
| 66 | 266 | 272 | 278 | 284 | 291 | 297 | 303 | 309 | 315 | 322 | 328 | 334 |
| 67 | 274 | 280 | 287 | 293 | 299 | 306 | 312 | 319 | 325 | 331 | 338 | 344 |
| 68 | 282 | 289 | 295 | 302 | 308 | 315 | 322 | 328 | 335 | 341 | 348 | 354 |
| 69 | 291 | 297 | 304 | 311 | 318 | 324 | 331 | 338 | 345 | 351 | 358 | 365 |
| 70 | 299 | 306 | 313 | 320 | 327 | 334 | 341 | 348 | 355 | 362 | 369 | 376 |
| 71 | 308 | 315 | 322 | 329 | 338 | 343 | 351 | 358 | 365 | 372 | 379 | 386 |
| 72 | 316 | 324 | 331 | 338 | 346 | 353 | 361 | 368 | 375 | 383 | 390 | 397 |
| 73 | 325 | 333 | 340 | 348 | 355 | 363 | 371 | 378 | 386 | 393 | 401 | 408 |
| 74 | 334 | 342 | 350 | 358 | 365 | 373 | 381 | 389 | 396 | 404 | 412 | 420 |
| 75 | 343 | 351 | 359 | 367 | 375 | 383 | 391 | 399 | 407 | 415 | 423 | 431 |
| 76 | 353 | 361 | 369 | 377 | 385 | 394 | 402 | 410 | 418 | 426 | 435 | 443 |

## Considering that 64 percent of all Americans are overweight, who is the **heaviest person** that ever lived?

John Brower Minnoch (1941–1983) of Bainbridge Island, Washington, weighed 976 pounds (443 kilograms) in 1976 and was estimated to have weighed more than 1,387 pounds (630 kilograms) when he was rushed to the hospital in 1978 with heart and respiratory failure. Much of his weight was due to fluid retention. After two years on a

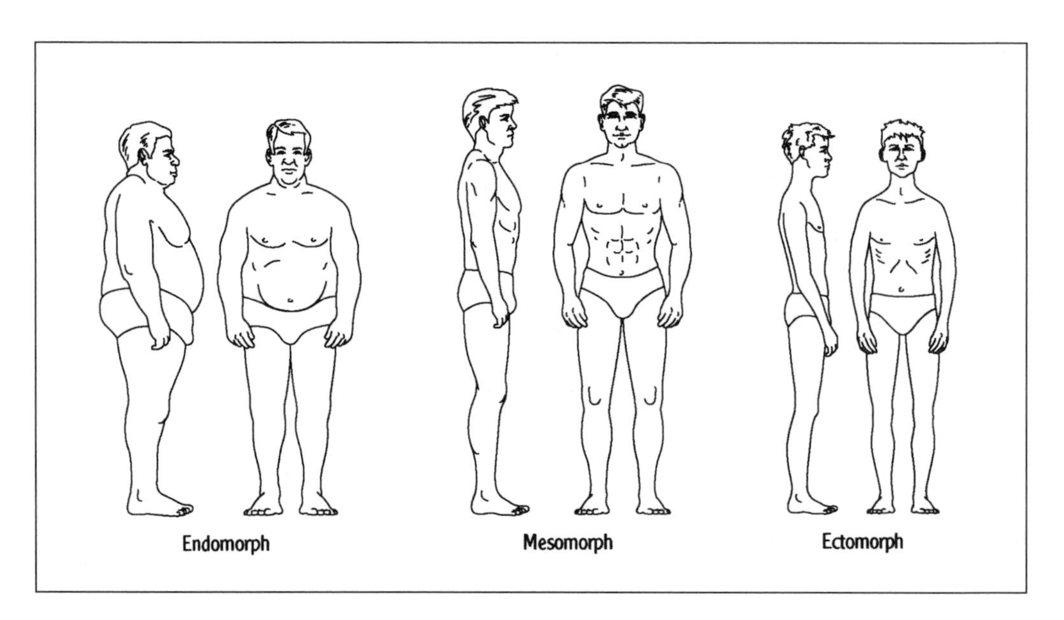


Classification of body types.

hospital diet, he was discharged at 476 pounds (216 kilograms). He had to be readmitted, however, after reportedly gaining 197 pounds (87 kilograms) in seven days. In 1983, when he died, he weighed 798 pounds (362 kilograms).

The heaviest woman ever recorded was Rosie Carnemolla (b. 1944) of Poughkeepsie, New York, who weighed 850 pounds (386 kilograms). When Mrs. Percy Pearl Washington, who suffered from polydipsia (excessive thirst), died in a Milwaukee hospital in 1972, the scales registered only 800 pounds (363 kilograms) maximum, but she was credited with weighing 880 pounds (400 kilograms).

## What are the types of human **body shapes**?

The best known example of body typing (classifying body shape in terms of physiological functioning, behavior, and disease resistance) was devised by American psychologist William Herbert Sheldon (1898–1977). Sheldon's system, known as somatotyping, distinguishes three types of body shapes, ignoring overall size: endomorph, mesomorph, and ectomorph. The extreme endomorph tends to be spherical: a round head, a large, fat abdomen, weak penguin-like arms and legs, with heavy upper arms and thighs but slender wrists and ankles. The extreme mesomorph is characterized by a massive cubical head, broad shoulders and chest, and heavy muscular arms and legs. The extreme ectomorph has a thin face, receding chin, high forehead, a thin, narrow chest and abdomen, and spindly arms and legs. In Sheldon's system there are mixed body types, determined by component ratings. Sheldon assumed a close relationship between body build and behavior and temperament. This system of body typing has many critics.

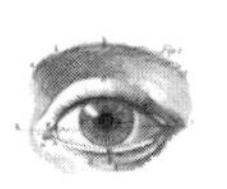



## Who were the congenitally joined twins who gave rise to the term **Siamese twins**?

The term "Siamese twins" originated with the appearance of Chang and Eng Bunker (1811–1874), conjoined Chinese twins born in Siam (now Thailand), who were used as a circus attraction by P. T. Barnum.

Siamese twins are identical twins joined at some point of their bodies, most commonly at the hip, chest, abdomen, buttocks, or head. Like other identical twins, they originate from a single fertilized egg; in the case of congenitally joined twins, however, the egg fails to split into two separate cell masses at the proper time. The condition is relatively rare; only about 500 cases have been reported worldwide. Surgery to separate Siamese twins is a complex task, and often results in the death of one or both of the twins.

Chang and Eng Bunker, the original "Siamese twins" made famous by P.T. Barnum.

## What causes **protruding ears**?

If a fold paralleling a part of the outer roll of the ear is sparse, but not absent, the ear will protrude. This trait usually runs in certain families.

## Who was the world's **oldest person**?

French actress Jeanne Louise Calment (1875–1997) was the oldest person to ever live with concrete proof of her age. She died at the age of 122 years and 164 days. At the age of 114, she played herself in the film *Vincent and Me* (1990).

### In addition to left- or right-handedness, what other left–right preferences do people have?

Most people have a preferred eye, ear, and foot. In one study, for example, 46 percent were strongly right-footed, while 3.9 percent were strongly left-footed; furthermore, 72 percent were strongly right-handed and 5.3 percent strongly left-handed. Estimates vary about the proportion of left-handers to right-handers, but it may be as high as one in ten. Some 90 percent of healthy adults use the right hand for writing; two-thirds favor the right hand for most activities requiring coordination and skill. There is no male-female difference in these proportions.

## How many people living in the United States are **centenarians**?

The number of centenarians in the United States has increased steadily over the decades. In the most recent census (2000), there were 68,000 centenarians, up from 48,000 in 1995. Although the Census Bureau asks for age information in more than one way, the U.S. Census Bureau cautions that the figures may be somewhat skewed since birth certificates were not common until the 1930s.

## How much of the body remains after **cremation**?

During cremation, the intense heat evaporates the large quantity of water that makes up the human body and burns the soft tissues and bone, reducing the body to four to eight pounds (1.8 to 3.6 kilograms) of ash and bone fragments. On average, adult male cremains weigh 7.4 pounds (3.4 kilograms), while adult female cremains weigh 5.8 pounds (2.6 kilograms). Most modern crematoria use electric processors to quickly pulverize residual bone fragments, which will fit into the average urn.

## What is **cryonic suspension**?

Suspended animation, as the long-term storage of humans is generally known, has been a topic of scientific speculation since the beginning of the modern era. Cryonic suspension, the controversial process of freezing and storing bodies for later revival, has been practiced since the late 1960s. Ordinarily, people are placed in cryonic suspension only after they are pronounced legally dead.

Opinion among scientists is divided over the feasibility of cryonic suspension. The necessary science and technology needed to revive a cryonically preserved body are not known at this time. The number of cryonics adherents is small. One cryonics group, Cryonics Institute, had 40 bodies in 2001—the same year it celebrated its 25th anniversary. The cost of cryonic suspension varies from $28,000 to $120,000.

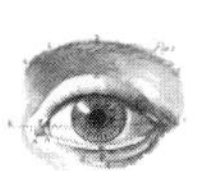



# BONES, MUSCLES, AND NERVES

## **How many bones** are in the human body?

Babies are born with about 300 to 350 bones, but many of these fuse together between birth and maturity to produce an average adult total of 206. Bone counts vary according to the method used to count them, because some systems treat as multiple bones a structure that other systems treat as a single bone with multiple parts.

| Location | Number |
|---|---|
| Skull | 22 |
| Ears (pair) | 6 |
| Vertebrae | 26 |
| Sternum | 3 |
| Throat | 1 |
| Pectoral girdle | 4 |
| Arms (pair) | 60 |
| Hip bones | 2 |
| Legs (pair) | 58 |
| *TOTAL* | *206* |

## What is the most commonly **broken bone**?

The clavicle (collar bone) is probably one of the most frequently fractured bones in the body. Fractured clavicles are caused by either a direct blow or a transmitted force resulting from a fall on the outstretched arm. Other commonly broken bones are the forearm (a Colles fracture) and hip fractures in individuals over 75 years old. A Colles fracture occurs when someone begins to fall and they often extend their hand to "catch the fall." The impact of the individual's weight upon the hand may cause the end of the lower arm bone (radius) to fracture just above the wrist.

## What is the only **bone** in the human body that **does not touch another bone**?

The hyoid bone is the only bone that does not touch another bone. Found above the larynx, it anchors the tongue muscles. It is also usually broken when a person is hanged or strangled, and therefore will often figure in trials concerning such crimes.

## What makes **knuckles crack**?

When a person pulls quickly on his or her finger, a vacuum is created in the joint space between the bones, displacing the fluid liquid normally found in the space. The popping sound occurs when the fluids rush back into the empty gap.

## Is it **harmful** to crack one's knuckles?

A study of 300 knuckle crackers found no apparent connection to between joint cracking and arthritis. Other damage was observed, including soft tissue damage to the joint capsule and a decrease in grip strength. The rapid, repeated stretching of the ligaments surrounding the joint is most likely the cause of damage to the soft tissue.

## What is the **funny bone**?

The funny bone is not a bone but part of the ulnar nerve located at the the back of the elbow. A bump in this area can cause a tingling sensation or it can make the forearm feel temporarily numb.

## What is the **hardest substance** in the body?

Tooth enamel is the hardest substance in the body. It is composed of 96 percent mineral salts and 4 percent organic matter and water.

## What is the purpose of **primary teeth**?

Primary teeth (also known as baby, deciduous, temporary, or milk teeth, for their milk-white color) serve many of the same purposes as permanent teeth. They are needed for chewing, they make the face more attractive, and they are necessary for speech development. They also prepare the mouth for the permanent teeth by maintaining space for the permanent teeth to emerge in proper alignment. Each individual has 20 primary teeth followed by 32 permanent teeth.

## Why do some dentists treat the **molars** and **premolars** of children with **sealants**?

Sealants, a soft plastic coating applied to the tooth surface, can protect a child's first and second permanent molars from decay by filling in the pits and fissures where food and bacteria might otherwise accumulate. The plastic is hardened with a special light or chemical.

## **How many muscles** are in the human body?

There are about 656 muscles in the body, although some authorities make this figure even as high as 850 muscles. No exact figure is available because authorities disagree about which are separate muscles and which ones slip off larger ones. Also, there is a wide variability from one person to another, though the general plan remains the same.

Muscles are used in three body systems. The skeletal muscles move various parts of the body, are striped or striated fibers, and are called voluntary muscles because the person controls their use. The second system includes smooth muscles found in the stom-

ach and intestinal walls, vein and artery walls, and in various internal organs. Called involuntary muscles, they are not generally controlled by the person. The last are the cardiac muscles, or the heart muscles, containing striped and involuntary muscles.

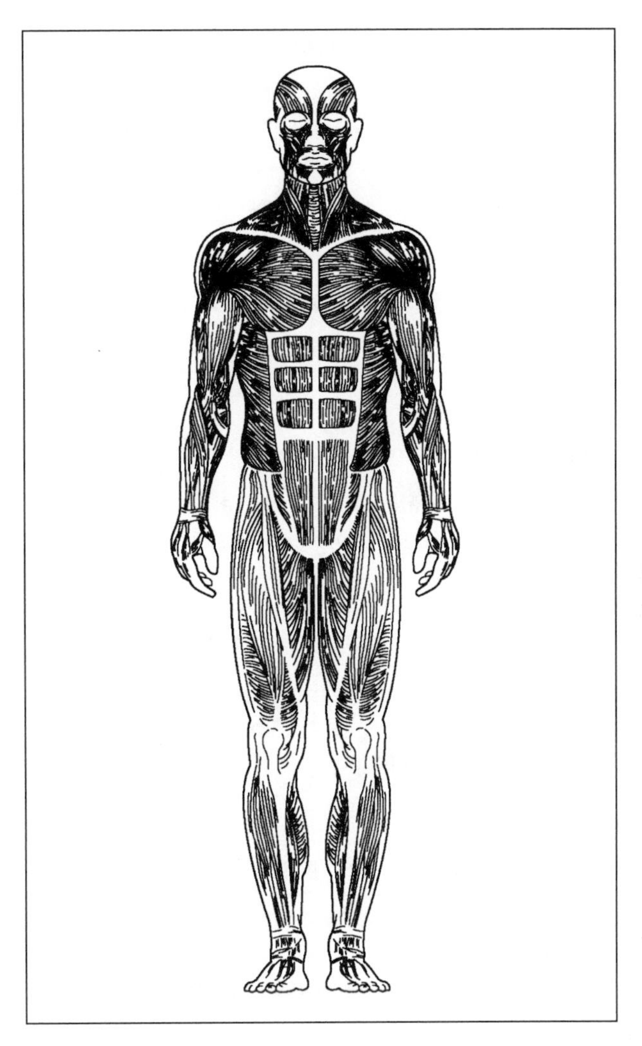

Illustration of an ecorche.

## What is an **ecorche**?

An ecorche is a flayed figure, a three-dimensional representation of the human body, usually made of plaster, with the envelope of skin and fat removed. Its purpose is to depict the surface muscles with precise anatomical correctness.

## Which **muscle** is the **most variable** among humans?

The platysma muscle in the side of the neck is probably the most variable. It can cover the whole region in some people while in others it is straplike or in a few situations it is missing completely.

## What is the **longest muscle** in the human body?

The longest muscle is the sartorius, which runs from the waist to the knee. Its purpose is to flex the hip and knee. The largest muscle is the gluteus maximus (buttock muscle), which moves the thighbone away from the body and straightens out the hip joint.

## What are the **hamstring muscles**?

There are three hamstring muscles, located at the back of the thigh. They flex the leg on the thigh, as in the process of kneeling.

## Why does excessive exercise cause **muscles** to become **stiff** and **sore**?

During vigorous exercise, the circulatory system cannot supply oxygen to muscle fibers quickly enough. In the absence of oxygen, the muscle cells begin to produce lactic acid, which accumulates in the muscle. It is this build-up of lactic acid that causes soreness and stiffness.

## How many **muscles** does it take to produce a **smile** and a **frown**?

Seventeen muscles are used in smiling. The average frown uses 43.

## What is the name of the small fleshy mass hanging from the **back of the mouth**?

The uvula is a small, soft structure, hanging from the free edge of the soft palate. It is composed of muscle, connective tissue, and mucous membrane.

## How much **force** does a **human bite** generate?

All the jaw muscles working together can close the teeth with a force as great as 55 pounds (25 kilograms) on the incisors or 200 pounds (90.7 kilograms) on the molars. A force as great as 268 pounds (122 kilograms) for molars has been reported.

## What is the **largest nerve** in the body?

The sciatic nerve is the largest in the human body—about as thick as a lead pencil—0.78 inch (1.98 centimeters). It is a broad, flat nerve composed of fibers that run from the spinal cord to the back of each leg.

# ORGANS AND GLANDS

## What is the **largest organ** in the human body?

The largest and heaviest human organ is the skin, with a total surface area of about 20 square feet (1.9 square meters) for an average person or 25 square feet (2.3 square meters) for a large person; it weighs 5.6 pounds (2.7 kilograms) on average. Although generally it is not thought of as an organ, medically it is. An organ is a collection of various tissues integrated into a distinct structural unit and performing specific functions.

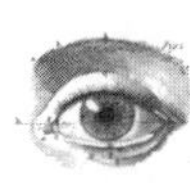



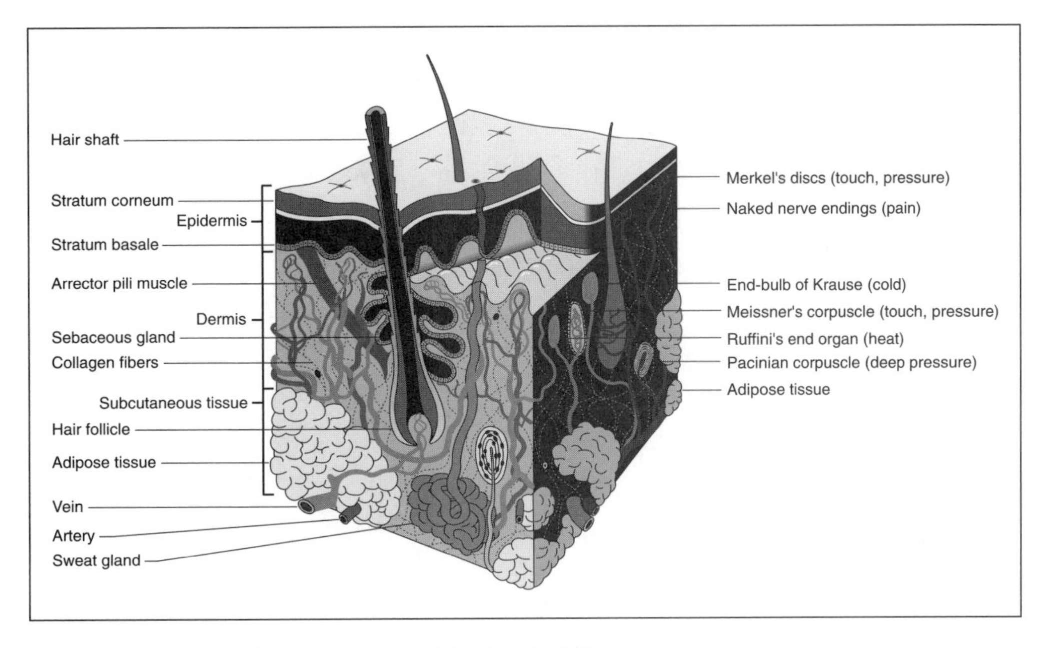


**A cross section of the skin. Sensory structures are labeled on the right.**

## What is the basic unit of the **brain**?

Neurons are the nerve cells that are the major constituent of the brain. At birth the brain has the maximum number of neurons—20 billion to 200 billion neurons. Thousands are lost daily, never to be replaced and apparently not missed, until the cumulative loss builds up in very old age.

## What is the average **weight** of the human **brain**?

The average human brain weighs three pounds (1.36 kilograms). The average female brain capacity is 79.3 cubic inches, slightly smaller than the male brain of 88.5 cubic inches. The largest human brains may be twice those of average size, but size has no relevance to brain performance.

## How hard does the **heart** work?

The heart squeezes out about two ounces (71 grams) of blood at every beat and daily pumps at least 2,500 gallons (9,450 liters) of blood. On the average, the adult heart beats 70 to 75 times a minute. The rate of the heartbeat is determined in part by the size of the organism. Generally, the smaller the size, the faster the heartbeat. Thus women's hearts beat six to eight beats per minute faster than men's hearts do. At birth the heart of a baby can beat as fast as 130 times per minute.

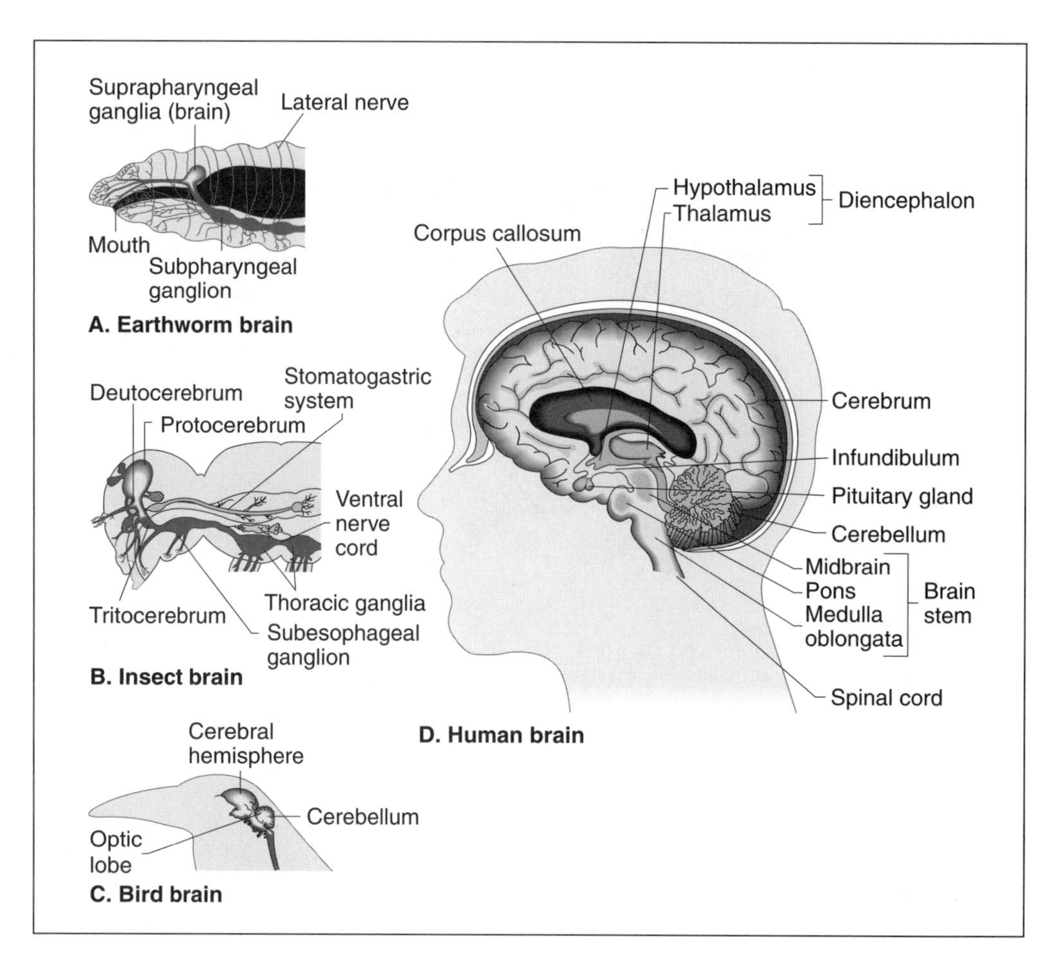


A comparison of the brains of an earthworm, an insect, a bird, and a human.

## Whose **brain** is **larger**: that of Neanderthal or modern humans?

The capacity of the skull of "classic" Neanderthal (or Neandertal) man was often larger than that of modern humans. The capacity was between 1,350 and 1,700 cubic centimeters with the average being 1,400 to 1,450 cubic centimeters. The mean cranial capacity of modern man is 1,370 cubic centimeters, with a range of 950 to 2,200 cubic centimeters. However, brain size alone is not an index of intelligence.

## **How much air** does a person breathe **in a lifetime**?

During his or her life, the average person will breathe about 75 million gallons (284 million liters) of air. Per minute, the human body needs two gallons (7.5 liters) of air when lying down, four gallons (15 liters) when sitting, six gallons (23 liters) when walking, and 12 gallons (45 liters) or more when running.

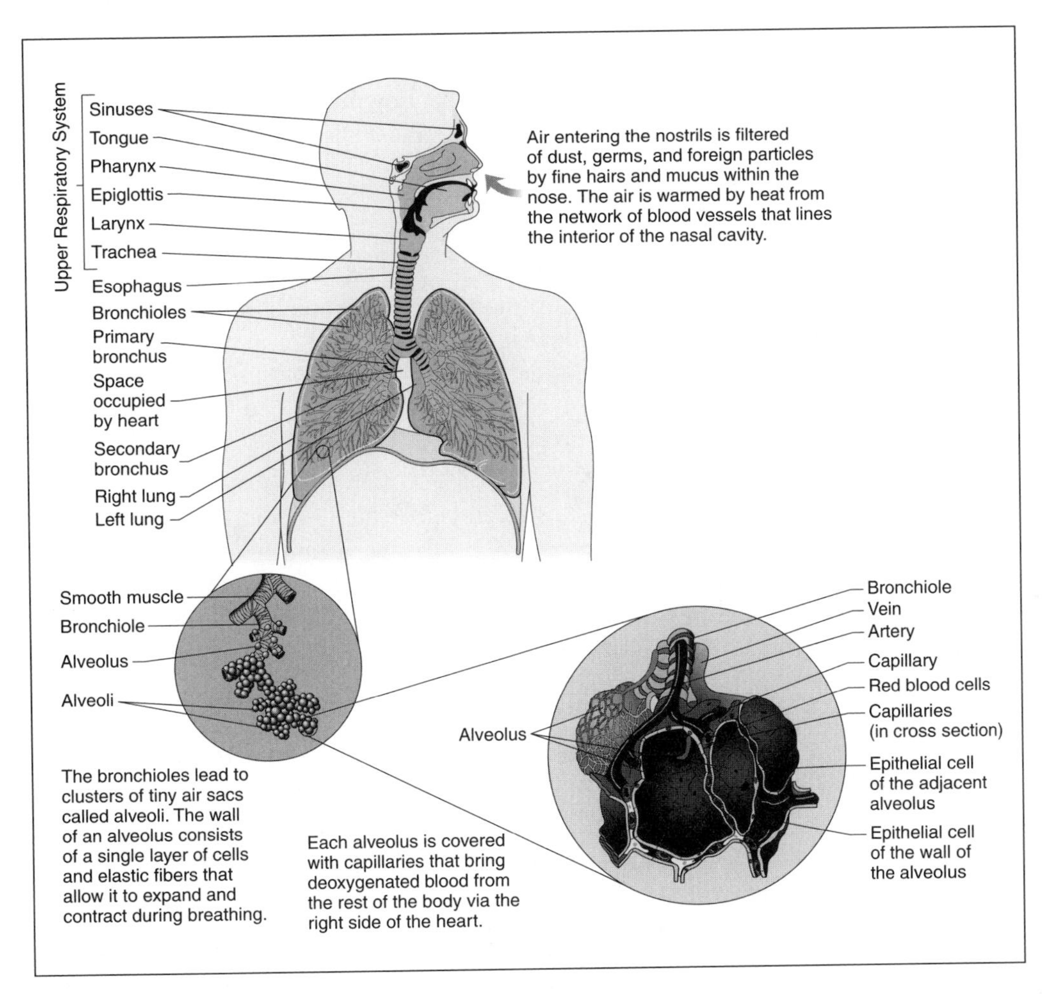


**The human respiratory system.**

## Are the **lungs** identical?

No, the right lung is shorter than the left by one inch (2.5 centimeters); however, its total capacity is greater. The right lung has three lobes, the left lung has two.

## How is there enough room in the **chest cavity** to accommodate a pair of lungs filled with air?

A pair of lungs filled with air contains the same volume of air as in eight large soda bottles—more than 1.5 gallons (6 liters). In order to accommodate this volume, muscles raise and expand the rib cage while the diaphragm, a domed muscular sheet under the lungs, flattens with each breath inhaled.

## What causes **hiccups** and how can they be cured?

Hiccupping is the involuntary contraction of the diaphragm. When the diaphragm contracts, the vocal cords close quickly, causing the familiar noise associated with hiccups. Hiccups help the stomach get rid of a bit of gas, relieve the esophagus of an irritation, or resolves a temporary loss of coordination between the nerves controlling the movement of the diaphragm. A bout of hiccups can be brought on by eating or drinking too fast or fatigue or nervousness. Although hiccups generally go away on their own after a few minutes, several remedies have been recommended to cure one of hiccups: a loud distraction to scare the sufferer of hiccups, swallowing a spoonful of sugar, blowing into a paper bag, or drinking from the opposite side of the glass all have their followers.

## When is the **earliest age** that a person can **hiccup**?

People from every age suffer from hiccups, even before birth—pregnant women have reported that fetuses hiccup in the womb. A hiccup happens when there is an involuntary contraction of the diaphragm, the large divider of muscle and tendons that separate the chest from the abdomen. An opening called the glottis near the top of the trachea between the vocal cords closes briskly. This closing, along with air trapped in the larynx, causes the distinct hiccup sound.

## What was the likely **purpose** of the human **appendix**?

Experts can only theorize on its use. It may have had the same purpose it does in present-day herbivores, where it harbors colonies of bacteria that help in the digestion of cellulose in plant material. Another theory suggests that tonsils and the appendix might manufacture the antibody-producing white blood cells called B lymphocytes; however, B lymphocytes could also be produced by the bone marrow. The third theory is that the appendix may "attract" body infections to localize the infection in one spot that is not critical to body functioning. The earliest surgical removal of the appendix was by Claudries Amyand (1680–1740) in England in 1736.

## What are the seven **endocrine glands**?

The major endocrine glands include the pituitary, thyroid, parathyroids, adrenals, pancreas, testes, and ovaries. These glands secrete hormones into the blood system, which generally stimulate some change in metabolic activity:

*Pituitary*—secretes ACTH to stimulate the adrenal cortex, which produces aldosterone to control sodium and potassium reabsorption by the kidneys; FSH to stimulate gonad function and prolactin to stimulate milk secretion of breasts; TSH to stimulate thyroid gland to produce thyroxin; LH to stimulate ovulation in females and testerone production in males; GH to stimulate general growth. Stores oxytocin for uterine contraction.

*Thyroid gland*—secretes triiodothyronine (T3) and thyroxine (T4) to stimulate metabolic rate, especially in growth and development, and secretes calcitonin to lower blood-calcium levels.

*Parathyroids*—secrete hormone PTH to increase blood-calcium levels; stimulates calcium reabsorption in kidneys.

*Adrenals*—secrete epinephrine and norepinephrine to help the body cope with stress, and raise blood pressure, heart rate, metabolic rate, blood sugar levels, etc. Aldosterone secreted by the adrenal cortex maintains sodium-potassium balance in kidneys and cortisol helps the body adapt to stress, mobilizes fat, and raises blood sugar level.

*Pancreas*—secretes insulin to control blood sugar levels, stimulates glycogen production, fat storage, and protein synthesis. Glucagon secretion raises blood sugar level and mobilizes fat.

*Ovaries and testes*—secrete estrogens, progesterone, or testosterone to stimulate growth and reproductive processes.

## Which **gland** is the **largest**?

The liver is the largest gland and the second largest organ after the skin. At 2.5 to 3.3 pounds (1.1 to 1.5 kilograms) the liver is seven times larger than it needs to be to perform its estimated 500 functions. It is the main chemical factory of the body. A ducted gland that produces bile to break down fats and reduce acidity in the digestive process, the liver is also a part of the circulatory system. It cleans poisons from the blood and regulates blood composition.

## What regulates **body temperature** in humans?

The hypothalamus controls internal body temperature by responding to sensory impulses from temperature receptors in the skin and in the deep body regions. The hypothalamus establishes a "set point" for the internal body temperature, then constantly compares this with its own actual temperature. If the two do not match, the hypothalamus activates either temperature-decreasing or temperature-increasing procedures to bring them into alignment.

# BODY FLUIDS

## What are the **four humors** of the body?

The four constituent humors of the body were identified as blood, phlegm, yellow bile, and black bile, originating in the heart, brain, liver, and spleen, respectively. Empedocles of Agrigentum (504–433 B.C.E.) probably originated the theory, in which he equated the body fluids to the four elements of nature: earth, fire, air, and water. These humors could determine the health of the body and the personality of the person as well. To be in good health the humors should be in harmony within the body. Ill health could be remedied by treatments to realign the humors and reestablish the harmony.

## What is the **normal pH** of blood, urine, and saliva?

Normal pH of arterial blood is 7.4; pH of venous blood is about 7.35. Normal urine pH averages about 6.0. Saliva has a pH between 6.0 and 7.4.

## How similar are **seawater** and **blood**?

| Component | Seawater (grams/liter) | Blood (grams/liter) |
|---|---|---|
| Na (sodium) | 10.7 | 3.2–3.4 |
| K (potassium) | 0.39 | 0.15–0.21 |
| Ca (calcium) | 0.42 | 0.09–0.11 |
| Mg (magnesium) | 1.34 | 0.012–0.036 |
| Cl (chloride) | 19.3 | 3.5–3.8 |
| $SO_4$ (sulfate) | 2.69 | 0.16–0.34 |
| $CO_3$ (carbonate) | 0.073 | 1.5–1.9 |
| Protein | | 70.0 |

## How does the body introduce **oxygen** to the **blood** and where does this happen?

Blood entering the right side of the heart (right auricle or atrium) contains carbon dioxide, a waste product of the body. The blood travels to the right ventricle, which pushes it through the pulmonary artery to the lungs. In the lungs, the carbon dioxide is removed and oxygen is added to the blood. Then the blood travels through the pulmonary vein carrying the fresh oxygen to the left side of the heart, first to the left auricle, where it goes through a one-way valve into the left ventricle, which must push the oxygenated blood to all portions of the body (except the lungs) through a network of arteries and capillaries. The left ventricle must contract with six times the force of the right ventricle, so its muscle wall is twice as thick as the right.

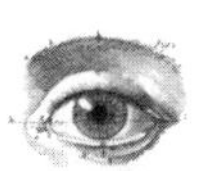



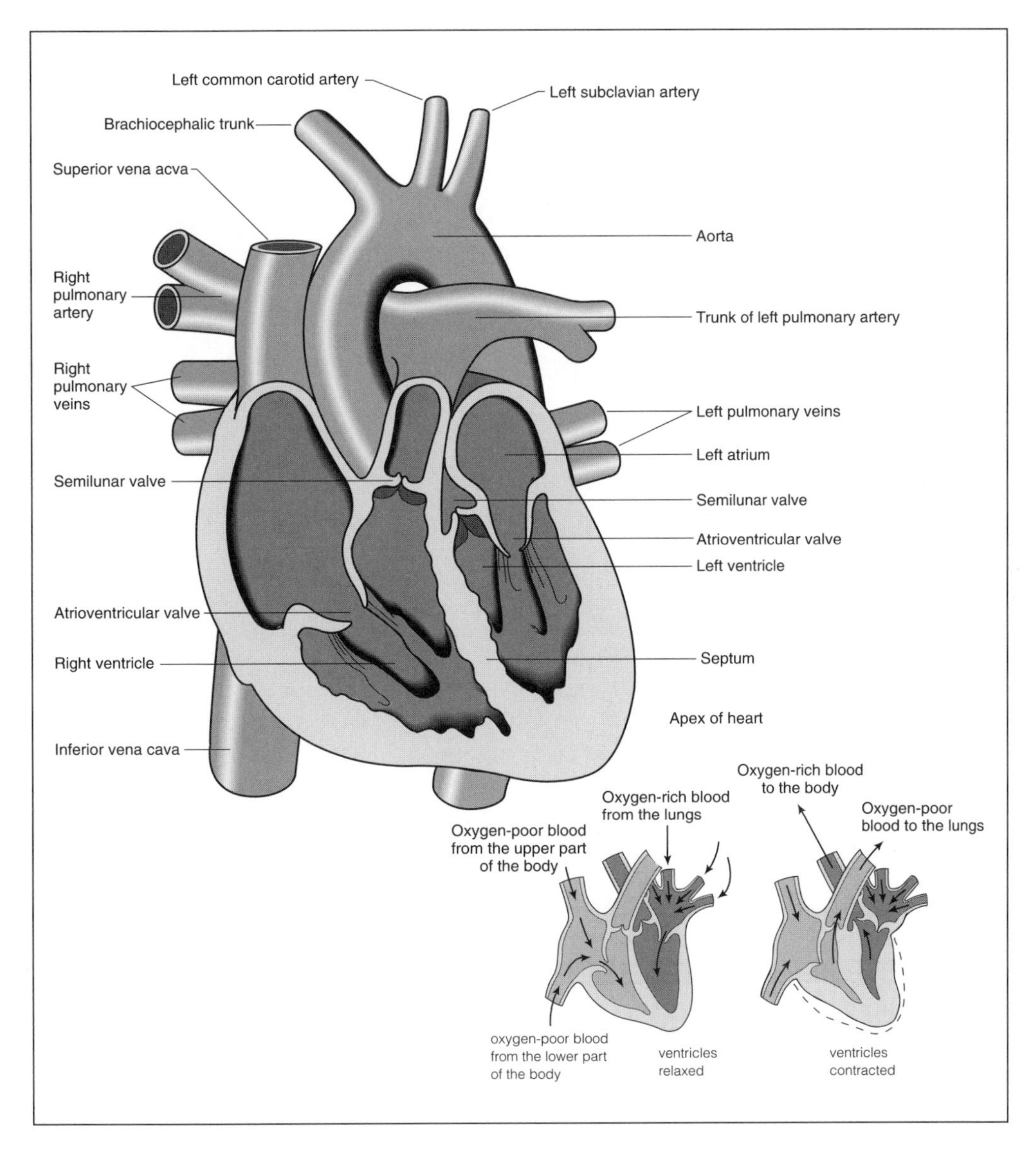


**A cutaway of the human heart (top) and a diagram showing blood flow in the heart during diastole (relaxation) and systole (contraction).**

## How much **saliva** does a person produce in a day?

Saliva is a mixture of mucus, water, salts, and the enzymes that break down carbohydrates. Awake individuals secrete saliva at a rate of approximately 0.5 milliliters per minute. Multiplying 0.5 milliliters by 960 minutes in an average 16-hour waking day amounts to 480 milliliters of saliva in 16 hours. This is only a basic estimate since various activities such as exercise, eating, drinking, and speaking all increase salivary volume.

## **How much blood** is in the average human body?

A man weighing 154 pounds (70 kilograms) would have about 5.5 quarts (5.2 liters) of blood. A woman weighing 110 pounds (50 kilograms) would have about 3.5 quarts (3.3 liters).

## How many miles of **blood vessels** are contained in the body?

If they could be laid end to end, the blood vessels would span about 60,000 miles (96,500 kilometers).

## What is the **largest artery** in the human body?

The aorta is the largest artery in the human body.

## Why do various parts of the body **fall asleep**?

The sensation of "pins and needles" when an arm or leg "falls asleep" is caused by impaired blood circulation to that limb.

## Who discovered the **ABO system** of typing blood?

The Austrian physician Karl Landsteiner (1868–1943) discovered the ABO system of blood types in 1909. Landsteiner had investigated why blood transfused from one individual was sometimes successful and other times resulted in the death of the patient. He theorized that there must be several different blood types. A person with one type of blood will have antibodies to the antigens in the blood type they do not have. If a transfusion occurs between two individuals with different blood types, the red blood cells will clump together, blocking the blood vessels.

## Which of the **major blood types** are the most common in the United States?

| Blood type | Frequency in U.S. |
|---|---|
| O+ | 37.4% |
| O– | 6.6% |
| A+ | 35.7% |
| A– | 6.3% |
| B+ | 8.5% |
| B– | 1.5% |
| AB+ | 3.4% |
| AB– | 0.6% |

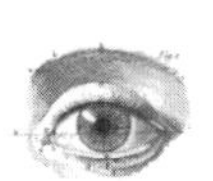



In the world, the preponderance of one blood group varies greatly by locality. Group O is generally the most common (46 percent), but in some areas Group A predominates.

## How often can a person **donate blood**?

Blood is one of the most easily donated tissues. According to the American Red Cross, a person in good health who weights at least 110 pounds (50 kilograms) can donate a unit of blood every eight weeks.

## What are **apheresis** and **plateletpheresis**?

Although most blood is donated as whole blood, it is also possible to donate only a portion of blood using a technique called apheresis. Blood is drawn from the vein of a donor into an apheresis instrument, which separates the blood into different portions by centrifugation. By appropriately adjusting the instrument, a selected portion of the blood, such as the platelets, can be recovered, while the rest of the blood is returned to the donor either into the same vein or into a vein in the other arm. This process takes more time than whole blood donation, but the yield of platelets is much greater. Platelets collected by apheresis are particularly useful for patients who require numerous platelet transfusions, for example cancer patients who have received chemotherapy.

## Which blood type is the **universal donor** and which is the **universal recipient**?

Persons with blood type O are universal donors. They are able to donate blood to anyone. Persons with blood type AB are universal recipients. They are able to receive blood from any donor.

## What is the **Rh factor**?

In addition to the ABO system of blood types, blood types can also be grouped by the Rhesus factor, or Rh factor, an inherited blood characteristic. Discovered independently in 1939 by Philip Levine (1900–1987) and R. E. Stetson and in 1940 by Karl Landsteiner (1868–1943) and A. S. Weiner, the Rh system classifies blood as either having the Rh factor or lacking it. Pregnant women are carefully screened for the Rh factor. If a mother is found to be Rh-negative, the father is also screened. Parents with incompatible Rh factors can lead to potentially fatal blood problems in newborn infants. The condition can be treated with a series of blood transfusions.

## What is the amount of **carbon dioxide** found in normal blood?

Carbon dioxide normally ranges from 19 to 50 millimeters per liter in arterial blood and 22 to 30 millimeters per liter in venous blood.

## Which **blood type** is the rarest?

The rarest blood type is Bombay blood (subtype h-h), found only in a Czechoslovakian nurse in 1961 and in a brother and sister named Jalbert living in Massachusetts in 1968.

## What are the **blood group combinations** that can normally be used to prove that a man is not the father of a particular child?

| If the mother is | and the child is | the father can be | but not |
|---|---|---|---|
| O | O | O, A, or B | AB |
| O | A | A or AB | O or B |
| O | B | B or AB | O or A |
| A | O | O, A, or B | AB |
| A | A | any group | |
| A | B | B or AB | O or A |
| A | AB | B or AB | O or A |
| B | O | O, A, or B | AB |
| B | B | any group | |
| B | A | A or AB | O or B |
| B | AB | A or AB | O or B |
| AB | AB | A, B, or AB | O |

No child can acquire a gene, and consequently a blood grouping, if it is not possessed by either parent.

## How do **wounds heal**?

Damage to tissue, such as a cut in the skin, begins to heal with the formation of a sticky lump known as a blood clot. Blood clots prevent blood and other fluids from leaking out. Microscopic sticky threads of the clotting protein fibrin make a tangled mesh that traps blood cells. Within a short time the clot begins to take shape, harden, and become more solid. The clot turns into a scab as it dries and hardens. Skin cells beneath the scab multiply to repair the damage. When the scab falls off the wound will be healed.

## What percent of human body weight is **water**?

The human body is 61.8 percent water by weight. Protein accounts for 16.6 percent, fat 14.9 percent, and nitrogen 3.3 percent. Other elements are present in lesser amounts.

## Why do **eyes tear** when we work with **onions**?

When an onion is cut, the pierced cells release a sulfur compound, thio-propanal-s-oxide, through a series of rapid chemical reactions. This substance is irritating to the eyes.

### What causes people to **sweat** when they eat **spicy foods**?

The chemical capsaicin, a component of spicy foods, causes sweating by stimulating nerve endings in the mouth and tongue that, normally, only respond to a rise in body temperature. As a result, the brain receives a false signal that body temperature has risen, and in turn launches the chain of physiological events that leads to facial sweating.

### Why do some people experience a **runny nose** while **eating**?

Called prandial rhinorrhea, this condition occurs when eating stimulates the autonomic nervous system to release the compound acetylcholine. This in turn prompts the increased production of saliva, stomach acid, and nasal mucus. Usually, the spicier the meal, the greater the reaction.

### Does your **heart stop beating** when you **sneeze**?

The heart does not stop when you sneeze. Sneezing, however, does affect the cardiovascular system. It causes a change in pressure inside the chest. This change in pressure affects the blood flow to the heart, which in turn affects the heart's rhythm. Therefore, a sneeze does produce a harmless delay between one heartbeat and the next, often misinterpreted as a "skipped beat."

## SKIN, HAIR, AND NAILS

### **How much skin** does an average person have?

The average human body is covered with about 20 square feet (two square meters) of skin. Weighing six pounds (2.7 kilograms), the skin is composed of two main layers: the epidermis (outer layer) and the dermis (inner layer). The epidermis layer is replaced continually as new cells, produced in the stratum basale, mature, and are pushed to the surface by the newer cells beneath; the entire epidermis is replaced in about 27 days. The dermis, the lower layer, contains nerve endings, sweat glands, hair follicles, and blood vessels. The upper portion of the dermis has small, fingerlike projections called "papillae," which extend into the upper layer. The patterns of ridges and grooves visible on the skin of the soles, palms, and fingertips are formed from the tops of the dermal papillae. The capillaries in these papillae deliver oxygen and nutrients to the epidermis cells and also function in temperature regulation.

## Who first used **fingerprints** as a means of identification?

It is generally acknowledged that Francis Galton (1822–1911) was the first to classify fingerprints. However, his basic ideas were further developed by Sir Edward Henry (1850–1931), who devised a system based on the pattern of the thumb print. In 1901 in England, Henry established the first fingerprint bureau with Scotland Yard called the Fingerprint Branch.

## Do identical **twins** have the **same fingerprints**?

No. Even identical twins have differences in their fingerprints, which, though subtle, can be discerned by experts.

## In a set of **identical twins** can each twin roll their tongue?

It has been postulated that the ability to roll one's tongue (roll up the lateral edges of the tongue) is a genetic trait. Results of studies of identical twins do not support the claim that tongue rolling is an inherited trait. In one study of 33 sets of identical twins, in only 18 pairs were both twins able to roll their tongue. In eight pairs of twins, neither one was able to roll his or her tongue. In the remaining seven pairs of identical twins, one twin was able to roll his or her tongue and the other was not able to do so.

## Are **freckles** dangerous?

Freckles, tan or brown spots on the skin, are small areas of increased skin pigment or melanin. Freckles are signs of sun damage to the skin. There is a genetic tendency to freckle, and often many members of the same family will have freckles. They are usually on the face, arms and other parts of the body that are exposed to the sun. Freckles appear in childhood, fade during the winter months, and reappear in the summer. Freckles themselves pose no health risks, but they are a marker of increased risk for all types of skin cancer.

## What is the purpose of **goose-bumps**?

The puckering of the skin that takes place when goose-flesh is formed is the result of contraction of the muscle fibers in the skin. This muscular activity will produce more heat, and raise the temperature of the body.

## How can **tattoos** be removed?

Tattoos can be removed by a dermatological surgeon on an outpatient basis with local anesthesia. The most common techniques used are:

*Laser Surgery*—The surgeon removes the tattoo by selectively treating the pigment colors with a high-intensity laser beam. Lasers have become the standard

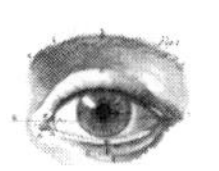



**How many hairs does the average person have on his or her head?**

The amount of hair covering varies from one individual to another. An average person has about 100,000 hairs on their scalp (blondes 140,000, brunettes 155,000, and redheads only 85,000). Most people shed between 50 to 100 hairs daily.

treatment because they offer a "bloodless," low-risk, highly effective approach with minimal side effects. The type of laser used generally depends upon the pigment colors. In many cases, multiple treatments may be required.

*Dermabrasion*—The surgeon "sands" the skin, removing the surface and middle layers of the tattoo. The combination of surgical and dressing techniques helps to raise and absorb the tattoo inks.

*Surgical Excision*—The surgeon removes the tattoo with a scalpel and closes the wound with stitches. This technique proves highly effective in removing some tattoos and allows the surgeon to excise inked areas with great control. Side effects are generally minor, but may include skin discoloration at the treatment site, infection of the tattoo site, lack of complete pigment removal, or some scarring. A raised or thickened scar may appear three to six months after the tattoo is removed.

## How much does human **hair grow** in a year?

Each hair grows about nine inches (23 centimeters) every year.

## Does human **hair grow faster** in summer or winter?

During the summertime, the rate of human hair growth increases by about 10 percent to 15 percent. This is because warm weather enhances blood circulation to the skin and scalp, which in turn nourishes hair cells and stimulates growth. In cold weather, when blood is needed to warm internal organs, circulation to the body surface slows and hair cells grow less quickly.

## Why does **hair turn gray** as part of the aging process?

The pigment in hair, as well as in the skin, is called melanin. There are two types of melanin: eumelanin, which is dark brown or black, and pheomelanin, which is reddish yellow. Both are made by a type of cell called a melanocyte that resides in the hair bulb and along the bottom of the outer layer of skin, or epidermis. The melanocytes pass this pigment to adjoining epidermal cells called keratinocytes, which produce the protein keratin—hair's chief component. When the keratinocytes undergo their scheduled death, they retain the melanin. Thus, the pigment that is visible in the hair and in the skin lies in these dead keratinocyte bodies. Gray hair is simply hair with less melanin, and white hair has no melanin at all. It remains unclear as to how hair loses its pigment. In the early stages of graying, the melanocytes are still present but inactive. Later they seem to decrease in number. Genes control this lack of deposition of melanin. In some families, many members' hair turns white they are still in their twenties. Generally speaking, among Caucasians 50 percent are gray by age 50. There is, however, wide variation.

### Why is some hair **curly** while some hair is **straight**?

The shape of the hair follicle determines how wavy a hair will be. Round follicles produce straight hair. Oval follicles produce wavy hair. Flat follies produce curly hair.

### What information can a **forensic scientist** determine from a **human hair**?

A single strand of human hair can identify the age and sex of the owner, drugs and narcotics the individual has taken, and, through DNA evaluation and sample comparisons, from whose head the hair came.

### Do the **nails and hair** of a dead person **continue to grow**?

No. Between 12 and 18 hours after death, the body begins to dry out. That causes the tips of the fingers and the skin of the face to shrink, creating the illusion that the nails and hair have grown.

### How fast do **fingernails** grow?

Healthy nails grow about 0.12 inch (3mm) each month or 1.4 inches (3.5 cm) each year. The middle fingernail grows the fastest, because the longer the finger the faster its nail growth.

## SENSES AND SENSE ORGANS

### What are the **floaters** that move around on the eye?

Floaters are semi-transparent specks perceived to be floating in the field of vision. Some originate with red blood cells that have leaked out of the retina. The blood cells

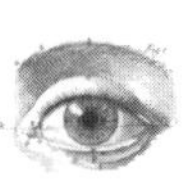

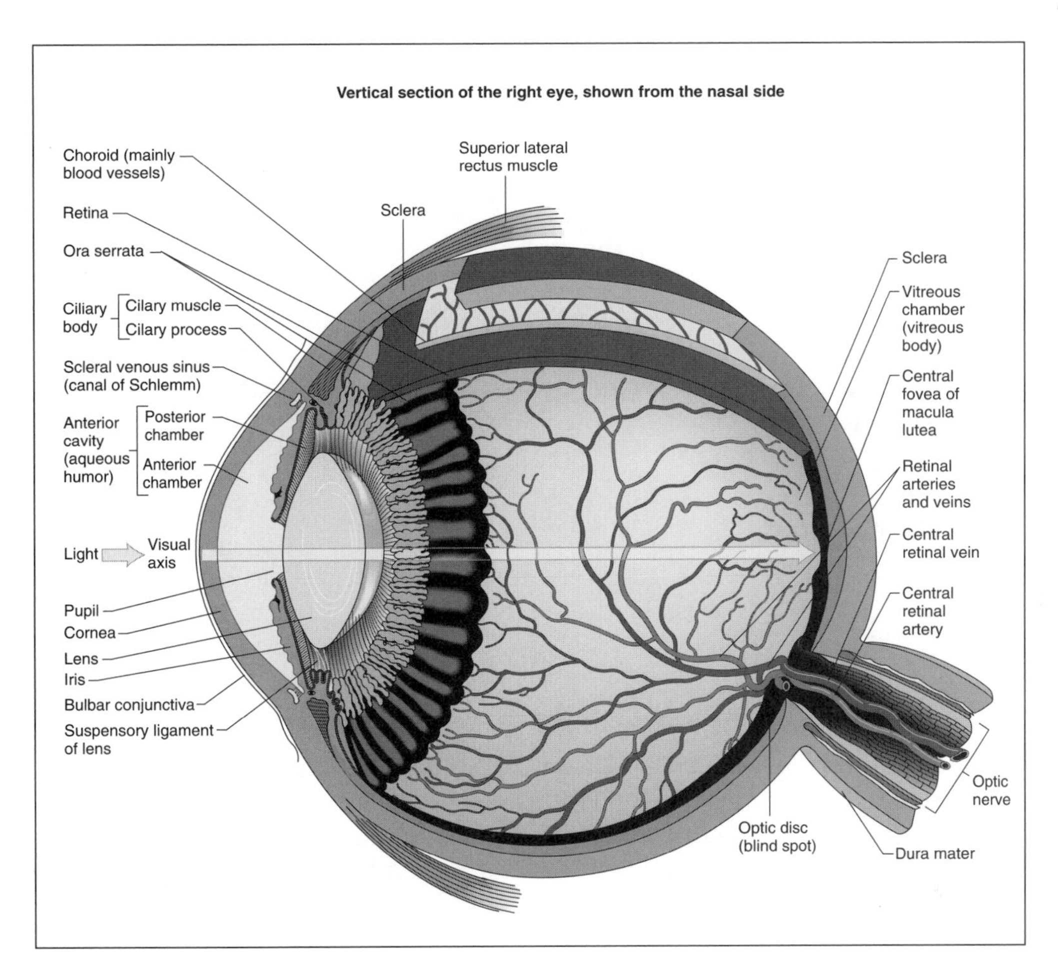


**The human eye.**

swell into spheres, some forming strings, and float around the areas of the retina. Others are shadows cast by the microscopic structures in the vitreous humor, a jellylike substructure located behind the retina. A sudden appearance of a cloud of dark floaters, if accompanied by bright light flashes, could indicate retinal detachment.

## What is the difference in the functions of the **rods and cones** found in the eyes?

Rods and cones contain photoreceptors that convert light first to chemical energy and then into electrical energy for transmission to the vision centers of the brain via the optic nerve. Rods are specialized for vision in dim light; they cannot detect color, but they are the first receptors to detect movement and register shapes. There are about 126 million rods in each eye. Cones provide acute vision, functioning best in bright daylight. They allow us to see colors and fine detail. Cones are divided into three different types, which absorb wavelengths in the short (blue), middle (green), and long (red) ranges. There are about six million cones in each eye.

## How often does the **human eye blink**?

The rate of blinking varies, but on the average the eye blinks once every five seconds (12 blinks per minute) or 17,000 times each day or 6.25 million times a year. In an award-winning science fair project, fourth-grader Holly Feldman observed that adults blink an average of 16 times per minute, or 5.48 million times a year. Miss Feldman based her calculations on a 16-hour day, concluding that humans don't blink for approximately eight hours each day, when they are asleep.

## What are **phosphenes**?

If the eyes are shut tightly, the lights seen are phosphenes. Technically, the luminous impressions are due to the excitation of the retina caused by pressure on the eyeball.

## What does it mean to have **20/20 vision**?

Many people think that with 20/20 vision the eyesight is perfect, but it actually means that the eye can see clearly at 20 feet (six meters) what a normal eye can see clearly at that distance. Some people can see even better—20/15, for example. With their eyes, they can view objects from 20 feet (six meters) away with the same sharpness that a normal-sighted person would have to move in to 15 feet (4.5 meters) to achieve.

## Are more people **nearsighted** or **farsighted**?

About 30 percent of Americans are nearsighted to some degree. About 60 percent are farsighted. If the light rays entering the pupils of the eye converge exactly on the retina, then a sharply focused picture is relayed to the brain. But if the eyeball is shaped differently, the focal point of the light rays is too short or too long, and vision is blurred. Convex lenses for farsightedness correct a too-long focal point. Concave lenses correct nearsightedness when the focal point of light rays, being too short, converge in front of the retina.

## Who invented **bifocal lenses**?

The original bifocal lens was invented in 1784 by Benjamin Franklin (1706–1790). At that time, the two lenses were joined in a metallic frame. In 1899, J. L. Borsch welded the two lenses together. One-part bifocal lenses were developed by Bentron and Emerson in 1910 for the Carl Zeiss Company.

### What is the rare condition called synesthesia?

Synesthesia, or cross perception, is a condition in which a person perceives stimuli not only with the sense for which it is intended, but with others as well. For example, a synesthete may see musical notes as color hues or feel flavors as different textures on the skin. Experiments have determined that the linking of the senses occurs because of some unique physical condition in the brains of these people. For example, blood flow to some parts of the brain, normally increased by sensory stimuli, decreases in synesthetes.

## Why do all **newborn babies** have **blue eyes**?

The color of the iris gives the human eye its color. The amount of dark pigment, melanin, in the iris is what determines its color. In newborns the pigment is concentrated in the folds of the iris. When a baby is a few months old, the melanin moves to the surface of the iris and gives the baby his or her permanent eye color.

## What causes eyes to appear as **red dots** in some **photographs**?

"Red eye" in some color flash photographs is caused by a reflection from the layer of blood vessels lying between the retina and sclera (whites) of the eye. Red eye usually occurs when the light level is relatively dim and the subject is looking directly at the camera. To minimize the effect, the photographer should move the flash away from the camera lens; if this cannot be done, turning on additional lights in the room may help.

## What is the **"sand"** that gathers in the **corners of our eyes** when we sleep?

The "sand" is dried mucus. Glands near the eye secrete the mucus to help the eye retain moisture and protect it against foreign particles. During sleep, the closed eyelids retain the moisture of the eyes, and the mucus may gather in the corners and dry out. When a person wakes, the dried mucus may feel like sand in the eyes.

## How do **colors** affect one's **moods**?

According to the American Institute for Biosocial Research, "colors are electromagnetic wave bands of energy." Each color has its own wavelength. The wave bands stimulate chemicals in your eye, sending messages to the pituitary and pineal glands. These master endocrine glands regulate hormones and other physiological systems in the body. Stimulated by response to colors, glandular activities can alter moods, speed up heart rates, and increase brain activity.

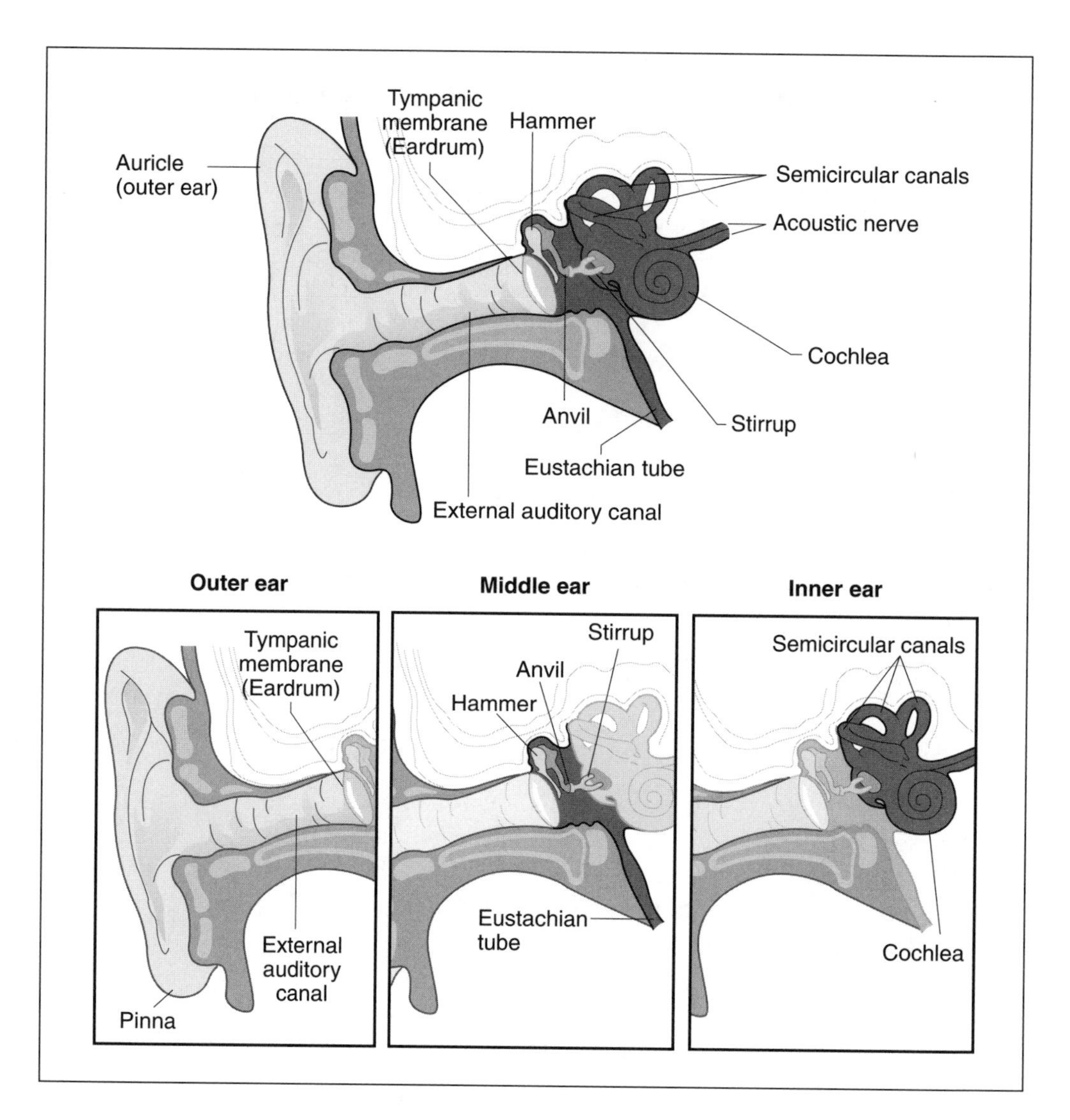


The anatomy of the human ear.

## How does an individual maintain their **balance**?

The ability to maintain one's balance depends on information the brain receives from three different sources—the inner ears, the special sensory receptors in muscles, and joints of the body and the eyes. The inner ears contain two sets of special vestibular sensory organs that are important for balance. Semicircular canals sense rotations of the head in space, and otolith organs sense both gravity and linear movements of the head in space. Sensory information from the muscles and joints gives information about the position of joints and forces acting on the body. The visual system gives important information about which direction is "up" and about

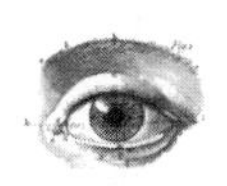



movements of the head; this is why an OmniMax theater can give the illusion that you are moving. These signals are processed by the brain to control balance, posture, and a sense of stability.

## What are the **three bones** in the **ear** called?

The three bones are the malleus, which means hammer; incus, which means anvil; and stapes, which means stirrup. The bones look somewhat like the objects for which they are named. The stapes is the smallest bone in the body, measuring 1.02 to 1.34 inches (2.6 to 3.4 centimeters) and weighing 0.00071 to 0.0015 ounce (0.002 to 0.004 gram). These three tiny bones in the middle ear conduct sound vibrations from the outer to inner ear.

## How broad is the **range of sound frequency** that most people can hear?

Most people can hear sounds with frequencies from about 20 to 20,000 hertz. A hertz is a measure of sound frequencies. Environmental sound is measured in decibels to calculate its loudness. Hearing starts with zero decibels; every increase of 10 units is equivalent to a tenfold increase. In comparison, leaves rustling = 10 decibels, a typical office is 50 decibels, pneumatic drills = 80 decibels, riveting machines = 110 decibels, and a jet takeoff at 200 feet (61 meters) measures 120 decibels. Noise above 70 decibels is harmful to hearing; noise at 140 decibels is physically painful.

## Why do humans **get old**?

Solving this question—along with finding ways to stay forever young—has vexed mankind since time immemorial. Scientists have tried to explain why people age with two broad theoretical directions: programmed theories and error theories. Programmed theories state that aging is "programmed" into humankind: we have a biological clock that simply runs down according to a set timeline. Examples of programmed theories include the Programmed Senescence theory, which holds that aging happens because of the switching off and on of genes, and the Immunological Theory, which holds that the human immune system declines at a set pace, leading to defenselessness from illnesses. Error theories define aging according to unavoidable failures and accumulating damages: we simply wear out. Examples of error theories include Somatic Mutation, which holds that genetic mutations cause cells to fail, and the Free Radical Theory, which holds that a buildup of damage by oxygen radicals created during normal metabolism causes cells to stop working.

Advances in society and knowledge have ameliorated aging's effects by allowing people to live longer. Average life expectancy in the United States has grown from about 47 years in 1900 to about 79 years at the beginning of 2000. But nothing, unfortunately, has enabled people to stop growing old.

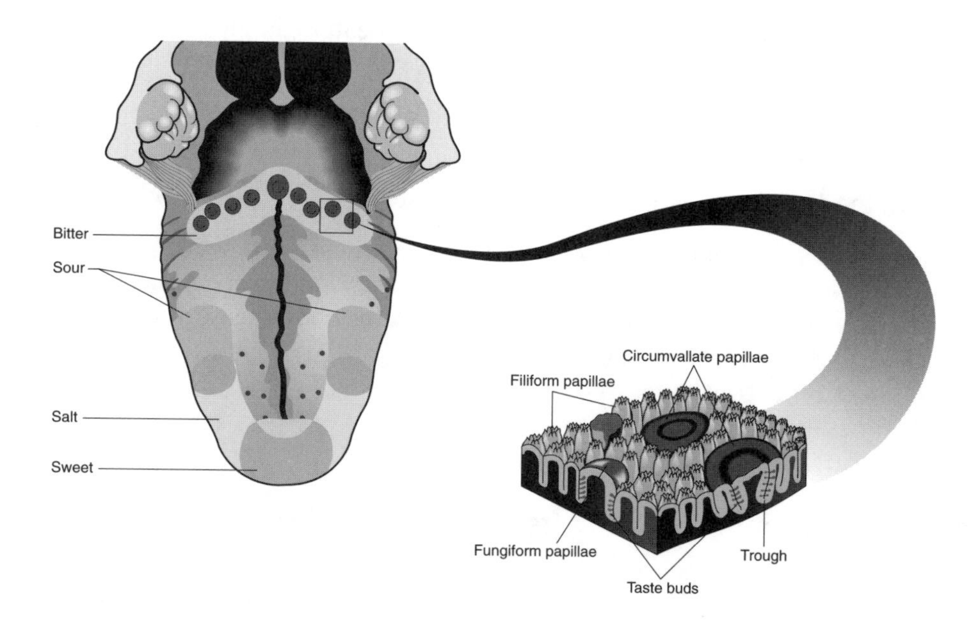


**Taste regions of the tongue (left) and taste bud anatomy (right).**

## What are the primary **sensations** of **taste**?

The four primary categories are sweet, sour, salty, and bitter. The sensitivity and location of these areas on the tongue varies from person to person. Some of the nine thousand taste buds are located in the other areas of the mouth as well. The lips (usually very salt-sensitive), the inner cheeks, the underside of the tongue, the back of the throat, and the roof of the mouth are some examples. The sense of taste is intimately associated with the sense of smell, so that foods taste bland to someone suffering from a cold. Also related are the appearance, texture, and temperature of food.

# HEALTH AND MEDICINE

## HEALTH HAZARDS, RISKS, ETC.

### Which **risk factors** affect one's health?

Such characteristics as age, gender, work, family history, behavior, and body chemistry are some of the factors to consider when deciding whether one is at risk for various conditions. Some risk factors are statistical, describing trends among large groups of people but not giving information about what will happen to individuals. Other risk factors might be described as causative—exposure to them has a direct effect on whether or not the person will become sick.

### What are the leading **causes of stress**?

In 1967, when they conducted a study of the correlation between significant life events and the onset of illness, Dr. Thomas H. Holmes and Dr. Richard H. Rahe from the University of Washington compiled a chart of the major causes of stress with assigned point values. They published their findings on stress effects as "The Social Readjustment Scale," printed in *The Journal of Psychosomatic Research*. The researchers calculated that a score of 150 points indicated a 50/50 chance of the respondent developing an illness or a "health change." A score of 300 would increase the risk to 90 percent. This type of rating scale continues to be used to help individuals determine their composite stress level within the last year. Since 1967 other researchers have adapted and modified the checklist, but the basic checklist has remained constant. Of course, many factors enter into an individual's response to a particular event, so this scale, partially represented below, can only be used as a guide.

| Event | Point value |
| --- | --- |
| Death of spouse | 100 |
| Divorce | 73 |
| Marital separation | 65 |
| Jail term or death of close family member | 63 |
| Personal injury or illness | 53 |
| Marriage | 50 |
| Fired at work | 47 |
| Marital reconciliation or retirement | 45 |
| Pregnancy | 40 |
| Change in financial state | 38 |
| Death of close friend | 37 |
| Change in employment | 36 |
| Foreclosure of mortgage or loan | 30 |
| Outstanding personal achievement | 28 |
| Trouble with boss | 23 |
| Change in work hours or conditions or change in residence or schools | 20 |
| Vacation | 13 |
| Christmas | 12 |
| Minor violations of the law | 11 |

## What are the **odds against** being **struck by lightning**?

The National Oceanic and Atmospheric Administration estimates the odds of being struck by lightning at one in 700,000. However, the odds drop to one in 240,000 based on the number of unreported lightning strikes. The odds of being struck by lightning in one's lifetime are 1 in 3,000.

## What are the **odds against** being **killed on a motorcycle**?

1,250 to one against.

## Is there **more violence** on the streets and in mental institutions when there is a **full moon**?

A review of 37 studies that attempted to correlate the phases of the moon with violent crime, suicide, crisis center hotline calls, psychiatric disorders, and mental hospital admissions found that there is absolutely no relation between the moon and the mind. Despite these findings, many people continue to believe that the moon is a powerful, and sometimes malevolent, force.

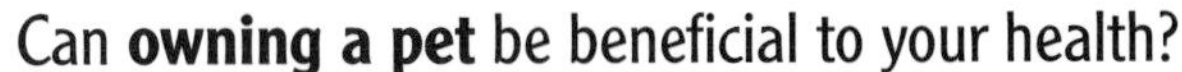

## Can **owning a pet** be beneficial to your health?

As a result of several studies, researchers now believe that regular contact with pets can reduce heart rate, blood pressure, and levels of stress. In a study of 93 heart attack patients, only one of 18 pet owners died compared to one of three patients who did not have pets. Pets offer constancy, stability, comfort, security, affection, and intimacy.

## What are the **leading causes of death** in the United States?

Of the 2,404,598 deaths in 2000, the leading cause of death was heart disease. Below are listed the four major causes of death in the United States.

| Rank | Cause of Death | Number | Percentage of Total |
|---|---|---|---|
| 1 | Heart disease | 709,894 | 29.5 |
| 2 | Cancer | 551,833 | 22.9 |
| 3 | Stroke | 166,028 | 6.9 |
| 4 | Chronic lower respiratory disease | 123,550 | 5.1 |

## Did raising the **speed limit** on rural interstate highways from 55 to 65 miles per hour have an effect on the **accident and death rate**?

There was an estimated 20 percent to 30 percent increase in deaths on those roads and a 40 percent increase in serious injuries when the speed limit was raised from 55 miles per hour (88.5 kilometers per hour) to 65 miles per hour (104.5 kilometers per hour) on rural interstate highways.

## Are men or women more **accident-prone**?

Women drive and even cross the street more safely than men. Men account for 70 percent of pedestrian fatalities since 1980. Between the ages of 18 and 45, males outnumber females as fatal crash victims by almost three to one, according to the National Highway Traffic Safety Administration. Accidental deaths of all types—from falls, firearms, drownings, fires, even food and other poisonings—also are more common among men than women.

## Which **sport** has the highest rate of **injuries** and what kind of injury is most common?

Football players suffer more injuries than other athletes, collectively. They have 12 times as many injuries as do basketball players, who have the next highest rate of injury. Knee problems are the most common type of injury, with two-thirds of basketball players' injuries and one-third of football players' injuries being knee-related.

## How many children sustain head injuries while riding bicycles?

Approximately 140,000 children are treated in emergency rooms for head injuries sustained while riding bikes each year. Many bike-related injuries and deaths could be avoided if riders wore helmets properly. Wearing a bike helmet reduces the risk of brain injury by as much as 88 percent and reduces the risk of injury to the face by 65 percent.

## Which direction of impact results in the greatest number of fatalities in **automobile crashes**?

Frontal crashes are responsible for the largest percentage of fatalities in passenger cars.

## What is "**boomeritis**"?

The American Academy of Orthopaedic Surgeons coined the term "boomeritis" to describe injuries to aging athletes. A 2000 study found sports injuries to baby boomers, individuals between the ages of 35 and 54, increased 33 percent between 1991 and 1998. According to the study, more than 365,000 baby boomers injured themselves badly enough while engaging in sports activities to require emergency room treatment. Including all injuries requiring medical attention, the total was more than one million.

## What are the most frequent **sports-related injuries** requiring visits to the emergency room for **children** and **young adults**?

Injuries associated with basketball and cycling are the most frequent sports-related injuries requiring visits to emergency rooms among children and young adults.

Approximate Number of Children's Sports Injuries per Year Requiring Visits to Emergency Rooms

| Sport | Number of Emergency Room Visits |
|---|---|
| Basketball and cycling* | 900,000* |
| Football | 250,000 |
| Baseball | 250,000 |
| Soccer | 100,000 |
| Ice or roller skating and skateboarding | 150,000 |
| Gymnastics and cheerleading | 146,000 |
| Playground injuries | 137,000 |
| Water and snow sports | 100,000 |

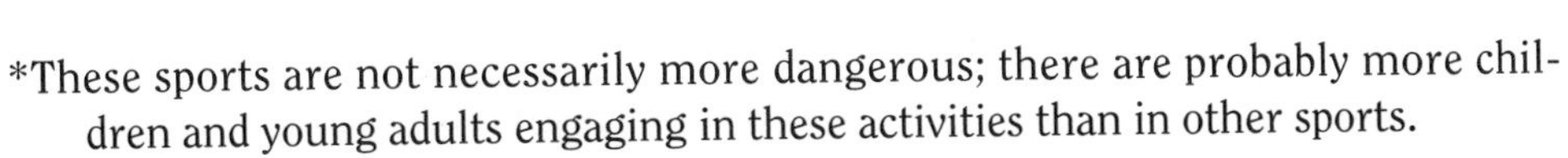

*These sports are not necessarily more dangerous; there are probably more children and young adults engaging in these activities than in other sports.

Sports-related injuries are more likely to be to the brain or skull and upper and lower extremities, more likely to be a fracture, strain or sprain, and more likely to have diagnostic and therapeutic services provided, especially orthopedic care.

## Do **electric** and **magnetic fields** produced by power transmission lines present a health hazard?

No study has produced evidence to allow a firm conclusion on this question. Scientists cannot agree on the significance of the inconsistent and puzzling findings. On electromagnetic fields, in general, a 1990 study by the United States Environmental Protection Agency (EPA) noted a possible statistically significant link between cancer and exposure to extremely low frequency (ELF) electromagnetic fields. ELF waves are non-ionizing electromagnetic radiation that closely resemble the body's micropulsations. Biological studies have yet to prove such a connection. So far the ELF risks are controversial. Effects (from those studies supporting ELF hazards) run the gamut from headaches, to miscarriages, to cancer.

In May 1995, the American Physical Society reported that it could find no evidence that the electromagnetic fields emanating from power lines can cause cancer. While acknowledging that studies are still underway, the society concluded that existing research does not substantiate any reported adverse health effects. More recent studies, including a recent California Department of Health Services, continue to be inconclusive.

In addition, electromagnetic radiation (EMR)—from devices such as electric blankets, video display terminals, microwaves, toasters, and hair dryers, powered by alternating current (AC) power lines—has been the target of study to better understand the health hazards that might be posed.

## Why is exposure to **asbestos** a health hazard?

Exposure to asbestos has long been known to cause asbestosis. This is a chronic, restrictive lung disease caused by the inhalation of tiny mineral asbestos fibers that scar lung tissues. Asbestos has also been linked with cancers of the larynx, pharynx, oral cavity, pancreas, kidneys, ovaries, and gastrointestinal tract. The American Lung Association reports that prolonged exposure doubles the likelihood that a smoker will develop lung cancer. It takes cancer 15 to 30 years to develop from asbestos. Mesothelioma is a rare cancer affecting the surface lining of the pleura (lung) or peritoneum (abdomen) that generally spreads rapidly over large surfaces of either the thoracic or abdominal cavities. No effective treatment exists for mesothelioma.

Asbestos fibers were used in building materials between 1900 and the early 1970s as insulation for walls and pipes, as fireproofing for walls and fireplaces, in sound-proofing and acoustic ceiling tiles, as a strengthener for vinyl flooring and joint compounds, and as a paint texturizer. Asbestos poses a health hazard only if the tiny fibers are released into the air, but this can happen with any normal fraying or cracking. Asbestos removal aggravates this normal process and multiplies the danger level—it should only be handled by a contractor trained in handling asbestos. Once released, the particles can hang suspended in the air for more than 20 hours.

## Can **ozone** be harmful to humans?

Ozone ($O_3$) in the lower atmosphere contributes to air pollution. It is formed by chemical reactions between sunlight and oxygen in the air in the presence of impurities, such as those found in automobile exhaust. Ozone can damage rubber, plastic, and plant and animal tissue. Exposure to certain concentrations can cause headaches, burning eyes, and irritation of the respiratory tract in many individuals. Asthmatics and others with impaired respiratory systems are particularly susceptible. Exposure to low concentrations for only a few hours can significantly affect normal persons while exercising. Symptoms include chest pain, coughing, sneezing, and pulmonary congestion.

## Why is **radon** a health hazard?

Radon is a colorless, odorless, tasteless, radioactive gaseous element produced by the decay of radium. It has three naturally occurring isotopes found in many natural materials, such as soil, rocks, well water, and building materials. Because the gas is continually released into the air, it makes up the largest source of radiation that humans receive. A 1999 National Academy of Sciences (NAS) report noted that radon was the second leading cause of lung cancer. It has been estimated that it may cause as much as 12 percent, or about 15,000 to 22,000 cases, of lung cancer deaths annually. Smokers seem to be at a higher risk than non-smokers. The U.S. Environmental Protection Agency (EPA) recommends that in radon testing the level should not be more than four picocuries per liter. The estimated national average is 1.5 picocuries per liter. Because EPA's "safe level" is equivalent to 200 chest X-rays per year, some experts believe that lower levels are appropriate. The American Society of Heating, Refrigeration, and Air-Conditioning Engineers (ASHRAE) recommends two picocuries/liter. The EPA estimates that nationally 8 percent to 12 percent of all houses are above the four picocuries/liter limit; whereas in another survey in 1987 it was estimated that 21 percent of homes were above this level.

## How is human exposure to **radiation** measured?

The radiation absorbed dose (rad) and the roentgen equivalent man (rem) were used for many years to measure the amount and effect of ionizing radiation absorbed by humans. While officially replaced by the gray and the sievert, both are still used in

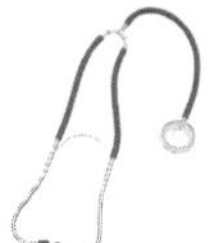



many reference sources. The rad equals the energy absorption of 100 ergs per gram of irradiated material (an erg is a unit of work or energy). The rem is the absorbed dose of ionizing radiation that produces the same biological effect as one rad of X-rays or gamma rays (which are equal). The rem of X-rays and gamma rays is therefore equal to the rad; for each type of radiation, the number of rads is multiplied by a specific factor to find the number of rems. The millirem, 0.001 rems, is also frequently used; the average radiation dose received by a person in the United States is about 360 millirems per year. Natural radiation accounts for about 82 percent of a person's yearly exposure, and manufactured sources for 18 percent. Indoor radon has only recently been recognized as a significant source of natural radiation, with 55 percent of the natural radiation coming from this source.

In the SI system (*Système International d'Unités,* or International System of Units), the gray and the sievert are used to measure radiation absorbed; these units have largely superseded the older rad and rem. The gray (Gy), equal to 100 rads, is now the base unit. It is also expressed as the energy absorption of one joule per kilogram of irradiated material. The sievert (Sv) is the absorbed dose of radiation that produces the same biological effect as one gray of X-rays or gamma rays. The sievert is equal to 100 rems, and has superseded the rem. The becquerel (Bq) measures the radioactive strength of a source, but does not consider effects on tissue. One becquerel is defined as one disintegration (or other nuclear transformation) per second.

## What is the **effect** of **radiation** on humans?

When ionizing radiation penetrates living tissue, random collisions with atoms and molecules in its path cause the formation of ions and reactive radicals. These ions and reactive radicals break chemical bonds and cause other molecular changes that produce biological injury. At the cellular level, radiation exposure inhibits cell division, and produces chromosomal damage and gene mutations as well as various other changes. Large enough doses of ionizing radiation will kill any kind of living cell.

## How much radiation does the average **dental X-ray** emit?

Dental examinations are estimated to contribute 0.15 millirems per year to the average genetically significant dose, a small amount when compared to other medical X-rays.

## How does the United States Environmental Protection Agency (EPA) classify **carcinogens**?

A carcinogen is an agent that can produce cancer (a malignant growth or tumor that spreads throughout the body, destroying tissue). The EPA classifies chemical and physical substances according to their toxicity to humans.

EPA classification system for carcinogens

*Group A. Human carcinogen*

This classification indicates that there is sufficient evidence from epidemiological studies to support a cause–effect relationship between the substance and cancer.

*Group B. Probable human carcinogen*

$B_1$: Substances are classified as $B_1$ carcinogens on the basis of sufficient evidence from animal studies, and limited evidence from epidemiological studies.

$B_2$: Substances are classified as $B_2$ carcinogens on the basis of sufficient evidence from animal studies, with inadequate or nonexistent epidemiological data.

*Group C. Possible human carcinogen*

For this classification, there is limited evidence of carcinogenicity from animal studies and no epidemiological data.

*Group D. Not classifiable as to human carcinogenicity*

The data from human epidemiological and animal studies are inadequate or completely lacking, so no assessment as to the substance's cancer-causing hazard is possible.

*Group E. Evidence of noncarcinogenicity for humans*

Substances in this category have tested negative in at least two adequate (as defined by the EPA) animal cancer tests in different species and in adequate epidemiological and animal studies. Classification in group E is based on available evidence; substances may prove to be carcinogenic under certain conditions.

## What is "good" and "bad" **cholesterol**?

Chemically a lipid, cholesterol is an important constituent of body cells. This fatty substance, produced mostly in the liver, is involved in bile salt and hormone formation, and in the transport of fats in the bloodstream to the tissues throughout the body. Both cholesterol and fats are transported as lipoproteins (units having a core of cholesterol and fats in varying proportions with an outer wrapping of carrier protein [phospholoids and apoproteins]). An overabundance of cholesterol in the bloodstream can be an inherited trait, can be triggered by dietary intake, or can be the result of a metabolic disease, such as diabetes mellitus. Fats (from meat, oil, and dairy products) strongly affect the cholesterol level. High cholesterol levels in the blood may lead to a narrowing of the inner lining of the coronary arteries from the build-up of a fatty tissue called atheroma. This increases the risk of coronary heart disease or stroke. However, if most cholesterol in the blood is in the form of high density lipoproteins (HDL),

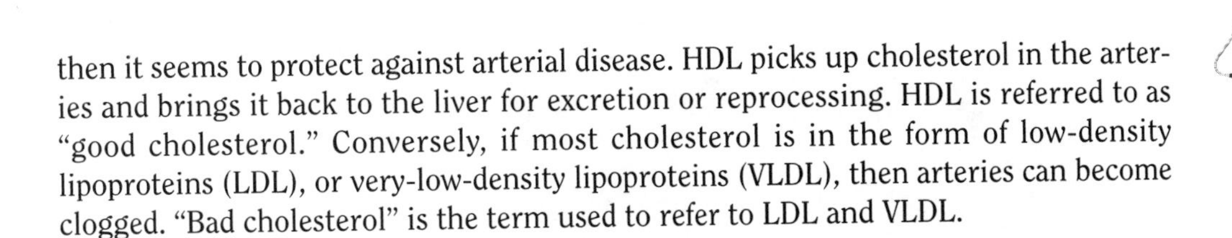

then it seems to protect against arterial disease. HDL picks up cholesterol in the arteries and brings it back to the liver for excretion or reprocessing. HDL is referred to as "good cholesterol." Conversely, if most cholesterol is in the form of low-density lipoproteins (LDL), or very-low-density lipoproteins (VLDL), then arteries can become clogged. "Bad cholesterol" is the term used to refer to LDL and VLDL.



## How does **blood alcohol level** affect the body and behavior?

The effects of drinking alcoholic beverages depend on body weight and the amount of actual ethyl alcohol consumed. The level of alcohol in the blood is calculated in milligrams (one milligram equals 0.035 ounce) of pure (ethyl) alcohol per deciliter (3.5 fluid ounces), commonly expressed in percentages.

| Number of drinks | Blood alcohol level | Effect of drinks |
|---|---|---|
| 1 | 0.02–.03% | Changes in behavior, coordination, and ability to think clearly |
| 2 | 0.05% | Sedation or tranquilized feeling |
| 3 | 0.08–0.10% | Legal intoxication in many states |
| 5 | 0.15–0.20% | Person is obviously intoxicated and may show signs of delirium |
| 12 | 0.30–0.40% | Loss of consciousness |
| 24 | 0.50% | Heart and respiration become so depressed that they cease to function and death follows |

## What percentage of the American public are **smokers**?

Smokers constitute a minority, and a shrinking one. The annual prevalence of cigarette smoking among adults in the United States declined 40 percent from 1965 to 1990 (from 42.4 percent to 25.5 percent), but was virtually unchanged from 1990 to 1992. In 1999, an estimated 48 million adults (23.5 percent) were current smokers. Epidemiologic data suggest that more than 70 percent of the nearly 50 million smokers in the United States today have made at least one prior attempt to quit, and approximately 46 percent try to quit each year. Most smokers make several quit attempts before they successfully kick the habit.

Studies have linked pipe and cigar smoking to oral cavity cancers and cigarette smoking to lung cancer and to respiratory diseases such as chronic bronchitis, emphysema, and coronary heart disease.

## What is the composition of **cigarette smoke**?

Cigarette smoke contains about 4,000 chemicals. Carbon dioxide, carbon monoxide, methane, and nicotine are some of the major components, with lesser amounts of ace-

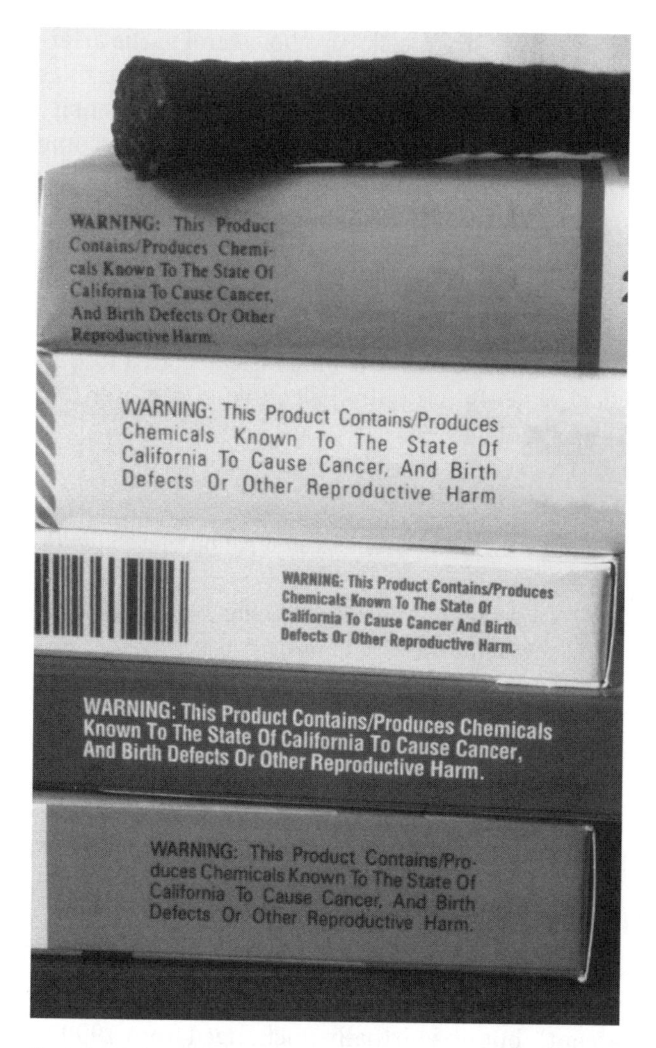


Cigarette smoke contains traces of nearly 4,000 different chemicals.

tone, acetylene, formaldehyde, propane, hydrogen cyanide, toluene, and many others.

### Why does the risk of **cancer** diminish rapidly after one quits the **cigarette smoking** habit?

Exposure of a premalignant cell to a promoter converts the cell to an irreversibly malignant state. Promotion is a slow process, and exposure to the promoter must be sustained for a certain period of time. This requirement explains why the risk of cancer diminishes rapidly after one quits smoking; both cancer initiators and promoters appear to be contained in tobacco smoke.

## FIRST AID, POISONS, ETC.

### Who discovered **cardiopulmonary resuscitation** (CPR) as a method to resuscitate an individual whose heart had stopped?

Cardiopulmonary resuscitation (CPR) is a first-aid technique that combines mouth-to-mouth resuscitation and rhythmic compression of the chest to a person whose heart has stopped. The Scottish surgeon William Tossach first performed mouth-to-mouth resuscitation in 1732. The technique was not further developed (or widely used) for many centuries until Dr. Edward Schafer developed a method of chest pressure to stimulate respiration. In 1910 the American Red Cross adopted and began to teach Schafer's method. A team of specialists at Johns Hopkins Medical School, O. R.

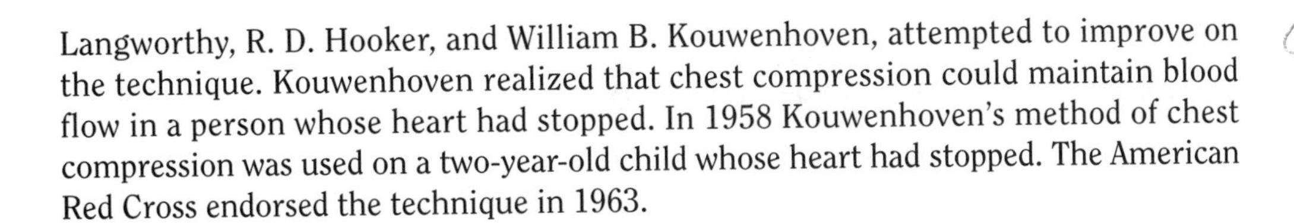

Langworthy, R. D. Hooker, and William B. Kouwenhoven, attempted to improve on the technique. Kouwenhoven realized that chest compression could maintain blood flow in a person whose heart had stopped. In 1958 Kouwenhoven's method of chest compression was used on a two-year-old child whose heart had stopped. The American Red Cross endorsed the technique in 1963.



## How long does it take to **bleed to death**?

Serious bleeding requires immediate attention and care. If a large blood vessel is severed or lacerated, a person can bleed to death in one minute or less. Rapid loss of one quart or more of the total blood volume often leads to irreversible shock and death.

## What is the **"ABCD" survey** first responders use to evaluate an emergency?

"A" stands for airway. It is important to be certain the airway from the mouth or nose to the lungs is clear. The airway can be opened by tilting the head back and lifting the chin.

"B" stands for breathing. Be certain the person is breathing or perform rescue breathing (CPR) to ensure a supply of oxygen.

"C" stands for circulation. If a pulse cannot be found, then there is no blood circulating. Emergency personnel can attempt to get the heart to resume breathing by performing rhythmic chest thrusts (CPR). Adults require 15 chest compressions for every two rescue breaths. It also means to check for profuse bleeding which must be controlled.

"D" stands for disability. It involves checking for consciousness and the likelihood of spinal cord or neck injury.

## What is the **Heimlich maneuver**?

This effective first-aid technique to resuscitate choking and drowning victims was introduced by Dr. Henry J. Heimlich (b. 1920) of Xavier University in Cincinnati, Ohio. It is a technique for removing a foreign body from the trachea or pharynx where it is preventing flow of air to the lungs. When the victim is in the vertical position, the maneuver consists of applying subdiaphragmatic pressure by wrapping one's arms around the victim's waist from behind, making a fist with one hand and placing it against the victim's abdomen between the navel and the rib cage, clasping one's fist with the other hand, and pressing in with a quick, forceful thrust. Repeat several times if necessary. When the victim is in the horizontal position (which some experts recommend), the rescuer straddles the victim's thighs.

## What safety rules should be observed during a **thunderstorm**?

These safety rules should be observed when lightning threatens:

1. Stay indoors. Seek shelter in buildings. If no buildings are available, the best protection is a cave, ditch, canyon, or under head-high clumps of trees in open forest glades. If there is no shelter, avoid the highest object in the area. Keep away from isolated trees.
2. Get out of the water and off small boats.
3. Do not use the telephone.
4. Do not use metal objects like fishing rods and golf clubs.
5. Stay in your automobile if you are traveling.
6. Do not use plug-in electrical equipment like hair dryers, electric razors, or electric toothbrushes during the storm.

## What items should be included in a household **first-aid kit**?

According to the American Medical Association and National Safety Council, a first-aid kit should contain:

Allergy and medication information for each family member
Antiseptic cream
Antiseptic wipes
Aspirin or an aspirin substitute such as acetaminophen or ibuprofen
Adhesive bandages
Elastic bandages
Emergency telephone numbers
First-aid manual
Gauze bandages
Triangular bandage
Calamine lotion
Flashlight
Foil blanket
Hydrogen peroxide or rubbing alcohol
Medical exam gloves
Roll of sterile cotton
Round-ended tweezers
Safety pins
Snub-nosed scissors
Syrup of ipecac

### How was Mr. Yuk, the symbol for dangerous or poisonous products, developed?

The Pittsburgh Poison Center developed the Mr. Yuk symbol in 1971. It is used by Children's Hospital to promote prevention education through affiliated hospitals and poison centers. In the testing program, children at day care centers selected Mr. Yuk as the symbol that represented the most unappealing product. Among the other symbols used were a red stop sign and a skull and crossbones. Interestingly, the children found the skull and crossbones the most appealing symbol.

## How is **activated charcoal** used medically?

Activated charcoal is an organic substance, such as burned wood or coal, that has been heated to approximately 1,000°F (537°C) in a controlled atmosphere. The result is a fine powder containing thousands of pores that have great absorbent qualities to rapidly absorb toxins and poisons. Activated charcoal is used medically in the treatment of drug overdoses and poisonings.

## What is the **deadliest natural toxin**?

Botulinal toxin, produced by the bacterium *Clostridium botulinum*, is the most potent poison of humans. It has an estimated lethal dose in the bloodstream of $10^{-9}$ milligrams per kilogram. It causes botulism, a severe neuroparalytic disease that travels to the junctions of skeletal muscles and nerves, where it blocks the release of the neurotransmitter acetylcholine, causing muscle weakness and paralysis, and impairing vision, speech, and swallowing. Death occurs when the respiratory muscles are paralyzed; this usually occurs during the first week of illness. Mortality from botulism is about 25 percent.

Because the bacterium can form the toxin only in the absence of oxygen, canned goods and meat products wrapped in airtight casings are potential sources of botulism. The toxin is more likely to grow in low-acid foods, such as mushrooms, peas, corn, or beans, rather than high-acid foods like tomatoes. However, some new tomato hybrids are not acidic enough to prevent the bacteria from forming the toxin. Foods being canned must be heated to a temperature high enough and for a long enough time to kill the bacteria present. Suspect food includes any canned or jarred food product with a swollen lid or can. Ironically, this dreaded toxin in tiny doses is being used to treat disorders that bring on involuntary muscle contractions, twisting, etc. The United States Food and Drug Administration (FDA) has approved the toxin for the treatment of strabismus (misalignment of the eyes), blepharospasm (forcible closure of eyelids), and hemifacial spasm (muscular contraction on one side of the face).

## What are some common causes of **poisoning**?

Poisoning is defined as the exposure to any substance in sufficient quantity to cause adverse health effects. Poisonings can be grouped into several different categories, including intentional, accidental, occupational and environmental, social and iatrogenic. Accidental poisonings are the most common, with more than 90 percent occurring in children at home. Intentional poisonings are usually suicide-related, with carbon monoxide being one of the most frequently used agents. Toxic chemical releases in industrial accidents are an occupational and environmental hazard.

## Have childhood deaths from poisoning decreased since the introduction of **childproof containers**?

After childproof packaging was required on all drugs and medications beginning in 1973, the childhood poisoning death rate declined dramatically. A 50 percent decrease was noted from 1973 to 1976, and the decline has continued. Other factors in this decline have been the development of poison control centers, changes in products to reduce poisonous agents, and the introduction of single-dose packages.

## Can toxins in plants in the **nightshade** family, such as potatoes, tomatoes, and eggplant, cause **arthritis** in some people?

No scientific evidence supports this belief. Studies have shown that in groups of people who eat lots of potatoes, there is no increased incidence of arthritis.

## How deadly is **strychnine**?

The fatal dose of strychnine or deadly nightshade (the plant from which it is obtained) is 0.0005 to 0.001 ounces (15 to 30 milligrams). It causes severe convulsions and respiratory failure. If the patient lives for 24 hours, recovery is probable.

## Which part of **mistletoe** is poisonous?

The white berries contain toxic amines, which cause acute stomach and intestinal irritation with diarrhea and a slow pulse. Mistletoe should be considered a potentially dangerous Christmas decoration, especially if children are around.

## What is the **poison on arrows** used by South American Indians to kill prey and enemies?

The botanical poison used by the Aucas and similar tribes in the South American jungles is curare. It is a sticky, black mixture with the appearance of licorice and is processed from either of two different vines. One is a liana (*Chondodendron tomentosum*); the other is a massive, tree-like vine (*Strychonos quianensis*).

## How deadly is ***Amanita phalloides***?

The poisonous mushroom *Amanita phalloides* has a fatality rate of about 50 percent. Ingestion of part of one mushroom may be sufficient to cause death. Over 100 fatalities occur each year from eating poisonous mushrooms, with more than 90 percent caused by the *Amanita phalloides* group.

## What first-aid remedies may be used for **bee stings**?

If a person is allergic to bee stings, he or she should seek professional medical care immediately. For persons not allergic to bee stings, the following steps may be taken: The stinger should be removed by scraping with a knife, a long fingernail, or a credit card, rather than by trying to pull it out. A wet aspirin may be rubbed on the area of the sting to help neutralize some of the inflammatory agents in the venom (unless the person is allergic or sensitive to aspirin taken by mouth).

A paste made of meat tenderizer (or other product that contains papain) mixed with water will relieve the pain. Adults may take an antihistamine along with a mild pain reliever such as aspirin, ibuprofen, or acetaminophen.

## Which first-aid measures can be used for a bite by a **black widow spider**?

The black widow spider (*Latrodectus mactans*) is common throughout the United States. Its bite is severely poisonous, but no first-aid measures are of value. Age, body size, and degree of sensitivity determine the severity of symptoms, which include an initial pinprick with a dull numbing pain, followed by swelling. An ice cube may be placed over the bite to relieve pain. Between 10 and 40 minutes after the bite, severe abdominal pain and rigidity of stomach muscles develop. Muscle spasms in the extremities, ascending paralysis, and difficulty in swallowing and breathing follow. The mortality rate is less than one percent, but anyone who has been bitten should see a doctor; the elderly, infants, and those with allergies are most at risk, and should be hospitalized.

## How can the amount of **lead in tap water** be reduced in an older house having lead-containing pipes?

The easiest way is to let the tap run until the water becomes very cold before using it for human consumption. By letting the tap run, water that has been in the lead-containing pipes for awhile is flushed out. Also, cold water, being less corrosive than warm, contains less lead from the pipes. Lead (Pb) accumulates in the blood, bones, and soft tissues of the body as well as the kidneys, nervous system, and blood-forming organs. Excessive exposure to lead can cause seizures, mental retardation, and behavior disorders. Infants and children are particularly susceptible to low doses of lead and suffer from nervous system damage.

Another source of lead poisoning is old, flaking lead paint. Lead oxide and other lead compounds were added to paints before 1950 to make the paint shinier and more durable. Fourteen percent of the lead ingested by humans comes from the seam soldering of food cans, according to the United States Food and Drug Administration (FDA). The FDA has proposed a reduction in this lead to 50 percent over the next five years. Improperly glazed pottery can be a source of poisoning, too. Acidic liquids such as tea, coffee, wine, and juice can break down the glazes so that the lead can leak out of the pottery. The lead is ingested little by little over a period of time. People can also be exposed to lead in the air. Lead gasoline additives, nonferrous smelters, and battery plants are the most significant contributors of atmospheric lead emissions.

## How did **lead** contribute to the **fall of the Roman Empire**?

Some believe Romans from the period around 150 B.C.E. may have been victims of lead poisoning. Symptoms of lead poisoning include sterility, general weakness, apathy, mental retardation, and early death. The lead could have been ingested in water taken from lead-lined water pipes or from food cooked in their lead-lined cooking pots or from wine served in lead-lined goblets. Unaware of its dangers, some ancient Romans unwittingly used lead as a sweetening agent or medical treatment for diarrhea. Lead poisoning could have caused infertility in women, leading to a subsequent long-term decline in the birth rate of the Roman upper classes. The effect of this inadvertent toxic food additive on Roman history, however, is only speculative.

## What often happened to hat makers that caused Lewis Carroll to use the expression **"mad as a hatter"** in his *Alice in Wonderland*?

In the 19th century, craftsmen who made hats were known to be excitable and irrational, as well as to tremble with palsy and mix up their words. Such behavior gave rise to familiar expression "mad as a hatter." The disorder, called hatter's shakes, was caused by chronic mercury poisoning from the solution used to treat the felt. Attacking the central nervous system, the toxin led to the behavioral problems.

## During what period of time may a **cut or wound** be **stitched**?

Stitches should be placed within six to eight hours of the injury. The time may be extended to twelve hours if the amount of contamination is small and the wound area is very vascular.

# DISEASES, DISORDERS, AND OTHER HEALTH PROBLEMS

## What is the difference between a **virus** and a **retrovirus**?

A virus is a rudimentary biosystem that has some of the aspects of a living system such as having a genome (genetic code) and the ability to adapt to its environment. A virus, however, cannot acquire and store energy and is therefore not functional outside of its hosts. Viruses and retroviruses infect cells by attaching themselves to the host cell and either entering themselves or injecting their genetic material into the cell and then reproducing its genetic material within the host cell. The reproduced virus then is released to find and attack more host cells. The difference between a virus and retrovirus is a function of how each replicates its genetic material. A virus has a single strand of genetic material—either DNA or RNA. A retrovirus consists of a single strand of RNA. Once a retrovirus enters a cell, it collects nucleotides and assembles itself as a double strand of DNA that splices itself into the host's genetic material. Retroviruses were first identified by David Baltimore (b. 1938) and Howard Temin (b. 1934). They were awarded the Nobel Prize in Medicine for their discovery.

## What was the **first retrovirus** discovered?

Dr. Robert Gallo (b. 1937) discovered the first retrovirus, human T cell lymphoma virus (HTLV), in 1979. The second human retrovirus to be discovered was human immunodeficiency virus, HIV.

## Which **disease** is the **most common**?

The most common noncontagious disease is periodontal disease, such as gingivitis or inflammation of the gums. Few people in their lifetime can avoid the effects of tooth decay. The most common contagious disease in the world is coryza or the common cold. There are nearly 62 million cases of the common cold in the United States annually.

## Which **disease** is the **deadliest**?

The most deadly infectious disease was the pneumonic form of the plague, the so-called Black Death of 1347–1351, with a mortality rate of 100 percent. Today, the disease with the highest mortality (almost 100 percent) is rabies in humans when it prevents the victim from swallowing water. This disease is not to be confused with being bitten by a rabid animal. With immediate attention, the rabies virus can be prevented from invading the nervous system and the survival rate in this circumstance is 95 percent. AIDS (acquired immunodeficiency syndrome), first reported in 1981, is caused by HIV (the human immunodeficiency virus). In 1993, HIV infection was the most common cause of death among persons aged 25 to 44 years. In 1999 alone, 14,802 U.S. residents died from the AIDS/HIV infection, according to the National Center for Health Statistics. Although still a significant cause of death among persons aged 25 to 44, it is no longer the most common cause of death.

## What is the difference between human immunodeficiency virus **(HIV)** and **AIDS**?

The term AIDS applies to the most advanced stages of HIV infection. The Center for Disease Control (CDC) definition of AIDS includes all HIV-infected people who have fewer than 200 CD4+ T cells per cubic millimeter of blood. (Healthy adults usually have CD4+ T cell counts of 1,000 or more.) The definition also includes 26 clinical conditions (mostly opportunistic infections) that affect people with advanced HIV disease.

## **How many individuals** are infected with **HIV/AIDS**?

At the end of 2001, it was estimated that 40 million people worldwide were living with HIV/AIDS. Adults accounted for 37.2 million cases of HIV/AIDS (48 percent or 17.6 million are women) and children under 15 accounted for 2.7 million cases.

## Which **metropolitan areas** in the United States have the **greatest incidence of HIV/AIDS**?

| Metropolitan area | # of cumulative AIDS cases |
|---|---|
| New York City | 122,062 |
| Los Angeles | 42,796 |
| San Francisco | 28,212 |
| Miami | 24,838 |
| Washington, DC | 24,029 |
| Chicago | 22,217 |
| Philadelphia | 19,605 |
| Houston | 19,582 |
| Newark | 17,472 |
| Atlanta | 16,423 |

## What are the symptoms and signs of **AIDS**?

The early symptoms (AIDS-related complex, or ARC, symptoms) include night sweats, prolonged fevers, severe weight loss, persistent diarrhea, skin rash, persistent cough, and shortness of breath. The diagnosis changes to AIDS (acquired immunodeficiency syndrome) when the immune system is affected and the patient becomes susceptible

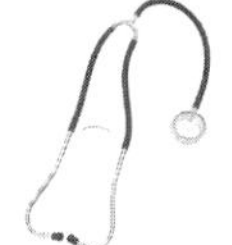



to opportunistic infections and unusual cancers, such as herpes viruses (herpes simplex, herpes zoster, cytomegalovirus infection), *Candida albicans* (fungus) infection, *Cryptosporidium enterocolitis* (protozoan intestinal infection), *Pneumocystis carinii* pneumonia (PCP, a common AIDS lung infection), toxoplasmosis (protozoan brain infection), progressive multifocal leukoencephalopathy (PML, a central nervous system disease causing gradual brain degeneration), *Mycobacterium avium intracellulare* infection (MAI, a common generalized bacterial infection), and Kaposi's sarcoma (a malignant skin cancer characterized by blue-red nodules on limbs and body, and internally in the gastrointestinal and respiratory tracts, where the tumors cause severe internal bleeding).

The signs of AIDS are generalized swollen glands, emaciation, blue or purple-brown spots on the body, especially on the legs and arms, prolonged pneumonia, and oral thrush.

## How is the term **zoonosis** defined?

A zoonosis is any infectious disease or parasitic disease of animals that can be transmitted to humans. Lyme disease and Rocky Mountain spotted fever, for example, are indirectly spread to humans from an animal through the bite of a tick. Common household pets also can directly transmit diseases to humans unless preventive measures are taken. Cat-scratch fever and toxoplasmosis may be contracted from cats. Wild animals and dogs can transmit rabies. However, most zoonosis diseases are relatively rare and can be treated once detected. Such sensible actions as regularly vaccinating pets and wearing long-sleeved shirts and pants when hiking can prevent the spread of most zoonoses.

## What is meant by **vectors** in medicine?

A vector is an animal that transmits a particular infectious disease. A vector picks up disease organisms from a source of infection, carries them within or on its body, and later deposits them where they infect a new host. Mosquitoes, fleas, lice, ticks, and flies are the most important vectors of disease to humans.

## Which species of mosquito causes **malaria** and **yellow fever** in humans?

The bite of the female mosquito of the genus *Anopheles* can contain the parasite of the genus *Plasmodium*, which causes malaria, a serious tropical infectious disease affecting 200 to 300 million people worldwide. More than one million African babies and children die from the disease annually. The *Aedes aegypti* mosquito transmits yellow fever, a serious infectious disease characterized by jaundice, giving the patient yellowish skin; 10 percent of the patients die.

## What was the contribution of **Dr. Gorgas** to the building of the **Panama Canal**?

Dr. William C. Gorgas (1854–1920) brought the endemic diseases of Panama under control by destroying mosquito breeding grounds, virtually eliminating yellow fever and malaria. His work was probably more essential to the completion of the canal than any engineering technique.

## What is **"mad cow disease"** and how does it affect humans?

Mad cow disease, bovine spongiform encephalopathy (BSE), is a cattle disease of the central nervous system. First identified in Britain in 1986, BSE is a transmissible spongiform encephalopathy (TSE), a disease characterized by the damage caused to the brain tissue. The tissue is pierced with small holes like a sponge. The disease is incurable, untreatable, and fatal. Researchers believe BSE is linked to Creutzfeldt-Jakob disease (CJD) in humans through the consumption of contaminated bovine products. CJD is a fatal illness marked by brain tissue deterioration and progressive degeneration of the central nervous system.

## How is **Lyme disease** carried?

The cause of Lyme disease is the spirochete *Borrelia burgdorferi* that is transmitted to humans by the small tick *Ixodes dammini* or other ticks in the Ixodidae family. The tick injects spirochete-laden saliva into the bloodstream or deposits fecal matter on the skin. This multisystemic disease usually begins in the summer with a skin lesion called erythema chronicum migrans (ECM), followed by more lesions, a malar rash, conjunctivitis, and urticaria. The lesions are eventually replaced by small red blotches. Other common symptoms in the first stage include fatigue, intermittent headache, fever, chills, and muscle aches.

In stage two, which can be weeks or months later, cardiac or neurologic abnormalities sometimes develop. In the last stage (weeks or years later) arthritis develops with marked swelling, especially in the large joints. If tetracycline, penicillin, or erythromycin is given in the early stages, the later complications can be minimized. High dosage of intravenously given penicillin can also be effective on the late stages.

## When were the first cases of **West Nile virus** reported in the United States?

The first cases of West Nile virus were identified in 1999 in the New York City area. West Nile virus is primarily a disease of birds found in Africa, West Asia, and the Middle East. It is transmitted to humans mainly via mosquito bites (mainly from the species *Culex pipiens*). The female mosquito catches the virus when it bites an infected bird and then passes it on when it later bites a human. In humans it causes encephalitis, an infection of the brain that can be lethal.

## Why is **Legionnaire's disease** known by that name?

Legionnaire's disease was first identified in 1976 when a sudden, virulent outbreak of pneumonia took place at a hotel in Philadelphia, Pennsylvania, where delegates to an American Legion convention were staying. The cause was eventually identified as a previously unknown bacterium that was given the name *Legionnella pneumophilia*. The bacterium probably was transmitted by an airborne route. It can spread through cooling tower or evaporation condensers in air-conditioning systems, and has been known to flourish in soil and excavation sites. Usually, the disease occurs in late summer or early fall, and its severity ranges from mild to life-threatening, with a mortality rate as high as 15 percent. Symptoms include diarrhea, anorexia, malaise, headache, generalized weakness, recurrent chills, and fever accompanied by cough, nausea, and chest pain. Antibiotics such as Erythroycin™ are administered along with other therapies (fluid replacement, oxygen, etc.) that treat the symptoms.

## Which name is now used as a synonym for **leprosy**?

Hansen's disease is the name of this chronic, systemic infection characterized by progressive lesions. Caused by a bacterium, *Mycobacterium leprae,* that is transmitted through airborne respiratory droplets, the disease is not highly contagious. Continuous close contact is needed for transmittal. Antimicrobial agents, such as sulfones (dapsone in particular), are used to treat the disease.

## Who was **Typhoid Mary**?

Mary Mallon (1855–1938), a cook who lived in New York City at the turn of the century, was identified as a chronic carrier of the typhoid bacilli. Immune to the disease herself, she was the cause of at least three deaths and 51 cases of typhoid fever. She was confined to an isolation center on North Brother Island, near the Bronx, from 1907 to 1910 and from 1914 to 1938. The New York City Health Department released her after the first confinement on the condition that she never accept employment that involved handling food. But when a later epidemic occurred at two places where she had worked as a cook, the authorities returned her to North Brother Island, where she remained until her death from a stroke in 1938.

## How many types of **herpes virus** are there?

There are five human herpes viruses:

*Herpes simplex type 1*—causes recurrent cold sores and infections of the lips, mouth, and face. The virus is contagious and spreads by direct contact with the lesions or fluid from the lesions. Cold sores are usually recurrent at the same sites and reoccur where there is an elevated temperature at the affected site, such as with a fever or prolonged sun exposure. Occasionally this virus may occur on the fingers with a rash of blisters. If the virus gets into the eye, it could cause conjunctivitis, or even a corneal ulcer. On rare occasions, it can spread to the brain to cause encephalitis.

*Herpes simplex type 2*—causes genital herpes and infections acquired by babies at birth. The virus is contagious and can be transmitted by sexual intercourse. The virus produces small blisters in the genital area that burst to leave small painful ulcers, which heal within 10 days to three weeks. Headache, fever, enlarged lymph nodes, and painful urination are the other symptoms.

*Varicella-zoster (Herpes zoster)*—causes chicken pox and shingles. Shingles can be caused by the dormant virus in certain sensory nerves that re-emerge with the decline of the immune system (because of age, certain diseases, and the use of immunosuppressants), excessive stress, or use of corticosteroid drugs. The painful rash of small blisters dry and crust over, eventually leaving small pitted scars. The rash tends to occur over the rib area or a strip on one side of the neck or lower body. Sometimes it involves the lower half of the face and can affect the eyes. Pain that can be severe and long-lasting affects about half of the sufferers and is caused by nerve damage.

*Epstein-Barr*—causes infectious mononucleosis (acute infection having high fever, sore throat and swollen lymph glands, especially in the neck, which occurs mainly during adolescence) and is associated with Burkitt's lymphoma (malignant tumors of the jaw or abdomen that occur mainly in African children and in tropical areas).

*Cytomegalovirus*—usually results in no symptoms but enlarges the cells it infects; it can cause birth defects when a pregnant mother infects her unborn child.

Three other human herpes viruses are also known: Human herpes virus 6 (HHV-6), commonly associated with roseola, and human herpes viruses 7 and 8 (HHV 7/8), whose disease association is not yet understood. Herpes gestationis is a rare skin-blister disorder occurring only in pregnancy and is not related to the herpes simplex virus.

## What is **necrotizing fasciitis**?

This very rare infection is caused by strains of Group A streptococcus, close relatives of the bacteria that cause strep throat and scarlet fever. When this chain-linked bacteria enters the body through a small cut, bite, or scratch, the infected skin becomes discolored, then blisters and cracks, exposing the destroyed tissue below. Within hours, an infected person can lose inches of flesh, or, in extreme cases, his or her life. If the infection is diagnosed early, antibiotics are generally enough to stop the infection. However, amputation of an affected limb may be the only means of curing an advanced case. The bacteria are often called "flesh-eating bacteria" in sensational media accounts.

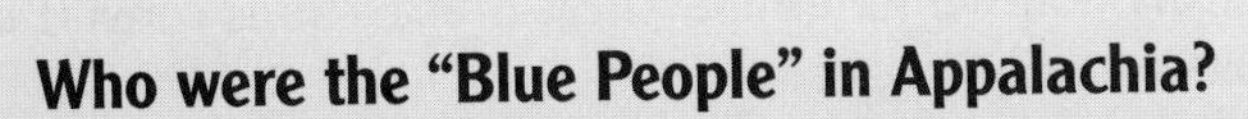

### Who were the "Blue People" in Appalachia?

The "Blue People" were descendants of Martin Fugate, a French immigrant to Kentucky. He had a recessive gene that limited or stopped the body's production of the enzyme diaphorase. Diaphorase breaks down methemoglobin into hemoglobin in red blood cells. When the enzyme is not present, a disproportionate amount of methemoglobin remains in the blood, giving the cells a bluish tint, rather than the normal pink associated with Caucasians. The condition is strictly one of pigment and does not deprive the person of oxygen.

Despite bluish color, there are no known health risks associated with the deficiency. Fugate's family suffered from the condition because of excessive inbreeding. When both spouses had the recessive gene, their children would be blue. As the family became mobile following World War II and moved out of their Kentucky valley, the inbreeding ceased. As of 1982 there were only two to three members of the family with the condition.

## How are **warts** caused?

A wart is a lump on the skin produced when one of the 30 types of papillomavirus invades skin cells and causes them to multiply rapidly. There are several different types of warts: common warts, usually on injury sites; flat warts on hands, accompanied by itching; digitate warts having fingerlike projections; filiform warts on eyelids, armpits, and necks; plantar warts on the soles of the feet; and genital warts, pink cauliflower-like areas that, if occurring in a woman's cervix, could predispose her to cervical cancer. Each is produced by a specific virus, and most are usually symptomless. Wart viruses are spread by touch or by contact with the skin shed from a wart.

## What is **lactose intolerance**?

Lactose, the principal sugar in cow's milk and found only in dairy products, requires the enzyme lactase for human digestion. Lactose intolerance occurs when the lining of the walls of a person's small intestine does not produce normal amounts of this enzyme. Lactose intolerance causes abdominal cramps, bloating, diarrhea, and excessive gas when more than a certain amount of milk is ingested. Most people are less able to tolerate lactose as they grow older.

A person having lactose intolerance need not eliminate dairy products totally from the diet. Decreasing the consumption of milk products, drinking milk only during meals, and getting calcium from cheese, yogurt, and other dairy products having lower lactose values are options. Another alternative is to buy a commercial lactose preparation that can be mixed into milk. These preparations convert lactose into simple sugars that can be easily digested.

## What is the difference between **Type I** and **Type II diabetes**?

Type I is insulin-dependent diabetes mellitus (IDDM) and Type II is non-insulin-dependent diabetes mellitus (NIDDM). In Type I diabetes there is an absolute deficiency of insulin. It accounts for approximately 10 percent of all cases of diabetes and has a greater prevalence is children. In Type II diabetes, insulin secretion may be normal, but the target cells for insulin are less responsive than normal. The incidence of Type II diabetes increases greatly after age 40 and is normally associated with obesity and lack of exercise as well as genetic predisposition. The symptoms of Type II diabetes are usually less severe than Type I, but long-term complications are similar in both types.

## What causes a **stomach ulcer**?

For decades, doctors thought that genetics or anxiety or even spicy foods caused stomach ulcers. Scientists now believe that stress and spicy foods only worsen the pain of an ulcer. The gastric ulcer itself is caused by a bacterium called *Helicobacter pylori*. Researcher Barry Marshall (1951–) of Australia observed that many ulcer patients had these bacteria present in their systems. So in 1984, to decide whether there was a link, he consumed a large amount of the bacteria. He developed ulcers 10 days later. Ulcers are now treated with antibiotics. In 1994, the *Helicobacter pylori* bacteria was classified as a carcinogen by the National Institutes of Health; ulcer sufferers would therefore do well to consult a doctor instead of ignoring or masking the pain with antacids.

## Which **medical condition** is associated with **Abraham Lincoln's** lanky appearance?

Abraham Lincoln (1809–1865) probably had Marfan's syndrome (Arachnodactyly), which abnormally lengthens the bones. It is a rare inherited degenerative disease of the connective tissue. Besides the excessively long bones, there are chest deformities, scoliosis (spine curvature), an arm span that can exceed height, eye problems (especially myopia or nearsightedness), abnormal heart sounds, and sparse subcutaneous fat. In 1991, researchers identified the gene behind the disease.

## What is **carpal tunnel syndrome**?

Carpal tunnel syndrome occurs when a branch of the median nerve in the forearm is compressed at the wrist as it passes through the tunnel formed by the wrist bones (or carpals), and a ligament that lies just under the skin. The syndrome occurs most often in middle age and more so in women than men. The symptoms are intermittent at

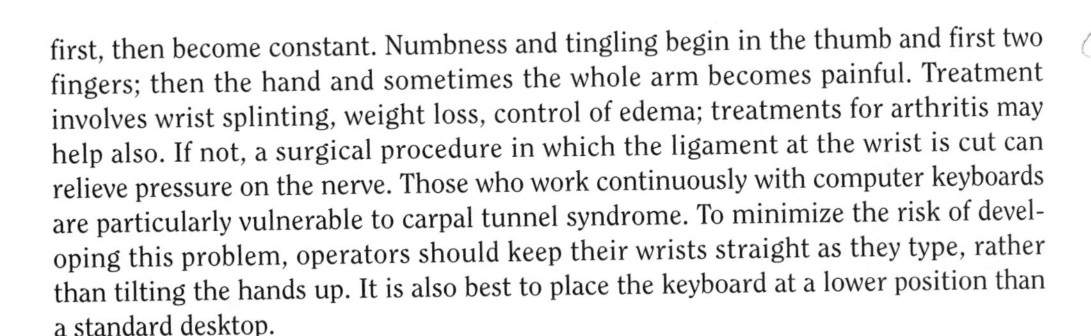

first, then become constant. Numbness and tingling begin in the thumb and first two fingers; then the hand and sometimes the whole arm becomes painful. Treatment involves wrist splinting, weight loss, control of edema; treatments for arthritis may help also. If not, a surgical procedure in which the ligament at the wrist is cut can relieve pressure on the nerve. Those who work continuously with computer keyboards are particularly vulnerable to carpal tunnel syndrome. To minimize the risk of developing this problem, operators should keep their wrists straight as they type, rather than tilting the hands up. It is also best to place the keyboard at a lower position than a standard desktop.

## What is the medical term for **tennis elbow**?

The technical term for tennis elbow is epicondylitis. A result of repeated strain on the forearm, it is a painful inflammation of the muscle and surrounding tissues of the elbow. A number of behaviors can cause its onset, ranging from playing tennis or golf to carrying a heavy load with the arm extended.

## What is **Lou Gehrig's disease**?

Sometimes called Lou Gehrig's disease, amyotrophic lateral sclerosis (ALS) is a motor neuron disease of middle or late life. It results from a progressive degeneration of nerve cells controlling voluntary motor functions that ends in death three to 10 years after onset. There is no cure for it. At the beginning of the disease, the patient notices weakness in the hands and arms, with involuntary muscle quivering and possible muscle cramping or stiffness. Eventually all four extremities become involved. As nerve degeneration progresses, disability occurs and physical independence declines until the patient, while mentally and intellectually aware, can no longer swallow or move.

## What is **narcolepsy**?

Although most people think of a narcoleptic as a person who falls asleep at inappropriate times, victims of narcolepsy also share other symptoms, including excessive daytime sleepiness, hallucinations, and cataplexy (a sudden loss of muscle strength following an emotional event). Persons with narcolepsy experience an uncontrollable desire to sleep, sometimes many times in one day. Episodes may last from a few minutes to several hours.

## How does **jet lag** affect one's body?

The physiological and mental stress encountered by airplane travelers when crossing four or more time zones is commonly called jet lag. Patterns of hunger, sleep, and elimination, along with alertness, memory, and normal judgment, may all be affected.

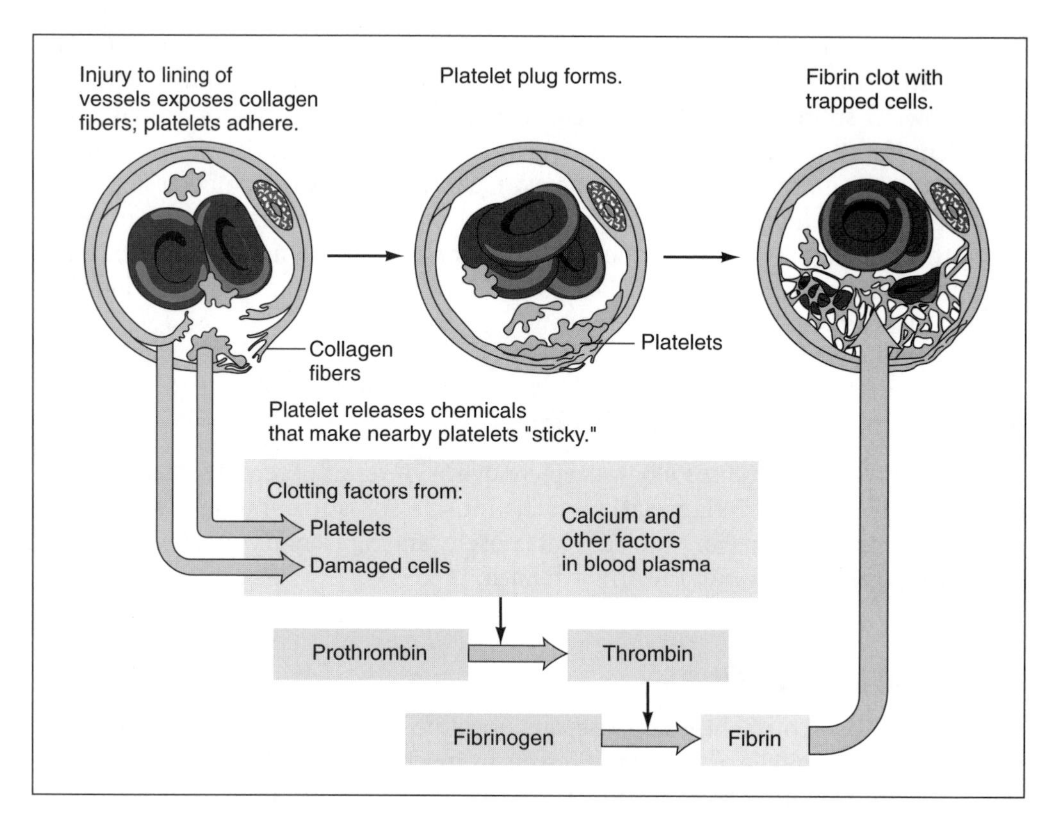


**The blood clotting process. Persons with Christmas disease lack a crucial clotting factor that keeps this process from proceeding properly.**

More than 100 biological functions that fluctuate during the 24-hour cycle (circadian rhythm) can become desynchronized. Most people's bodies adjust at a rate of about one hour per day. Thus after four time zone changes, the body will require about four days to return to its usual rhythms. Flying eastward is often more difficult than flying westward, which adds hours to the day.

## What is **factor VIII**?

Factor VIII is one of the enzymes involved in the clotting of blood. Hemophiliacs lack this enzyme and are at high risk of bleeding to death unless they receive supplemental doses of factor VIII. The lack of factor VIII is due to a defective gene, which shows a sex-linked inherited pattern (it affects about one in 10 thousand males). Females can carry the gene. Hemorrhage into joints and muscles usually makes up the majority of bleeding episodes.

## What is the **"Christmas factor"**?

In the clotting of blood, factor IX, or the Christmas factor, is a coagulation factor present in normal plasma, but deficient in the blood of persons with hemophilia B

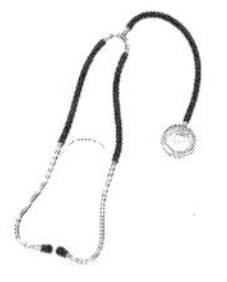



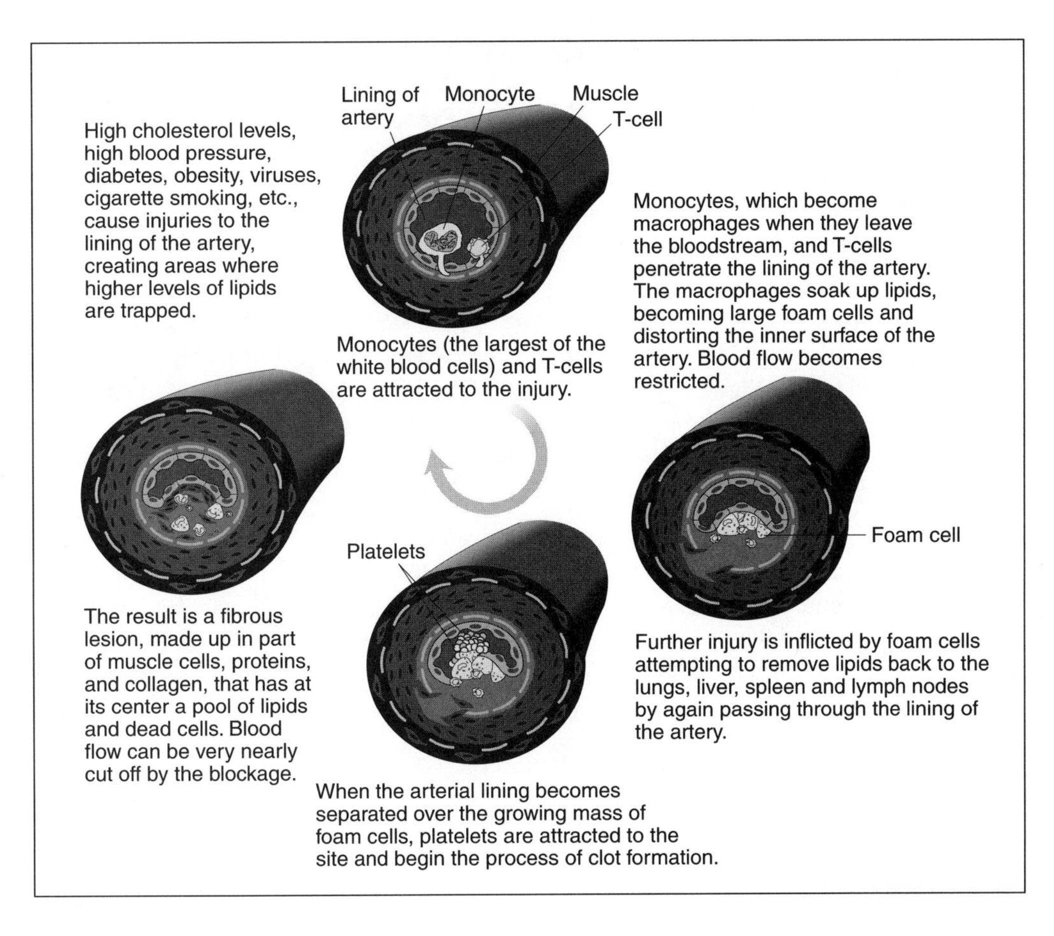


**The progression of arteriosclerosis, which may in turn lead to a heart attack.**

or Christmas disease. It was named after a man named Christmas who, in 1952, was the first patient in whom this genetic disease was shown to be distinct from hemophilia (another genetic blood-clotting disease in which the blood does not have factor VIII).

## What is the medical term for a **heart attack**?

Myocardial infarction is the term used for a heart attack in which part of the heart muscle's cells die as a result of reduced blood flow through one of the main arteries (many times due to arteriosclerosis). The outlook for the patient is dependant on the size and location of the blockage and extent of damage, but 33 percent of patients die within 20 days after the attack; it is a leading cause of death in the United States. Also, almost half of sudden deaths due to myocardial infarction occur before hospitalization. However, the possibility of recovery improves if vigorous treatment begins immediately.

## What is the difference between **heatstroke** and **heat exhaustion**?

| Heatstroke | Heat Exhaustion |
| --- | --- |
| **Caused by:** Body cannot regulate its own temperature due to intensive sweating under conditions of high heat and humidity. Advanced age can be a factor. | **Caused by:** Person doesn't get enough liquid and salt in very hot, humid weather. |
| **Symptoms:** Weakness, vertigo, nausea, headache, heat cramps, mild heat exhaustion, excessive sweating. Sweating stops just before heatstroke. Temperature rises rapidly, as high as 106°F (41°C); blood pressure is elevated. Skin is flushed at first, then turns ashen or purplish. Delirium or coma is common. | **Symptoms:** Excessive sweating, weakness, vertigo, and sometimes heat cramps. Skin is cold and pale, clammy with sweat; pulse is thready and blood pressure is low. Body temperature is normal or sub-normal. Vomiting may occur. Unconsciousness is rare. |
| **First Aid:** Heatstroke is a medical emergency. Call for medical assistance. Move person to a cool, indoor place. Loosen or remove clothing. Primary objective is to reduce body temperature, preferably by iced bath or sponging down with cool water until pulse lowers to below 110 per minute and body temperature is below 103°F (39.4°C). Caution is necessary. | **First Aid:** Lay person in cool place. Loosen clothing. Give water to drink with 1 tsp. salt to each quart of water. Fluid intake usually brings about full recovery. Seek medical assistance if severe. |

## How are the forms of **cancer** classified?

The over 150 different types of cancer are classified into four major groups:

1. Carcinomas—Nine in 10 cancers are carcinomas, which involve the skin and skin-like membranes of the internal organs.
2. Sarcomas—Involve the bones, muscles, cartilage, fat, and linings of the lungs, abdomen, heart, central nervous system, and blood vessels.
3. Leukemias—Develop in blood, bone marrow, and the spleen.
4. Lymphomas—Involve the lymphatic system.

## What adverse effects may a person who is allergic to **sulfites** experience?

Sulfites are chemical agents used to prevent discoloration in dried fruits and freshly cut vegetables. They are also used by winemakers to inhibit bacterial growth and fermentation. If a person is allergic to sulfites, he or she can develop breathing difficulties within minutes of consuming food or drink containing sulfites. Reactions to sulfites can include acute asthma attacks, loss of consciousness, and anaphylactic shock.

## What are **HeLa cells**?

HeLa cells, used in many biomedical experiments, were obtained from a cervical carcinoma in a woman named Henriette Lacks. Epithelial tissue obtained by biopsy became the first continuously cultured human malignant cells.

## What are **dust mites**?

Dust mites are microscopic arachnids (members of the spider family) commonly found in house dust. Dust mite allergen is probably one of the most important causes of asthma (breathlessness and wheezing caused by the narrowing of small airways of the lungs) in North America, as well as the major cause of common allergies (exaggerated reactions of the immune system to exposure of offending agents).

Thorough, regular cleaning of the home, including the following measures, will help control dust mites:

1. Clean all major appliances such as furnaces and air conditioners, and change filters as recommended by the manufacturer.
2. Launder bedding every seven to 10 days in hot water. Use synthetic or foam rubber mattress pads and pillows. Cover mattresses with dust-proof covers. Clean or replace pillows regularly.
3. Keep moist surfaces in kitchen and bathroom clean and free of mold.
4. Vacuum and dust often. A high-efficiency particulate air filter (HEPA) vacuum is especially effective.

## If the sap of the **poison ivy** plant touches the skin, will a rash develop?

Studies show that 85 percent of the population will develop an allergic reaction if exposed to poison ivy, but this sensitivity varies with each individual according to circumstance, age, genetics, and previous exposure. The poison comes mainly from the leaves whose allergens touch the skin. A red rash with itching and burning will develop, and skin blisters will usually develop within six hours to several days after exposure. Washing the affected area thoroughly with mild soap within five minutes of exposure can be effective; sponging with alcohol and applying a soothing and drying lotion, such as calamine lotion, is the prescribed treatment for light cases. If the affected area is large, fever, headache, and generalized body weakness may develop. For severe reactions, a physician should be consulted to prescribe a corticosteroid drug. Clothing that touched the plants should also be washed.

## What is **dyslexia** and what causes it?

Dyslexia covers a wide range of language difficulties. In general, a person with dyslexia cannot grasp the meaning of sequences of letters, words, or symbols or the concept of direction. The condition can affect people of otherwise normal intelligence. Dyslexic children may reverse letter and word order, make bizarre spelling errors, and may not be able to name colors or write from dictation. It may be caused by minor visual defects, emotional disturbance, or failure to train the brain. New evidence shows that a neurological disorder may be the underlying cause. Approximately 90 percent of dyslexics are male.

The term *dyslexia* (of Greek origin) was first suggested by Professor Rudolph Berlin of Stuttgart, Germany, in 1887. The earliest references to the condition date as far back as 30 C.E. when Valerius Maximus and Pliny described a man who lost his ability to read after being struck on the head by a stone.

**A person with dyslexia may reverse the order of words or letters, as illustrated here.**

## What is **anorexia**?

Anorexia simply means a loss of appetite. Anorexia nervosa is a psychological disturbance that is characterized by an intense fear of being fat. It usually affects teenage or young adult women. This persistent "fat image," however untrue in reality, leads the patient to self-imposed starvation and emaciation (extreme thinness) to the point where one-third of the body weight is lost. There are many theories on the causes of this disease, which is difficult to treat and can be fatal. Between five and 10 percent of patients hospitalized for anorexia nervosa later die from starvation or suicide. Symptoms include a 25 percent or greater weight loss (for no organic reason) coupled with a morbid dread of being fat, an obsession with food, an avoidance of eating, compulsive exercising and restlessness, binge eating followed by induced vomiting, and/or use of laxatives or diuretics.

## Why do deep-sea divers get **the bends**?

Bends is a painful condition in the limbs and abdomen. It is caused by the formation and enlargement of bubbles of nitrogen in blood and tissues as a result of rapid reduction of pressure. This condition can develop when a diver ascends too rapidly after being exposed to increased pressure. Severe pain will develop in the muscles and joints of the arms and legs. More severe symptoms include vertigo, nausea, vomiting, choking, shock, and sometimes death. Bends is also known as decompression sickness, caisson disease, tunnel disease, and diver's paralysis.



## What is **progeria**?

Progeria is premature old age. There are two distinct forms of the condition, both of which are extremely rare. In Hutchinson-Gilford syndrome, aging starts around the age four, and by 10 or 12, the affected child has all the external features of old age, including gray hair, baldness, and loss of fat, resulting in thin limbs and sagging skin on the trunk and face. There are also internal degenerative changes, such as atherosclerosis (fatty deposits lining the artery walls). Death usually occurs at puberty. Werner's syndrome, or adult progeria, starts in early adult life and follows the same rapid progression as the juvenile form. The cause of progeria is unknown.

## How is an **iatrogenic illness** defined?

It is an adverse mental or physical condition caused by the effects of treatment by a physician or surgeon. The term implies that it could have been avoided by judicious care on the part of the physician.

## How many people in America are estimated to have **Alzheimer's disease**?

Researchers have concluded that three percent of men and women ages 65 to 74 have Alzheimer's disease, including nearly 50 percent of people over 85. The National Institute on Aging estimated that four million Americans have Alzheimer's disease. Some experts, however, do not agree with these figures.

Alzheimer's disease is a progressive condition in which nerve cells in the brain degenerate and the brain substance shrinks. Although the cause is unknown, some theorize that it is toxic poisoning by a metal such as aluminum. Others believe it to be of genetic origin. There are three stages; in the first, the person becomes forgetful; in the second, the patient experiences severe memory loss and disorientation, lack of concentration, loss of ability to calculate and find the right word to use (dysphasia), anxiety, and sudden personality changes. In the third stage the patient is severely disoriented and confused, suffers from hallucinations and delusions, and has severe memory loss; the nervous system also declines, with regression into infantile behavior, violence, etc., often requiring hospital care.

### What is pica?

Pica refers to the craving for unnatural or non-nutritious substances. It is named after the magpie (*Pica pica*), which has a reputation for sticking its beak into all kinds of things to satisfy its hunger or curiosity. It can happen in both sexes, all races, and in all parts of the world, but is especially noted in pregnant women.

Alzheimer's disease is emerging as one of the most costly medical and social problems in the United States. In 1999, 44,536 deaths (1.9 percent of all deaths) were attributed to Alzheimer's disease.

## What is **pelvic inflammatory disease**?

Pelvic inflammatory disease (PID) is a term used for a group of infections in the female organs, including inflammations of the Fallopian tubes, cervix, uterus, and ovaries. It is the most common cause of female infertility today. PID is most often found in sexually active women under the age of 25 and almost always results from gonorrhea or chlamydia, but women who use IUDs are also at risk. A variety of organisms have been shown to cause PID, including *Neisseria gonorrhoeae* and such common bacteria as staphylococci, chlamydiae, and coliforms (*Pseudomonas* and *Escherichia coli*). Signs and symptoms of PID vary with the site of the infection, but usually include profuse, purulent vaginal discharge, low-grade fever and malaise (especially with *N. gonorrhoeae* infections), and lower abdominal pain. PID is treated with antibiotics, and early diagnosis and treatment will prevent damage to the reproductive system. Severe, untreated PID can result in the development of a pelvic abscess that requires drainage. A ruptured pelvic abscess is a potentially fatal complication, and a patient who develops this complication may require a total hysterectomy.

## What is **Chinese restaurant syndrome**?

Monosodium glutamate (MSG), a commonly used flavor-enhancer, is thought to cause flushing, headache, and numbness about the mouth in susceptible people. Because many Chinese restaurants use MSG in food preparation, these symptoms may appear after a susceptible person eats Chinese food.

## What is the chemical composition of **kidney stones**?

About 80 percent are calcium, mainly calcium oxalate and/or phosphate; five percent are uric acid; two percent are amino acid cystine; the remainder are magnesium ammonium phosphate. About 20 percent of these stones are infective stones, linked to chronic urinary infections, and contain a combination of calcium, magnesium, and ammonium phosphate produced from the alkalinity of the urine and bacteria action on urea (a substance in urine).

## How are **burns** classified?

| Type | Causes and effects |
|---|---|
| First-degree | Sunburn; steam. Reddening and peeling. Affects epidermis (top layer of skin). Heals within a week. |
| Second-degree | Scalding; holding hot metal. Deeper burns causing blisters. Affects dermis (deep skin layer). Heals in two to three weeks. |
| Third-degree | Fire. A full layer of skin is destroyed. Requires a doctor's care and grafting. |
| Circumferential | Any burns (often electrical) that completely encircle a limb or body region (such as the chest), which can impair circulation or respiration; requires a doctor's care; fasciotomy (repair of connective tissues) is sometimes required. |
| Chemical | Acid, alkali. Can be neutralized with water (for up to half an hour). Doctor's evaluation recommended. |
| Electrical | Destruction of muscles, nerves, circulatory system, etc., below the skin. Doctor's evaluation and ECG monitoring required. |

If more than 10 percent of body surface is affected in second-and third-degree burns, shock can develop when large quantities of fluid (and its protein) are lost. When skin is burned, it cannot protect the body from airborne bacteria.

## What is the **phobia** of number 13 called?

Fear of the number 13 is known as tridecaphobia, tredecaphobia, or triskaidekaphobia. Persons may fear any situation involving this number, including a house number, the floor of a building, or the 13th day of the month. Many buildings omit labeling the 13th floor as such for this reason. A phobia can develop for a wide range of objects, situations, or organisms. The list below demonstrates the variety:

| Phobia subject | Phobia term |
|---|---|
| Animals | Zoophobia |
| Beards | Pogonophobia |
| Books | Bibliophobia |
| Churches | Ecclesiaphobia |
| Dreams | Oneirophobia |
| Flowers | Anthophobia |
| Food | Sitophobia |

| Phobia subject | Phobia term |
|---|---|
| Graves | Taphophobia |
| Infection | Nosemaphobia |
| Lakes | Limnophobia |
| Leaves | Phyllophobia |
| Lightning | Astraphobia |
| Men | Androphobia |
| Money | Chrometophobia |
| Music | Musicophobia |
| Sex | Genophobia |
| Shadows | Sciophobia |
| Spiders | Arachnophobia |
| Sun | Heliophobia |
| Touch | Haptophobia |
| Trees | Dendrophobia |
| Walking | Basiphobia |
| Water | Hydrophobia |
| Women | Gynophobia |
| Work | Ergophobia |
| Writing | Graphophobia |

### Was **Napoleon** poisoned?

The most common opinion today is that Napoleon Bonaparte (1769–1821), Emperor of France from 1804 to 1815, died of a cancerous, perforated stomach. A significant minority of doctors and historians have made other claims ranging from various diseases to benign neglect to outright homicide. A Swedish toxicologist, Sten Forshufvud, advanced the theory that Napoleon died from arsenic administered by an agent of the French Royalists who was planted in Napoleon's household during his final exile on the island of St. Helena.

## HEALTH CARE

### Which **symbol** is used to represent **medicine**?

The staff of Aesculapius has represented medicine since 800 B.C.E. It is a single serpent wound around a staff. The caduceus, the twin-serpent magic wand of the god Hermes or Mercury, came into use after 1800 and is commonly used today. The serpent has

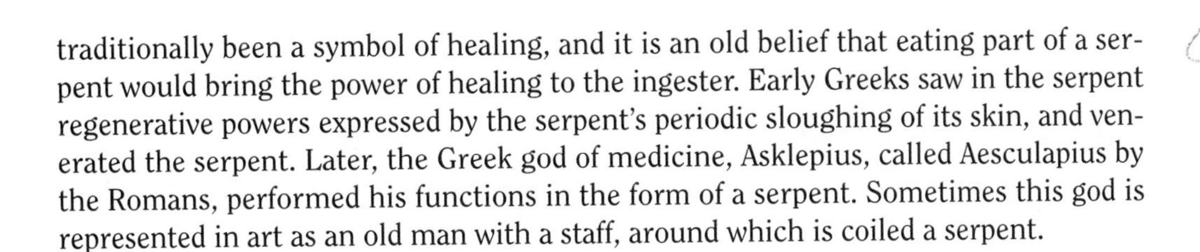

traditionally been a symbol of healing, and it is an old belief that eating part of a serpent would bring the power of healing to the ingester. Early Greeks saw in the serpent regenerative powers expressed by the serpent's periodic sloughing of its skin, and venerated the serpent. Later, the Greek god of medicine, Asklepius, called Aesculapius by the Romans, performed his functions in the form of a serpent. Sometimes this god is represented in art as an old man with a staff, around which is coiled a serpent.

## What is the **Hippocratic Oath**?

The Hippocratic Oath is an oath demanded of physicians who are entering practice and can be traced back to the Greek physician and teacher Hippocrates (ca. 460–ca. 377 B.C.E.). The oath reads as follows:

"I swear by Apollo the physician, by Aesculapius, Hygeia, and Panacea, and I take to witness all the gods, all the goddesses, to keep according to my ability and my judgment the following Oath:

"To consider dear to me as my parents him who taught me this art; to live in common with him and if necessary to share my goods with him; to look upon his children as my own brothers, to teach them this art if they so desire without fee or written promise; to impart to my sons and the sons of the master who taught me and the disciples who have enrolled themselves and have agreed to rules of the profession, but to these alone, the precepts and the instruction. I will prescribe regiment for the good of my patients according to my ability and my judgment and never to harm anyone. To please no one will I prescribe a deadly drug, nor give advice which may cause his death. Nor will I give a woman a pessary to procure abortion. But I will preserve the purity of my life and my art. I will not cut for stone, even for patients in whom the disease is manifest; I will leave this operation to be performed by practitioners (specialists in this art). In every house where I come I will enter only for the good of my patients, keeping myself far from all intentional ill-doing and all seduction, and especially from the pleasures of love with women or with men, be they free or slaves. All that may come to my knowledge in the exercise of my profession or outside of my profession or in daily commerce with men, which ought not to be spread abroad, I will keep secret and will never reveal. If I keep this oath faithfully, may I enjoy life and practice my art, respected by all men and in all times; but if I swerve from it or violate it, may the reverse be my lot."

The oath varies slightly in wording among different sources.

## Who was the founder of **modern medicine**?

Thomas Sydenham (1624–1689), who was also called the English Hippocrates, reintroduced the Hippocratic method of accurate observation at the bedside, and recording of observations, to build up a general clinical description of individual diseases. He

### Who is generally regarded as the father of medicine?

Hippocrates (ca. 460–ca. 377 B.C.E.), a Greek physician, holds this honor. Greek medicine, previous to Hippocrates, was a mixture of religion, mysticism, and necromancy. Hippocrates established the rational system of medicine as a science, separating it from religion and philosophy. Diseases had natural causes and natural laws: they were not the "wrath of the gods." Hippocrates believed that the four elements (earth, air, fire, and water) were represented in the body by four body fluids (blood, phlegm, black bile, and yellow bile) or "humors." When they existed in harmony within the body, the body was in good health. The duty of the physician was to help nature to restore the body's harmony. Diet, exercise, and moderation in all things kept the body well, and psychological healing (good attitude toward recovery), bed rest, and quiet were part of his therapy. Hippocrates was the first to recognize that different diseases had different symptoms; and he described them in such detail that the descriptions generally would hold today. His descriptions not only included diagnosis but prognosis.

is also considered one of the founders of epidemiology and was among the first to describe scarlet fever and Sydenham's chorea.

## What was the **first medical college** in the United States?

The College of Philadelphia Department of Medicine, now the University of Pennsylvania School of Medicine, was established on May 3, 1765. The first commencement was held June 21, 1768, when medical diplomas were presented to the ten members of the graduating class.

## Who was the **first woman physician** in the United States?

Elizabeth Blackwell (1821–1910) received her degree in 1849 from Geneva Medical College in New York. After overcoming many obstacles, she set up a small practice that expanded into the New York Infirmary for Women and Children, which featured an all-female staff.

## When and where did the **first blood bank** open?

Several sites claim the distinction. Some sources list the first blood bank as opening in 1940 in New York City under the supervision of Dr. Richard C. Drew (1904–1950). Others list an earlier date of 1938 in Moscow at the Sklifosovsky Institute (Moscow's central emergency service hospital) founded by Professor Sergei Yudin. The term blood bank was coined by Bernard Fantus (1974–1940), who set up a centralized storage depot for blood in 1937 at the Cook County Hospital in Chicago, Illinois.

## What is the difference between **homeopathy**, **osteopathic medicine**, **naturopathy**, and **chiropractic** medicine?

Developed by a German physician, Christian F. S. Hahnemann (1755–1843), the therapy of homeopathy treats patients with small doses of 2,000 substances. Based on the principle that "like cures like," the medicine used is one that produces the same symptoms in a healthy person that the disease is producing in the sick person.

Chiropractic medicine is based on the belief that disease results from the lack of normal nerve function. Relying on physical manipulation and adjustment of the spine for therapy, rather than on drugs or surgery, this therapy, used by the ancient Egyptians, Chinese, and Hindus, was rediscovered in 1895 by American osteopath Daniel David Palmer (1845–1913).

Developed in the United States by Dr. Andrew Taylor Still (1828–1917), osteopathic medicine recognizes the role of the musculoskeletal system in healthy function of the human body. The physician is fully licensed and uses manipulation techniques as well as traditional diagnostic and therapeutic procedures. Osteopathy is practiced as part of standard Western medicine.

Naturopathy is based on the principle that disease is due to the accumulation of waste products and toxins in the body. Practitioners believe that health is maintained by avoiding anything artificial or unnatural in the diet or in the environment.

## **How many physicians** are there in the United States?

In 2000 there were 618,223 male physicians and 195,537 female physicians in the United States.

## How many **nursing homes** are in the United States?

In 1999 there were 18,000 nursing homes in the United States. Approximately 1.6 million individuals, or five percent, of the elderly people in the United States are residents of nursing homes.

## What is the difference between a **skilled nursing facility**, an **intermediate care facility**, and an **assisted living facility**?

A skilled nursing facility is required to provide twenty-four-hour nursing supervision by registered or licensed nurses. Residents of skilled nursing facilities need assistance with many aspects of day-to-day living and daily care. Skilled nursing facilities provide medical, nursing, dietary, pharmacy, and activity services.

An intermediate care facility is required to provide eight hours of nursing supervision per day. Residents of an intermediate care facility are usually ambulatory and require less assistance with day-to-day living.

An assisted living facility is a home-like environment that provides individualized health and personal care assistance. Assisted living facilities offer personal care assistance, health care monitoring, limited health care services, and/or the dispensing of medications. Assisted living facilities promote independence by meeting residents' supportive needs.

## What is the difference between the degrees **doctor of dental surgery** (DDS) and **doctor of medical dentistry** (DMD)?

The title depends entirely on the school's preference in terminology. The degrees are equivalent.

## How many people **visit a dentist** regularly?

Only about half the population visits a dentist as often as once a year. Children, ages two to 17, visit the dentist more frequently than adults.

| Age | Percent who visited a dentist in the last year |
|---|---|
| 2–17 | 72.6% |
| 18–64 | 64.6% |
| 65 and over | 55.0% |

## What is the difference between an **ophthalmologist**, **optometrist**, and **optician**?

An ophthalmologist is a physician who specializes in care of the eyes. Ophthalmologists conduct examinations to determine the quality of vision and the need for corrective glasses or contact lenses. They also check for the presence of any disorders, such as glaucoma or cataracts. Ophthalmologists may perform surgery or prescribe glasses, contact lenses, or medication, as necessary.

An optometrist is a specialist trained to examine the eyes and to prescribe, supply, and adjust glasses or contact lenses. Because they are not physicians, optometrists may not prescribe drugs or perform surgery. An optometrist refers patients requiring these types of treatment to an ophthalmologist.

An optician is a person who fits, supplies, and adjusts glasses or contact lenses. Because their training is limited, opticians may not examine or test eyes or prescribe glasses or drugs.

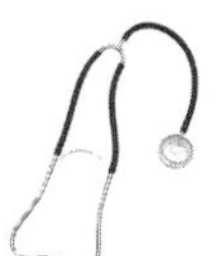



# DIAGNOSTIC EQUIPMENT, TESTS, ETC.

## What is the meaning of the **medical abbreviation NYD**?

Not yet diagnosed.

## How is **blood pressure** measured?

A sphygmomanometer is the device used to measure blood pressure. It was invented in 1881 by an Austrian named Von Bash. It consists of a cuff with an inflatable bladder that is wrapped around the upper arm, a rubber bulb to inflate the bladder, and a device that indicates the pressure of blood. Measuring arterial tension (blood pressure) of a person's circulation is achieved when the cuff is applied to the arm over the artery and pumped to a pressure that occludes or blocks it. This gives the systolic measure, or the maximum pressure of the blood, which occurs during contraction of the ventricles of the heart. Air is then released from the cuff until the blood is first heard passing through the opening artery (called Korotkoff sounds). This gives diastolic pressure, or the minimum value of blood pressure that occurs during the relaxation of the arterial-filling phase of the heart muscle.

## What is the meaning of the numbers in a **blood pressure reading**?

When blood is forced into the aorta, it exerts a pressure against the walls; this is referred to as blood pressure. The upper number, the systolic, measures the pressure during the period of ventricular contraction. The lower number, the diastolic, measures the pressure when blood is entering the relaxed chambers of the heart. While these numbers can vary due to age, sex, weight, and other factors, the normal blood pressure is around 110/60 to 140/90 millimeters of mercury.

## Is there a name for the **heart-monitoring machine** that people sometimes wear for a day or two while carrying on their normal activities?

A portable version of the electrocardiograph (ECG) designed by J. J. Holter is called a Holter monitor. Electrodes attached to the chest are linked to a small box containing a recording device. The device records the activity of the heart.

## Who invented the **pacemaker**?

Paul Zoll (1911–1999) invented an electric stimulator device to deliver electrical impulses to the heart externally. In 1958, biomedical engineer Wilson Greatbatch (1919–), in cooperation with doctors William M. Chardack and Andres A. Gage, invented the first internal pacemaker. It was a small, flat, plastic disk powered by a battery. It was implanted into the body and connected by wires sewn directly onto the heart. The wires emitted rhythmic electric impulses to trigger the heart's action. Pacemaker batteries now last from six to ten years.

## What are the normal **test ranges** for total **cholesterol**, low-density lipoproteins (**LDL**), and high-density lipoproteins (**HDL**)?

The National Cholesterol Education Program has drawn up these guidelines:

| | **Desirable** | **Borderline** | **High Risk** |
|---|---|---|---|
| Total Blood Cholesterol | Less than 200 mg/dl | 200–239 mg/dl | 240 mg/dl or more |
| LDL | Less than 130 mg/dl | 130–159 mg/dl | 160 mg/dl or more |
| HDL | 45–65 mg/dl | 35–45 mg/dl | Below 35 mg/dl |

mg/dl = milligrams per deciliter

## What is **nuclear magnetic resonance imaging**?

Magnetic resonance imaging (MRI), sometimes called nuclear magnetic resonance imaging (NMR), is a non-invasive, non-ionizing diagnostic technique. It is useful in detecting small tumors, blocked blood vessels, or damaged vertebral disks. Because it does not involve the use of radiation, it can often be used where X-rays are dangerous. Large magnets beam energy through the body causing hydrogen atoms in the body to resonate. This produces energy in the form of tiny electrical signals. A computer detects these signals, which vary in different parts of the body and according to whether an organ is healthy or not. The variation enables a picture to be produced on a screen and interpreted by a medical specialist.

What distinguishes MRI from computerized X-ray scanners is that most X-ray studies cannot distinguish between a living body and a cadaver, while MRI "sees" the difference between life and death in great detail. More specifically, it can discriminate between healthy and diseased tissues with more sensitivity than conventional radiographic instruments like X-rays or CAT scans. CAT (computerized axial tomography) scanners have been around since 1973 and are actually glorified X-ray machines. They offer three-dimensional viewing but are limited because the object imaged must remain still.

The concept of using MRI to detect tumors in patients was proposed by Raymond Damadian (1936–) in a 1972 patent application. The fundamental MRI imaging concept used in all present-day MRI instruments was proposed by Paul Lauterbar in an article published in *Nature* in 1973. The main advantages of MRI are that it not only gives superior images of soft tissues (like organs), but can also measure dynamic physiological changes in a non-invasive manner (without penetrating the body in any way). A disadvantage of MRI is that it cannot be used for every patient. For example, patients with implants, pacemakers, or cerebral aneurysm clips made of metal cannot be exam-

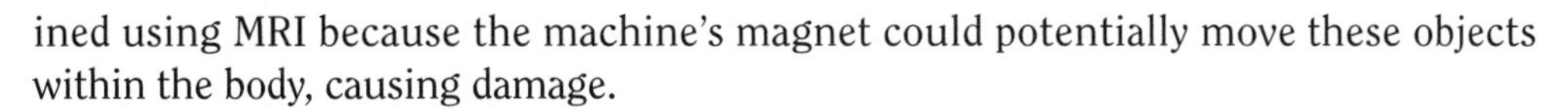

ined using MRI because the machine's magnet could potentially move these objects within the body, causing damage.

Ultrasound is another type of 3-D computerized imaging. Using brief pulses of ultrahigh frequency acoustic waves (lasting 0.01 second), it can produce a sonar map of the imaged object.

### When were **hearing aids** first invented?

Specially designed hearing aids were described as early as 1588 by Giovanni Battista Porta (1535–1615) in his book *Natural Magick*. These hearing aids were made out of wood in the shape of the ears of animals with a sharp sense of hearing. During the 1700s, speaking tubes and ear trumpets were developed. Bone conduction devices to transmit sound vibrations from outside to the bones of the ear, first suggested in 1550 by Gerolamo Cardano (1501–1576), were further developed during the 1800s. The first battery-powered hearing aid in the United States was made by the Dictagraph Company in 1898. Miller Reese Hutchison (1876–1944) filed the patent for the first electric hearing aid in 1901. Refinements were made to the hearing aids throughout the twentieth century. Miniaturization and microchips made it possible for hearing aids to become so small they now fit invisibly inside the ear.

### What is the instrument a doctor uses to check **reflexes**?

A plessor or plexor or percussor is a small hammer, usually with a soft rubber head, used to tap the part directly. Also called a reflex hammer or a percussion hammer, it is used by a doctor to elicit reflexes by tapping on tendons. In the most common test, the patient sits on a surface high enough to allow his legs to dangle freely, and the physician lightly taps the patellar tendon, just below the kneecap. This stimulus briefly stretches the quadriceps muscle on top of the thigh. The stretch causes the muscle to contract, which makes the leg kick forward. The time interval between the tendon tap and the start of the leg extension is about 50 microseconds. That interval is too short for the involvement of the brain and is totally reflexive. This test indicates the status of an individual's reflex control of movement.

## DRUGS, MEDICINES, ETC.

### What is **pharmacognosy**?

It is the science of natural drugs and their physical, botanical, and chemical properties. Natural products derived from plant, vegetable, animal, and mineral sources have been a part of medical practice for thousands of years. Today about 25 percent of all prescriptions dispensed in pharmacies contain active ingredients that are extracted from higher plants, and many more are found in over-the-counter products.

## Why do physicians use the symbol **Rx** when they write their prescriptions?

There are several explanations for the symbol Rx. One common explanation is that it comes from the Latin word *recipi* or *recipere*, which means "take" and is abbreviated as Rx. The symbol can also be traced to the sign of Jupiter, which was found on ancient prescriptions to appeal to the Roman god Jupiter. In ancient medical books the crossed R has been found wherever the letter R occurred.

Others believe that the origin of the Rx symbol can be found in the Egyptian myth about two brothers, Seth and Horus, who ruled over Upper and Lower Egypt as gods. Horus's eye was injured in a battle with Seth and healed by another god, Thoth. The eye of Horus consisted of the sun and the moon, and it was the moon eye that was damaged. This explained the phases of the moon—the waning of the moon was the eye being damaged and the waxing, the healing. The eye of Horus became a powerful symbol of healing in the eyes of the Egyptians. In Egyptian art, the eye of Horus strongly resembles the modern Rx of the physician.

## What is the meaning of the **abbreviations** often used by a doctor when writing a **prescription**?

| Latin phrase | Shortened form | Meaning |
|---|---|---|
| quaque hora | qh | every hour |
| quaque die | qd | every day |
| bis in die | bid | twice a day |
| ter in die | tid | three times a day |
| quarter in die | qid | four times a day |
| pro re nata | prn | as needed |
| ante cibum | a.c. | before meals |
| post cibum | p.c. | after meals |
| per os | p.o. | by mouth |
| nihil per os | n.p.o. | nothing by mouth |
| signetur | sig | let it be labeled |
| statim | stat | immediately |
| ad libitum | ad lib | at pleasure |
| hora somni | h.s. | at bedtime |
| cum | c | with |
| sine | s | without |
| guttae | gtt | drops |
| semis | ss | a half |
| et | et | and |

## What are the common **medication measures**?

Approximate equivalents of apothecary measures are given below.

| Volume | Apothecary volume equivalent |
|---|---|
| 1 minim | 0.06 milliliters or 0.02 fluid drams or 0.002 fluid ounces |
| 1½ minims | 0.1 milliliters |
| 15 minims | 1 milliliter |
| 480 minims | 1 fluid ounce |
| 1 dram | 3.7 milliliters or 60 minims |
| 1 t (teaspoon) | 60 drops |
| 3 t (teaspoons) | ½ ounce |
| 1 T (tablespoon) | ½ ounce |
| 2 T (tablespoons) | 1 ounce |
| 1 C (cup) | 8 ounces or 30 milliliters |

**Apothecary weights**

| Weight | Equivalent |
|---|---|
| 1 grain | 60 milligrams or ½ dram |
| 60 grains | 1 dram or 3.75 grams |
| 8 drams | 1 ounce or 30 grams |

## How long can a **prescription drug** be kept?

Generally, a prescription drug should not be more than one year old. Some over-the-counter medications have an expiration date on their box or container. A cream should not be used if it has separated into its components. Although some general guidelines are given below, "When in doubt, throw it out."

| Remedy | Maximum shelf life (years) |
|---|---|
| Cold tablets | 1 to 2 |
| Laxatives | 2 to 3 |
| Minerals | 6 or more |
| Nonprescription painkiller tablets | 1 to 4 |
| Prescription antibiotics | 2 to 3 |
| Prescription antihypertension tablets | 2 to 4 |
| Travel sickness tablets | 2 |
| Vitamins (protected from heat, light, and moisture) | 6 or more |

## Which drugs are **the most frequently prescribed** at medical provider office visits?

In 2000, the drugs most frequently prescribed were:

| Brand Name | Therapy |
|---|---|
| Claritin | Antihistamines |
| Lipitor | Hyperlipidemia |
| Synthroid | Thyroid hormone replacement agents |
| Premarin | Symptoms of menopause |
| Amoxicillin | Penicillins |
| Tylenol | Nonnarcotic analgesics |
| Lasix | Diuretics |
| Celebrex | Nonsteroidal anti-inflammatory drugs |
| Glucophage | Blood glucose regulators |
| Albuterol sulfate | Antiasthmatics/bronchodilators |

## What is a **double-blind study**?

In drug tests a double-blind study is a study in which neither the investigator administering the drug nor the subject taking it knows if the patient is receiving the experimental drug or a neutral substitute called a placebo. In this manner, bias, either on the part of the administrator or the subject, can be eliminated from the study.

## What is meant by the term **orphan drugs**?

Orphan drugs are intended to treat diseases that affect fewer than 200,000 Americans. With little chance of making money, a drug company is not likely to undertake the necessary research and expense of finding drugs that might treat these diseases. Also, if the drug is a naturally occurring substance, it cannot be patented in the United States, and companies are reluctant to invest money in such a medication when it cannot be protected against exploitation by competing drug companies. Encouragingly, the Orphan Drug Act of 1983 offers a number of incentives to drug companies to encourage development of these drugs. The act has provided hope for millions of people with rare and otherwise untreatable conditions.

## How long does it take for a **new drug** to be **developed**?

Pharmaceutical research is a long process. Only about one out of every 10,000 investigated substance reaches the market in the form of a new drug. Drug development research can take eight to fifteen years. The cost to develop a new drug can amount to about $300-500 million for each new drug.

## How many of the **medications** used today are **derived from plants**?

Of the more than 250,000 known plant species, less than one percent have been thoroughly tested for medical applications. Yet out of this tiny portion have come 25 percent of our prescription medicines. The United States National Cancer Institute has

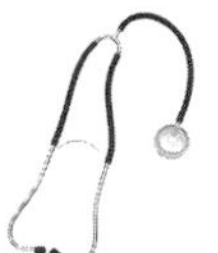



identified 3,000 plants from which anti-cancer drugs are or can be made. This includes ginseng (*Panax quinquefolius*), Asian mayapple (*Podophyllum hexandrum*), western yew (*Taxus brevifolia*), and rosy periwinkle. Seventy percent of these 3,000 come from rain-forests, which also are a source of countless other drugs for diseases and infections. Rainforest plants are rich in so-called secondary metabolites, particularly alkaloids, which biochemists believe the plants produce to protect them from disease and insect attack. However, with the current rate of rainforest destruction, raw materials for future medicines are certainly being lost. Also, as tribal groups disappear, their knowledge of the properties and uses of these plants species will be lost.

## What are some **medications** that have been obtained from **rain forest** plants, animals, and microorganisms?

| Drug | Medicinal Use | Source |
|---|---|---|
| Allantoin | Wound healer | Blowfly larva |
| Atropine | High blood pressure | Bee venom |
| Cocaine | Analgesic | Coca bush |
| Cortisone | Anti-inflammatory | Mexican yam |
| Cytarabine | Leukemia | Sponge |
| Diosgenin | Birth control | Mexican yam |
| Erythromycin | Antibiotic | Bacterium |
| Morphine | Analgesic | Opium poppy |
| Quinine | Malaria | Chincona tree bark |
| Reserpine | Hypertension | Rauwolfia plant |
| Tetracycline | Antibiotic | Bacterium |
| Vinblastine | Hodgkin's disease and leukemia | Rosy periwinkle plant |

## From what plant is **taxol** extracted?

Taxol is produced from the bark of the western or Pacific yew (*Taxus brevifolia*). It has been shown to inhibit the growth of HeLa cells (human cancer cells) and is a promising new treatment for several kinds of cancer. Originally it was a scarce drug, but in 1994 two groups of researchers announced its synthesis. The synthesis is a formidable challenge, and better procedures and modifications remain to be developed. Since taxol is developed now from needles instead of tree bark, the natural source is more available, but the synthetic version will be needed to devise modified or "designer" taxols whose cancer-fighting ability may prove more effective.

## What is **herbal medicine**?

Herbal medicine treats disease and promotes health with plant material. For centuries herbal medicines were the primary methods to administer medicinally active compounds.

## What are some of the most **common herbal remedies**?

| **Herb** | **Botanical name** | **Common use** |
|---|---|---|
| Aloe | *Aloe vera* | Skin, gastritis |
| Black cohosh | *Cimicifuga racemosa* | Menstrual, menopause |
| Dong quai | *Angelica sinensis* | Menstrual, menopause |
| Echinacea | *Echinacea angustifolia* | Colds, immunity |
| Ephedra (ma huang) | *Ephedra sinica* | Asthma, energy, weight loss |
| Evening primrose oil | *Oenothera biennis* | Eczema, psoriasis, premenstrual syndrome, breast pain |
| Feverfew | *Tanacetum parthenium* | Migraine |
| Garlic | *Allium sativum* | Cholesterol, hypertension |
| Ginger | *Zingiber officinale* | Nausea, arthritis |
| Ginkgo biloba | *Ginkgo biloba* | Cerebrovascular insufficiency, memory |
| Ginseng | *Panax ginseng, Panax quinquifolius, Panax pseudoginseng, Eleutherococcus senticosus* | Energy, immunity, mentation, libido |
| Goldenseal | *Hydrastis candensis* | Immunity, colds |
| Hawthorne | *Crateaegus laeviagata* | Cardiac function |
| Kava kava | *Piper methysticum* | Anxiety |
| Milk thistle | *Silybum marianum* | Liver disease |
| Peppermint | *Mentha piperita* | Dyspepsia, irritable bowel syndrome |
| Saw palmetto | *Serona repens* | Prostate problems |
| St. John's Wort | *Hypericum perforatum* | Depression, anxiety, insomnia |
| Tea tree oil | *Melaleuca alternifolia* | Skin infections |
| Valerian | *Valeriana officinalis* | Anxiety, insomnia |

## What is the **Dietary Supplement Health and Education Act**?

The Dietary Supplement Health and Education Act was passed by Congress in 1994. The law bars the Food and Drug Administration (FDA) from regulating dietary supplements. It allows manufacturers to make claims about the benefits of an herbal medicine. Further, it requires the FDA to prove an herbal is harmful before the sale of the herbal can be restricted.

## What is a **dietary supplement**?

According to the Dietary Supplement Health and Education Act (DSHEA), a dietary supplement is a product taken by mouth that contains a "dietary ingredient" intended to supplement the diet. The dietary ingredient may include: vitamins, minerals, herbs or other botanicals, amino acids, a dietary substance to supplement the diet by increasing the total dietary intake or a concentrate, metabolite, constituent, or extract. They may be in any form—tablets, capsules, liquids, gelcaps, or powders. The DSHEA places dietary supplements in a special category under the general umbrella of "foods," not drugs, and requires that every supplement be labeled a dietary supplement.

## How long is a **tetanus shot** effective?

In the United States, infants are vaccinated against tetanus at two months, four months, and six months. The vaccination is part of the DPT shot, which protects against diphtheria, tetanus, and pertussis (whooping cough). In order to insure immunity, booster doses are given every 10 years or at the time of a major injury if it occurs more than five years after a dose.

## Who discovered **penicillin**?

British bacteriologist Sir Alexander Fleming (1881–1955) discovered the bacteria-killing property of penicillin in 1928. Fleming noticed that no bacteria grew around bits of the *Penicillium notatum* fungus that accidentally fell into a bacterial culture in his laboratory. However, although penicillin was clinically used, it was not until 1941 that Dr. Howard Florey (1898–1968) purified and tested it. The first large-scale plant to produce penicillin was constructed under the direction of Dr. Ernest Chain (1906–1979). By 1945, penicillin was commercially available. In that year, Chain, Florey, and Fleming received a joint Nobel Prize for their work on penicillin.

Sir Alexander Fleming, the British bacteriologist who discovered the medicinal properties of penicillin in 1928.

Today penicillin is still used successfully in the treatment of many bacterial diseases, including pneumonia, strep throat, scarlet fever, gonorrhea, and impetigo. Its discovery also led to the development of other antibiotics that are useful in destroying a broad spectrum of pathogenic bacteria.

## Who discovered the antibiotic **streptomycin**?

The Russian-born microbiologist Selman A. Waksman (1888–1973) coined the term "antibiotic" and subsequently discovered streptomycin in 1943. In 1944 Merck and Company agreed to produce it to be used against tuberculosis and tuberculosis meningitis.

Streptomycin ultimately proved to have some human toxicity and was supplanted by other antibiotics, but its discovery changed the course of modern medicine. In addition to its use in treating tuberculosis, it was also used to treat bacterial meningitis, endocarditis, pulmonary and urinary tract infections, leprosy, typhoid fever, bacillary dysentery, cholera, and bubonic plague. Streptomycin saved countless lives, and its development led scientists to search the microbial world for other antibiotics and medicines.

## Who developed the **poliomyelitis vaccine** in America?

Immunologist Jonas E. Salk (1914–1995) developed the first vaccine (made from a killed virus) against poliomyelitis. In 1952 he prepared and tested the vaccine, and in 1954 massive field tests were successfully undertaken. Two years later immunologist Albert Sabin (1906–1993) developed an oral vaccine made from inactivated live viruses of three polio strains. Because of its easy administration and the fact that it requires fewer booster inoculations, Sabin's vaccine has replaced the Salk vaccine as the one to prevent polio. However, Salk remains known as the man who defeated polio.

## Who was the first to use **chemotherapy**?

Chemotherapy is the use of chemical substances to treat diseases, specifically malignant diseases. The drug must interfere with the growth of bacterial, parasitic, or tumor cells, without significantly affecting host cells. Especially effective in types of cancer such as leukemia and lymphoma, chemotherapy was introduced in medicine by the German physician Paul Ehrlich (1854–1915).

## What are **monoclonal antibodies**?

Monoclonal antibodies are artificially produced antibodies designed to neutralize a specific foreign protein (antigen). Cloned cells (genetically identical) are stimulated to produce antibodies to the target antigen. Most monoclonal antibody work so far has used cloned cells from mice infected with cancer. In some cases they are used to destroy cancer cells directly; in others they carry other drugs to combat the cancer cells.

## How are **anabolic steroids** harmful to those who use them?

Anabolic (protein-building) steroids are drugs that mimic the effects of testosterone and other male sex hormones. They can build muscle tissue, strengthen bone, and speed muscle recovery following exercise or injury. They are sometimes prescribed to treat osteoporosis in postmenopausal women and some types of anemia. Some athletes use anabolic steroids to build muscle strength and bulk and to allow a more rigorous training schedule. Weightlifters, field event athletes, and body-builders are most likely to use anabolic steroids. The drugs are banned from most organized competitions because of the dangers they pose to health and to prevent an unfair advantage.

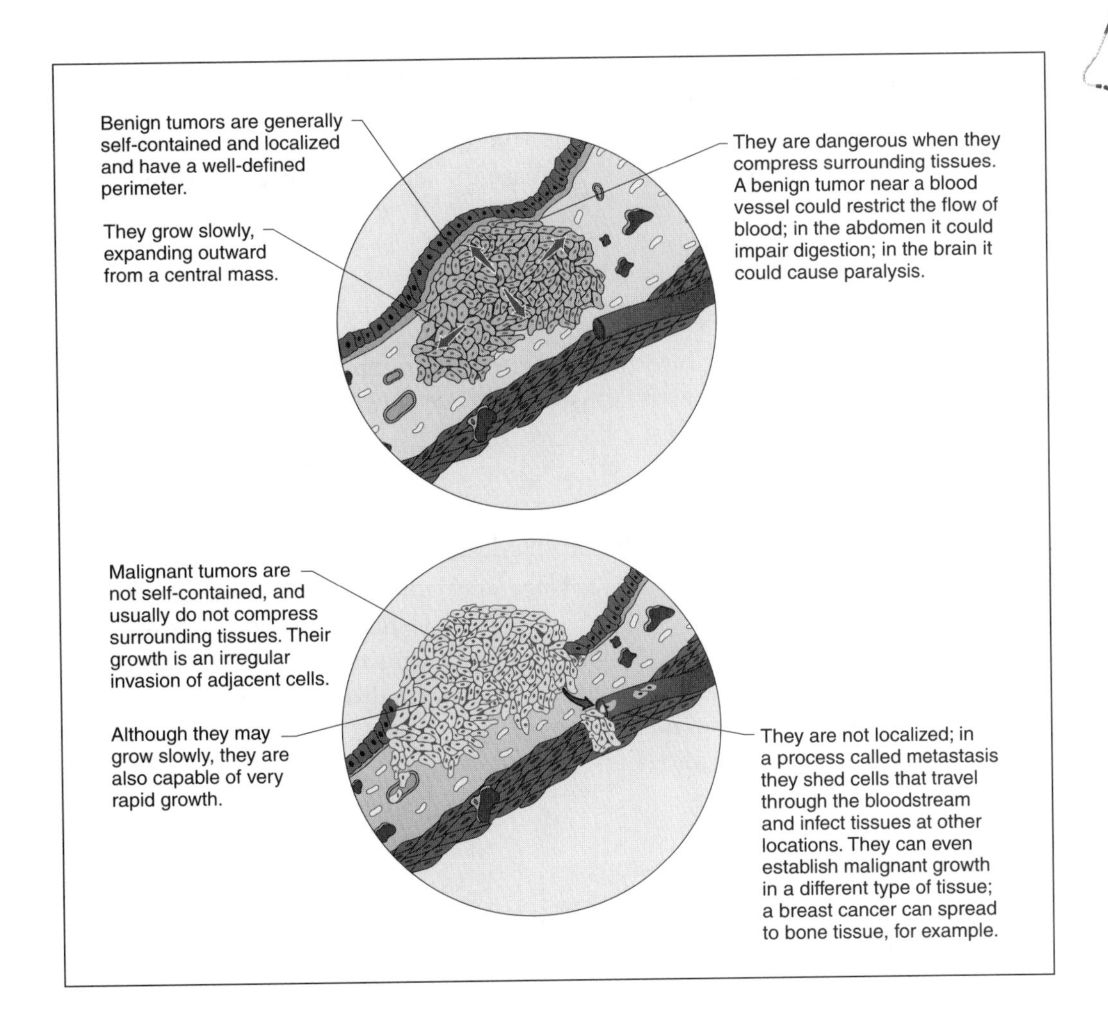


**A comparison of benign and malignant tumor characteristics.**

Adverse effects include hypertension, acne, edema, and damage to liver, heart, and adrenal glands. Psychiatric symptoms can include hallucinations, paranoid delusions, and manic episodes. In men, anabolic steroids can cause infertility, impotence, and premature balding. Women can develop masculine characteristics, such as excessive hair growth, male-pattern balding, disruption of menstruation, and deepening of the voice. Children and adolescents can develop problems in growing bones, leading to short stature.

## How is **patient-controlled analgesia** (PCA) administered?

This is a drug delivery system that dispenses a preset intravenous (IV) dose of a narcotic analgesic for reduction of pain whenever a patient pushes a switch on an electric cord. The device consists of a computerized pump with a chamber containing a syringe holding up to 60 milliliters of a drug. The patient administers a dose of narcotic when the need for pain relief arises. A lockout interval device automatically inactivates the system if the patient tries to increase the amount of narcotic within a preset time period.

### What is a **Brompton's cocktail**?

Named for the Brompton Chest Hospital in England, Brompton's cocktail is a mixture of cocaine, morphine, and antiemetics used to reduce pain and induce euphoria, particularly in terminally ill cancer patients.

### What were the **birth defects** caused by the drug **Thalidomide**?

In the early 1960s, Thalidomide was marketed as a sedative and anti-nausea drug. It was found to cause birth defects in babies whose mothers had taken the drug for morning sickness. Some babies were born without arms or legs. Others were born blind or deaf or with heart defects or intestinal abnormalities. Although some were mentally retarded, most were of normal intelligence. This tragedy led to much stricter laws regulating the sale and testing of new drugs.

### How does **RU-486** cause an abortion?

A pill containing RU-486 (mifepristone) deprives a fertilized embryo of a compatible uterine environment, terminating a pregnancy within 49 days of fertilization. It was approved for use in the United States on September 28, 2000.

### What are **designer drugs**, such as **China White**?

Designer drugs are synthesized chemicals that resemble such available narcotics as fentanyl and meperidine. China white (3-methyl-fentanyl) is one of these drugs and is an analogue of fentanyl. It is 3,000 times more potent than morphine. Even small amounts can be fatal, and it has been responsible for more than 100 overdose deaths in California.

### What is a **controlled substance**?

The Comprehensive Drug Abuse Prevention and Control Act of 1971 was designed to control the distribution and use of all depressant and stimulant drugs and other drugs of abuse or potential abuse. Centrally acting drugs are divided into five classes called Schedule I through V.

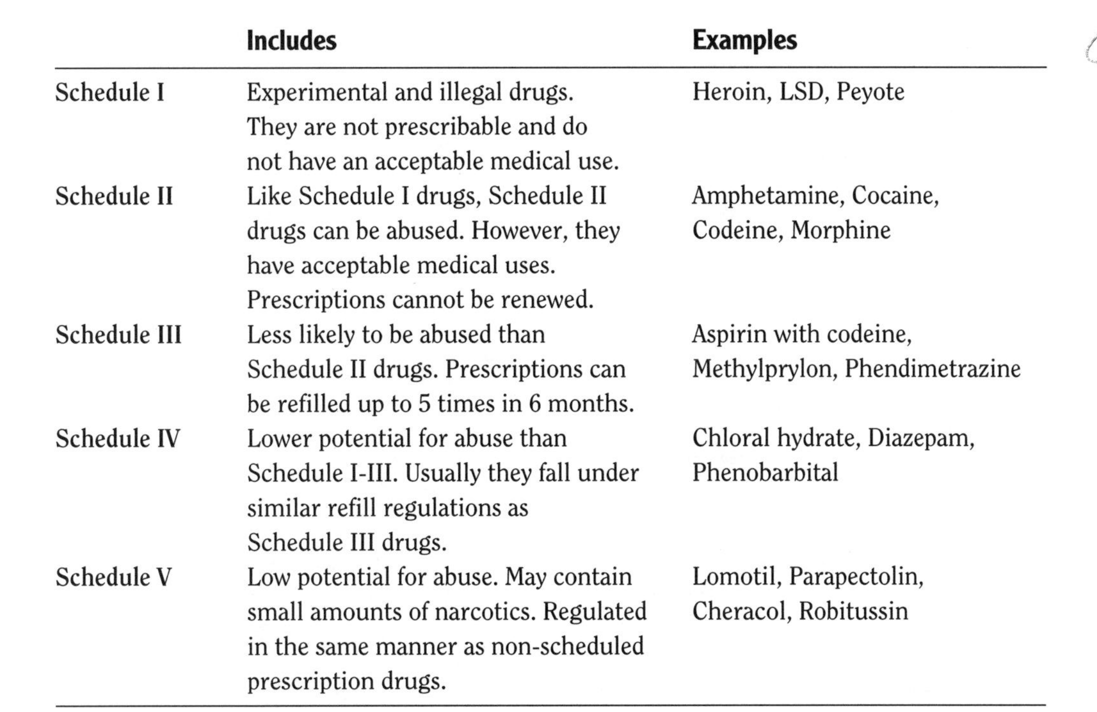

| | Includes | Examples |
|---|---|---|
| Schedule I | Experimental and illegal drugs. They are not prescribable and do not have an acceptable medical use. | Heroin, LSD, Peyote |
| Schedule II | Like Schedule I drugs, Schedule II drugs can be abused. However, they have acceptable medical uses. Prescriptions cannot be renewed. | Amphetamine, Cocaine, Codeine, Morphine |
| Schedule III | Less likely to be abused than Schedule II drugs. Prescriptions can be refilled up to 5 times in 6 months. | Aspirin with codeine, Methylprylon, Phendimetrazine |
| Schedule IV | Lower potential for abuse than Schedule I-III. Usually they fall under similar refill regulations as Schedule III drugs. | Chloral hydrate, Diazepam, Phenobarbital |
| Schedule V | Low potential for abuse. May contain small amounts of narcotics. Regulated in the same manner as non-scheduled prescription drugs. | Lomotil, Parapectolin, Cheracol, Robitussin |



## What is the difference between **cocaine, freebase**, and **crack**?

Cocaine comes in several different forms. Coca paste is widely used in South America and is usually added to and smoked in tobacco or marijuana cigarettes. Cocaine hydrochloride, the form most common in the United States, is a white powder. It is inhaled through the nose ("snorted") from tiny spoons, rolled-up dollar bills, or straws. It can also be mixed with water and injected.

Freebase is a purified form of cocaine that is smoked in a water pipe. It is prepared by applying ether, baking powder, or other solvent to cocaine powder and heating the mixture.

Crack is freebase that comes in ready-to-smoke chunks or "rocks." Crack is usually smoked in a water pipe. Some users apply it to tobacco or marijuana cigarettes.

## Which tests detect **illegal drug use**?

Blood samples are rarely if ever used in routine screening of individuals for drug abuse. Blood samples provide little valuable information unless the drug was consumed a short time before the blood was drawn. Urine, on the other hand, is easily collected and can be analyzed and transported cheaply. Samples of skin, saliva, and hair can also be tested but with greater difficulty.

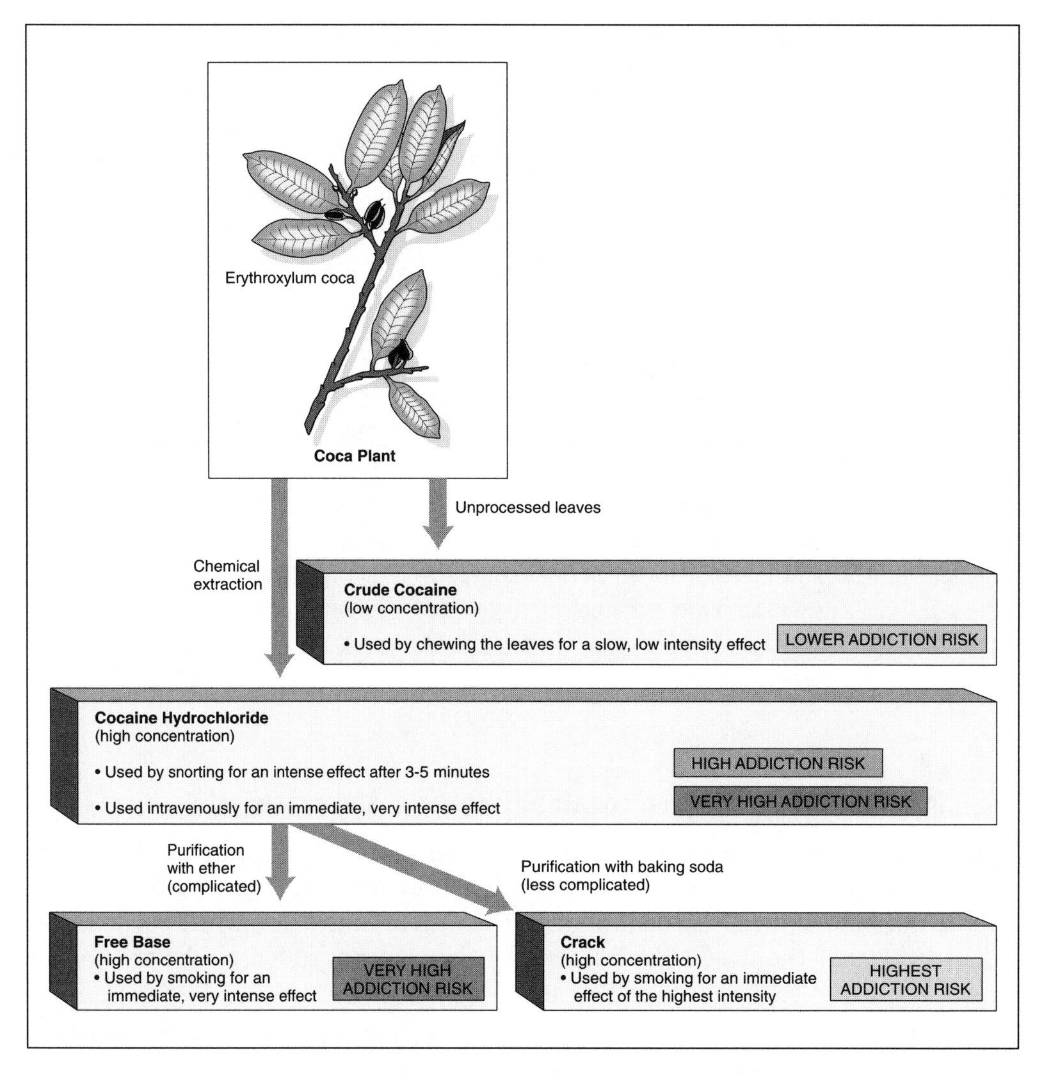


**Various forms of cocaine and the addiction risks associated with them.**

Urine tests detect cocaine used by a person within the past 24 to 36 hours. Hair analysis can detect cocaine used more than a year ago. Other drugs that can be detected by urine testing are:

PCP (phencylidine)—up to 7–8 days

Barbiturates—up to 72 hours

Morphine—1–2 days

Heroin—2–3 days or 4–5 days if larger doses are taken

Methaqualone (Quaalude)—up to 10 days

Cannabis (marijuana)—up to 5 days for infrequent users; up to 10 days for heavy smokers



### Where is **marijuana grown legally** in the United States?

The federal marijuana farm at the University of Mississippi was established in 1968. This farm supplies most of the marijuana (*Cannabis sativa*) used for official medical research in America.

### How long do **chemicals from marijuana** stay in the body?

When marijuana is smoked, tetrahydrocannabinol (THC), its active ingredient, is absorbed primarily in the fat tissues. The body transforms the THC into metabolites, which can be detected by urine tests for up to a week. Tests involving radioactively labeled THC have traced the metabolites for up to a month. The retention of labeled THC in humans is about 40 percent at three days and 30 percent at one week.

## SURGERY AND OTHER NON-DRUG TREATMENTS

### How does **minimally invasive surgery** differ from traditional, major, open surgery?

Traditional, major, open surgery requires a major incision in the body, often several inches long, allowing surgeons to physically place their hands inside the body to work. In minimally invasive surgery the incision is very small and surgeons do not place their hands inside the body. Using a laparoscope, a narrow wand containing a video camera, surgeons are able to insert tools into tiny surgical openings to remove diseased tissue. Laparoscopy was introduced in the 1970s for gynecological treatment and gall bladder removal. At least half of all surgeries are now minimally invasive (laparoscopic or arthroscopic) with a wide range of applications—gall bladder removal, appendix removal, hernia repair, gynecological, colon removal, partial lung removal, spleen removal, and surgery for chronic heartburn or reflux disease.

The major advantages of minimally invasive surgery is that it is less traumatic to the patient. There is less scarring, and recovery time is much quicker. Hospital stays are shorter.

## Who was the first African American surgeon to perform heart surgery?

Dr. Daniel Hale Williams (1858–1931) was a pioneer in open heart surgery. In 1893, he was able to save a knifing victim by opening up the patient's chest with the help of a surgical team, exposing the beating heart. He sewed the knife wound a fraction of an inch from the heart without the aid of X-rays, blood transfusions, or anesthetics.

## How many **surgical procedures** are performed each year?

In 1999, 41.3 million surgical procedures were performed on an inpatient basis. Obstetrical procedures were the most common with 6,174,000 procedures, followed closely by operations on the cardiovascular system with 6,133,000 procedures.

## Does **catgut** really come from cats?

Catgut, an absorbable sterile strand, is obtained from collagen derived from healthy mammals. It was originally prepared from the submucosal layer of the intestines of sheep. It is used as a surgical ligature.

## Who received the **first heart transplant**?

On December 3, 1967, in Capetown, South Africa, Dr. Christiaan Barnard (1922–2001) and a team of 30 associates performed the first heart transplant. In a five-hour operation the heart of Denise Ann Darval, age 25, an auto accident victim, was transplanted into the body of Louis Washansky, a 55-year-old wholesale grocer. Washansky lived for 18 days before dying from pneumonia.

The first heart transplant performed in the United States was on a 2½-week-old baby boy at Maimonides Hospital, Brooklyn, New York, on December 6, 1967, by Dr. Adrian Kantrowitz (b. 1918). The baby boy lived 6½ hours. The first adult to receive a heart transplant in the United States was Mike Kasperak, age 54, at the Stanford Medical Center in Palo Alto, California, on January 6, 1968. Dr. Norman Shumway (b. 1923) performed the operation. Mr. Kasperak lived 14 days.

From December 1967 until March 31, 1993, 14,085 heart transplants have been performed. Interestingly, almost no transplants were done in the 1970s because of the problem of rejection of the new heart by the recipient's immune system. In 1969, Jean-François Borel discovered the anti-rejection drug cyclosporine, but it was not widely used until 1983 when the FDA granted approval. Today heart transplantation is an established medical procedure.

## Is is possible to use **animal organs** as replacements for human organ transplantation?

In 1984, a 12-day-old infant, Baby Fae, received a baboon's heart. She lived for 20 days before her body rejected the transplanted heart. Since the supply of human organs is far less than the need for human organs, researchers continue to search for alternative sources of organs. In 1999, the Food and Drug Administration (FDA) prohibited the used of nonhuman primate organs for human transplantation unless researchers could assess the risk of disease. A major concern had been that viruses and other diseases that were harmful to the animals would be deadly to humans. Researchers continue to search for alternate sources of organs, including pigs.

## When was the **first artificial heart** used?

On December 2, 1982, Dr. Barney B. Clark (1921–1983), a 61-year-old retired dentist, became the first human to receive a permanently implanted artificial heart. It was known as the Jarvik-7 after its inventor, Dr. Robert Jarvik (b. 1946). The 7.5-hour operation was performed by Dr. William DeVries (b. 1943), a surgeon at the University of Utah Medical Center. Dr. Clark died on March 23, 1983, 112 days later. In Louisville, Kentucky, William Schroeder (1923–1986) survived 620 days with an artificial heart (November 25, 1984 to August 7, 1986). On January 11, 1990, the United States Food and Drug Administration (FDA) recalled the Jarvik-7, which had been the only artificial heart approved by the FDA for use.

The Jarvik-7 artificial heart—invented by Dr. Robert Jarvik—was first implanted into a human being in 1982.

## How much does the **cardiac pacemaker** weigh?

The modern pacemaker generator is a hermetically sealed titanium metal can weighing from one to 4.5 ounces (30 to 130 grams) and powered by a lithium battery that can last two to 15 years. Used to correct an insufficient or irregular heartbeat, the cardiac pacemaker corrects the low cardiac rhythm through electrical stimulation, which increases the contractions of the heart muscle. The contraction and expansion of the heart muscle produces a heartbeat, one of three billion in an average lifetime, which pumps blood throughout the body. A pacemaker has gold or platinum electrodes, conducting wires, and a pacing box (miniature generator). In both kinds of pacemakers—

those implanted in the chest or externally—the electrode is attached to the heart's right ventricle (chamber), either directly through the chest or by being threaded through a vein.

## Why are **eye transplants** not available?

It is because the eye's retina is part of the brain, and the retina's cells are derived from brain tissue. Retinal cells and the cells that connect them to the brain are the least amenable to being manipulated outside the body.

## Who was the **first test-tube baby?**

Louise Brown, born on July 25, 1978, is the first baby produced from fertilization done in vitro—outside the mother's body. Patrick Steptoe (1913–1988) was the obstetrician and Robert Edwards (1939–) the biologist who designed the method for in-vitro fertilization and early embryo development.

In-vitro fertilization occurs in a glass dish, not a test tube, where eggs from the mother's ovary are combined with the father's sperm (in a salt solution). Fertilization should occur within 24 hours, and when cell division begins these fertilized eggs are placed in the mother's womb (or possibly another woman's womb).

## What is **lithotripsy**?

Lithotripsy is the use of ultrasonic or shock waves to pulverize kidney stones (calculi), allowing the small particles to be excreted or removed from the body. There are two different methods: extracorporeal shock wave lithotripsy (ESWL) and percutaneous lithotripsy. The ESWL method, used on smaller stones, breaks up the stones with external shock waves from a machine called a lithotripter. This technique has eliminated the need for more invasive stone surgery in many cases. For larger stones, a type of endoscope, called a nephroscope, is inserted into the kidney through a small incision. The ultrasonic waves from the nephroscope shatter the stones, and the fragments are removed through the nephroscope.

## Prior to the condom, what was the main **contraceptive** practice?

Contraceptive devices have been used throughout recorded history. The most traditional of such devices was a sponge soaked in vinegar. The condom was named for its English inventor, the personal physician to Charles II (1630–1685), who used a sheath of stretched, oiled sheep intestine to protect the king from syphilis. Previously penile sheaths were used, such as the linen one made by Italian anatomist Gabriel Fallopius (1523–1562), but they were too heavy to be successful.

## What is **synthetic skin**?

The material consists of a very porous collagen fiber bonded to a sugar polymer (glycoaminoglycan) obtained from shark cartilage. It is covered by a sheet of silicon rubber. It was developed by Ioannis V. Yannas (1935–) and his colleagues at Massachusetts Institute of Technology around 1985. Soon after its introduction, synthetic skin was used to successfully treat more than 100 severely burned victims.

## How is a **mustard plaster** made?

Mix one tablespoon of powdered mustard with four tablespoons of flour. Add enough warm water to make a runny paste. Put the mixture in a folded cloth and apply to the chest. Add olive oil first to the skin if the patient has a delicate skin.

## What is a **negative ion generator**?

A negative ion generator is an electrostatic air cleaner that sprays a continuous fountain of negatively charged ions into the air. Some researchers say that the presence of these ions causes a feeling of well-being, increased mental and physical energy, and relief from some of the symptoms of allergies, asthma, and chronic headaches. It may also aid in healing of burns and peptic ulcers.

## What is **reflexology**?

Reflexology is the application of specific pressures to reflex points in the hands and feet. The reflex points relate to every organ and every part of the body. Massaging of the reflex points is done to prevent or cure diseases. Believed to have been used in Asian cultures as long as 2,000 years ago, reflexology was introduced to the United States at the turn of the century by Dr. William Fitzgerald and Eunice D. Ingham. Today nearly 25,000 certified practitioners can be found throughout the world.

## What is **iridology**?

Iridology is the study of the iris of the eye with the intent of diagnosing weaknesses in the body. Iridologists believe that areas of the iris correspond with different body parts. Among the conditions they monitor are color, clarity, texture, fibers, rings, and spots on the iris. Once the diagnosis is made, they recommend natural methods of healing and conditioning of weak areas.

## What is **aromatherapy**?

Aromatherapy involves using particular scents derived from essential oils to influence emotions and to treat and cure minor ailments. Rene Maurice Gatlefosse, a French cosmetic chemist, introduced the theory in the late 1970s. It is based on the fact that the olfactory and emotional centers of the body are connected. By inhaling different aromas, emotional concerns as well as physical complaints are said to be eased.

## Who developed **psychodrama**?

Psychodrama was developed by Jacob L. Moreno (1892–1974) , a psychiatrist born in Romania, who lived and practiced in Vienna until he came to the United States in 1927. He soon began to conduct psychodrama sessions and founded a psychodrama institute in Beacon, New York, in 1934. Largely because of Moreno's efforts it is practiced worldwide.

Psychodrama is defined as a method of group psychotherapy in which personality makeup, interpersonal relationships, conflicts, and emotional problems are explored by means of special dramatic methods.

# WEIGHTS, MEASURES, TIME, TOOLS, AND WEAPONS

## WEIGHTS, MEASURES, AND MEASUREMENT

*See also: Physics and Chemistry—Measurement, Methodology, etc.; Space—Observation and Measurement; Earth—Observation and Measurement; Energy—Measures and Measurement; Health and Medicine—Drugs, Medicines, etc.*

### How much does the **biblical shekel** weigh in modern units?

A shekel is equal to 0.497 ounces (14.1 grams). Below are listed some ancient measurements with some modern equivalents given.

Biblical

*Volume*

| | | |
|---|---|---|
| omer | = | 4.188 quarts (modern) or 0.45 peck (modern) or 3.964 liters (modern) |
| 9.4 omers | = | 1 bath |
| 10 omers | = | 1 ephah |

*Weight*

| | | |
|---|---|---|
| shekel | = | 0.497 ounces (modern) or 14.1 grams (modern) |

*Length*

| | | |
|---|---|---|
| cubit | = | 21.8 inches (modern) |

Egyptian

*Weight*

| | | |
|---|---|---|
| 60 grams | = | 1 shekel |
| 60 shekels | = | 1 great mina |
| 60 great minas | = | 1 talent |

Greek

*Length*

| | | |
|---|---|---|
| cubit | = | 18.3 inches (modern) |
| stadion | = | 607.2 or 622 feet (modern) |

*Weight*

| | | |
|---|---|---|
| obol or obolos | = | 715.38 milligrams (modern) or. 0.04 ounces (modern) |
| drachma | = | 4.2923 grams (modern) or 6 obols |
| mina | = | 0.9463 pounds (modern) or 96 drachmas |
| talent | = | 60 mina |

Roman

*Length*

| | | |
|---|---|---|
| cubit | = | 17.5 inches (modern) |
| stadium | = | 202 yards (modern) or 415.5 cubits |

*Weight*

| | | |
|---|---|---|
| denarius | = | 0.17 ounces (modern) |

*Volume*

| | | |
|---|---|---|
| amphora | = | 6.84 gallons (modern) |

## What is the **SI system** of measurement?

French scientists as far back as the 17th and 18th centuries questioned the hodgepodge of the many illogical and imprecise standards used for measurement, and so began a crusade to make a comprehensive, logical, precise, and universal measurement system called Système Internationale d'Unités, or SI for short. It uses the metric system as its base. Since all the units are in multiples of 10, calculations are simplified. Today all countries except the United States, Burma, and Liberia use this system. However, some elements within American society do use SI—scientists, exporting/importing industries, and federal agencies (as of November 30, 1992).

The SI or metric system has seven fundamental standards: the meter (for length), the kilogram (for mass), the second (for time), the ampere (for electric current), the kelvin (for temperature), the candela (for luminous intensity), and the mole (for amount of substance). In addition, two supplementary units, the radian (plane angle) and steradian (solid angle), and a large number of derived units compose the current system, which is still evolving. Some derived units, which use special names, are the hertz, newton, pascal, joule, watt, coulomb, volt, farad, ohm, siemens, weber, tesla, henry, lumen, lux, becquerel, gray, and sievert. Its unit of volume or capacity is the cubic decimeter, but many still use "liter" in its place. Very large or very small dimensions are expressed through a series of prefixes, which increase or decrease in multiples of ten. For example, a decimeter is 1/10 of a meter; a centimeter is 1/100 of a meter, and a millimeter is 1/1000 of a meter. A dekameter is 10 meters, a hectometer is 100 meters, and a kilometer is 1,000 meters. The use of these prefixes enable the system to express these units in an orderly way and avoid inventing new names and new relationships.

## How was the length of a **meter** originally determined?

It was originally intended that the meter should represent one ten-millionth of the distance along the meridian running from the North Pole to the equator through Dunkirk, France, and Barcelona, Spain. French scientists determined this distance, working nearly six years to complete the task in November 1798. They decided to use a platinum-iridium bar as the physical replica of the meter. Although the surveyors made an error of about two miles, the error was not discovered until much later. Rather than change the length of the meter to conform to the actual distance, scientists in 1889 chose the platinum-iridium bar as the international prototype. It was used until 1960. Numerous copies of it are in other parts of the world, including the United States National Bureau of Standards.

## How is the **length of a meter** presently determined?

The meter is equal to 39.37 inches. It is presently defined as the distance traveled by light in a vacuum during 1/299,792,458 of a second. From 1960 to 1983, the length of a meter had been defined as 1,650,763.73 times the wavelength of the orange light emitted when a gas consisting of the pure krypton isotope of mass number 86 is excited in an electrical discharge.

## How did the **yard** as a unit of measurement originate?

In early times the natural way to measure lengths was to use various portions of the body (the foot, the thumb, the forearm, etc.). According to tradition, a yard measured on King Henry I (1068–1135) of England became the standard yard still used today. It was the distance from his nose to the middle fingertip of his extended arm.

Other measures were derived from physical activity such as a pace, a league (distance that equaled an hour's walking), an acre (amount plowed in a day), a furlong (length of a plowed ditch), etc., but obviously these units were unreliable. The ell, based on the distance between the elbow and index fingertip, was used to measure out cloth. It ranged from 0.513 to 2.322 meters depending on the locality where it was used and even on the type of goods measured.

Below are listed some linear measurements that evolved from this old reckoning into U.S. customary measures:

**U.S. customary linear measures**

| | | |
|---|---|---|
| 1 hand | = | 4 inches |
| 1 foot | = | 12 inches |
| 1 yard | = | 3 feet |
| 1 rod (pole or perch) | = | 16.5 feet |
| 1 fathom | = | 6 feet |

**U.S. customary linear measures**

| | | |
|---|---|---|
| 1 furlong | = | 220 yards or 660 feet or 40 rods |
| 1 (statute) mile | = | 1,760 yards or 5,280 feet or 8 furlongs |
| 1 league | = | 5,280 yards or 15,840 feet or 3 miles |
| 1 international nautical mile | = | 6,076.1 feet |

**Conversion to metric**

| | | |
|---|---|---|
| 1 inch | = | 2.54 centimeters |
| 1 foot | = | 0.304 meters |
| 1 yard | = | 0.9144 meters |
| 1 fathom | = | 1.83 meters |
| 1 rod | = | 5.029 meters |
| 1 furlong | = | 201.168 meters |
| 1 league | = | 4.828 kilometers |
| 1 mile | = | 1.609 kilometers |
| 1 international nautical mile | = | 1.852 kilometers |

## Why is a **nautical mile** different from a **statute mile**?

Queen Elizabeth I established the statute mile as 5,280 feet (1,609 meters). This measure, based on walking distance, originated with the Romans, who designated 1,000 paces as a land mile.

The nautical mile is not based on human locomotion, but on the circumference of the Earth. There was wide disagreement on the precise measurement, but by 1954 the United States adopted the international nautical mile of 1,852 meters (6,076 feet). It is the length on the Earth's surface of one minute of arc.

1 nautical mile (Int.) = 1.1508 statute miles

1 statute mile = 0.868976 nautical miles

## How are U.S. customary measures **converted** to metric measures and vice versa?

Listed below is the conversion process for common units of measure:

| To convert from | To | Multiply by |
|---|---|---|
| acres | meters, square | 4,046.856 |
| centimeters | inches | 0.394 |
| centimeters | feet | 0.0328 |
| centimeters, cubic | inches, cubic | 0.06 |
| centimeters, square | inches, square | 0.155 |
| feet | meters | 0.305 |

## How can one use U.S. money as a measuring device?

U.S. paper currency is 6 1/8 inches wide by 2 5/8 inches long. The diameter of a quarter is approximately one inch. The diameter of a penny is approximately three-quarters of an inch.

| To convert from | To | Multiply by |
|---|---|---|
| feet, square | meters, square | 0.093 |
| gallons, U.S. | liters | 3.785 |
| grams | ounces (avoirdupois) | 0.035 |
| hectares | kilometers, square | 0.01 |
| hectares | miles, square | 0.004 |
| inches | centimeters | 2.54 |
| inches | millimeters | 25.4 |
| inches, cubic | centimeters, cubic | 16.387 |
| inches, cubic | liters | 0.016387 |
| inches, cubic | meters, cubic | 0.0000164 |
| inches, square | centimeters, square | 6.4516 |
| inches, square | meters, square | 0.0006452 |
| kilograms | ounces, troy | 32.15075 |
| kilograms | pounds (avoirdupois) | 2.205 |
| kilograms | tons, metric | 0.001 |
| kilometers | feet | 3,280.8 |
| kilometers | miles | 0.621 |
| kilometers, square | hectares | 100 |
| knots | miles per hour | 1.151 |
| liters | fluid ounces | 33.815 |
| liters | gallons | 0.264 |
| liters | pints | 2.113 |
| liters | quarts | 1.057 |
| meters | feet | 3.281 |
| meters | yards | 1.094 |
| meters, cubic | yards, cubic | 1.308 |
| meters, cubic | feet, cubic | 35.315 |
| meters, square | feet, square | 10.764 |
| meters, square | yards, square | 1.196 |
| miles, nautical | kilometers | 1.852 |
| miles, square | hectares | 258.999 |
| miles, square | kilometers, square | 2.59 |
| miles (statute) | meters | 1,609.344 |

| To convert from | To | Multiply by |
|---|---|---|
| miles (statute) | kilometers | 1.609344 |
| ounces (avoirdupois) | grams | 28.35 |
| ounces (avoirdupois) | kilograms | 0.0283495 |
| ounces, fluid | liters | 0.03 |
| pints, liquid | liters | 0.473 |
| pounds (avoirdupois) | grams | 453.592 |
| pounds (avoirdupois) | kilograms | 0.454 |
| quarts | liters | 0.946 |
| tons (short/U.S.) | tonne | 0.907 |
| ton, long | tonne | 1.016 |
| tonne | ton(short/U.S.) | 1.102 |
| tonne ton, | long | 0.984 |
| yards | meters | 0.914 |
| yards, square | meters, square | 0.836 |
| yards, cubic | meters, cubic | 0.765 |

## Which **countries** of the world have not formally begun **converting** to the metric system?

The United States, Burma, and Liberia are the only countries that have not formally converted to the metric system. As early as 1790, Thomas Jefferson, then Secretary of State, proposed adoption of the metric system. It was not implemented because Great Britain, America's major trading source, had not yet begun to use the system.

## What are the equivalents for **dry** and **liquid measures**?

**U.S. customary dry measures**

| | | |
|---|---|---|
| 1 pint | = | 33.6 cubic inches |
| 1 quart | = | 2 pints or 67.2006 cubic inches |
| 1 peck | = | 8 quarts or 16 pints or 537.605 cubic inches |
| 1 bushel | = | 4 pecks or 2,150.42 cubic inches or 32 quarts |
| 1 barrel | = | 105 quarts or 7,056 cubic inches |
| 1 pint, dry | = | 0.551 liter |
| 1 quart, dry | = | 1.101 liters |
| 1 bushel | = | 35.239 liters |

**U.S. customary liquid measures**

| | | |
|---|---|---|
| 1 tablespoon | = | 4 fluid drams or 0.5 fluid ounce |
| 1 cup | = | 0.5 pint or 8 fluid ounces |
| 1 gill | = | 4 fluid ounces |

| | | |
|---|---|---|
| 4 gills | = | 1 pint or 28.875 cubic inches |
| 1 pint | = | 2 cups or 16 fluid ounces |
| 2 pints | = | 1 quart or 57.75 cubic ounces |
| 1 quart | = | 2 pints or 4 cups or 32 fluid ounces |
| 4 quarts | = | 1 gallon or 231 cubic inches or 8 pints or 32 gills or 0.833 British quart |
| 1 gallon | = | 16 cups or 231 cubic inches or 128 fluid ounces |
| 1 bushel | = | 8 gallons or 32 quarts |

**Conversion to metric**

| | | |
|---|---|---|
| 1 fluid ounce | = | 29.57 milliliters or 0.029 liter |
| 1 gill | = | 0.118 liter |
| 1 cup | = | 0.236 liter |
| 1 pint | = | 0.473 liter |
| 1 U.S. quart | = | 0.833 British quart or 0.946 liter |
| 1 U.S. gallon | = | 0.833 British gallon or 3.785 liters |

## What is the difference between a **short ton**, a **long ton**, and a **metric ton**?

A short ton or U.S. ton or net ton (sometimes just "ton") equals 2,000 pounds; a long ton or an avoirdupois ton is 2,240 pounds; and a metric ton or tonne is 2,204.62 pounds. Other weights are compared below:

**U.S. customary measure**

| | | |
|---|---|---|
| 1 ounce | = | 16 drams or 437.5 grains |
| 1 pound | = | 16 ounces or 7,000 grains or 256 drams |
| 1 (short) hundredweight | = | 100 pounds |
| 1 long hundredweight | = | 112 pounds |
| 1 (short) ton | = | 20 hundredweights or 2,000 pounds |
| 1 long ton | = | 20 long hundredweights or 2,240 pounds |

**Conversion to metric**

| | | |
|---|---|---|
| 1 grain | = | 65 milligrams |
| 1 dram | = | 1.77 grams |
| 1 ounce | = | 28.3 grams |
| 1 pound | = | 453.5 grams |
| 1 metric ton (tonne) | = | 2,204.6 pounds |

## What are the U.S. and metric units of measurement for **area**?

**U.S. customary area measures**

| | | |
|---|---|---|
| 1 square foot | = | 144 square inches |
| 1 square yard | = | 9 square feet or 1,296 square miles |

| | | |
|---|---|---|
| 1 square rod or pole or perch | = | 30.25 square yards or 272.5 square feet |
| 1 rood | = | 40 square rods |
| 1 acre | = | 160 square rods or 4,840 square yards or 43,460 square feet |
| 1 section | = | 1 mile square or 640 acres |
| 1 township | = | 6 miles square or 36 square miles or 36 sections |
| 1 square mile | = | 640 acres or 4 roods, or 1 section |

**International area measures**

| | | |
|---|---|---|
| 1 square millimeter | = | 1,000,000 square microns |
| 1 square centimeter | = | 100 square millimeters |
| 1 square meter | = | 10,000 square centimeters |
| 1 are | = | 100 square meters |
| 1 hectare | = | 100 ares or 10,000 square meters |
| 1 square kilometer | = | 100 hectares or 1,000,000 square meters |

## How much does **water** weigh?

**U.S. customary measures**

| | | |
|---|---|---|
| 1 gallon | = | 4 quarts |
| 1 gallon | = | 231 cubic inches |
| 1 gallon | = | 8.34 pounds |
| 1 gallon | = | 0.134 cubic foot |
| 1 cubic foot | = | 7.48 gallons |
| 1 cubic inch | = | .0360 pound |
| 12 cubic inches | = | .433 pound |

**British measures**

| | | |
|---|---|---|
| 1 liter | = | 1 kilogram |
| 1 cubic meter | = | 1 tonne (metric ton) |
| 1 imperial gallon | = | 10.022 pounds |
| 1 imperial gallon salt water | = | 10.3 pounds |

## How are **avoirdupois** measurements converted to **troy measurements** and how do the measures differ?

Troy weight is a system of mass units used primarily to measure gold and silver. A troy ounce is 480 grains or 31.1 grams. Avoirdupois weight is a system of units that is used to measure mass, except for precious metals, precious stones, and drugs. It is based on the pound, which is approximately 454 grams. In both systems, the weight of a grain

### Which is heavier—a pound of gold or a pound of feathers?

A pound of feathers is heavier than a pound of gold because gold is measured in troy pounds, while feathers are measured in avoirdupois pounds. Troy pounds have 12 ounces; avoirdupois pounds have 16 ounces. A troy pound contains 372 grams in the metric system; an avoirdupois pound contains 454 grams. Each troy ounce is heavier than an avoirdupois ounce.

is the same—65 milligrams. The two systems do not contain the same weights for other units, however, even though they use the same name for the unit.

*Troy*

1 grain = 65 milligrams
1 ounce = 480 grains = 31.1 grams
1 pound = 12 ounce = 5760 grains = 373 grams

*Avoirdupois*

1 grain = 65 milligrams
1 ounce = 437.5 grains = 28.3 grams
1 pound = 16 ounces = 7000 grains = 454 grams

| To convert from | To | Multiply by |
|---|---|---|
| pounds avdp | ounces troy | 14.583 |
| pounds avdp | pounds troy | 1.215 |
| pounds troy | ounces avdp | 1.097 |
| pounds troy | pounds avdp | 0.069 |
| ounces avdp | ounces troy | 0.911 |
| ounces troy | ounces avdp | 1.097 |

## How is the distance to the **horizon** measured?

Distance to the horizon depends on the height of the observer's eyes. To determine that, take the distance (in feet) from sea level to eye level and multiply by three, then divide by two and take the square root of the answer. The result is the number of miles to the horizon. For example, if eye level is at a height of six feet above sea level, the horizon is almost three miles away. If eye level were exactly at sea level, there would be no distance seen at all; the horizon would be directly in front of the viewer.

## What is a **bench mark**?

A bench mark is a permanent, recognizable point that lies at a known elevation. It may be an existing object, such as the top of a fire hydrant, or it may be a brass plate placed on top of a concrete post. Surveyors and engineers use bench marks to find the elevation of objects by reading, through a level telescope, the distance a point lies above some already established bench mark.

### What is a **theodolite**?

This optical surveying instrument used to measure angles and directions is mounted on an adjustable tripod and has a spirit level to show when it is horizontal. Similar to the more commonly used transit, the theodolite gives more precise readings; angles can be read to fractions of a degree. It is comprised of a telescope that sights the main target, a horizontal plate to provide readings around the horizon, and a vertical plate and scale for vertical readings. The surveyor uses the geometry of triangles to calculate the distance from the angles measured by the theodolite. Such triangulation is used in road- and tunnel-building and other civil engineering work. One of the earliest forms of this surveying instrument was described by Englishman Leonard Digges (d. 1571?) in his 1571 work, *Geometrical Treatise Named Pantometria*.

## TIME

*See also: Biology—Classification, Measurement, and Terms*

### How is **time** measured?

The passage of time can be measured in three ways. Rotational time is based on the unit of the mean solar day (the average length of time it takes the Earth to complete one rotation on its axis). Dynamic time, the second way to measure time, uses the motion of the moon and planets to determine time and avoids the problem of the Earth's varying rotation. The first dynamic time scale was Ephemeris Time, proposed in 1896 and modified in 1960.

Atomic time is a third way to measure time. This method, using an atomic clock, is based on the extremely regular oscillations that occur within atoms. In 1967, the atomic second (the length of time in which 9,192,631,770 vibrations are emitted by a hot cesium atom) was adopted as the basic unit of time. Atomic clocks are now used as international time standards.

But time has also been measured in other less scientific terms. Listed below are some other "timely" expressions.

*Twilight*—The first soft glow of sunlight; the sun is still below the horizon; also the last glow of sunlight.

*Midnight*—12 a.m.; the point of time when one day becomes the next day and night becomes morning.

*Daybreak*—The first appearance of the sun.

*Dawn*—A gradual increase of sunlight.

*Noon*—12 p.m.; the point of time when morning becomes afternoon.

*Dusk*—The gradual dimming of sunlight.

*Sunset*—The last diffused glow of sunlight; the sun is below the horizon.

*Evening*—A term with wide meaning, evening is generally the period between sunset and bedtime.

*Night*—The period of darkness, lasting from sunset to midnight.

## What is the basis for **modern timekeeping**?

Mankind has typically associated the durations of the year, month, week, and day on the earth and moon cycles. The modern clock, however, is based on the number 60. The Sumerians around 3000 B.C.E. employed a base ten counting system and also a base 60 counting system. The timekeeping system inherited this pattern with 60 seconds per minute and 60 minutes per hour. Ten and sixty fit together to form the notion of time: 10 hours is 600 minutes; 10 minutes is 600 seconds; one minute is 60 seconds. One second is not related to any of these factors; instead, scientists based the duration of one second on cesium-133, an isotope of the metal cesium. Officially, one second is the amount of time it takes for a cesium-133 atom to vibrate 9,192,631,770 times.

## What is the exact length of a **calendar year**?

The calendar year is defined as the time between two successive crossings of the celestial equator by the sun at the vernal equinox. It is exactly 365 days, 5 hours, 48 minutes, 46 seconds. The fact that the year is not a whole number of days has affected the development of calendars, which over time generate an accumulative error. The current calendar used, the Gregorian calendar, named after Pope Gregory XII (1502–1585), attempts to compensate by adding an extra day to the month of February every four years. A "leap year" is one with the extra day added.

## When does a **century begin**?

A century has 100 consecutive calendar years. The first century consisted of years 1 through 100. The 20th century started with 1901 and ended with 2000. The 21st century began on January 1, 2001.

## When did **January 1** become the first day of the **new year**?

When Julius Caesar (100–44 B.C.E.) reorganized the Roman calendar and made it solar rather than lunar in the year 45 B.C.E., he moved the beginning of the year to January 1. When the Gregorian calendar was introduced in 1582, January 1 continued to be recognized as the first day of the year in most places. In England and the American colonies, however, March 25, intended to represent the spring equinox, was the beginning of the year. Under this system, March 24, 1700, was followed by March 25, 1701. In 1752, the British government changed the beginning date of the year to January 1.

## Besides the Gregorian calendar, what **other kinds of calendars** have been used?

*Babylonian calendar*—A lunar calendar composed of alternating 29-day and 30-day months to equal roughly 354 lunar days. When the calendar became too misaligned with astronomical events, an extra month was added. In addition, three extra months were added every eight years to coordinate this calendar with the solar year.

*Chinese calendar*—A lunar month calendar of 12 periods having either 29 or 30 days (to compensate for the 29.5 days from new moon to new moon). The new year begins on the first new moon over China after the sun enters Aquarius (between January 21st and February 19th). Each year has both a number and a name (for example, the year 1992, or 4629 in the Chinese era, is the year of the monkey). The calendar is synchronized with the solar year by the addition of extra months at fixed intervals.

*Muslim calendar*—A lunar 12-month calendar with 30 and 29 days alternating every month for a total of 354 days. The calendar has a 30-year cycle with designated leap years of 355 days (one day added to the last month) in the 30-year period. The Islamic year does not attempt to relate to the solar year (the season). The dating of the beginning of the calendar is 622 C.E. (the date of Mohammed's flight from Mecca to Medina).

*Jewish calendar*—A blend of the solar and lunar calendar, this calendar adds an extra month (Adar Sheni, or the second Adar or Veadar) to keep the lunar and solar years in alignment. This occurs seven times during a 19-year cycle. When the extra 29-day month is inserted, the month Adar has 30 days instead of 29. In a usual year the 12 months alternately have 30, then 29 days.

*Egyptian calendar*—The ancient Egyptians were the first to use a solar calendar (about 4236 B.C.E. or 4242 B.C.E.), but their year started with the rising of Sirius, the brightest star in the sky. The year, composed of 365 days, was one-quarter day short of the true solar year, so eventually the Egyptian calendar did not coincide with the seasons. It used twelve 30-day months with five-day weeks and five dates of festival.

*Coptic calendar*—Still used in areas of Egypt and Ethiopia, it has a similar cycle to the Egyptian calendar: 12 months of 30 days followed by five complementary

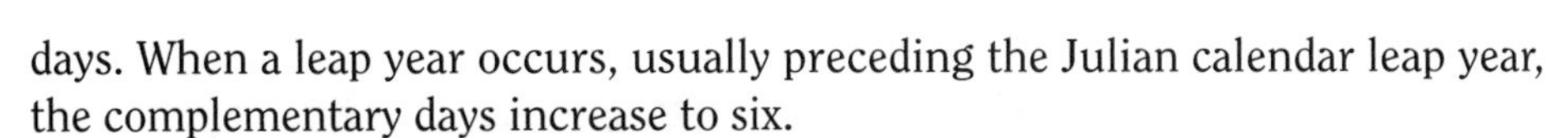

days. When a leap year occurs, usually preceding the Julian calendar leap year, the complementary days increase to six.

*Roman calendar*—Borrowing from the ancient Greek calendar, which had a four-year cycle based on the Olympic Games, the earliest Roman calendar (about 738 B.C.E.) had 304 days with 10 months. Every second year a short month of 22 or 23 days was added to coincide with the solar year. Eventually two more months were added at the end of the year (Januarius and Februarius) to increase the year to 354 days. The Roman republican calendar replaced this calendar during the reign of Tarquinius Priscus (616–579 B.C.E.). This new lunar calendar had 355 days, with the month of February having 28 days. The other months had either 29 or 31 days. To keep the calendar aligned with the seasons, an extra month was added every two years. By the time the Julian calendar replaced this one, the calendar was three months ahead of the season schedule.

*Julian calendar*—Julius Caesar, (100–44 B.C.E.) in 46 B.C.E., wishing to have one calendar in use for all the empire, had the astronomer Sosigenes develop a uniform solar calendar with a year of 365 days with one day ("leap day") added every fourth year to compensate for the true solar year of 365.25 days. The year had 12 months with 30 or 31 days except for February, which had 28 days (or 29 days in a leap year). The first of the year was moved from March 1 to January 1.

*Gregorian calendar*—Pope Gregory XIII (1502–1585) instituted calendrical reform in 1582 to realign the church celebration of Easter with the vernal equinox (the first day of spring). To better align this solar calendar with the seasons, the new calendar would not have leap year in the century years that were not divisible by 400. Because the solar year is shortening, today a one-second adjustment is made (usually on December 31 at midnight) when necessary to compensate.

*Japanese calendar*—It has the same structure as the Gregorian calendar in years, months, and weeks. But the years are enumerated in terms of the reigns of emperors as epochs. The last epoch (for Emperor Akihito) is Epoch Heisei, which started January 8, 1989.

*Hindu calendars*—The principal Indian calendars reckon their epochs from historical events, such as rulers' accessions or death dates, or a religious founder's dates. The Vikrama era (originally from Northern India and still used in western India) dates from February 23, 57 B.C.E., in the Gregorian calendar. The Saka era dates from March 3, 78 C.E., in the Gregorian calendar and is based on the solar year with 12 months of 365 days and 366 days in leap years. The first five months have 31 days; the last seven have 30 days. In leap years, the first six months have 31 days and the last six have 30. The Saka era is the national calendar of India (since 1957). The Buddhist era starts with 543 B.C.E. (believed to be the date of Buddha's death).

Three other secular calendars of note are the *Julian Day calendar* (a calendar astronomers use, which counts days within a 7,980-year period and must be used with a table), the *perpetual calendar* (which gives the days of the week for the Julian and

## What is the World Calendar?

Following World War II, the United States led a movement in the United Nations to encourage the international community to adopt a common calendar. Called the World Calendar, it would self-adjust the irregular months, equalize the quarterly divisions of the years, and fix the sequence of weekdays and month dates so that days of the month would always fall on the same day of the week. This calendar would make the normal year 364 days long, divided into four quarters of 91 days each. The quarters could be segmented into three months of 31, 30, and 30 days. Every year would have 52 whole weeks; every quarter would begin with a Sunday and end with a Saturday. An extra day would need to be added to each year. This day, "World's Day," is formally called "W December," observed following the last day of December. Leap Year Day would be a similar insertion in the calendar every fourth year, following June 30, and would be called "W June." Although it was predicted that the calendar would be adopted in 1961, it has repeatedly failed passage in the United Nations.

Gregorian calendars as well), and the *world calendar*, which is similar to the perpetual calendar, having 12 months of 30 or 31 days, a year-day at the end of each year, and a leap-year-day before July 1 every four years.

There have been attempts to reform and simplify the calendar. One such example is the Thirteen-Month or International Fixed calendar, which would have 13 months of four weeks each. The month *Sol* would come before July; there would be a year-day at the end of each year and a leap-year-day every four years just before July 1. A radical reform was made in France when the French republican calendar (1793–1806) replaced the Gregorian calendar after the French Revolution. It had 12 months of 30 days and five supplementary days at the end of the year (six in a leap-year), and weeks were replaced with 10-day decades.

## What is the **Julian Day Count**?

This system of counting days rather than years was developed by Joseph Justus Scaliger (1540–1609) in 1583. Still used by astronomers today, the Julian Day Count (named after Scaliger's father, Julius Caesar Scaliger) Julian Day (JD) 1 was January 1, 4113 B.C.E. On this date the Julian calendar, the ancient Roman tax calendar, and the lunar calendar all coincided. This event would not occur again until 7,980 years later. Each day within this 7,980-year period is numbered. December 31, 1991, at noon is the beginning of JD 2,448,622. The figure reflects the number of days that have passed since its inception. To convert Gregorian calendar dates into Julian Day, simple JD conversion tables have been devised for astronomers.

## Do all calendars other than the Gregorian calendar use a **twelve-month cycle**?

No. Some calendars have a varying cycle of months, with their first month falling at a different time. Listed below are some variations with the beginning month of the year in italics:

**Months of the year**

| Gregorian | Hebrew | Hindu |
|---|---|---|
| January | Shebat | Magha |
| February | Adar | Phalguna |
| March | Nisan | Caitra |
| April | Iyar | Vaisakha |
| May | Sivan | Jyaistha |
| June | Tammuz | Asadhe |
| July | Ab | Sravana |
| August | Elul | Bhadrapada |
| September | Tishri | Asvina |
| October | Heshvan | Karttika |
| November | Kislev | Margasivsa |
| December | Tebet | Pansa |
| | Ve-Adar (13th month every three years) | |

| Muslim | Chinese | |
|---|---|---|
| Muharram | Li Chun | Li Qui |
| Safar | Yu Shui | Chu Shu |
| Rabi I | Jing Zhe | Bai Lu |
| Rabi II | Chun Fen | Qui Fen |
| Jumada I | Qing Ming | Han Lu |
| Jumada II | Gu Yu | Shuang Jiang |
| Rajab | Li Xia | Li Dong |
| Sha'ban | Xiao Man | Xiao Xue |
| Ramadan | Mang Zhong | Da Xue |
| Shawwal | Xia Zhi | Dong Zhi |
| Dhu'l-Qa'da | Xiao Shu | Xiao Han |
| Dhu'l-hijja | Da Shu | Da Han |

## Which animal designations have been given to the **Chinese years**?

There are 12 different names, always used in the same sequence: Rat, Ox, Tiger, Hare (Rabbit), Dragon, Snake, Horse, Sheep (Goat), Monkey, Rooster, Dog, and Pig. The following table shows this sequence.

**Chinese year cycle**

| Rat | Ox | Tiger | Hare (Rabbit) | Dragon | Snake |
|---|---|---|---|---|---|
| 1996 | 1997 | 1998 | 1999 | 2000 | 2001 |
| 2008 | 2009 | 2010 | 2011 | 2012 | 2013 |
| 2020 | 2021 | 2022 | 2023 | 2024 | 2025 |
| **Horse** | **Sheep (Goat)** | **Monkey** | **Rooster** | **Dog** | **Pig** |
| 2002 | 2003 | 2004 | 2005 | 2006 | 2007 |
| 2014 | 2015 | 2016 | 2017 | 2018 | 2019 |
| 2026 | 2027 | 2028 | 2029 | 2030 | 2031 |

The Chinese year of 354 days begins three to seven weeks into the western 365-day year, so the animal designation changes at that time, rather than on January 1.

## When does **leap year** occur?

A leap year occurs when the year is exactly divisible by four, except for centenary years. A centenary year must be divisible by 400 to be a leap year. There was no February 29 in 1900, which was not a leap year. The year 2000 was a centenary leap year, and the next one will be 2400.

## What is a **leap second**?

The Earth's rotation is slowing down, and to compensate for this lagging motion a leap second is added to a specified day. One leap second was used in 1992 to keep the calendar in close alignment with international atomic time. To complete this change, 23h 59m 59s universal time on June 30, 1992 was followed by 23h 59m 60s and this in turn was followed by 0h 0m 0s on July 1.

## What were the **longest** and **shortest** years on record?

The longest year was 46 B.C.E. when Julius Caesar (100–44 B.C.E.) introduced his calendar, called the Julian calendar, which was used until 1582. He added two extra months and 23 extra days to February to make up for accumulated slippage in the Egyptian calendar. Thus, 46 B.C.E. was 455 days long. The shortest year was 1582, when Pope Gregory XIII (1502–1585) introduced his calendar, the Gregorian calendar. He decreed that October 5 would be October 15, eliminating 10 days, to make up for the accumulated error in the Julian calendar. Not everyone changed over to this new calendar at once. Catholic Europe adopted it within two years of its inception. Many Protestant continental countries did so in 1699–1700; England imposed it on its colonies in 1752

and Sweden in 1753. Many non-European countries adopted it in the 19th century, with China doing so in 1912, Turkey in 1917, and Russia in 1918. To change from the Julian to the Gregorian calendar, 10 days are added to dates October 5, 1582, through February 28, 1700; after that date add 11 days through February 28, 1800; 12 days through February 28, 1900; and 13 days through February 28, 2100.

## Where do the names of the **days of the week** come from?

The English days of the week are named for a mixture of figures in Anglo-Saxon and Roman mythology.

| Day | Named after |
|---|---|
| Sunday | The sun |
| Monday | The moon |
| Tuesday | Tiu (the Anglo-Saxon god of war, equivalent to the Norse Tyr or the Roman Mars) |
| Wednesday | Woden (the Anglo-Saxon equivalent of Odin, the chief Norse god) |
| Thursday | Thor (the Norse god of thunder) |
| Friday | Frigg (the Norse god of love and fertility, the equivalent of the Roman Venus) |
| Saturday | Saturn (the Roman god of agriculture) |

## What is the **origin** of the **week**?

The week originated in the Babylonian calendar, where one day out of seven was devoted to rest.

## How were the **months** of the year **named**?

The English names of the months of the current (Gregorian) calendar come from the Romans, who tended to honor their gods and commemorate specific events by designating them as month names:

*January* (*Januarius* in Latin), named after Janus, a Roman two-faced god, one face looking into the past, the other into the future.

*February* (*Februarium*) is from the Latin word *Februare*, meaning "to cleanse." At the time of year corresponding to our February, the Romans performed religious rites to purge themselves of sin.

*March* (*Martius*) is named in honor of Mars, the god of war.

*April* (*Aprilis*), after the Latin word *Aperio*, meaning "to open," because plants begin to grow in this month.

*May* (*Maius*), after the Roman goddess Maia, as well as from the Latin word *Maiores* meaning "elders," who were celebrated during this month.

*June* (*Junius*), after the goddess Juno and Latin word *iuniores*, meaning "young people."

*July* (*Iulius*) was, at first, known as *Quintilis* from the Latin word meaning five, since it was the fifth month in the early Roman calendar. Its name was changed to July, in honor of Julius Caesar (100–44 B.C.E.).

*August* (*Augustus*) is named in honor of the Emperor Octavian (63–14 B.C.E.), first Roman emperor, known as Augustus Caesar. Originally the month was known as *Sextilis* (sixth month of early Roman calendar).

*September* (*September*) was once the seventh month and accordingly took its name from *septem*, meaning "seven."

*October* (*October*) takes its name from *octo* (eight); at one time it was the eighth month.

*November* (*November*) from *novem*, meaning "nine," once the ninth month of the early Roman calendar.

*December* (*December*) from *decem*, meaning "ten," once the tenth month of the early Roman calendar.

## Why are the **lengths of the seasons** not equal?

The lengths of the seasons are not exactly equal because the orbit of the Earth around the sun is elliptic rather than a circular. When the Earth is closest to the sun in January, gravitational forces cause the planet to move faster than it does in the summer months when it is far away from the sun. As a result, the autumn and winter seasons in the Northern Hemisphere are slightly shorter than spring and summer. The duration of the northern seasons are:

| | |
|---|---|
| spring | 92.76 days |
| summer | 93.65 days |
| autumn | 89.84 days |
| winter | 88.99 days |

## What are the **beginning dates of the seasons** for 2002 to 2009?

The four seasons in the northern hemisphere coincide with astronomical events. The spring season starts with the vernal equinox (around March 21) and lasts until the summer solstice (June 21 or 22); summer is from the summer solstice to the autumnal equinox (about September 21); autumn, from the autumnal equinox to the winter solstice (December 21 or 22); and winter from the winter solstice to the vernal equinox. In the southern hemisphere the seasons are reversed; autumn corresponds to spring, winter to summer. The seasons are caused by the tilt of the Earth's axis, which changes the position of the sun in the sky. In winter the sun is at its lowest position (or angle of declination) in the sky; in summer it is at its highest position.

| Year | Spring Equinox | Summer Solstice | Fall Equinox | Winter Solstice |
|---|---|---|---|---|
| 2002 | Mar 20 | June 21 | Sept 22 | Dec 21 |
| 2003 | Mar 20 | June 21 | Sept 23 | Dec 22 |
| 2004 | Mar 19 | June 20 | Sept 22 | Dec 21 |
| 2005 | Mar 20 | June 20 | Sept 22 | Dec 21 |
| 2006 | Mar 20 | June 21 | Sept 23 | Dec 22 |
| 2007 | Mar 21 | June 21 | Sept 22 | Dec 22 |
| 2008 | Mar 20 | June 21 | Sept 22 | Dec 21 |
| 2009 | Mar 20 | June 21 | Sept 22 | Dec 21 |

## How is the date for **Easter** determined?

The rule for establishing the date in the Christian Churches is that Easter always falls on the first Sunday after the first full moon that occurs on or just after the vernal equinox. The vernal equinox is the first day of spring in the Northern Hemisphere. Because the full moon can occur on any one of many dates after the vernal equinox, the date of Easter can be as early as March 22 and as late as April 25. For the period 2002–2012 Easter will occur on the following dates:

| Year | Date |
|---|---|
| 2002 | March 31 |
| 2003 | April 20 |
| 2004 | April 11 |
| 2005 | March 27 |
| 2006 | April 16 |
| 2007 | April 8 |
| 2008 | March 23 |
| 2009 | April 12 |
| 2010 | April 4 |
| 2011 | April 24 |
| 2012 | April 8 |

## How is the date for **Passover** determined?

Passover (Pesah) begins at sundown on the evening before the 15th day of the Hebrew month of Nisan, which falls in March or April. Celebrated as a public holiday in Israel, the Passover commemorates the exodus of the Israelites from Egypt in 1290 B.C.E. The term "Passover" is a reference to the last of the 10 plagues that "passed over" the Israelite homes at the end of their captivity in Egypt.

## Occasionally one sees a date expressed as "**B.P.** 6500" instead of "B.C. 6500". What does this signify?

Archaeologists use the B.P. or BP as an abbreviation for years before the present. This date is a rough generalization of the number of years before 1950, and is a date not necessarily based on radiocarbon dating methods.

## How is **"local noon"** calculated?

Local noon, also called "the time of solar meridian passage," is the time of day when the sun is at its highest point in the sky. This is different from noon on clocks. To calculate local noon, first find the sunrise and sunset (readily available in larger newspapers), add the total amount of sunlight, divide this number by two, and add that to the sunrise. For instance, if sunrise is at 7:30 a.m. and sunset is at 8:40 p.m., the total sunlight is 13 hours and ten minutes, divided by two is 6 hours and 40 minutes (13 hours/2 = 6.5 hours plus 10 minutes), added to 7:30 a.m. equals a local noon of 1:40 p.m.

## When is **Daylight Savings Time** observed in the United States?

In 1967, all states and possessions of the United States were to begin observing Daylight Savings Time at 2 a.m. on the first Sunday in April of every year. The clock would advance one hour at that time until 2 a.m. of the last Sunday of October, when the clock would be turned back one hour. In the intervening years, the length of this time period changed, but on July 8, 1986, the original starting and ending dates were reinstated. A 1972 amendment allowed some areas to be exempt.

This time change was enacted to provide more light in the evening hours. The phrase "fall back, spring forward" indicates the direction in which the clock setting moved during these seasons. Other countries have adopted DST as well. For instance, in western Europe the period is from the last Sunday in March to the last Sunday in September, with the United Kingdom extending the range to the last Sunday in October. Many countries in the Southern Hemisphere generally maintain DST from October to March; countries near the equator maintain standard time.

## Which states and territories of the United States are **exempt from Daylight Savings Time**?

Arizona, Hawaii, Puerto Rico, the U.S. Virgin Islands, American Samoa, and most of Indiana are exempt from Daylight Savings Time.

### Why do the minute and hour hands on a clock go "clockwise"?

Henry Fried, an expert on timekeeping, postulates that the clockwise motion of minute and hour hands has its origin in the use of sundials before the development of clocks. In the northern hemisphere, the shadow rotates in a clockwise direction, and clock inventors built the hands to copy the natural movement of the sun.

## What are some **other names** for **Daylight Saving Time**?

It is also known as fast time or summer time.

## How many **time zones** are there in the world?

There are 24 standard time zones that serially cover the Earth's surface at coincident intervals of 15 degrees longitude and 60 minutes Universal Time (UT), as agreed at the Washington Meridian Conference of 1884, thus accounting, respectively, for each of the 24 hours in a calendar day.

## What is unusual about the observance of **time zones** in **Russia** and **China**?

Russia, which spans 11 time zones, is on "advanced time" year round, meaning that it maintains its Standard Time one hour faster than the zone designation. The country also observes Daylight Saving Time from the fourth Sunday in March until the fourth Sunday in September. China, though it lies across five time zones, keeps one time, which is eight hours faster than Greenwich Time.

## When traveling from Tokyo to Seattle and crossing the **International Date Line**, what day is it?

Traveling from west to east (Tokyo to Seattle), the calendar day is set back (e.g., Sunday becomes Saturday). Traveling east to west, the calendar day is advanced (e.g., Tuesday becomes Wednesday). The International Date Line is a zigzag line at approximately the 180th meridian, where the calendar days are separated.

## What is meant by **Universal Time**?

On January 1, 1972, Universal Time (UT) replaced Greenwich Mean Time (GMT) as the time reference coordinate for scientific work. Universal Time is measured by an atomic clock and is seen as the logical development of the adoption of the atomic second in 1968. An advantage of UT is that the time at which an event takes place can be determined very readily without recourse to the time-consuming astronomical observations and calculations that were necessary before the advent of atomic clocks. Universal Time is also referred to as International Atomic Time. GMT is measured according to when the sun crosses the Greenwich Meridian (zero degrees longitude, which passes through the Greenwich Observatory).

### Who establishes the **correct time** in the United States?

The United States National Institute of Standards and Technology (NIST) uses a cesium beam clock as its NIST atomic frequency standard to determine atomic time. It is possible to check the time by accessing the NIST Web site at http://www.nist.gov and selecting "check time" from the options. The clock is accurate to plus or minus one second. The atomic second was officially defined in 1967 by the 13th General Conference of Weights and Measures as 9,192,631,770 oscillations of the atom of cesium-133. This cesium-beam clock of the NIST is referred to as a primary clock because, independently of any other reference, it provides highly precise and accurate time.

### What is the **United States Time Standard** signal?

Universal Time is announced in International Morse Code each five minutes by radio stations WWV (Fort Collins, Colorado) and WWVH (Puuene, Maui, Hawaii). These stations are under the direction of the U.S. National Institute of Standards and Technology (formerly called National Bureau of Standards). They transmit 24 hours a day on 2.5, 5, 10, 15, and 20 megahertz. The first radio station to transmit time signals regularly was the Eiffel Tower Radio Station in Paris in 1913.

### What do the initials **a.m.** and **p.m.** mean?

The initials a.m. stand for *ante meridian,* Latin for "before noon." The initials p.m. stand for *post meridian,* Latin for "after noon."

### Why do **clocks and watches** with **Roman numerals** on their faces use "IIII" for the number 4, rather than "IV"?

A clock with a "IV" has a somewhat unbalanced appearance. Thus tradition has favored the strictly incorrect "IIII," which more nearly balances the equally heavy "VIII."

### How does a **sundial** work?

The sundial, one of the first instruments used in the measurement of time, works by simulating the movement of the sun. A gnomon is affixed to a time (or hour) scale and the time is read by observing the shadow of the gnomon on the scale. Sundials gener-

ally use the altitude of the sun to give the time, so certain modifications and interpretations must be made for seasonal variance.

## What is a **floral clock**?

In the Middle Ages it was believed that one could tell the hour of day by observing flowers, which were believed to open and close at specific times. These would be planted in flower dials. The first hour belonged to the budding rose, the fourth to hyacinths and the twelfth to pansies. The unreliability of this method of timekeeping must have soon become apparent, but floral clocks are still planted in parks and gardens today. The world's largest clock face is that of the floral clock located inside the Rose Building in Hokkaido, Japan. The diameter of the clock is almost 69 feet (21 meters), and the large hand of the clock is almost 28 feet (8.5 meters) in length.

## How is **time** denoted at **sea**?

The day is divided into watches and bells. A watch equals four hours except for the time period between 4 p.m. and 8 p.m., which has two short watches. Within each watch there are eight bells—one stroke for each half hour, so that each watch ends on eight bells except the dog watches, which end at four bells. New Year's Day is marked with 16 bells.

| Bell | Time equivalent | | |
|---|---|---|---|
| 1 bell | 12:30 or | 4:30 | or 8:30 A.M. or P.M. |
| 2 bells | 1:00 | 5:00 | 9:00 |
| 3 bells | 1:30 | 5:30 | 9:30 |
| 4 bells | 2:00 | 6:00 | 10:00 |
| 5 bells | 2:30 | 6:30 | 10:30 |
| 6 bells | 3:00 | 7:00 | 11:00 |
| 7 bells | 3:30 | 7:30 | 11:30 |
| 8 bells | 4:00 | 8:00 | 12:00 |

## Where did the term **grandfather clock** come from?

The weight-and-pendulum clock was invented by Dutch scientist Christiaan Huygens (1629–1695) around the year 1656. In the United States, Pennsylvania German settlers considered such clocks, often called long-case clocks, to be a status symbol. In 1876 American songwriter Henry Clay Work (1832–1884) referred to long-case clocks in his song "My Grandfather's Clock," and the nickname stuck.

## What is the difference between a **quartz** watch and a **mechanical** watch?

Quartz watches and mechanical watches use the same gear mechanism for turning the hour and minute gear wheels. Mechanical watches are powered by a coiled spring known as the mainspring and regulated by a system called a lever escarpment. As the watch runs, the mainspring unwinds. Quartz watches are powered by a battery-powered electronic integrated circuit on a tiny piece of silicon chip and regulated by quartz crystals that vibrate and produce electric pulses at a fixed rate.

## Who invented the **alarm clock**?

Levi Hutchins of Concord, New Hampshire, invented an alarm clock in 1787. His alarm clock, however, rang at only one time—4 a.m. He invented this device so that he would never sleep past his usual waking time. He never patented or manufactured it. The first modern alarm clock was made by Antoine Redier (1817–1892) in 1847. It was a mechanical device; the electric alarm clock was not invented until around 1890. The earliest mechanical clock was made in 725 C.E. in China by Yi Xing and Liang Lingzan.

## What is **military time**?

Military time divides the day into one set of 24 hours, counting from midnight (0000) to midnight of the next day (2400). This is expressed without punctuation.

Midnight becomes 0000 (or 2400 of the next day)

1:00 a.m. becomes 0100 (pronounced oh–one hundred)

2:10 a.m. becomes 0210

Noon becomes 1200

6:00 p.m. becomes 1800

9:45 p.m. becomes 2145

To translate 24-hour time into familiar time, the a.m. times are obvious. For p.m. times, subtract 1200 from numbers larger than 1200; e.g., 1900 – 1200 is 7 p.m.

## Who set a **doomsday clock** for nuclear annihilation?

The clock first appeared on the cover of the magazine *Bulletin of the Atomic Scientists* in 1947 and was set at 11:53 p.m. The clock, created by the magazine's board of directors, represents the threat of nuclear annihilation, with midnight as the time of destruction. The clock has been reset eighteen times in fifty-five years. In 1953, just after the United States tested the hydrogen bomb, it was set at 11:58 p.m., the closest to midnight ever. In 1991, following the collapse of the Soviet Union, it was moved back to 11:43 p.m., the farthest from midnight the clock has ever been. However, in 1995, the clock was shifted forward to 11:47, reflecting the instability of the post-Cold-War world. The clock was set at seven minutes to midnight (11:53 p.m.) in 2002 because little progress is being made on global nuclear disarmament, and because terrorists are seeking to acquire and use nuclear and biological weapons.

# TOOLS, MACHINES, AND PROCESSES



## What are some of the **earliest agricultural tools**?

The Natufians of Palestine, circa 8000 B.C.E., are believed to have used simple digging and harvesting tools. At this early time in human agricultural history, digging sticks or hoes were used to break the ground. They also had a form of sickle to harvest sown and wild grain.

Later, around 6500 B.C.E., the first rather primitive plow, called the *ard*, was used in the Near East. This important agricultural implement evolved from a simple digging stick—a plow with a handle used by humans to push or pull along the ground. This in turn evolved into the Egyptian plow, with stag antlers or a forked branch tied to a pole to break the ground for planting.

## Which **tools** did **Neanderthal man** use?

Neanderthal tool kits are termed *Mousterian* from finds at LeMoustier in France (traditionally dated to the early Fourth Glacial Period, around 40,000 B.C.E.). Using fine-grained glassy stone like flint and obsidian, Neanderthals improved on the already old, established *Levallois* technique for striking one or two big flakes of predetermined shape from a prepared core. They made each core yield many small, thin, sharp-edged flakes, which were then trimmed to produce hide scrapers, points, backed knives, stick sharpeners, tiny saws, and borers. These could have served for killing, cutting, and skinning prey, and for making wooden tools and clothing.

## What are the six **simple machines**?

All machines and mechanical devices, no matter how complicated, can be reduced to some combinations of six basic, or simple, machines. The lever, the wheel and axle, the pulley, the inclined plane, the wedge, and the screw were all known to the ancient Greeks, who learned that a machine works because an "effort," which is exerted over an "effort distance," is magnified through "mechanical advantage" to overcome a "resistance" over a "resistance distance." Some consider there to be only five simple machines, and regard the wedge as a moving inclined plane.

## How does a **four-stroke** differ from a **two-stroke** engine?

A four-stroke engine functions by going through four cycles: 1) the intake stroke draws a fuel-air mixture in on a down stroke; 2) the mixture is compressed on an upward stroke; 3) the mixture is ignited causing a down stroke; and 4) the mixture is exhausted on an up stroke. A two-stroke engine combines the intake and compression

### How did the monkey wrench get its name?

This type of wrench, on which the jaws are at right angles to the handle, derives its name from its inventor, Charles Moncky, whose last name was eventually corrupted into "monkey."

strokes (1 and 2) and the power and exhaust strokes (3 and 4) by covering and uncovering ports and valves in the cylinder wall. Two-stroke engines are typically used in small displacement applications, such as chain saws, some motorcycles, etc.

## What does "cc" mean in **engine sizes**?

Cubic centimeters (cc), as applied to an internal combustion engine, is a measure of the combustion space in the cylinders. The size of an engine is measured by theoretically removing the top of a cylinder, pushing the piston all the way down, and then filling the cylinder with liquid. The cubic centimeters of liquid displaced (spilled out) when the piston is returned to its high position is the measure of the combustion volume of the cylinder. If a motorcycle has four cylinders, each displacing 200 ccs, it has an 800 cc engine. Automobile engines are sized essentially the same way.

## What is a **donkey engine**?

A donkey engine is a small auxiliary engine that is usually portable or semi-portable. It is powered by steam, compressed air, or other means. It is often used to power a windlass or lift cargo on shipboard.

## What is a **power take-off**?

The standard power take-off is a connection that will turn a shaft inserted through the rear wall of a gear case. It is used to power accessories such as a cable control unit, a winch, or a hydraulic pump. Farmers use the mechanism to pump water, grind feed, or saw wood. A power take-off drives the moving parts of mowing machines, hay balers, combines, and potato diggers.

## Who invented the **compass**?

The first compass has no known inventor. The Chinese, in the first century B.C.E., discovered that lodestone, an iron mineral, always pointed north when placed on a surface. The first Chinese compass was a spoon made of lodestone with markings indicating the four directions. Later, the Chinese enclosed the lodestone in a decorative

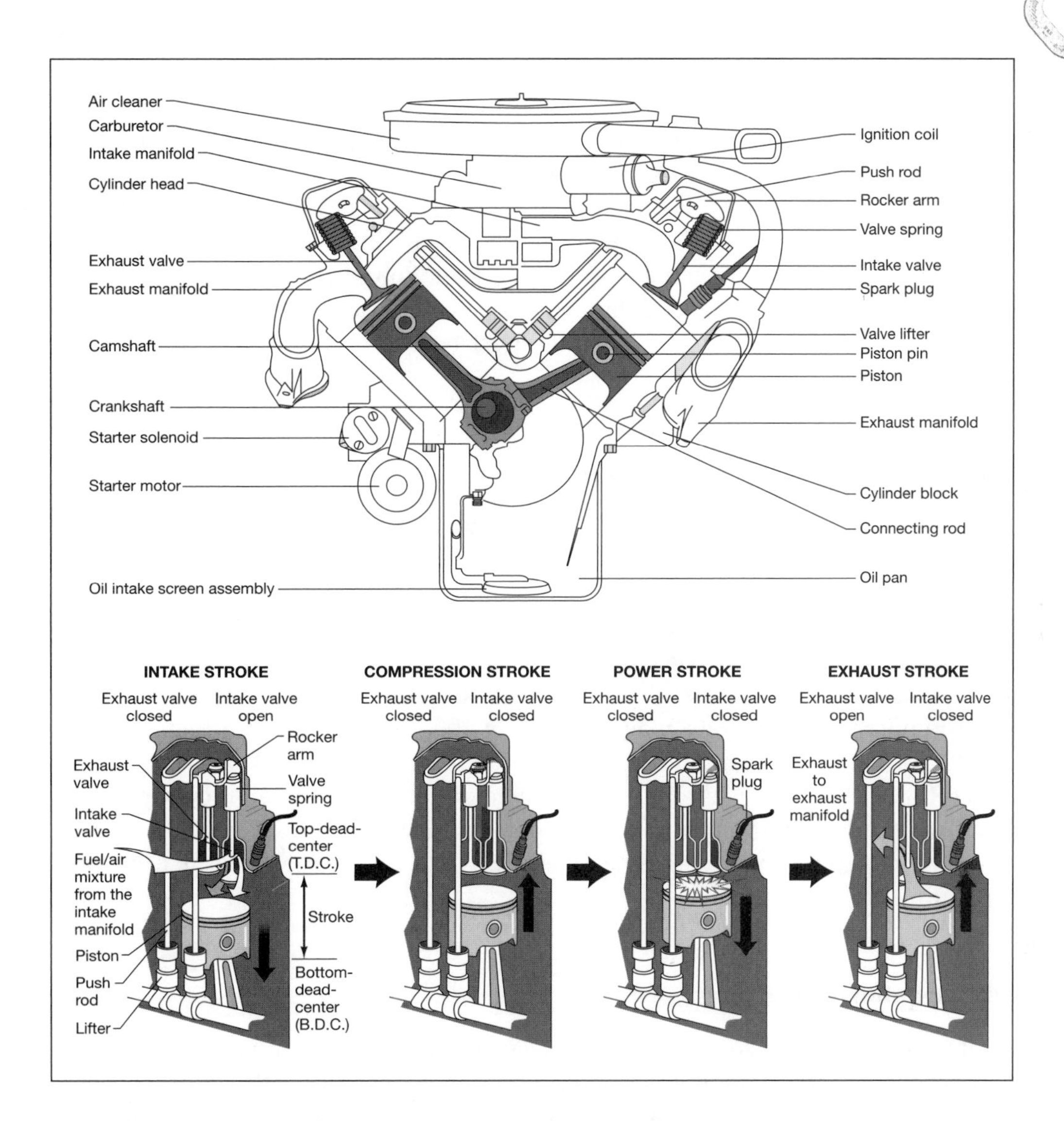


**The major components of an internal combustion engine (top) and the four strokes of its combustion sequence (bottom).**

casing, and used a needle to indicate which direction was north. The problem with lodestone, however, was that it easily lost its magnetic properties. After much experimentation with different metals, the Chinese discovered that adding carbon to iron to make steel gave the needle a strong magnetic charge that lasted for a long time.

## Who invented the **compound microscope**?

The principle of the compound microscope, in which two or more lenses are arranged to form an enlarged image of an object, occurred independently, at about the same time, to more than one person. Certainly many opticians were active at the end of the 16th century, especially in Holland, in the construction of telescopes, so that it is likely that the idea of the microscope may have occurred to several of them independently. In all probability, the date may be placed within the period 1590–1609, and the credit should go to three spectacle makers in Holland. Hans Janssen, his son Zacharias (1580–1638), and Hans Lippershey (1570–1619) have all been cited at various times as deserving chief credit. An Englishman, Robert Hooke (1635–1703), was the first to make the best use of a compound microscope, and his book *Micrographia*, published in 1665, contains some of the most beautiful drawings of microscopic observations ever made.

## Who invented the **electron microscope**?

The theoretical and practical limits to the use of the optical microscope were set by the wavelength of light. When the oscilloscope was developed, it was realized that cathode-ray beams could be used to resolve much finer detail because their wavelength was so much shorter than that of light. In 1928, Ernst Ruska (1906–1988) and Max Knoll, using magnetic fields to "focus" electrons in a cathode-ray beam, produced a crude instrument that gave a magnification of 17, and by 1932 they developed an electron microscope having a magnification of 400. By 1937 James Hillier (b. 1915) advanced this magnification to 7,000. The 1939 instrument Vladimir Zworykin (1889–1982) developed gave 50 times more detail than any optical microscope ever could, with a magnification up to two million. The electron microscope revolutionized biological research: for the first time, scientists could see the molecules of cell structures, proteins, and viruses.

## What is **holography**?

Hungarian-born scientist Dennis Gabor invented the technique of holography (image in the round) in 1947, but it was not until 1961 when Emmet Leith and Juris Upatnieks produced the modern hologram using a laser, which gave the hologram the strong, pure light it needed. Three dimensions are seen around an object because light waves are reflected from all around it, overlapping and interfering with each other. This interaction of these collections of waves, called wave fronts, give an object its light, shade, and depth. A camera cannot capture all the information in these wave fronts, so it produces two-dimensional objects. Holography captures the depth of an object by measuring the distance light has traveled from the object.

A simple hologram is made by splitting a laser light into two beams through a silvered mirror. One beam, called the object beam, lights up the subject of the hologram. These light waves are reflected onto a photographic plate. The other beam, called a reference beam, is reflected directly onto the plate itself. The two beams coincide to create, on the plate, an "interference pattern." After the plate is developed, a laser light

is projected through this developed hologram at the same angle as the original reference beam, but from the opposite direction. The pattern scatters the light to create a projected, three-dimensional, ghost-like image of the original object in space.

## In which industries are **robots** being used?

A robot is a device that can execute a wide range of maneuvers under the direction of a computer. Reacting to feedback from sensors or by reprogramming, a robot can alter its maneuvers to fit a changed task or situation. The worldwide population of robots is about 250,000, with approximately 65 percent being used in Japan and 14 percent being used in the United States (the second largest user).

Robots are used to do welding, painting, drilling, sanding, cutting, and moving tasks in manufacturing plants. Automobile factories account for more than 50 percent of robot use in the United States. Robots can work in environments that are extremely threatening to humans or in difficult physical environments. They clean up radioactive areas, extinguish fires, disarm bombs, and load and unload explosives and toxic chemicals. Robots can process light-sensitive materials, such as photographic films that require near darkness—a difficult task for human workers. They can perform a variety of tasks in underwater exploration and in mines (a hazardous, low-light environment). There are only 100 robots in mines, but they produce about one-third of the coal coming from these mines. In the military and security fields, robots are used to sense targets or as surveillance devices. In the printing industry, robots perform miscellaneous tasks, such as sorting and tying bundles of output material, delivering paper to the presses, and applying book covers. In research laboratories, small desktop robots prepare samples and mix compounds.

## What **film** was the first to feature a **robot**?

In 1886, the French movie *l'Eve Futur* had a Thomas Edison-like mad scientist building a robot in the likeness of a woman. A British lord falls in love with her in a variation on the Pygmalion theme. True working robots, in contrast to these entertainment devices, are strictly functional in appearance, looking more like machines rather than human beings or animals. However, one early exception to this is the very human-looking "Scribe" built in 1773 by two French inventors, Pierre and Henri-Louis Jacquel-Droz, a father and son team, who produced "The Automaton" to dip a quill pen into an inkwell and write a text of 40 characters maximum.

## Who was the founder of **cybernetics**?

Norbert Wiener (1894–1964) is considered the creator of cybernetics. Derived from the Greek word kubernetes, meaning steerman or helmsman, cybernetics is concerned with the common factors of control and communication in living organisms, automatic machines, or organizations. These factors are exemplified by the skill used in steering a boat, in which the helmsman uses continual judgment to maintain control. The principles of cybernetics are used today in control theory, automation theory, and computer programs to reduce many time-consuming computations and decision-making processes formerly done by people.

## When was the **jackhammer** invented?

In 1861 a French engineer, German Sommelier, came up with the idea of the pneumatic pick or jackhammer while working on the Mont Cenis tunnel, which would link Italy with France. Engineers at the time said the tunnel project would take thirty years to complete. Sommelier worked with steam-powered drills and compressed air to invent the more efficient pneumatic drill pick that became know as the jackhammer. The tunnel was completed in 1871, some twenty years ahead of schedule.

## When was the **road roller** invented?

A Frenchman, Louis Lemoine, invented the steam-driven road roller in 1859. His invention revolutionized road construction, greatly improving the quality of the roadbed. Prior to the road roller, the roadbed was packed manually. Later, a roller harnessed to oxen or horses was introduced.

## Is there significance to the color of **fire hydrant** nozzles?

Color coding of fire hydrants has significance to water and fire departments. The color indicates the water flow available from the hydrant.

| Class | Flow gallons per minute | Color of hydrant bonnets & nozzle caps |
|---|---|---|
| AA | 1500 or greater | Light Blue |
| A | 1000–1499 | Green |
| B | 500–999 | Orange |
| C | Less than 500 | Red |

## Who brought about the **standardization** of **screw threads**?

Two Englishmen are usually credited with pursuing standards for nuts, bolts and screws. H. Mandslay, working in the machine construction industry, tried to introduce a standard between 1800 and 1810. The number of different sized nuts and bolts was relatively small at this time. Mandslay influenced his apprentice J. Withworth to continue to develop standards, and in 1841 the Withworth thread was adopted in England as a standard.

### How does a breathalyzer determine the alcohol level of the breath?

Breathalyzers used by police are usually electronic, using the alcohol blown in through the tube as fuel to produce electric current. The more alcohol the breath contains, the stronger the current. If it lights up a green light, the driver is below the legal limit and has passed the test. An amber light means the alcohol level is near the limit; a red light above the limit. The device has a platinum anode, which causes the alcohol to oxidize into acetic acid with its molecules losing some electrons. This sets up the electric current. Earlier breathalyzers detected alcohol by color change. Orange-yellow crystals of a mixture of sulfuric acid and potassium dichromate in a blowing tube turn to blue-green chromium sulfate and colorless potassium sulphate when the mixture reacts with alcohol, which changes into acetic acid (vinegar). The more crystals that change color, the higher the alcohol level in the body.

## What is the **lost wax** process?

This process is used for making valve parts, small gears, magnets, surgical tools, and jewelry. It involves the use of a wax pattern between a two-layered mold. The wax is removed by melting and replaced with molten metal; hence the name "lost wax."

## What is **sintering**?

Sintering is the bonding together of compacted powder particles at temperatures below the melting point. This bonding produces larger forms such as cakes and pellets. Sintering is used in powder metallurgy technology, which is the production of useful artifacts from metal powder without passing through the molten state. The components produced are referred to as sintered parts. These parts are normally quite small, and some typical examples made today are shock absorber pistons, belt pulleys, small helical gears, drive gears for chainsaws, and automotive pump gears. Because they are molded, sintered parts can have extremely complex shapes, and do not require machining. The toughness and high strength properties of sintered parts make them especially good for today's high technology systems.

## What are **Lichtenberg figures**?

Lichtenberg figures are patterns appearing on a photographic plate or on a plate coated with fine dust when the plate is placed between electrodes and a high voltage is applied between them. The figures were first produced by George Christopher Litchenberg in 1777, marking the discovery of electrostatic recording.

# WEAPONS

*See also: Boats, Trains, Cars, and Planes—Military Vehicles*

## What is an **onager**?

The onager was the simplest of the early catapults. One type of onager twisted a mass of human hair or animal sinew with one wooden beam inserted into it. Geared winches were used to twist the hair or sinew without letting it unwind. To load it, soldiers manned a windlass, which pulled the beam down until it was horizontal, which added more twist to the fiber. A stone was attached to the end of the beam, and this weapon was fired when a soldier pulled a rope that released the beam from its mooring.

## What is the name of the historic weapon consisting of a **round, spiked ball** attached to a chain?

A flail is a weapon consisting of a stout handle fitted with a short iron-shod bar or wooden rod with iron spikes. The "morning star" flail or mace featured one or more spiked balls on a chain. The flexibility of the chain made the weapon harder to parry.

## What is **napalm** and how has it been used?

Napalm was invented in 1943 by the Harvard chemist Louis Feiser in cooperation with the U.S. Army. It's composed of gasoline (33 percent), benzene (21 percent), and polystyrene (46 percent). It was first used in World War II. It burned more slowly and at higher temperatures than pure gasoline and sticks to anything it touches. Napalm depletes the air of oxygen, and, when dropped on bunkers or into caves, it asphyxiates the occupants without burning them. It was used in the Korean War, the Vietnam War, and in Desert Storm, where canisters were dropped on Iraqi fortifications and tank obstacles.

## Is **chemical warfare** a twentieth-century phenomenon?

Chemical warfare can be traced back to ancient and medieval times, especially siege warfare. Assaults on castles and walled cities could take months, even years, so innovative means were sought to end these stalemates. Incendiaries and toxic or greasy smokes were common weapons in both attacking and defending besieged castles or cities. Possibly the first recorded use of poison gas occurred in the wars between Athens and Sparta (431–404 B.C.). Flaming pitch or tar and sulfur mixtures were used during the Peloponnesian War. These smoke-generating mixtures could be quite toxic when inhaled. The Spartan armies are said to have deployed a toxic metal arsenic in vapor clouds.

## What is **mustard gas** and what happens upon exposure?

Mustard gas is made from a variety of chemicals, including sulfur mustard. It is actually a clear liquid, but when mixed with other chemicals, it looks brown and smells like garlic. Mustard gas was used heavily as a chemical warfare agent in World Wars I and II; it causes skin burns, blisters, and damage to the respiratory tract. Exposure to large amounts can cause death.

## Can **anthrax** be used as a **weapon**?

A mere two grams of dried anthrax spores, distributed evenly as a powder over a population of 500,000 in an urban setting, could cause death or serious illness for perhaps 200,000 people. However, many comments about the potential devastation of a massive anthrax attack are misleading and unnecessarily alarming. The attackers would have to overcome a series of very challenging technical problems.

First, the attackers would have to have access to a very virulent strain of the bacterium, because spores harvested from the ground or from farm animals could not be spread as a powder. Next, they would have to grind down the supply of spores so that the particles would be small enough to be deeply inhaled, but not so small that they wouldn't be immediately exhaled. Lastly, the attackers would have to add an anti-static element to the milled particles, because in the process of milling the spores become electrostatically charged, thus forming large clumps of spores that then would drop harmlessly to the ground.

## What other **viruses** are similar to anthrax and could be used in **bio-terrorism**?

Other toxins can be introduced and used in biological warfare. Examples of other toxins are:

*Ricin* is one of the most toxic naturally occurring substances known. It is derived from the seeds of castor bean plants—the same plants that are used to make castor oil.

*Botulism*, like anthrax, is bacteria found in soil. It occasionally strikes people who eat poorly processed canned food or fish in which the bacteria has grown. The bacteria produce an extremely toxic substance, botulinum, that causes blurred vision, dry mouth, difficulty in swallowing or speaking, weakness, and other symptoms. Paralysis, respiratory failure, and death may follow.

*Aflatoxin* and *mycotixn* are both common in crops. Aflatoxin B1 is frequently found in molds that grow on nuts. Iraq and Iran are two of the world's largest producers of pistachio nuts, but the toxin can also be cultivated from molds that grow on corn and other crops. These toxins destroy the immune system in animals and are carcinogenic over a long period of time in humans.

## Which warfare innovations were introduced during the American Civil War?

Barbed wire, trench warfare, hand grenades, land mines, armored trains, ironclad ships, aerial reconnaissance, submarine vessels, machine guns, and even a primitive flamethrower were products of the American Civil War, 1861–1865.

*Clostridium perfringens* is a common source of food poisoning. It is similar to anthrax since it also forms spores that can live in soil. Though its spores are less nasty in food, the organism causes gas gangrene when it finds its way into open battlefield wounds. Gas gangrene produces pain and swelling as the infected area bloats with gas. It later causes shock, jaundice, and death.

Little data exists on *Camelpox,* a virus being developed by Iraq. It is classified as one of the riskiest illegal animal pathogens.

## Who invented the **Bowie knife**?

A popular weapon of the American West, the Bowie knife was named after Jim Bowie (1796–1836), who was killed at the Alamo. According to most reliable sources, his brother Rezin Bowie might have been the actual inventor. The knife's blade measured up to two inches (five centimeters) wide and its length varied from nine to 15 inches (23 to 38 centimeters).

## Who invented **mine barrage**?

In 1777, David Bushnell (1742?–1824) conceived the idea of floating kegs containing explosives that would ignite upon contact with ships.

## When was the **Colt revolver** patented?

The celebrated six-shooter of the American West was named after its inventor, Samuel Colt (1814–1862). Although he did not invent the revolver, he perfected the design, which he first patented in 1835 in England and then in the United States the following year. Colt had hoped to mass-produce this weapon, but failed to get enough backing to acquire the necessary machinery. Consequently, the guns, made by hand, were expensive and attracted only limited orders. In 1847, the Texas Rangers ordered 1,000 pistols, enabling Colt to finally set up assembly-line manufacture in a plant in Hartford, Connecticut.

An early example of a gatling gun—the first successful machine gun—and Richard J. Gatling's claim to fame.

## Why is a **shot tower** used in shot making?

For shot to be accurate and of high velocity, it should be perfectly round. However, early methods for molding lead for shot often resulted in flawed or misshapen balls. In 1872, a British plumber, William Watts, solved this problem when he devised a simple method. This method consisted of pouring the molten lead through a sieve and then allowing the resulting drops to fall from a great height into a pool of water. The air cooled the drops and the water cushioned their fall, preventing them from being deformed. This new technology spread rapidly and shot towers, from 150 to 215 feet (46 to 65 meters) high appeared across Europe and America. Although steel shot has largely replaced lead due to environmental concerns, about 30 towers still drop lead shot worldwide, including five in the United States. The essentials of the 1782 design remain unchanged.

## Who invented the **machine gun**?

The first successful machine gun, invented by Richard J. Gatling (1818–1903), was patented in 1862 during the Civil War. Its six barrels were revolved by gears operated by a hand crank, and it fired 1,200 rounds per minute. Although there had been several partially successful attempts at building a multi-firing weapon, none were able to overcome the many engineering difficulties until Gatling. In his gear-driven machine, cocking and firing were performed by cam action. The United States Army officially adopted the gun on August 24, 1866.

The first automatic machine gun was a highly original design by Hiram S. Maxim (1840–1915). In 1884, the clever Maxim designed a portable, single-barreled automatic weapon that made use of the recoil energy of a fired bullet to eject the spent cartridge and load the next.

The original "tommy gun" was the Thompson Model 1928 SMG. This 45-caliber machine gun, designed in 1918 by General John Taliaferro Thompson (1860–1940), was to be used in close-quarter combat. The war ended before it went into production, however, and Thompson's Auto Ordnance Corporation did not do well, until the gun was adopted by American gangsters during Prohibition. The image of a reckless criminal spraying his enemies with bullets from his hand-held "tommy gun" became a symbol of the Depression years. The gun was modified several times and was much used during World War II.

## How did the **bazooka** get its name?

It was coined by American comedian Bob Burns (1893–1956). As a prop in his act he used a unique musical instrument that was long and cylindrical and resembled an oboe. When United States Army soldiers in World War II were first issued hollow-tube rocket launchers, they named them bazookas because of their similarity to Burns' instrument.

## Who was known as the **Cannon King**?

Alfred Krupp (1812–1887), whose father Friedrich (1787–1826) established the family's cast-steel factory in 1811, began manufacturing guns in 1856. Krupp supplied large weapons to so many nations that he became known as the "Cannon King." Prussia's victory in the Franco-German War of 1870–1871 was largely the result of Krupp's field guns. In 1933 when Hitler came to power, this family business began manufacturing a wide range of artillery. Alfred Krupp (1907–1967), the great-grandson of Alfred, supported the Nazis in power and accrued staggering wealth for the company. The firm seized property in occupied countries and used slave labor in its factories. After the war, Alfred was imprisoned for 12 years and had to forfeit all his property. Granted amnesty in 1951, he restored the business to its former position by the early 1960s. On his death in 1967, however, the firm became a corporation and the Krupp family dynasty ended.

## What is considered the most **technologically advanced handgun**?

Metal Storm's O'Dwyer Variable Lethality Law Enforcement pistol is cited as the most sophisticated handgun. It has no moving parts, and its firing system is totally electronic. The seven-shot single barrel can fire multiple rounds with a single pull of the trigger. It has an in-built electronic security system to limit its use to authorized users. The O'Dwyer is intended for specialist police teams and the military.

## What does a **crystal ball** refer to in the military?

"Crystal ball" is slang used for a radar scope by pilots.

## How did **Big Bertha** get its name?

This was the popular name first applied to the 16.5-inch (42-centimeter) howitzers used by the Germans and Austrians in 1914, the year World War I started. Subsequently, the term included other types of huge artillery pieces used during World Wars I and II. The large gun was built by the German arms manufacturer Friedrich A. Krupp (1854–1902), and was named after his only child, Bertha Krupp (1886–1957).

These large guns were used to destroy the concrete and steel forts defending Belgium in 1914, and their shells weighed 205 pounds (930 kilograms) and were nearly as tall as a man. Since it was slow and difficult to move such guns, the use of them was practical only in static warfare. In World War II, bomber aircraft took away this role of long-range bombardment from these cumbersome guns.

## What was the **Manhattan Project**?

The Manhattan Engineer District was the formal code name for the United States government project to develop an atomic bomb during World War II. It soon became known as the Manhattan Project—a name taken from the location of the office of Colonel James C. Marshall, who had been selected by the United States Army Corps of Engineers to build and run the bomb's production facilities. When the project was activated by the United States War Department in June 1942, it came under the direction of Colonel Leslie R. Groves (1896–1970).

The first major accomplishment of the project's scientists was the successful initiation of the first self-sustaining nuclear chain reaction, done at a University of Chicago laboratory on December 2, 1942. The project tested the first experimental detonation of an atomic bomb in a desert area near Alamogordo, New Mexico, on July 16, 1945. The test site was called Trinity, and the bomb generated an explosive power equivalent to between 15,000 and 20,000 tons (15,240 to 20,320 tonnes) of TNT. Two of the project's bombs were dropped on Japan the following month (Hiroshima on August 6 and Nagasaki on August 9, 1945) resulting in the Japanese surrender that ended World War II.

## What were **"Little Boy"** and **"Fat Man"**?

"Little Boy" was the code name for the first atomic weapon dropped by the United States on Hiroshima. Little Boy was designed at Los Alamos, New Mexico, by a team of scientists headed by J. Robert Oppenheimer. Little Boy measured about 10 feet long and weighed 8,000 pounds with an explosive yield of a minimum of 12,000 tons of

TNT. It was fueled by uranium 235. "Fat Man" was the code name of the second atomic bomb dropped by the United States on Nagasaki. Fat Man was also designed at Los Alamos by a team of scientists headed by Oppenheimer. Like Little Boy, it measured about 10 feet long, but it weighed 9,000 pounds and had an explosive yield of 20,000 tons of TNT. Fat Man was fueled by plutonium. The least powerful nuclear weapons available today are more powerful than either Little Boy or Fat Man.

### How does a **smart bomb** work?

"Smart bomb" is slang for a bomb that can be accurately guided to its target by a laser beam, radar, radio control, or electro-optical system. Although pilots must accurately release the bomb in the general direction of the target, steerable tail fins allow minor adjustments in the bomb's glide path to be made in order to more accurately reach the target. An advantage of smart bombs is that they can be released further away from the target, thus reducing the aircraft's vulnerability to enemy ground fire.

### How far can a **Pershing missile** travel?

The surface-to-surface nuclear missile, which is 34.5 feet (10.5 meters) long and weighs 10,000 pounds (4,536 kilograms), has a range of about 1,120 miles (1,800 kilometers). It was developed by the United States Army in 1972. Other surface-to-surface missiles are the Polaris, with a range of 2,860 miles (4,600 kilometers); the Minuteman, with a range of 1,120 miles (1,800 kilometers); the Tomahawk, 2,300 miles (3,700 kilometers); the Trident, 4,600 miles (7,400 kilometers); and the Peacemaker, 6,200 miles (10,000 kilometers).

### How does a **cruise missile** work?

Cruise missiles are highly accurate, long-range guided missiles that fly at low altitudes at high subsonic speeds. They are guided by Global Positioning System (GPS) or Terrain Contour Matching (TERCOM) or Digital Scene Matching Area Correlation (DSMAC) guidance systems. Cruise missiles are difficult to detect on radar because of their small size and low-altitude flying capability.

### What is a **nuclear winter**?

The term "nuclear winter" was coined by American physicist Richard P. Turco in a 1983 article in the journal *Science*, in which he describes a hypothetical post-nuclear war scenario having severe worldwide climatic changes: prolonged periods of darkness, below-freezing temperatures, violent windstorms, and persistent radioactive fallout. This would be caused by billions of tons of dust, soot, and ash being tossed into the atmosphere, accompanied by smoke and poisonous fumes from firestorms. In the case

of a severe nuclear war, within a few days, the entire northern hemisphere would be under a blanket so thick that as little as 1/10 of 1 percent of available sunlight would reach the Earth. Without sunlight, temperatures would drop well below freezing for a year or longer, causing dire consequences for all plant and animal life on Earth.

Reaction to this doomsday prediction led critics to coin the term "nuclear autumn," which downplayed such climatic effects and casualties. In January 1990, the release of *Climate and Smoke: An Appraisal of Nuclear Winter*, based on five years of laboratory studies and field experiments, reinforced the original 1983 conclusions.

## What is the most destructive **non-lethal weapon**?

Known as the "Blackout Bomb", the BLU-114/B graphite bomb was used by NATO against Serbia in May 1999, disabling 70 percent of its power grid with minimal casualties. The bomb was specifically designed to attack electrical power infrastructures. A similar bomb was used against Iraq in the Gulf War of 1990–1991. It knocked out 85 percent of Iraq's electricity generating capacity. It works by exploding a cloud of ultra-fine carbon-fiber wires over electrical installations, causing a temporary short-circuiting and disruption of electrical supply.

## What is a **radiological dispersal device**?

Better known as a "dirty bomb," this device is designed to release massive amounts of radioactive material as widely as possible. Unlike a nuclear bomb, which is designed to release large amounts of energy such as heat and radiation by splitting or fusing atoms, the dirty bomb is a conventional bomb packed with radioactive materials. Components include materials used in medicine and research as well as low-grade, unenriched uranium. Dirty bombs are not considered true weapons of mass destruction; instead they are intended to cause mass disruption. The economic and psychological consequences of having an important urban area contaminated with radiation could be severe.

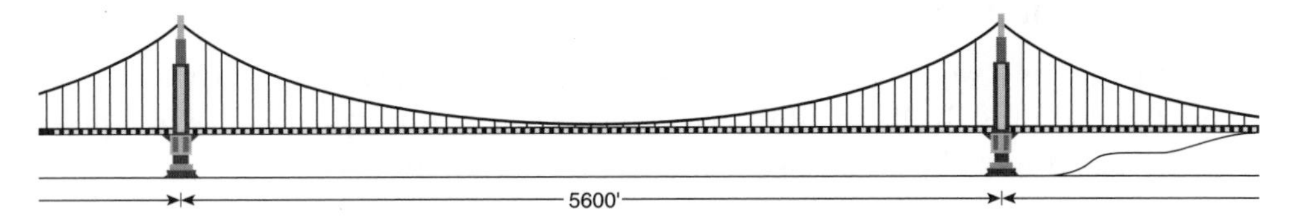


# BUILDINGS, BRIDGES, AND OTHER STRUCTURES

## BUILDINGS AND BUILDING PARTS

### How much **wood** is used in the construction of a **2,000-square-foot house**?

It takes about 15,000 board feet of lumber to build a 2,000-square-foot house.

### How does a **chimney** differ from a **flue**?

A chimney is a brick-and-masonry construction that contains one or more flues. A flue is a passage within a chimney through which smoke, fumes, and gases ascend. A flue is lined with clay or steel to contain the combustion wastes. By channeling the warm, rising gases, a flue creates a draft that pulls the air over the fire and up the flue. Each heat source needs its own flue, but one chimney can house several flues.

### How tall is the world's **tallest chimney**?

The number two stack of the Ekibastuz, Kazakhstan, power plant is 1,378 feet (420 meters) tall.

### Which part of a door is a **doorjamb**?

The doorjamb is not part of the door, but is the surrounding case into and out of which the door opens and closes. It consists of two upright pieces, called side jambs, and a horizontal head jamb.

## What is **crown molding**?

Crown molding is a wood, metal, or plaster finishing strip placed on the wall where it intersects the ceiling. If the molding has a concave face, it is called a cove molding. In inside corners it must be cope-jointed to insure a tight joint.

## What is **R-value**?

An R-value, or resistance to the flow of heat, is a special measurement of insulation that represents the difficulty with which heat flows through an insulating material. The higher the R-value, the greater the insulating value of the material. A wall's component R-values can be added up to get its total R-value:

| Wall component | R-value |
|---|---|
| Inside Air Film | 0.7 |
| 1/2" Gypsum Wallboard | 0.5 |
| R-13 Insulation | 13.0 |
| 1/2" Wood Fiber Sheathing | 1.3 |
| Wood Siding | 0.8 |
| Outside Air Film | 0.2 |
| Total R-value | 16.5 |

| Building component | R-value |
|---|---|
| Standard attic ceiling | 19 |
| Standard 4-inch-thick insulated wall | 11 |
| Typical single-pane glass window | 1 |
| Double-glazed window | 2 |
| "Superwindows" (inner surface of a pane coated with an infrared radiation reflector material such as tin oxide and argon gas filling in the space between the panes of a double-glazed window.) | 4 |

## Why are **Phillips screws** used?

The recessed head and cross-shaped slots of a Phillips screw are self-centering and allow a closer, tighter fit than conventional screws. Straight-slotted screws can allow the screwdriver to slip out of the groove, ruining the wood.

Screws were used in carpentry as far back as the 16th century, but slotted screws with a tapering point were made at the beginning of the 19th century. The great advantage of screws over nails is that they are extremely resistant to longitudinal tension. Larger-size screws that require considerable force to insert have square heads

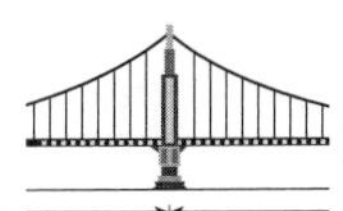



that can be tightened with a wrench. Screws provide more holding power than nails and can be withdrawn without damaging the material. Types of screws include wood screws, lag screws (longer and heavier than wood screws), expansion anchors (usually for masonry), and sheet metal screws. The screw sizes vary from 0.25 to six inches (six millimeters to 15 centimeters).

## What is the **BOCA code**?

The Building Officials and Code Administrators (BOCA) International is a service organization that issues a series of model regulatory construction codes for the protection of public health, safety, and welfare. The codes are published in sections, such as the National Building Code, National Plumbing Code, National Fire Prevention Code, etc. They are designed for adoption by state or local governments and may be amended or modified to accomplish desired local requirements.

## What is an **STC rating** and what does it mean?

The STC (Sound Transmission Class) tells how well a wall or floor assembly inhibits airborne sound. The higher the number, the greater the sound barrier. Typical STC ratings are as follows:

| STC number | What it means |
|---|---|
| 25 | Normal speech can be easily understood |
| 30 | Loud speech can be understood |
| 35 | Loud speech is audible, but not intelligible |
| 42 | Loud speech audible as a murmur |
| 45 | Must strain to hear loud speech |
| 48 | Some loud speech barely audible |
| 50 | Loud speech not audible |

The degree to which sound barriers work depends greatly on the elimination of air gaps under doors, through electrical outlets, and around heating ducts.

## Why is the term **penny** used in nail sizes?

The term "penny" originated in England and is a measurement relating to the length of nails. One explanation is that it refers to cost, with the cost of 100 nails of a certain size being 10 pence or 10d ("d" being the British symbol for a penny). Another explanation suggests that the term refers to the weight of 1,000 nails, with "d" at one time being used as an abbreviation for a pound in weight.

People have been using nails for the last 5,000 years or so. Nails are known to have been used in Ur (ancient Iraq) to fasten together sheet metal. Before 1500, nails were made by hand by drawing small pieces of metal through a succession of graded holes in a metal plate. In 1741, 60,000 people were employed in England making nails.

The first nail-making machine was invented by the American Ezekiel Reed. In 1851, Adolphe F. Brown of New York invented a wire-nail-making machine. This enabled nails to be mass-produced cheaply.

## What is **pisé** or rammed earth?

Rammed earth is an ancient building technique in which moist earth is compacted into a rough approximation of sedimentary rock. Using forms, rammed earth may be shaped into bricks or entire walls. Dating back as far as 7000 B.C.E., it was used in portions of the 2,000-year-old Great Wall of China as well as temples in Mali and in Morocco. Romans and Phoenicians introduced the technique to Europeans, and it became a popular building technique in France, where it became known as *pisé de terre*. In the United States, it was used for both gracious Victorian homes and low-cost housing.

Today, builders add cement to the earthen mix, which results in stronger, waterproof walls. In traditional rammed-earth construction, hand or compressor-powered tampers are used to compact a mixture of moistened earth and cement between double-sided wooden forms to about 60 percent of its original volume. In a newer technique, a high-pressure hose sprays the mixture against a single-sided form. Sometimes steel-reinforced bars are also used.

## What is a **yurt**?

Originally a Mongolian hut, the yurt has been adapted in the United States as a low-cost structure that can be used as a dwelling. The foundation is built of wood on a hexagonal frame. The wooden lattice-work side walls have a tension cable sandwiched between the wall pieces at the top to keep the walls from collapsing. The walls are insulated and covered with boards, log slabs, canvas, or aluminum siding. A shingle roof, electricity, plumbing, and a small heating stove may be installed. The interior can be finished as desired with shelves, room dividers, and interior siding. The yurt is beautiful, practical, and relatively inexpensive to erect.

## When was the first **skyscraper** built?

Designed by William Le Baron Jenney (1832–1907), the first skyscraper, the 10-story Home Insurance Company Building in Chicago, Illinois, was completed in 1885. A skyscraper—a very tall building supported by an internal frame (skeleton) of iron and steel rather than by load-bearing walls—maximizes floor space on limited land. Three technological developments made skyscrapers feasible: a better understanding of how materials behave under stress and load (from engineering and bridge design); the use of steel or iron framing to create a structure, with the outer skin "hung" on the frame;

### How much does the leaning tower of Pisa lean?

The leaning tower of Pisa, 184.5 feet (56 meters) tall, is about 17 feet (five meters) out of perpendicular, increasing by about 0.2 inch (1.25 millimeters) a year. The Romanesque-style tower was started by Bonanno Pisano in 1173 as a campanile or bell tower for the nearby baptistry, but was not completed until 1372. Built entirely of white marble, with eight tiers of arched arcades, it began to lean during construction. Although the foundation was dug down to 10 feet (three meters), the builders did not reach bedrock for a firm footing. Ingenious attempts were made to compensate for the tilt by straightening up the subsequent stories and making the pillars higher on the south side than on the north. The most recent attempt at compensation for the tilt, completed in 2001, removed earth from the north side of the foundation—the side opposite the tilt. It is hoped that this "fix" will hold for about 300 years.

and the introduction of the first "safety" passenger elevator, invented by Elisha Otis (1811–1861) and patented on January 15, 1861 (No. 31,128).

## When was the **first shopping center** built?

The first shopping center in the world was built in 1896 at Roland Park, Baltimore, Maryland. One of the world's largest shopping centers is the West Edmonton Mall in Alberta, Canada, which covers 5.2 million square feet (480,000 square meters) on a 121-acre (49-hectare) site. It has 828 stores and services, with parking for 20,000 vehicles.

## What material was used to construct the exterior of the **Empire State Building**?

The exterior of the Empire State Building is made of Indiana limestone and granite with vertical strips of stainless steel.

## What is the ground area of the **Pentagon Building** in Washington, D.C.?

The Pentagon, the headquarters of the United States Department of Defense, is the world's largest office building in terms of ground space. Its construction was completed on January 15, 1943—in just over 17 months. With a gross floor area of over 6.5 million square feet (604,000 square meters), this five-story, five-sided building has three times the floor space of the Empire State Building and is one and a half times larger than the Sears Tower in Chicago. The World Trade Center complex in New York City, completed in 1973 and destroyed in 2001, was larger with over nine million square feet (836,000 square meters), but it consisted of two structures (towers). Each of the Pentagon's five sides is 921 feet (281 meters) long with a perimeter of 4,610 feet (1,405 meters). The secretary of defense, the secretaries of the three military departments, and the military heads of the army, navy, and air force are all located in the Pentagon. The National Military Command Center, which is the nation's military communications hub, is located where the Joint Chiefs of Staff convene. It is commonly called the "war room."

The largest commercial building in the world under one roof is the flower auction building of the cooperative VBA in Aalsmeer, the Netherlands. In 1986, the floor plan was extended to 91 acres (37 hectares). It measures 2,546 × 2,070 feet (776 × 639 meters). The largest building in the United States, and the largest assembly plant in the world, is the Boeing 747 assembly plant in Everett, Washington, with a capacity of 200 million cubic feet (5.5 million cubic meters) and covering 47 acres (19 hectares).

## What **criteria** are used to determine the **tallest building**?

The Council on Tall Buildings and Urban Habitat, representing world leaders in the field of the built environment, measures the height (structural top) of a building from the sidewalk level of the main entrance to the structural top of the building. The measurement includes spire, but does not include television antennas, radio antennas or flagpoles. Other measures of height include the height to the highest occupied floor of building, height to the top of the roof, and height to the tip of a spire, pinnacle antenna, mast or flagpole.

## What are the **tallest buildings** in the United States?

The tallest buildings in the United States are:

| Building | Location | Year built | Height | | Stories |
|---|---|---|---|---|---|
| | | | Feet | Meters | |
| Sears Tower | Chicago, Illinois | 1974 | 1454 | 443 | 110 |
| plus two TV towers | | | 1707 | 520.9 | |
| Empire State Bldg. | New York, New York | 1931 | 1250 | 381 | 102 |
| plus two TV towers | | | 1414 | 431 | |

Prior to its destruction in 2001, the World Trade Center in New York City was taller than the Empire State Building. The North Tower, completed in 1972, stood 1,368 feet (417 meters) high, and the South Tower, completed in 1973, stood 1,362 feet (415 meters) high. Both structured were 110 stories tall.

## What are the **tallest buildings** in the **world**?

The Petronas Tower 1 and Petronas Tower 2, located in Kuala Lumpur, Malaysia, are 1,483 feet (452 meters) tall. The buildings were completed in 1998.

## What is the **tallest self-supporting structure** in the world?

The tallest self-supporting structure in the world is the CN Tower in Toronto, Canada, at 1,815 feet (533 meters), built in 1975.

The CN Tower in Toronto, Canada, is among the world's tallest self-supporting structures.

## What is the **tallest structure** in the world?

The tallest structure in the world is the KVLY-TV transmitting tower in Blanchard, North Dakota, at 2,063 feet (629 meters) tall.

## Who invented the **geodesic dome**?

A geodesic line is the shortest distance between two points across a surface. If that surface is curved, a geodesic line across it will usually be curved as well. A geodesic line on the surface of a sphere will be part of a great circle. Buckminster Fuller (1895–1983) realized that the surface of a sphere could be divided into triangles by a network of geodesic lines, and that structures could be designed so that their main elements either followed those lines or were joined along them.

This is the basis of his very successful geodesic dome: a structure of generally spherical form constructed of many light, straight structural elements in tension and arranged in a framework of triangles to reduce stress and weight. These contiguous tetrahedrons are made from lightweight alloys with high tensile strength.

An early example at a geodesic structure is Britain's Dome of Discovery, built in 1951. It was the first dome ever built with principal framing members intentionally aligned along great circle arcs. The ASM (American Society for Metals) dome, which was built east of Cleveland, Ohio, in 1959–1960, is an open lattice-work geodesic dome. Built in 1965, the Houston Astrodome also forms a giant geodesic dome.

## When was the first **retractable roof stadium** opened?

The first retractable roof stadium, the Skydome in Toronto, was opened in 1989. Unlike previous stadiums that had removable tops, the Skydome had a fully retractable roof. The roof consists of four panels, weighing 11,000 tons (9,979 tonnes),

Buckminster Fuller stands before a geodesic dome structure—reflecting his unique design.

and runs on a system of steel tracks and bogies. One panel remains fixed while two of the other panels move forward and backward at a rate of 71 feet (21 meters) per minute to open or close. The fourth panel rotates 180 degrees to completely open or close the roof. It takes 20 minutes for the roof to open or close. Four other retractable roof stadiums have opened since the Skydome—Bank One Ballpark in Phoenix, Arizona (1998), Safeco Field in Seattle, Washington (1999), Enron Field in Houston, Texas (2000), and Reliant Stadium in Houston, Texas (2002).

## Which building has the largest **non-air-supported clear span roof**?

The Suncoast Dome in St. Petersburg, Florida, completed in 1990, has a clear span of 688 feet (210 meters). Covering an area of 372,000 square feet (34,570 square meters), the fabric-covered dome is a cable roof structure—the newest structural system for trussed domes. Its structural behavior is exactly opposite that of the traditional dome: the base ring is in compression rather than tension, and the ring at the crown is in tension rather than compression. Also, the space enclosed is not totally free and unobstructed, but includes structural, top-to-bottom members. To retain its characteristic lightness, the roofing surface employs flexible fabric membranes. These must be flexible since cable structures undergo major structural distortions, and they actually change shape under different load conditions. The largest air-supported building is the 80,600-capacity octagonal Pontiac Silverdome Stadium in Pontiac, Michigan. 522 feet (159 meters) wide and 722 feet (220 meters) long, it is supported by a 10-acre (1.62-hectare) translucent fiberglass roof.

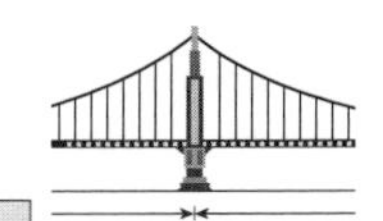



### What is the "topping out" party that iron workers have?

When the last beam is placed on a new bridge, skyscraper, or building, ironworkers hoist up an evergreen tree, attach a flag or a handkerchief, and brightly paint the final beam and autograph it.

This custom of raising an evergreen tree goes back to Scandinavia in 700 C.E., when attaching the tree to the building's ridge pole signaled to all who helped that the celebration of its completion would begin.

### When did Sears, Roebuck and Company sell **homes by mail order**?

Between 1908 and 1940, Sears manufactured and sold houses in approximately 450 ready-to-assemble designs from mansions to bungalows. Ordered by mail and sent by rail, these popular houses were complete with plumbing and electricity. Over 100,000 were sold at prices ranging from $595 to $5,000.

Sears' success in selling houses was tied, in large part, to its attractive financing plans. By 1911 the company had begun to offer loans for the purchase of materials, and by 1918 it would sometimes advance a portion of the capital required for labor.

Although most homes were sold to individuals, Sears also sold houses to companies for company towns near their factories. Standard Oil Company in Illinois, and Bethlehem Steel in Hellertown, Pennsylvania, were corporate customers.

## ROADS, BRIDGES, AND TUNNELS

### Who is known as the founder of **civil engineering**?

Thomas Telford (1757–1834), the first president of the Institute of Civil Engineers, is the founder of the British civil engineering profession. Telford established the professional ethos and tradition of the civil engineer—a tradition followed by all engineers today. He built bridges, roads, harbors, and canals. His greatest works include the Menai Strait Suspension Bridge and Pontcysyllte Aqueduct, the Gotha Canal in Sweden, the Caledonian Canal, and many Scottish roads. He was the first and greatest master of the iron bridge.

### Which state has the **most miles of roads**?

Texas lays claim to the most road mileage with 296,651 miles (474,642 kilometers). California is a distant second with 170,601 miles (272,962 kilometers) of roads. On the other end of the scale, only three states have less than 10,000 miles (16,090 kilometers) of roads: Delaware, Hawaii, and Rhode Island. The total mileage count for the United States is 3,944,721 miles (6,311,361 kilometers).

## When was the **first U.S. coast-to-coast highway** built?

Completed in 1923, after ten years of planning and construction, the Lincoln Highway was the first transcontinental highway connecting the Atlantic coast (New York) to the Pacific (California). Sometimes called the "Main Street of the United States," the highway was proposed by Carl G. Fisher to a group of automobile manufacturers, who formed the Lincoln Highway Association and promoted the idea. Fisher thought that "Lincoln" would be a good, patriotic name for the road.

Its original length of 3,389 miles (5,453 kilometers) was later shortened by relocations and improvements to 3,143 miles (5,057 kilometers). Crossing 12 states—New York, New Jersey, Pennsylvania, Ohio, Indiana, Illinois, Nebraska, Colorado, Wyoming, Utah, Nevada, and California—in 1925, the Lincoln Highway became, for most of its length, U.S. Route 30.

## How are U.S. **highways numbered**?

The main north–south interstate highways always have odd numbers of one or two digits. The system, beginning with Interstate 5 on the West Coast, increases in number as it moves eastward, and ends with Interstate 95 on the East Coast.

The east–west interstate highways have even numbers. The lowest-numbered highway begins in Florida with Interstate 4, increasing in number as it moves northward, and ends with Interstate 96. Coast-to-coast east–west interstates, such as Routes 10, 40, and 80, end in zero. An interstate with three digits is either a beltway or a spur route.

U.S. routes follow the same numbering system as the interstate system, but increase in number from east to west and from north to south. U.S. Route 1, for example, runs along the East Coast; U.S. Route 2 runs along the Canadian border. U.S. route numbers may have anywhere from one to three digits.

## Which road has the **most lanes**?

There are 23 lanes (17 westbound) at the San Francisco–Oakland Bay Bridge Toll Plaza in Oakland, California.

## Which city had the **first traffic light**?

On December 10, 1868, the first traffic light was erected on a 22-foot (6.7-meter) high cast-iron pillar at the corner of Bridge Street and New Palace Yard off Parliament

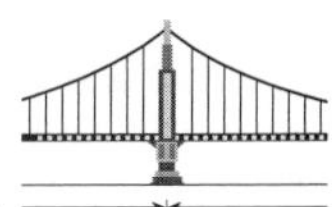



Square in London, England. Invented by J. P. Knight, a railway signaling engineer, the light was a revolving lantern illuminated by gas, with red and green signals. It was turned by hand using a lever at the base of the pole.

Cleveland, Ohio, installed an electric traffic signal at Euclid Avenue and 105th Street on August 5, 1914. It had red and green lights with a warning buzzer as the color changed.

Around 1913, Detroit, Michigan, used a system of manually operated semaphores. Eventually, the semaphores were fitted with colored lanterns for night traffic. New York City installed the first three-color light signals in 1918; these signals were still operated manually.

## What is a **Jersey barrier**?

Jersey barriers are concrete highway barriers developed by the New Jersey Department of Transportation. Originally only 12 to 18 inches (30 to 46 centimeters) high, they were designed to prevent left turns at certain intersections. Later barriers, reinforced concrete usually poured on the site, were used as temporary traffic safeguards where construction required motorists to cross over into a lane normally used by oncoming traffic. These barriers were 32 inches (81 centimeters) high. Now the barriers line thousands of miles of American highways as permanent fixtures. Barriers are now 54 inches (137 centimeters) high, thus blocking out the glare of oncoming headlights.

## How do cars, buses, and trucks travel in the **English Channel tunnel**?

Cars, buses, and trucks are carried on 2,500-foot-long (762-meter-long) trains traveling at 90 miles per hour (145 kilometers per hour) in the two 31-mile (50-kilometer) long tunnels under the English Channel. Automobile operators drive their vehicles onto the trains and remain inside until the passage is complete. The trains carry cars and buses in large wagons. Truck drivers travel separately from their vehicles, which ride in simpler, semi-open wagons. Interspersed with the shuttles are high-speed passenger trains designed to run on the different railroad systems of Britain, Continental Europe, and the tunnel.

## What is the world's **longest road tunnel**?

The world's longest road tunnel runs between Laerdal and Aurland in Norway. Completed in 2000, the tunnel is 15.2 miles (24.5 kilometers) long. The world's second longest road tunnel is the 10.14-mile (16-kilometer), two-lane St. Gotthard tunnel from Göschenen to Airolo, Switzerland, which opened to traffic on September 5, 1980. In the United States the longest road tunnel is the Anton Anderson Memorial Tunnel near the city of Whittier in Alaska. It is 2.5 miles (4,184 meters) long and was completed in June 2000.

### Why are manhole covers round?

Circular covers are almost universally used on sewer manholes because they cannot drop through the opening. The circular cover rests on a lip that is smaller than the cover, and the cover and the lip are manufactured together, making for a particularly tight fit. Any other shape—such as a square or rectangle—could slip into the manhole opening. In addition, round manhole covers can be machined more accurately than other shapes, and, once removed, round covers can be rolled rather than having to be lifted.

## Where is the **longest bridge-tunnel** in the world?

Completed in 1964 after 42 months' work and $200 million and spanning a great distance of open sea, the Chesapeake Bay Bridge-Tunnel is a 17.5-mile (28-kilometer) combination of trestles, bridges, and tunnels that connect Norfolk with Cape Charles in Virginia. Its only rival for crossing so much open and deep water is the Zuider Zee Dam in Holland—a road-carrying structure of similar length, but without the same water depth and same length of open sea.

## What are the various **types of bridge** structures?

Four basic types of structures can be used to bridge a stream or similar obstacle: rigid beam, cantilever, arch, and suspension systems.

The *rigid beam bridge*, the simplest and most common form of bridging, has straight slabs or girders carrying the roadbed. The span is relatively short and its load rests on its supports or piers.

The *arch bridge* is in compression, and thrusts outward on its bearings at each end.

In the *suspension bridge*, the roadway hangs on steel cables, with the bulk of the load carried on cables anchored to the banks. It can span a great distance without intermediate piers.

Each arm of a *cantilever bridge* is, or could be, free-standing, with the load of the short central truss span pushing down through the piers of the outer arms and pulling up at each end. The outer arms are usually anchored at the abutments and project into the central truss.

## What is a **"kissing bridge"**?

Covered bridges with roofs and wooden sides are called "kissing bridges," because people inside the bridge could not be seen from outside. Such bridges can be traced back

to the early 19th century. Contrary to folk wisdom, they were not designed to produce rural "lovers' lanes," but were covered to protect the structures from deterioration.

## How many **covered bridges** are there in the United States?

More than 10,000 covered bridges were built across the United States between 1805 (when the first was erected in Philadelphia) and the early 20th century. As of January 1980, only 893 of these covered bridges remained—231 in Pennsylvania, 157 in Ohio, 103 in Indiana, 100 in Vermont, 54 in Oregon, and 52 in New Hampshire. Three interstate bridges link New Hampshire and Vermont. The remainder are scattered throughout the country. Non-authentic covered bridges—built or covered for visual effect—appear in each state.

## What are the world's **longest bridge spans**?

The world's longest main-span in a suspension bridge is the Akashi-Kaikyo bridge in Japan at 6,066 feet (1,991 meters), linking the Kobe and Awajishima Islands. The overall length of the bridge is 12,825 feet (3,911 meters). Construction began in 1988, and the bridge was opened in 1998. When the Messina Bridge linking Sicily with Calabria on the Italian mainland is completed it will become the longest by far, with a single span of 10,892 feet (3,320 meters).

Major Suspension Bridges

| Bridge name | Location | Year opened | Length between spans feet | meters |
|---|---|---|---|---|
| Akashi-Kaikyo | Japan | 1998 | 6,066 | 1,991 |
| Great Belt Link | Denmark | 1997 | 5,328 | 1,624 |
| Humber Estuary | England | 1981 | 4,626 | 1,410 |
| Verrazano-Narrows | New York | 1964 | 4,260 | 1,298 |
| Golden Gate | California | 1937 | 4,200 | 1,280 |

The world's longest cable-stayed bridge span is Tatara in Japan, with a span of 2,920 feet (888 meters) completed in 1999.

The world's longest cantilever bridge is the Quebec Bridge over the St. Lawrence River in Canada, opened in 1917, which has a span of 1,800 feet (549 meters) between piers, with an overall length of 3,239 feet (987 meters).

The world's longest steel-arch bridge is the New River Gorge Bridge near Fayetteville, West Virginia, completed in 1977, with a span of 1,700 feet (518 meters).

The world's longest concrete-arch bridge is the Jesse H. Jones Memorial Bridge, which spans Houston Ship Canal in Texas. Completed in 1982, the bridge measures 1,500 feet (457 meters).

The world's longest stone-arch bridge is the 3,810-foot (1,161-meter) Rockville Bridge, completed in 1901, north of Harrisburg, Pennsylvania, with 48 spans containing 216 tons (219 tonnes) of stone.

### Where is the **longest suspension bridge** in the United States?

Spanning New York (City) Harbor, the Verrazano-Narrows Bridge is the longest suspension bridge in the United States. With a span of 4,260 feet (1,298 meters) from tower to tower, its total length is 7,200 feet (2,194 meters). Named after Giovanni da Verrazano (1485–1528), the Italian explorer who discovered New York Harbor in April 1524, it was erected under the direction of Othmar H. Ammann (1879–1965) and completed in 1964. To avoid impeding navigation in and out of the harbor, it provides a clearance of 216 feet (66 meters) between the water level and the bottom of the bridge deck. Like other suspension bridges, the bulk of the load of the Verrazano-Narrows Bridge is carried on cables anchored to the banks.

### Which place in the United States has the **most bridges**?

With the possible exception of Venice, Italy, Pittsburgh, Pennsylvania, is the bridge capital of the world. There are more than 1,900 bridges in Allegheny County's 731 square miles (1,892 square kilometers). The county has 2.6 bridges per square mile, or one bridge for every mile of highway. Nationwide, there are approximately 589,250 bridges.

### Are there any **floating bridges** in the United States?

The four floating pontoon bridges in the United States are all located in the state of Washington. The Lacy V. Murrow–Lake Washington Bridge (1993), The Evergreen Point Bridge (1963), and the Third Lake Washington Bridge (1989) in Seattle are 6,543 feet (1,994 meters), 7,518 feet (2,291 meters), and 6,130 feet (1,868 meters) long respectively. The Hood Canal Bridge (1961) in Port Gamble is 6,471 feet (1,972 meters) long.

### Who built the **Brooklyn Bridge**?

John A. Roebling (1806–1869), a German-born American engineer, constructed the first truly modern suspension bridge in 1855. Towers supporting massive cables, tension anchorage for stays, a roadway suspended from the main cables, and a stiffening deck below or beside the road deck to prevent oscillation are all characteristics of Roebling's suspension bridge. In 1867, Roebling was given the ambitious task of constructing the Brooklyn Bridge. In his design he proposed the revolutionary idea of using steel wire for cables rather than the less-resilient iron. Just as construction began, Roebling died of tetanus when his foot was crushed in an accident, and his son, Washington A. Roebling (1837–1926), assumed responsibility for the bridge's construction. Fourteen

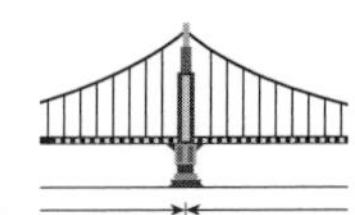



years later, in 1883, the bridge was completed. At that time, it was the longest suspension bridge in the world, spanning the East River and connecting New York's Manhattan with Brooklyn. The bridge has a central span of 1,595 feet (486 meters), with its masonry towers rising 276 feet (841 meters) above high water. Today, the Brooklyn Bridge is among the best-known of all American civil engineering accomplishments.

### What is the longest **causeway**?

The Lake Pontchartrain Causeway in Louisiana is the longest overwater highway bridge. The causeway consists of two twin spans 80 feet (24 meters) apart. The northbound span is 23.87 miles (38.19 kilometers) long and the southbound span is 23.86 miles (38.2 kilometers) long. The first span was opened in 1956 and thirteen years later, in 1969, the second span was opened.

### How was the safety of the **first bridge across the Mississippi River** at St. Louis proved in 1874?

According to a historical appraisal of the bridge by Howard Miller, "Progressively heavier trains shuttled back and forth across the bridge as its engineer, James B. Eads, took meticulous measurements. However, the general public was probably more reassured by a nonscientific test. Everyone knew that elephants had canny instincts and would not set foot on an unsound bridge. The crowd cheered as a great beast from a local menagerie mounted the approach without hesitation and lumbered placidly across to the Illinois side."

### Who designed the **Golden Gate Bridge**?

Joseph B. Strauss (1870–1938), formally named chief engineer for the project in 1929, was assisted by Charles Ellis and Leon Moissieff in the design. An engineering masterpiece that opened to traffic in May 1937, this suspension bridge spans San Francisco Bay, linking San Francisco with Marin County, California. It has a central span of 4,200 feet (1,280 meters) with towers rising 746 feet (227 meters).

## MISCELLANEOUS STRUCTURES

### What is the **tallest national monument**?

The Gateway Arch in St. Louis, Missouri, at 630 feet (192 meters), is 75 feet (23 meters) taller than the Washington Monument and the tallest national monument. It

### What is a Texas tower?

A Texas tower is an off-shore platform used as a radar station. Built on pilings sunk into the ocean floor, it resembles the offshore oil derricks or rigs first developed in the Gulf of Mexico off the coast of Texas. In addition to the radar, the components of a typical tower may include crew quarters, a helicopter landing pad, fog signals, and oceanographic equipment.

was designed by Eero Saarinen (1910–1961) in the shape of an inverted catenary curve using stainless steel. Some interesting facts and figures about the Gateway Arch are:

#### General Facts

Outer Width, Outside North Leg to Outer South Leg: 630 feet (192 meters)
Maximum Height: 630 feet (192 meters)
Shape of Arch Section: Equilateral Triangle
Dimension of Arch at Base: 54 feet (16.46 meters)
Deflection of Arch: 18 inches in 150 mph wind (0.46 meters in 240 km/h wind)
Number of Sections in Arch: 142
Thickness of Plates for Outer Skin: 0.25 inch (6.3 millimeters)
Type of Material Used in Arch: Exterior Stainless Steel; #3 Finish Type 304

#### Weight of Steel in Arch

Stainless Steel Plate Exterior Skin: 886 tons (804 metric tons)
Carbon Steel Plate Interior Skin: 3/8 inch (9.5 millimeters) 2,157 tons (1,957 metric tons)
Steel Stiffeners: 1,408 tons (1,277 metric tons)
Interior Steel Members, Stairs, Trains, etc.: 300 tons (272 metric tons)
Total Steel Weight: 5,199 tons (4,644 metric tons)

#### Weight of Concrete in Arch

Between Skins to 300 feet (91 meters): 12,127 tons (11,011 metric tons)
In Foundation Below Ground: 25,980 tons (23,569 metric tons)
Total Concrete Weight: 38,107 tons (34,570 metric tons)
External Protection: Six 0.50 inch x 20 inch (13 x 510 millimeter) lightning rods and one aircraft obstruction light

## Will building a **seawall** protect a beach?

It may for a while, but during a storm the sand cannot follow its natural pattern of allowing waves to draw the sand across the lower beach, making the beach flatter. With a seawall, the waves carry off more sand, dropping it into deeper water. A better

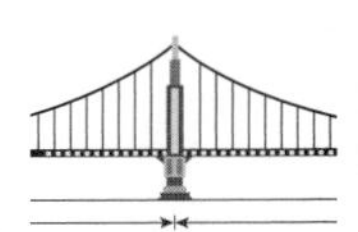



alternative to the seawall is the revetment. This is a wall of boulders, rubble, or concrete block, tilted back away from the waves. It imitates the way a natural beach flattens out under wave attack.

## What are the dimensions of the **Eiffel Tower**?

| | |
|---|---|
| Number of iron structural components in tower | 15,000 |
| Number of rivets | 2,500,000 |
| Weight of foundations | 306 tons (277,602 kilograms) |
| Weight of iron | 8,092 tons (7,341,214 kilograms) |
| Weight of elevator systems | 1,042 tons (946,000 kilograms) |
| Total weight | 9,441 tons (8,564,816 kilograms) |
| Pressure on foundations | 58–64 pounds per square inch (4–4.5 kilograms per square centimeter), depending on pier |
| Height of first platform | 189 feet (58 meters) |
| Height of second platform | 379 feet 8 inches (116 meters) |
| Height of third platform | 905 feet 11 inches (276 meters) |
| Total height in 1889 | 985 feet 11 inches (300.5 meters) |
| Total height with television antenna | 1,052 feet 4 inches (320.75 meters) |
| Number of steps to the top | 1,671 |
| Maximum sway at top caused by wind | 4.75 inches (12 centimeters) |
| Maximum sway at top caused by metal dilation | 7 inches (18 centimeters) |
| Size of base area | 2.54 acres (10,282 square meters) |
| Dates of construction | January 26, 1887 to March 31, 1889 |
| Cost of construction | 7,799,401.31 francs ($1,505,675.90) |

## How many locks are on the **Panama Canal**?

The 40-mile-long (64-kilometers-long) Panama Canal was completed in 1914, connecting the Atlantic and Pacific Oceans. There are three locks between Gatun Lake and Limon Bay at Colon on the Atlantic side and three locks between Gaillard Cut and Balboa on the Pacific side. Ships are raised 85 feet (26 meters) in their passage from one ocean to the other.

## Where is the **highest dam** in the United States and what is its capacity?

Oroville, the highest dam in the United States, is an earth-fill dam that rises 754 feet (230 meters) and extends more than a mile across the Feather River, near Oroville, California. Built in 1968, it forms a reservoir containing about 3.5 million acre-feet (4.3 million cubic meters) of water. The next highest dam in the United States is the Hoover Dam on the Colorado River, on the Nevada–Arizona border. It is 726 feet (221 meters) high, and was, for 22 years, the world's highest. Presently, nine dams are higher than the Oroville—the highest currently is the 1,098-foot (335-meter) Rogun(skaya) earth-fill dam that crosses the Vaksh River in Tadzhikistan. Built between 1981 and 1987, this dam has a volume of 92.9 million cubic yards (71 million cubic meters).

## How big is the **Hoover Dam**?

Formerly called Boulder Dam, the Hoover Dam is located between Nevada and Arizona on the Colorado River. The highest concrete-arch dam in the United States, the dam is 1,244 feet (379 meters) long and 726 feet (221 meters) high. It has a base thickness of 660 feet (201 meters) and a crest thickness of 45 feet (13.7 meters). It stores 21.25 million acre-feet of water in the 115-mile-long (185-kilometer-long) Lake Mead reservoir.

The dam was built because the Southwest was faced with constantly recurring cycles of flood and drought. Uncontrolled, the Colorado River had limited value, but once regulated the flow would assure a stabilized, year-round water supply, and the low-lying valleys would be protected against floods. On December 21, 1928, the Boulder Canyon Project Act became law, and the project was completed on September 30, 1935—two years ahead of schedule. For 22 years, Hoover Dam was the highest dam in the world.

## How tall is the figure on the **Statue of Liberty**, and how much does it weigh?

The Statue of Liberty, conceived and designed by the French sculptor Frédéric-Auguste Bartholdi (1834–1904), was given to the United States to commemorate its first centennial. Called (in his patent, U.S. Design Patent no. 11,023, issued February 18, 1879) "Liberty Enlightening the World," it is 152 feet (46 meters) high, weighs 225 tons (204 metric tons), and stands on a pedestal and base that are 151 feet (46 meters) tall. Her flowing robes are made from more than 300 sheets of hand-hammered copper over a steel frame. Constructed and finished in France in 1884, the statue's exterior and interior were taken apart piece by piece, packed into 200 mammoth wooden crates, and shipped to the United States in May 1885. The statue was placed by Bartholdi on Bedloe's Island, at the mouth of New York City Harbor. On October 28, 1886, ten years after the centennial had passed, the inauguration celebration was held.

It was not until 1903 that the inscription "Give me your tired, your poor, / Your huddled masses yearning to breathe free ..." was added. The verse was taken from *The New Colossus* composed by New York City poet Emma Lazarus in 1883. The statue, the tallest in the United States, and the tallest metal statue in the world, was refurbished for its own centennial, at a cost of $698 million, reopening on July 4, 1986. One visible difference is the flame of her torch is now 24-karat gold-leaf, just as in the original design. In 1916 the flame was redone into a lantern of amber glass. Concealed within the rim of her crown is the observation deck that can be reached by climbing 354 steps or by taking the more recently installed hydraulic elevator.

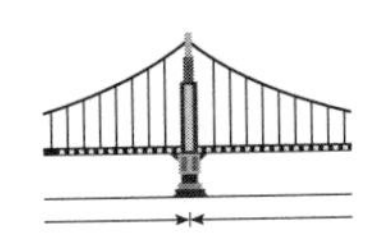



## How many **stairs** are in the **Statue of Liberty**?

Visitors to the Statue of Liberty must climb 354 steps (22 stories) to reach the Statue's crown.

## Who invented the **Ferris Wheel** and when?

Originally called "pleasure wheels," the first such rides were described by English traveler Peter Mundy in 1620. In Turkey he saw a ride for children consisting of two vertical wheels, 20 feet (six meters) across, supported by a large post on each side. Such rides were called "ups-and-downs" at the St. Bartholomew Fair of 1728 in England, and in 1860 a French pleasure wheel was turned by hand and carried 16 passengers. They were also in use in the United States by then, with a larger, wooden wheel operating at Walton Spring, Georgia.

Wanting a spectacular attraction to rival that of the 1889 Paris Centennial celebration—the Eiffel Tower—the directors of the 1893 Columbian Exposition had a design competition. The prize was won by the American bridge builder George Washington Gale Ferris (1859–1896). In 1893 he designed and erected a gigantic revolving steel wheel whose top reached 264 feet (80.5 meters) above ground. The wheel—825 feet (251.5 meters) in circumference, 250 feet (76 meters) in diameter, and 30 feet (nine meters) wide—was supported by two 140-foot (43-meter) towers. Attached to the wheel were 36 cars, each able to carry 60 passengers. Opening on June 21, 1893, at the exposition in Chicago, Illinois, it was extremely successful. Thousands lined up to pay 50 cents for a 20-minute ride—a large sum in those days, considering that a merry-go-round ride cost only four cents. In 1904 it was moved to St. Louis, Missouri, for the Louisiana Purchase Exposition. It was eventually sold for scrap. The largest-diameter wheel currently operating is the Cosmoclock 21 at Yokohama City, Japan. It is 345 feet (105 meters) high and 328 feet (100 meters) in diameter.

## How many **roller coasters** are there world?

There were 1,455 roller coasters in the world in 2001.

| Continent | No. of roller coasters |
|---|---|
| Africa | 17 |
| Asia | 314 |
| Australia | 22 |
| Europe | 410 |
| North America | 644 |
| South America | 48 |

Roller coasters have had a long history of thrill-giving. During the 15th and 16th centuries, the first known gravity rides were built in St. Petersburg and were called "Russian Mountains." A wheeled roller coaster, called the "Switchback," was used as early as 1784 in Russia. By 1817, the first roller coaster with cars locked to the tracks operated in France. The first United States roller coaster patent was granted to J. G. Taylor in 1872, and LaMarcus Thompson built the first known roller coaster in the United States at Coney Island in Brooklyn, New York, in 1884. The present record holder for the longest roller coaster in the world is the "Steel Dragon 2000" at Nagashima Spa Land. It is 8,133 feet (2,479 meters) in length.

# BOATS, TRAINS, CARS, AND PLANES

## BOATS AND SHIPS

### What is **dead reckoning**?

Dead reckoning is the determination of a craft's current latitude and longitude by advancing its previous position to the new one on the basis of assumed distance and direction traveled. The influences of current and wind as well as compass errors are taken into account in this calculation, all done without the aid of any celestial or physical observation. This is a real test of a navigator's skill.

Nuclear-powered submarines, which must retain secrecy of movement and cannot ascend to the surface, use the SINS system (Ship's Inertial Navigation System) developed by the United States Navy. It is a fully self-contained system, which requires no receiving or transmitting apparatus and thus involves no detectable signals. It consists of accelerometers, gyroscopes, and a computer. Together they produce inertial navigation, which is a sophisticated form of dead reckoning.

### What is the name of the carved **wooden figure of a woman** on a **sailing ship**?

A carved wooden figure at the top of the stem of a sailing ship, usually in the shape of a woman, is called a figurehead.

### What is the **nautical meaning** of the phrase **"by and large"**?

On a sailing ship or sailboat, new sailors at the helm are usually ordered to sail "by and large," meaning to sail into the wind at a slightly larger angle than those with more experience might choose. Sailing almost directly into the wind is most efficient, but

## Why is the right side of a ship called "starboard"?

In the time of the Vikings, ships were steered by long paddles or boards placed over the right side. They were known in Old English as steorbords, evolving into the word starboard. The left side of a ship, looking forward, is called port. Formerly the left side was called larboard, originating perhaps from the fact that early merchant ships were always loaded from the left side. Its etymology is Scandinavian, being lade (load) and bord (side). The British Admiralty ordered port to be used in place of larboard to prevent confusion with starboard.

doing so may cause the sail to flap back against the mast, resulting in loss of speed and control. Sailing "by and large" thus meant they were on the right, if not the perfect, course. Eventually the phrase became generally used as a synonym for approximately.

## Which type of **wood** was used to build **Noah's ark**?

According to the Bible, Noah's ark was made of gopher wood. This is identified as *Cupressus sempervirens*, one of the most durable woods in the world. Also called the Mediterranean cypress, the tree is native throughout southern Europe and western Asia. It grows up to 80 feet (24 meters) tall. Similar to this tree is the Monterey cypress (*Cupressus macrocarpa*), which is restricted to a very small area along the coast of central California. It can become as tall as 90 feet (27 meters) with horizontal branches that support a broad, spreading crown. When old, this tree looks very much like the aged cedars of Lebanon.

## Where does the term **"mark twain"** originate?

Mark twain is a riverboat term meaning two fathoms (a depth of 12 feet or 3.6 meters). A hand lead is used for determining the depth of water where there is less than 20 fathoms. The lead consists of a lead weight of seven to 14 pounds (three to six kilograms) and a line of hemp or braided cotton, 25 fathoms (150 feet or 46 meters) in length. The line is marked at 2, 3, 5, 7, 10, 15, 17, and 20 fathoms. The soundings are taken by a leadsman who calls out the depths while standing on a platform projecting from the side of the ship, called "the chains." The number of fathoms always forms the last part of the call. When the depth corresponds to any mark on the lead line, it is reported as "By the mark 7," "By the mark 10," etc. When the depth corresponds to a fathom between the marks on the line, it is reported as "By the deep 6," etc. When the line is a fraction greater than a mark, it is reported as "And a half 7," "And a quarter 5"; a fraction less than a mark is "Half less 7," "Quarter less 10," etc. If bottom is not reached, the call is "No bottom at 20 fathoms."

"Mark Twain" was also the pseudonym chosen by American humorist Samuel L. Clemens. Supposedly, he chose the name because of its suggestive meaning, since it was a riverman's term for water that was just barely safe for navigation. One implication of this "barely safe water" meaning was, as his character Huck Finn would later remark, "Mr. Mark Twain ... he told the truth, mostly." Another implication was that "barely safe water" usually made people nervous, or at least uncomfortable.



## How is a ship's **tonnage** calculated?

Tonnage of a ship is not necessarily the number of tons that the ship weighs. There are at least six different methods of rating ships; the most common are:

*Displacement tonnage*—used especially for warships and U.S. merchant ships—is the weight of the water displaced by a ship. Since a ton of sea water occupies 35 cubic feet (one cubic meter), the weight of water displaced by a ship can be determined by dividing the cubic footage of the submerged area of the ship by 35. The result is converted to long tons (2,240 pounds or 1,017 kilograms). Loaded displacement tonnage is the weight of the water displaced when a ship is carrying its normal load of fuel, cargo, and crew. Light displacement tonnage is the weight of water displaced by the unloaded ship.

*Gross tonnage* (GRST) or *gross registered tonnage* (GRT)—used to rate merchant shipping and passenger ships—is a measure of the enclosed capacity of a vessel. It is the sum in cubic feet of the vessel's enclosed space divided by 100 (100 such cubic feet is considered one ton). The result is gross (registered) tonnage. For example, the old *Queen Elizabeth* did not weigh 83,673 tons, but had a capacity of 8,367,300 cubic feet (236,878 cubic meters).

*Deadweight tonnage* (DWT)—used for freighters and tankers—is the total weight in long tons (2,240 pounds or 1,017 kilograms) of everything a ship can carry when fully loaded. It represents the amount of cargo, stores, bunkers, and passengers that are required to bring a ship down to her loadline, i.e., the carrying capacity of a ship.

*Net registered tonnage* (NRT)—used in merchant shipping—is the gross registered tonnage minus the space that cannot be utilized for paying passengers or cargo (crew space, ballast, engine room, etc.)

## What does the term **loadline** mean in shipping?

A loadline or load waterline is an immersion mark on the hull of a merchant ship. This indicates her safe load limit. The lines vary in height for different seasons of the year and areas of the world. Also called the "Plimsoll line" or "Plimsoll mark," it was accepted as law by the British Parliament in the Merchant Shipping Act of 1875, primarily at the instigation of Samuel Plimsoll (1824–1898). This law prevented unscrupulous owners from sending out unseaworthy and overloaded, but heavily insured, vessels (so-called "coffin ships"), which risked the crew's lives.

## Who created the **Liberty ship**?

The Liberty ship of World War II was the brainchild of Henry J. Kaiser (1882–1967), an American industrialist who had never run a shipyard before 1941. The huge loss of merchant tonnage during the war created an urgent need to protect merchant vessels transporting weapons and supplies, and the Liberty ship was born. It was a standard merchant ship with a deadweight tonnage of 10,500 long tons and a service speed of 11 knots. They were built to spartan standards, and production was on a massive scale. Simplicity of construction and operations, rapidity of building, and large cargo carrying capacity were assets. To these, Kaiser added prefabrication and welding instead of riveting. The ships were a deciding factor on the side of the Allies. In four years 2,770 ships with a deadweight tonnage of 29,292,000 long tons were produced.

## When was the first **hospital ship** built?

It is believed that the Spanish Armada fleet of 1587–1588 included hospital ships. England's first recorded hospital ship was the *Goodwill* in 1608, but it was not until after 1660 that the Royal Navy made it a regular practice to set aside ships for hospital use. The United States government outfitted six hospital ships, some of which were permanently attached to the fleet, during the Spanish-American War of 1898. Congress authorized the construction of the USS *Relief* in August 1916. It was launched in 1919 and delivered to the navy in December 1920.

## Why did the ***Titanic*** sink?

On its maiden voyage from Southampton, England, to New York, the British luxury liner *Titanic* sideswiped an iceberg at 11:40 p.m. on Sunday, April 14, 1912, and was badly damaged. The 882-foot (269-meter) long liner, whose eight decks rose to the height of an 11-story building, sank two hours and forty minutes later. Of the 2,227 passengers and crew, 705 escaped in 20 lifeboats and rafts; 1,522 drowned.

Famous as the greatest disaster in transatlantic shipping history, circumstances made the loss of life in the sinking of the *Titanic* exceptionally high. Although Capt. E. J. Smith was warned of icebergs in shipping lanes, he maintained his speed of 22 knots, and did not post additional lookouts. Later inquiries revealed that the liner *Californian* was only 20 miles (32 kilometers) away and could have helped, had its radio operator been on duty. The *Titanic* had an insufficient number of lifeboats, and those available for use were badly managed, with some leaving the boat only half full. The only ship responding to distress signals was the ancient *Carpathia*, which saved 705 people.

Contrary to a long-held belief, the *Titanic* had not been sliced open by the iceberg. When Dr. Robert Ballard (b. 1942) from Woods Hole Oceanographic Institution descended to the site of the sunken vessel in the research vessel *Alvin* in July 1986, he found that the ship's starboard bow plates had buckled under the impact of the collision. This caused the ship to be opened up to the sea.

Ballard found the bow and the stern more than 600 yards (548 meters) apart on the ocean's floor, and speculated on what happened after the collision with the iceberg: "Water entered six forward compartments after the ship struck the iceberg. As the liner nosed down, water flooded compartments one after another, and the ship's stern rose even higher out of the water, until the stress amidships was more than she could bear. She broke apart ..." and the stern soon sank by itself.

More recent investigations have shown that defective rivets resulted in a structural weakness in the *Titanic*. Rivets from the hull were recently analyzed by a corrosion laboratory and were found to contain unusually high concentrations of slag, making them brittle and prone to failure. The weakened rivets popped and the plates separated.

## What is the world's **largest ship**?

Among passenger liners, the cruise ship *Voyager of the Seas* is the biggest in the world. The 142,000-ton ship is not unlike a small city, containing a four-story-high shopping and entertainment street, an ice rink, a rock climbing wall, a 1,350-seat theatre, numerous cafes ... in short, all the amenities of a small, sophisticated city.

The largest ship of any kind currently in service is the tanker *Jahre Viking* (ex-*Happy Giant*, ex-*Seawise Giant*). This ship measures 260,851 gross register tons, with a cargo capacity of 564,763 tons (deadweight tonnage).

## When were the first **nuclear-powered vessels** launched?

A controlled nuclear reaction generates tremendous heat, which turns water into steam for running turbine engines. The USS *Nautilus* was the first submarine to be propelled by nuclear power, making her first sea run on January 17, 1955. It has been called the first true submarine since it can remain underwater for an indefinite period of time. The *Nautilus*, 324 feet (99 meters) long, has a range of 2,500 miles (4,023 kilometers) submerged, a diving depth of 700 feet (213 meters), and can travel submerged at 20 knots.

The first nuclear warship was the 14,000-ton cruiser USS *Long Beach*, launched on July 14, 1959. The USS *Enterprise* was the first nuclear-powered aircraft carrier. Launched on September 24, 1960, the *Enterprise* was 1,101.5 feet (336 meters) long and was designed to carry 100 aircraft.

The first nuclear-powered merchant ship was the *Savannah*, a 20,000-ton vessel, launched in 1962. The United States built it largely as an experiment and it was never operated commercially. In 1969, Germany built the *Otto Hahn*, a nuclear-powered ore carrier. The most successful use of nuclear propulsion in non-naval ships has been as icebreakers. The first nuclear-powered icebreaker was the Soviet Union's *Lenin*, commissioned in 1959.

# TRAINS AND TROLLEYS

## What is a **standard gauge** railroad?

The first successful railroads in England used steam locomotives built by George Stephenson (1781–1848) to operate on tracks with a gauge of four feet, 8.5 inches (1.41 meters), probably because that was the wheel spacing common on the wagons and tramways of the time. Stephenson, a self-taught inventor and engineer, had developed in 1814 the steam-blast engine that made steam locomotives practical. His railroad rival, Isambard K. Brunel (1806–1859), laid out the line for the Great Western Railway at seven feet, 0.25 inches (2.14 meters), and the famous "battle of the gauges" began. A commission appointed by the British Parliament decided in favor of Stephenson's narrower gauge, and the Gauge Act of 1846 prohibited using other gauges. This width eventually became accepted by the rest of the world. The distance is measured between the inner sides of the heads of the two rails of the track at a distance of 5/8 inch (16 millimeters) below the top of the rails.

## What is the **fastest train**?

The French SNCF high-speed train TGV (*Train à Grande Vitesse*) *Atlantique* achieved the fastest speed—320.2 miles per hour (515.2 kilometers per hour)—recorded on any national rail system on May 18, 1990, between Courtalain and Tours.

The fastest passenger train in North America is Amtrak's *Metroliner*. On the route between New York City and Washington, D.C., it can reach 125 mph (201 kph).

MAGLEV (*mag*netic *lev*itation) trains under development in Japan and Germany can travel 250 to 300 miles per hour (402 to 483 kilometers per hour) or more. These trains run on a bed of air produced from the repulsion or attraction of powerful magnetic fields (based on the principle that like poles of magnets repel and unlike poles [north and south] attract). The German Transrapid uses conventional magnets to levitate the train. The principle of attraction in magnetism, the employment of winglike flaps extending under the train to fold under a T-shaped guideway, and the use of electromagnets on board (that are attracted to the non-energized magnetic surface) are the guiding components. Interaction between the train's electromagnets and those built on top of the T-shaped track lift the vehicle 3/8 inch (one centimeter) off the guideway. Another set of magnets along the rail sides provides lateral guidance. The train rides on electromagnetic waves. Alternating current in the magnet sets in the guideway changes their polarity to alternately push and pull the train along. Braking is done by reversing the direction of the magnetic field (caused by reversing the magnetic poles). To increase train speed, the frequency of current is raised.

The Japanese MLV002 uses the same propulsion system, but the difference is in the levitation design, in which the train rests on wheels until it reaches a speed of 100 miles (161 kilometers) per hour. Then it levitates four inches (10 centimeters) above the guideway. The levitation depends on superconducting magnets and a repulsion system (rather than the attraction system that the German system uses).

## What is the world's **longest railway**?

The Trans-Siberian Railway, from Moscow to Vladivostok, is 5,777 miles (9,297 kilometers) long. If the spur to Nakhodka is included, the distance becomes 5,865 miles (9,436 kilometers). It was opened in sections, and the first goods train reached Irkutsk on August 27, 1898. The Baikal-Amur Northern Main Line, begun in 1938, shortens the distance by about 310 miles (500 kilometers). The journey takes approximately seven days, two hours, and crosses seven time zones. There are nine tunnels, 139 large bridges or viaducts, and 3,762 smaller bridges or culverts on the whole route. Nearly the entire line is electrified.

In comparison, the first American transcontinental railroad, completed on May 10, 1869, is 1,780 miles (2,864 kilometers) long. The Central Pacific Railroad was built eastward from Sacramento, California, and the Union Pacific Railroad was built westward to Promontory Point, Utah, where the two lines met to connect the line.

## When was the **first U.S. railroad** chartered?

The first American railroad charter was obtained on February 6, 1815, by Colonel John Stevens (1749–1838) of Hoboken, New Jersey, to build and operate a railroad between the Delaware and Raritan rivers near Trenton and New Brunswick. However, lack of financial backing prevented its construction. The Granite Railway, built by Gridley Bryant, was chartered on October 7, 1826. It ran from Quincy, Massachusetts, to the Neponset River—a distance of three miles (4.8 kilometers). The main cargo was granite blocks used in building the Bunker Hill Monument.

## When and why were **cabooses** eliminated from railroads?

The once familiar sight of a red caboose at the end of a train is mostly a historical memory now. Cabooses were home to conductors, brakemen and flagmen on a train. In 1972, the Florida East Coast Railway uncoupled and eliminated its cabooses. In the early 1980s, the United Transportation Union agreed to the elimination of cabooses on many trains. Technology has replaced many of the functions of railroad employees. Computers have replaced the conductor's record-keeping; an electronic "end-of-train device" monitors brake pressure, eliminating one of the rear brakeman's jobs; and trackside scanners have been installed to detect overhead axle bearings and report problems to the engineers.

## What was the railroad **velocipede**?

In the 19th century, railroad track maintenance workers used a three-wheeled handcar to speed their way along the track. The handcar was used for interstation express and package deliveries and for delivery of urgent messages between stations that could not wait until the next train. Also called an "Irish Mail," this 150-pound (68-kilogram) three-wheeler resembled a bicycle with a sidecar. The operator sat in the middle of the two-wheel section and pushed a crank back and forth, which propelled the triangle-shaped vehicle down the tracks. This manually powered handcar was replaced after World War I by a gasoline-powered track vehicle. This, in turn, was replaced by a conventional pickup truck fitted with an auxiliary set of flanged wheels.

## What does the term **gandy dancer** mean?

A track laborer. The name derived from the special tools used for track work made by the Gandy Manufacturing Company of Chicago, Illinois. These tools were used during the 19th century almost universally by section gangs.

## What was the route of the ***Orient Express***?

This luxury train service was inaugurated in June, 1883, to provide through connection between France and Turkey. It was not until 1889 that the complete journey could be made by train. The route left Paris and went via Chalons, Nancy, and Strasbourg into Germany (via Karlsruhe, Stuttgart, and Munich), then into Austria (via Salzburg, Linz, and Vienna), into Hungary (through Gyor and Budapest), south to Belgrade, Yugoslavia, through Sofia, Bulgaria and finally to Istanbul (Constantinople), Turkey. It ceased operation in May 1977. In 1982, part of the line, the Venice-Simplon-Orient Express went back into operation.

## How does a **cable car**, like those in San Francisco, move?

A cable runs continuously in a channel, between the tracks located just below the street. The cable is controlled from a central station, and usually moves about nine miles (14.5 kilometers) per hour. Each cable car has an attachment, on the underside of the car, called a grip. When the car operator pulls the lever, the grip latches onto the moving cable and is pulled along by the moving cable. When the operator releases the lever, the grip disconnects from the cable and comes to a halt when the operator applies the brakes. Also called an endless ropeway, it was invented by Andrew S. Hallidie (1836–1900), who first operated his system in San Francisco in 1873.



### What is a **funicular railway**?

A funicular railway is a type of railway used on steep grades, such as on a mountainside. Two counterbalanced cars or trains are linked by a cable, and when one moves down, the other moves up.

### Where is the **highest cable car** in the world?

The highest cable car, which is also the longest, is in the Venezuelan Andes. It begins in the mile-high (5,280 feet) city of Merida and ends after traversing 7.8 miles at Pico Espejo, Mirror Peak (15,720 feet).

## MOTOR VEHICLES

*See also: Energy—Consumption and Conservation, Buildings, Bridges, and Other Structures—Roads, Bridges, and Tunnels*

### How did the term **horsepower** originate?

Horsepower is the unit of energy needed to lift 550 pounds (247.5 kilograms) the distance of one foot (30.48 centimeters) in one second. Near the end of the 18th century the Scottish engineer James Watt (1736–1819) made improvements in the steam engine and wished to determine how its rate of pumping water out of coal mines compared with that of horses, which had previously been used to operate the pumps. In order to define a horsepower, he tested horses and concluded that a strong horse could lift 150 pounds (67.5 kilograms) 220 feet (66.7 meters) in one minute. Therefore, one horsepower was equal to 150 × 220/1 or 33,000 foot-pounds per minute (also expressed as 745.2 joules per second, 7,452 million ergs per second, or 745.2 watts).

The term horsepower was frequently used in the early days of the automobile because the "horseless carriage" was generally compared to the horse-drawn carriage. Today this inconvenient unit is still used routinely to express the power of motors and engines, particularly of automobiles and aircraft. A typical automobile requires about 20 horsepower to propel it at 50 miles (80.5 kilometers) per hour.

Karl Benz sits at the wheel of his original gasoline-powered automobile in 1885.

## How does a **gasoline engine** differ from one using **ethanol** or **gasohol**?

A gasoline engine uses only gasoline as its fuel. The ethanol or gasohol engine uses either ethanol (a fuel derived from plant sources) or a combination of gasoline and alcohol as its fuel source. The fuel system must have gaskets and components compatible with the fuel that will be used.

## Who invented the **automobile**?

Although the idea of self-propelled road transportation originated long before, Karl Benz (1844–1929) and Gottlieb Daimler (1834–1900) are both credited with the invention of the gasoline-powered automobile, because they were the first to make their automotive machines commercially practicable. Benz and Daimler worked independently, unaware of each other's endeavors. Both built compact, internal-combustion engines to power their vehicles. Benz built his three-wheeler in 1885; it was steered by a tiller. Daimler's four-wheeled vehicle was produced in 1887.

Earlier self-propelled road vehicles include a steam-driven contraption invented by Nicolas-Joseph Cugnot (1725–1804), who rode the Paris streets at 2.5 miles (four kilometers) per hour in 1769. Richard Trevithick (1771–1833) also produced a steam-driven vehicle that could carry eight passengers. It first ran on December 24, 1801, in Camborne, England. Londoner Samuel Brown built the first practical four-horsepower gasoline-powered vehicle in 1826. The Belgian engineer J. J. Etienne Lenoire (1822–1900)

built a vehicle with an internal combustion engine that ran on liquid hydrocarbon fuel in 1862, but he did not test it on the road until September 1863, when it traveled a distance of 12 miles (19.3 kilometers) in three hours. The Austrian inventor Siegfried Marcus (1831–1898) invented a four-wheeled, gasoline-powered handcart in 1864 and a full-size car in 1875; the Viennese police objected to the noise that the car made, and Marcus did not continue its development. Edouard Delamare-Deboutteville invented an eight-horsepower vehicle in 1883, which was not durable enough for road conditions.



## What was the first **mass produced alternative fuel vehicle** in the United States?

The Honda Civic GX Natural Gas Vehicle began production in 1998. This vehicle was rated as having the cleanest-burning internal combustion engine of its time.

## How does an **electric car** work?

An electric car uses an electric motor to convert electric energy stored in batteries into mechanical work. Various combinations of generating mechanisms (solar panels, generative braking, internal combustion engines driving a generator, fuel cells) and storage mechanisms are used in electric vehicles.

## Is the **electric automobile** a **recent** idea?

During the last decade of the 19th century, the electric vehicle became especially popular in cities. People had grown familiar with electric trolleys and railways, and technology had produced motors and batteries in a wide variety of sizes. The Edison Cell, a nickel-iron battery, became the leader in electric vehicle use. By 1900, electric vehicles nearly dominated the pleasure car field. In that year, 4,200 automobiles were sold in the United States. Of these, 38 percent were powered by electricity, 22 percent by gasoline, and 40 percent by steam. By 1911, the automobile starter motor did away with hand-cranking gasoline cars, and Henry Ford (1863–1947) had just begun to mass-produce his Model T's. By 1924, not a single electric vehicle was exhibited at the National Automobile Show, and the Stanley Steamer was scrapped the same year.

Because of the energy crises of the 1970s and 1990s concern for the environment (as well as "Clean Air" legislation), automobile manufacturers have marketed several all-electric and hybrid vehicles. General Motors marketed the Impact, an electric vehicle. Honda offers the Insight and a Civic sedan, both hybrids that use a gasoline engine and an electric motor. Toyota offers the Prius, which is also a hybrid.

## Who started the first **American automobile company**?

Charles Duryea (1861–1938), a cycle manufacturer from Peoria, Illinois, and his brother, Frank (1869–1967), founded America's first auto-manufacturing firm and became the first to build cars for sale in the United States. The Duryea Motor Wagon

Henry Ford drives his first car through the streets of Detroit in 1893.

company, set up in Springfield, Massachusetts, in 1895, built gasoline-powered horseless carriages similar to those built by Benz in Germany.

However, the Duryea brothers did not build the first automobile factory in the United States. Ransom Eli Olds (1864–1950) built it in 1899 in Detroit, Michigan, to manufacture his Oldsmobile. More than 10 vehicles a week were produced there by April 1901, for a total of 433 cars produced in 1901. In 1902 Olds introduced the assembly-line method of production and made over 2,500 vehicles in 1902 and 5,508 in 1904. In 1906, 125 companies made automobiles in the United States. In 1908, the American engineer Henry Ford (1863–1947) improved the automobile assembly-line techniques by adding the conveyor belt system that brought the parts to the workers on the production line; this made automotive manufacture quick and cheap, cutting production time to 93 minutes. His company sold 10,660 vehicles that year.

## How **many hours** does it take to **build a car**?

The amount of time varies depending on the vehicle, but industry averages for 2000 are:

| | |
|---|---|
| Nissan | 27.6 |
| Honda | 29.1 |
| Toyota | 31.1 |
| Ford | 39.9 |
| General Motors | 40.5 |
| DaimlerChrysler | 44.8 |

## What **materials** are used to manufacture an automobile?

**Materials Found in a Typical 1999 Automobile**

| Material | Pounds | Percent |
|---|---|---|
| Regular Steel, Sheet, Strip, Bar and Rod | 1,299.0 | 42.7% |
| High and Medium Strength Steel | 328.0 | 10.0% |
| Stainless Steel | 50.5 | 1.5% |
| Other Steels | 25.0 | 0.8% |
| Iron | 355.0 | 10.8% |
| Plastics and Plastic Composites | 245.0 | 7.5% |
| Aluminum | 236.0 | 7.2% |
| Copper and Brass | 45.5 | 1.4% |
| Powder Metal Parts | 35.0 | 1.1% |
| Zinc Die Castings | 12.0 | 0.4% |
| Magnesium Castings | 7.0 | 0.2% |
| Fluids and Lubricants | 194.0 | 5.9% |
| Rubber | 142.0 | 4.3% |
| Glass | 97.0 | 2.0% |
| Other Materials | 103.0 | 3.1% |
| Total | 3,274.0 | 100.0% |

## When was the **Michelin tire** introduced?

The first pneumatic (air-filled) tire for automobiles was produced in France by André (1853–1931) and Edouard Michelin (1859–1940) in 1885. The first radial-ply tire, the Michelin X, was made and sold in 1948. In radial construction, layers of cord materials called plies are laid across the circumference of the tire from bead to bead (perpendicular to the direction of the tread centerline). The plies can be made of steel wires or belts that circle the tire. Radial tires are said to give longer tread life, better handling, and a softer ride at medium and high speeds than bias or belted bias tires (both of which have plies laid diagonally). Radials give a firm, almost hard, ride at low speeds.

## What is a **rumble seat**?

A rumble seat is a folding external seat situated in the rear deck of some older two-door coupes, convertibles, and roadsters.

## When were **tubeless automobile tires** first manufactured?

In Akron, Ohio, the B. F. Goodrich Company announced the manufacture of tubeless tires on May 11, 1947. Dunlop was the first British firm to make tubeless tires in 1953.

## What do the **numbers** mean on **automobile tires**?

The numbers and letters associated with tire sizes and types are complicated and confusing. The "Metric P" system of numbering is probably the most useful method of indicating tire sizes. For example, if the tire had P185/75R-14, then "P" means the tire is for a passenger car; the number 185 is the width of the tire in millimeters; 75 indicates the aspect ratio, i.e., that the height of the tire from the rim to the road is 75 percent of the width; R indicates that it is a radial tire; 14 is the wheel diameter in inches (13 and 15 inches are also common sizes). Speed ratings are indicated by a letter located in the size markings:

| Symbol | Maximum speed MPH | KPH |
|---|---|---|
| S | 112 | 180 |
| T | 118 | 190 |
| U | 124 | 200 |
| H | 130 | 210 |
| V | 149 | 240 |
| W | 168 | 270 |
| Y | 186 | 300 |
| Z | 186+ | 300+ |

## Which vehicle had the first modern automobile **air conditioner**?

The first air-conditioned automobile was manufactured by the Packard Motor Car Company in Detroit, Michigan, and was exhibited publicly November 4–12, 1939, at the 40th Automobile Show in Chicago, Illinois. Air in the car was cooled to the temperature desired, dehumidified, filtered, and circulated. The first fully automatic air conditioning system was Cadillac's "Climate Control," introduced in 1964.

## What was the first car manufactured with an **automatic transmission**?

The first of the modern generation of automatic transmissions was General Motors' Hydramatic, first offered as an option on the Oldsmobile during the 1940 season. Between 1934 and 1936, a handful of 18 horsepower Austins were fitted with the American-designed Hayes infinitely variable gear. The direct ancestor of the modern automatic gearbox was patented in 1898.

## Has there ever been a **nuclear-powered automobile**?

In the 1950s, Ford automotive designers envisioned the Ford Nucleon, which was to be propelled by a small atomic reactor core located under a circular cover at the rear of the car. It was to be recharged with nuclear fuel. The car was never built.

## Where was the first automobile **license plate** issued?

Leon Serpollet of Paris, France, obtained the first license plate in 1889. They were first required in the United States by New York State in 1901. Registration was required within 30 days. Owners had to provide their names and addresses as well as a description of their vehicles. The fee was one dollar. The plates bore the owner's initials and were required to be over three inches (7.5 centimeters) high. Permanent plates made of aluminum were first issued in Connecticut in 1937.

## What information is available from the **vehicle identification number (VIN), body number plate,** and **engine** on a car?

These coded numbers reveal the model and make, model year, type of transmission, plant of manufacture, and sometimes even the date and day of the week a car was made. The form and content of these codes is not standardized (with one exception) and often changes from one year to the next for the same manufacturer. The exception (starting in 1981) is the tenth digit, which indicates the year. For example: W = 1998, x = 1999, 2 = 2002, etc. Various components of a car may be made in different plants, so a location listed on a VIN may differ from one on the engine number. The official shop manual lists the codes for a particular make of car.

## **How many motor vehicles** are registered in the United States?

The total United States registration of motor vehicles in 1999 is estimated to be 216,308,623. Of the total, 132,432,044 were automobiles and 83,876,579 were trucks and buses. Worldwide, in 1999 there were 681,799,000 total motor vehicles registered (491,597,000 passenger cars).

## How much does it **cost to operate** an automobile in the United States?

Below is listed the average cost per mile to operate an automobile in the United States in cents per mile. Figures are given for 12,500 miles in suburban driving conditions:

| | Medium | Large | Luxury | SUV | Minivan |
|---|---|---|---|---|---|
| Gas/oil | 5.0 | 6.3 | 7.4 | 7.2 | 6.8 |
| Maintenance | 2.9 | 3.1 | 3.2 | 3.4 | 3.2 |
| Tires | 1.3 | 1.4 | 1.4 | 1.4 | 1.3 |
| Annual costs: | | | | | |
| Insurance | $912 | $856 | $933 | $1,312 | $950 |
| License/Registration | $175 | $223 | $279 | $396 | $379 |
| Depreciation | $2,819 | $3,294 | $3,979 | $3,556 | $3,409 |
| Financing | $598 | $802 | $1,040 | $929 | $885 |
| Average cost per mile: | $0.45 | $0.52 | $0.62 | $0.62 | $0.56 |

## How do **antilock braking systems** (ABS) work?

The antilock brake system (ABS) was first developed and patented in 1936, and the term antilock brake is derived from the German term, "antiblockiersystem." ABS prevents the wheels on a vehicle from locking. Locked wheels create vehicle instability and can cause skidding. The ABS utilizes a computer that automatically modulates brake pressure, preventing the wheels from locking.

## What is the **braking distance** for an automobile at different speeds?

Average stopping distance is directly related to vehicle speed. On a dry, level, concrete surface, the minimum stopping distances are as follows (including driver reaction time to apply brakes):

| Speed | | Reaction time distance | | Braking distance | | Total distance | |
|---|---|---|---|---|---|---|---|
| Mph | Kph | Feet | Meters | Feet | Meters | Feet | Meters |
| 10 | 16 | 11 | 3.4 | 9 | 2.7 | 20 | 6.1 |
| 20 | 32 | 22 | 6.7 | 23 | 7.0 | 45 | 13.7 |
| 30 | 48 | 33 | 10.1 | 45 | 13.7 | 78 | 23.8 |
| 40 | 64 | 44 | 13.4 | 81 | 24.7 | 125 | 38.1 |
| 50 | 80 | 55 | 16.8 | 133 | 40.5 | 188 | 57.3 |
| 60 | 97 | 66 | 20.1 | 206 | 62.8 | 272 | 82.9 |
| 70 | 113 | 77 | 23.5 | 304 | 92.7 | 381 | 116.1 |

## How far will a car **skid** on **various road surfaces** at different speeds?

| Speed (mph) | Asphalt | Concrete | Snow | Gravel |
|---|---|---|---|---|
| 30 | 40 feet | 33 feet | 100 feet | 60 feet |
| 40 | 71 feet | 59 feet | 178 feet | 107 feet |
| 50 | 111 feet | 93 feet | 278 feet | 167 feet |
| 60 | 160 feet | 133 feet | 400 feet | 240 feet |

## How can one find out about **safety recalls** on automobiles?

The National Highway Traffic Safety Administration (NHTSA) keeps records of recalls and takes reports of safety problems experienced by consumers. You can call their 24-hour hotline at 1-800-424-9393, or check their Web site at http://www.nhtsa.dot.gov/, or write to NHTSA, Department of Transportation, Washington, DC 20590. Be sure to include the make, model, year, and vehicle identification number of the vehicle, and a description of the problem or part in question. You will get any recall information the NHTSA has, either on the phone or by a mailed printout.

## What day of the week do most **fatal automotive accidents** occur?

Ongoing studies have shown that Saturday continues to be the day of the week with the most fatal accidents. The National Highway Traffic Safety Administration collected these data for 1998:

| Hour of Day | Sun | Mon | Tues | Wed | Thurs | Fri | Sat | Total |
|---|---|---|---|---|---|---|---|---|
| 12–3 a.m. | 1,208 | 400 | 322 | 480 | 506 | 530 | 1,218 | 4,564 |
| 3–6 a.m. | 641 | 269 | 256 | 267 | 332 | 329 | 630 | 2,724 |
| 6–9 a.m. | 382 | 569 | 554 | 560 | 518 | 503 | 494 | 3,580 |
| 9–Noon | 479 | 543 | 560 | 526 | 494 | 558 | 611 | 3,771 |
| Noon–3 p.m. | 645 | 719 | 681 | 705 | 701 | 803 | 716 | 4,970 |
| 3–6 p.m. | 885 | 840 | 887 | 822 | 894 | 1,015 | 869 | 6,213 |
| 6–9 p.m. | 848 | 685 | 721 | 710 | 821 | 984 | 1,028 | 5,797 |
| 9 p.m.–12 a.m. | 581 | 561 | 575 | 593 | 678 | 1,099 | 1,047 | 5,135 |
| Total | 5,734 | 4,608 | 4,595 | 4,593 | 4,985 | 5,864 | 6,686 | 37,081 |

## When did **seatbelts** become mandatory equipment on United States motor vehicles?

The U.S. National Highway Safety Bureau first required the installation of lap belts for all seats and shoulder belts in the front seats of cars in 1968. However, most Americans did not regularly use safety belts until after 1984, when the first state laws were introduced that penalized drivers and passengers who did not use the device. As of the late 1990s, 68 percent of automobile occupants regularly use their seat belts.

## Which **colors** of cars are the **safest**?

Tests at the University of California concluded that either blue or yellow is the best color for car safety. Blue shows up best during daylight and fog; yellow is best at night. The worst color from the visibility standpoint is gray. In another study by Mercedes-Benz in Germany, white ranked the highest in all-around visibility, except in situations of completely snow-covered roads or white sand. In such extremes, bright yellow and bright orange ranked second and third respectively in visibility. The least visible car color in the Mercedes-Benz test was dark green.

## How did Ralph Nader's book *Unsafe at Any Speed* contribute to the demise of the **Corvair** automobile?

Nader intended the book as an indictment of all the sins of the Detroit automobile manufacturers, and principally of General Motors. Actually, the Corvair is discussed only in the first chapter. Nader believed that General Motors executives had marketed a car they knew to be unsafe because their desire for profit outweighed all other considerations.

Nader, at the time of the book's publication, was working on the staff of Senator Abraham Ribicoff, who chaired a Senate subcommittee that was crafting a bill to establish standards for automobile design. Thus the book received widespread attention, and Nader was called on as an expert witness during hearings on various automobile concerns, including the safety of the Corvair. Nader's testimony, along with some well-timed publicity, set the stage for passage of a strong National Traffic and Motor Vehicle Safety Act in September 1966.

Negative publicity about the car had done its damage, and even though design modifications were made, sales dropped catastrophically. Production was discontinued in 1969.

## What **car colors** are the most **popular**?

According to a 1998 study, these are the most popular colors for various sized vehicles:

**Automobile Paint Color Popularity by Vehicle Type**

| Luxury Cars | | Full Size/Intermediate Cars | | Compact/Sports Cars | |
|---|---|---|---|---|---|
| **Color** | **Percent** | **Color** | **Percent** | **Color** | **Percent** |
| Light Brown | 17.7 | Medium/Dark Green | 16.4 | Medium/Dark Green | 15.9 |
| White Metallic | 12.3 | White | 15.6 | Black | 15.0 |
| Black | 12.3 | Light Brown | 14.1 | White | 14.7 |
| White | 11.3 | Silver | 11.0 | Silver | 10.4 |
| Medium/Dark Green | 10.0 | Black | 8.9 | Bright Red | 9.5 |
| Silver | 9.2 | Medium Red | 6.5 | Light Brown | 7.0 |
| Medium Red | 7.5 | Medium/Dark Blue | 6.0 | Medium Red | 6.4 |
| Medium/Dark Gray | 5.3 | Dark Red | 4.9 | Medium/Dark Blue | 5.3 |
| Medium/Dark Blue | 4.8 | Light Blue | 3.8 | Teal/Aqua | 4.0 |
| Gold | 4.8 | White Metallic | 3.2 | Purple | 3.4 |
| Other | 4.8 | Other | 9.6 | Other | 8.4 |

## Which states allow a **right** or **left turn** on a **red light**?

All states permit drivers to turn right on a red signal after a complete stop if the intersection is not designated otherwise by posted signs. New York City now is the only major jurisdiction that prohibits the turn. According to the Federal Highway Administration, fewer accidents occur when drivers turn right on a red light than when they turn right on a green light. The statute also saves each driver an average of 14 seconds at a turn, cuts gasoline consumption and exhaust emissions, and allows intersections to handle more traffic.

### When was a speed trap first employed to apprehend speeding automobile drivers?

In 1905, William McAdoo, police commissioner of New York City, was stopped for traveling at 12 miles per hour (19 kilometers per hour) in an 8-miles-per-hour (13-kilometers-per-hour) zone in rural New England. The speed detection device consisted of two lookout posts, camouflaged as dead tree trunks, spaced one mile (1.6 kilometers) apart. A deputy with a stopwatch and a telephone kept watch for speeders. When a car appeared to be traveling too fast, the deputy pressed his stopwatch and telephoned ahead to his confederate who immediately consulted a speed-mileage chart and phoned ahead to another constable manning a road block to apprehend the speeder. McAdoo invited the New England constable to set up a similar device in New York City.

One of the most famous speed traps was in the Alabama town of Fruithurst, on the Alabama–Georgia border. In one year, this town of 250 people collected over $200,000 in fines and forfeitures from unwary "speeders."

Forty-one states permit left turns on a red signal, but only after a complete stop and only from a one-way street into another one-way street. Those states that prohibit the turn are: Connecticut, Maryland, Mississippi, Missouri, New Jersey, North Carolina, Rhode Island, Vermont, and Wisconsin. The District of Columbia and New York City also prohibit the turn.

## How does **VASCAR** work?

Invented in 1965, VASCAR (Visual Average Speed Computer and Recorder) is a calculator that determines a car's speed from two simple measurements of time and distance. No radar is involved. VASCAR can be used at rest or while moving to clock traffic in both directions. The patrol car can be behind, ahead of, or even perpendicular to the target vehicle. The device measures the length of a speed trap and then determines how long it takes the target car to cover that distance. An internal calculator does the math and displays the average speed on an LED readout. Most police departments now use several forms of moving radar, which are less detectable and more accurate.

## How does **police radar** work?

The Austrian physicist Christian Doppler (1803–1853) discovered that the reflected waves bouncing off a moving object are returned at a different frequency (shorter or longer waves, cycles, or vibrations). This phenomenon, called the Doppler effect, is the basis of police radar. Directional radio waves are transmitted from the radar device. The waves bounce off the targeted vehicle and are received by a recorder. The recorder compares the difference between the sent and received waves, translates the information into miles per hour, and displays the speed on a dial.

## How do **laser speed guns** differ from radar?

Laser speed guns rely on the reflection time of light rather than the Doppler effect. They measure the round-trip time for light to reach a vehicle and reflect back. Laser speed guns shoot a very short burst of infrared laser light at the target (a moving vehicle) and wait for it to reflect off the vehicle. The advantages of a laser speed gun are that the cone of light the gun emits is very small so they can target a specific vehicle and they are very accurate. The disadvantage is that they require better aim.

## How does an **air bag** work to prevent injury in an automobile crash?

When a frontal collision occurs, sensors trigger the release and reaction of sodium azide with iron, which produces large quantities of nitrogen gas. This gas fully inflates the bag in about two-tenths of a second after impact to create a protective cushion. The air bag deflates immediately thereafter, and the harmless nitrogen gas escapes through holes in the back.

An air bag will deploy only after the car has an impact speed of 11 to 14 miles per hour (17 to 22 kilometers per hour) or greater. It will not be set off by a minor fender bender, by hitting a cement stop in a parking space, or if someone kicks the bumpers. The National Highway Traffic Safety Administration estimates that between 1987 and July 1, 2001, air bags saved 7,224 lives. Federal safety officials recommended that owners of vehicles with air bags not use rear-facing infant seats in the front passenger seat. In that location, an inflating air bag can strike the child seats with enough force to cause injury.

## When was the **air bag invented**?

Patented ideas on air bag safety devices began appearing in the early 1950s. U.S. patent 2,649,311 was granted on August 18, 1953, to John W. Hetrick for an inflated safety cushion to be used in automotive vehicles. The Ford Motor Company studied the use of air bags around 1957, and other undocumented work was carried out by Mr. Assen Jordanoff before 1956. There are other earlier uses of an air bag concept, including a rumored method of some World War II pilots inflating their life vests before a crash.

In the mid-1970s, General Motors geared up to sell 100,000 air bag-equipped cars a year in a pilot program to offer them as a discounted option on luxury models. GM dropped the option after only 8,000 buyers ordered air bags in three years. As of September 1, 1989, all new passenger cars produced for sale in the United States are required to be equipped with passive restraints (either automatic seatbelts or air bags).

## When was the parking meter introduced?

Carlton C. Magee, editor of the Oklahoma City *Daily News* and a member of the Chamber of Commerce traffic committee, became concerned about the parking problem in larger cities. He proposed a device to charge people for parking spaces. He entered into a partnership with Gerald A. Hale, a professor at Oklahoma Agricultural and Mechanical College, to perfect the mechanism. In 1932, Magee applied for a patent on a parking meter. In July 1935, meters were installed on some streets in Oklahoma City. These machines now help solve traffic and parking problems in major cities throughout the world.

Today, nearly 20 million cars and trucks on the road have air bags. Federal law required dual air bags in all cars by 1998 and on all light trucks by 1999.

## Which automobile is the one **most often stolen**?

Automobile theft has been occurring since an automobile mechanic stole Baron de Zuylen's Peugeot in Paris in June 1896. In general, cars with the lowest overall theft losses are small and midsize four-door cars and station wagons. Sports and luxury models, especially convertibles, have the highest losses.

The Highway Loss Data Institute published the following data for 1999–2001 passenger vehicles:

**Most likely to be stolen**

Acura Integra
Jeep Wrangler
Jeep Cherokee (4WD)
Honda Prelude
Mitsubishi Mirage (2 door)
Chrysler 300M
Hyundai Tiburon
Dodge Intrepid
Mitsubishi Mirage (4 door)
Chrysler LHS

## What is the difference between a **medium truck** and a **heavy truck**?

Medium trucks weigh 14,001 to 33,000 pounds (6,351 to 14,969 kilograms). They span a wide range of sizes and have a variety of uses, from step-van route trucks to truck tractors. Common examples include beverage trucks, city cargo vans, and garbage trucks. Heavy trucks weigh 33,001 pounds (14,969 kilograms) or greater. Heavy trucks include over-the-road 18-wheelers, dump trucks, concrete mixers, and fire trucks. These trucks have come a long way from the first carrying truck, built in 1870 by John Yule, which moved at a rate of three-quarters of a mile (1.2 kilometers) per hour.

### What is the origin of the term **taxicab**?

The term taxicab is derived from two words—*taximeter* and *cabriolet*. The taximeter, an instrument invented by Wilhelm Bruhn in 1891, automatically recorded the distance traveled and/or the time consumed. This enabled the fare to be accurately measured. The cabriolet is a two-wheeled, one-horse carriage, which was often rented.

The first taxicabs for hire were two Benz-Kraftdroschkes operated by "Droschkenbesitzer" Dütz in the spring of 1896 in Stuttgart, Germany. In May 1897 Friedrich Greiner started a rival service. In a literal sense, Greiner's cabs were the first "true" taxis because they were the first motor cabs fitted with taximeters.

## AIRCRAFT

*See also: Boats, Trains, Cars, and Planes—Military Vehicles*

### Why did the dirigible ***Hindenburg*** explode?

Despite the official United States and German investigations into the explosion, it still remains a mystery today. The most plausible explanations are structural failure, St. Elmo's Fire, static electricity, or sabotage. The *Hindenburg*, built following the great initial success of the *Graf Zeppelin*, was intended to exceed all other airships in size, speed, safety, comfort, and economy. At 803 feet (245 meters) long, it was 80 percent as long as the ocean liner *Queen Mary*, 135 feet (41 meters) in diameter, and could carry 72 passengers in its spacious quarters.

In 1935, the German Air Ministry virtually took over the Zeppelin Company to use it to spread Nazi propaganda. After its first flight in 1936, the airship was very popular with the flying public. No other form of transport could carry passengers so swiftly, reliably, and comfortably between continents. During 1936, 1,006 passengers flew over the North Atlantic Ocean in the *Hindenburg*. On May 6, 1937, while landing at Lakehurst, New Jersey, its hydrogen burst into flames, and the airship was completely destroyed. Of the 97 people aboard, 62 survived.

Amazingly, 62 of the 97 people aboard the *Hindenburg* survived the fiery crash.

## How does the **wing** on an aircraft **generate lift**?

Bernoulli's principle states that an increase in a fluid medium's velocity results in a decrease in pressure. An aircraft's wing is shaped so that the air (a fluid medium) flows faster past the upper surface than the lower surface, thus generating a difference in pressures, causing lift.

## What is a **hovercraft**?

A hovercraft is a vehicle supported by a cushion of air, thus capable of operating on land or water. In 1968 regular commercial hovercraft service was established across the English Channel.

## What was the name of the **Wright brothers' airplane**?

The name of the Wright brothers' plane was the *Flyer*. A wood and fabric biplane, the *Flyer* was originally used by the brothers as a glider and measured 40 feet, 4 inches (12 meters) from wing-tip to wing-tip. For their historic flight, Wilbur and Orville Wright outfitted it with a four-cylinder, 12-horsepower gasoline engine and two propellers, all of their own design. On December 17, 1903, at Kitty Hawk, North Carolina, Orville Wright made the first engine-powered, heavier-than-air craft flight lying in the middle of the lower wing to pilot the craft, which flew 120 feet (37 meters) in 12 seconds. The brothers made three more flights that day, with Wilbur Wright completing the longest one—852 feet (260 meters) in 59 seconds.

### Who made the first **nonstop transatlantic flight**?

The first nonstop flight across the Atlantic Ocean, from Newfoundland, Canada, to Ireland, was made by two British aviators, Capt. John W. Alcock (1892–1919) and Lt. Arthur W. Brown (1886–1948), on June 14–15, 1919. The aircraft, a converted twin-engined Vickers Vimy bomber, took 16 hours, 27 minutes to fly 1,890 miles (3,032 kilometers). Later, Charles A. Lindbergh (1902–1974) made the first solo crossing flight on May 20–21, 1927, in the single-engined Ryan monoplane *Spirit of St. Louis*, with a wing spread of 46 feet (15 meters) and a chord of seven feet (2.2 meters). His flight from New York to Paris covered a distance of 3,609 miles (5,089 kilometers) and lasted 33.5 hours. The first woman to fly solo across the Atlantic was Amelia Earhart (1897–1937), who flew from Newfoundland to Ireland May 20–21, 1932.

### Who made the **first supersonic flight**?

Supersonic flight is flight at or above the speed of sound. The speed of sound is 760 miles (1,223 kilometers) per hour in warm air at sea level. At a height of about 37,000 feet (11,278 kilometers), its speed is only 660 miles (1,062 kilometers) per hour. The first person credited with reaching the speed of sound (Mach 1) was Major Charles E. (Chuck) Yeager (b. 1923) of the United States Air Force. In 1947, he attained Mach 1.45 at 60,000 feet (18,288 meters) while flying the Bell X-1 rocket research plane designed by John Stack and Lawrence Bell. This plane had been carried aloft by a B-29 and released at 30,000 feet (9,144 meters). It is, however, highly likely that the sound barrier was broken on April 9, 1945, by Hans Guido Mutke in a Messerschmitt Me262, the world's first operational jet aircraft. It is also probable that Chalmers Goodlin broke the barrier in the Bell X-1 six months prior to Yeager's flight, and the barrier was broken again, by George Welch, in a North American XP-86 *Sabre* shortly before Yeager's flight. In 1949, the Douglas *Skyrocket* was credited as the first supersonic jet-powered aircraft to reach Mach 1 when Gene May flew at Mach 1.03 at 26,000 feet (7,925 meters). The Me262 and the XP-86 were both jet-powered aircraft.

### When was the first **nonstop, unrefueled, round-the-world** airplane flight?

Dick Rutan (b. 1943) and Jeana Yeager (b. 1952) flew the *Voyager*, a trimaran monoplane, in a closed circuit loop westbound and back to Edwards Air Force Base, California, December 14–23, 1986. The flight lasted nine days, three minutes, 44 seconds, and covered 24,986.7 miles (40,203.6 kilometers). The first successful round-the-world flight was made by two Douglas World Cruisers between April 6 and September 28, 1924. Four aircraft originally left Seattle, Washington, and two went down. The two successful planes completed 27,553 miles (44,333 kilometers) in 175 days—with 371 hours, 11 minutes being their actual flying time. Between June 23 and July 1, 1931, Wiley Post (1900–1935) and Harold Gatty (1903–1957) flew around the world, starting from New York, in their Lockheed Vega *Winnie Mae*.



## Who was the first to fly **around the world** in a **balloon**?

Steve Fossett (1944–) was the first to fly solo around the world in a balloon. He departed western Australia on June 18, 2002, and returned on July 4, 2002, exactly 13 days, 23 hours, 16 minutes, and 13 seconds later. Fossett had previously made five attempts to circumnavigate the world in a balloon.

## Who made the first **parachute jump**?

The first successful parachute jump from a great height was made by French aeronaut Jacques Garnerin in 1797. He jumped 3,000 feet (914 meters) from a hot-air balloon.

## What is **avionics**?

Avionics, a term derived by combining aviation and electronics, describes all of the electronic navigational, communications, and flight management aids with which airplanes are equipped today. In military aircraft it also covers electronically controlled weapons, reconnaissance, and detection systems. Until the 1940s, the systems involved in operating aircraft were purely mechanical, electric, or magnetic, with radio apparatus being the most sophisticated instrumentation. The advent of radar and the great advance made in airborne detection during World War II led to the general adoption of electronic distance-measuring and navigational aids. In military aircraft such devices improve weapon delivery accuracy, and in commercial aircraft they provide greater safety in operation.

## Where is the **black box** carried on an airplane?

Actually painted bright orange to make it more visible in an aircraft's wreckage, the black box is a tough metal-and-plastic case containing two recorders. Installed in the rear of the aircraft—the area most likely to survive a crash—the case has two shells of stainless steel with a heat-protective material between the shells. The case must be able to withstand a temperature of 2,000°F (1,100°C) for 30 minutes. Inside it, mounted in a shockproof base, is the aircraft's flight data and cockpit voice recorders. The flight data recorder provides information about airspeed, direction, altitude, acceleration, engine thrust, and rudder and spoiler positions from sensors that are located around the aircraft. The data is recorded as electronic pulses on stainless steel tape, which is about as thick as aluminum foil. When the tape is played back, it gener-

**Why don't tires on airplanes blow out when the airplane lands?**

The Federal Aviation Agency requires airplane tires to meet rigid FAA standards to prevent accidental blowouts. A plug built into the tire will pop out when the air pressure gets too high, and the tire will deflate slowly rather than blow out.

ates a computer printout. The cockpit voice recorder records the previous 30 minutes of the flight crew's conversation and radio transmission on a continuous tape loop. If a crash does not stop the recorder, vital information can be lost.

### When was the first **full-scale wind tunnel** for testing airplanes used?

It began operations on May 27, 1931, at the Langley Research Center of the National Advisory Committee for Aeronautics, Langley Field, Virginia. This tunnel, still in use, is 30 feet (nine meters) high and 60 feet (18 meters) wide. A wind tunnel is used to simulate air flow for aerodynamic measurement; it consists essentially of a closed tube, large enough to hold the airplane or other craft being tested, through which air is circulated by powerful fans.

### What is the world's **largest wind tunnel**?

The largest wind tunnel in the world is the National Full-Scale Aerodynamics Complex at NASA Ames. The complex consists of two test areas, one 40 feet by 80 feet, the other 80 by 120 feet. The 80-by-120-foot area can generate wind speeds of up to 115 miles per hour.

### What is a **bird shot test**?

Bird strikes—incidents where a bird collides with an aircraft—are not uncommon. To test components such as windshields, a small, dead bird, usually a chicken, is fired from a device into the windshield at an appropriate velocity.

### Who designed the ***Spruce Goose***?

Howard Hughes (1905–1976) designed and built the all-wood H-4 Hercules flying boat, nicknamed the *Spruce Goose*. The aircraft had the greatest wingspan ever built and was powered by eight engines. It was flown only once—covering a distance of less than one mile at Los Angeles harbor on November 2, 1947, lifting only 33 feet (10.6 meters) off the surface of the water.

After the attack on Pearl Harbor on December 7, 1941, and the subsequent entry of the United States into World War II, the United States government needed a large,

Howard Hughes's *Spruce Goose*—the largest aircraft ever built—took its first and only flight in 1947.

cargo-carrying airplane that could be made from noncritical wartime materials, such as wood. Henry J. Kaiser (1882–1967), whose shipyards were producing Liberty ships at the rate of one per day, hired Howard Hughes to build such a plane. Hughes eventually produced a plane that weighed 400,000 pounds (181,440 kilograms) and had a wingspan of 320 feet (97.5 meters). Unfortunately, the plane was so complicated that it was not finished by the end of the war. In 1947, Hughes flew the plane himself during its only time off the ground—supposedly just to prove that something that big could fly. The plane was on public display in Long Beach, California, but was sold in 1992 to aviation enthusiast Delford Smith and shipped to McMinnville, Oregon, where it is now on display at the Evergreen Aviation Museum.

## What is a **Mach number**?

A Mach number is the equivalent of the speed of sound; therefore Mach 2 is twice the speed of sound, Mach 0.5 is half the speed of sound, etc.

## What is the world's **fastest aircraft**?

The world's highest and fastest flying airplane is the North American X-15. The X-15 reached a peak altitude of 354,200 feet (67 miles, 108 kilometers). The X-15A-2 reached a speed of Mach 6.72 (4,534 mph, 7,295 kph). The X-15 is a rocket-powered aircraft that was launched/dropped by a modified B-52 bomber. It was designed in 1954 and first flown in 1959.

## What is the maximum **seating capacity** in a Boeing 747?

The seating capacity of the 747 and some other jets servicing cities in the United States are listed below.

| Airplane | Maximum seating capacity |
|---|---|
| Boeing 707 | 179 |
| Boeing 707–320, 707–420 | 189 |
| Boeing 720 | 149 |
| Boeing 727 | 125 |
| Boeing 737 | 149 |
| Boeing 747 | 498 |
| Boeing 757 | 196 |
| Boeing 767 | 289 |
| Boeing 777 | 375 |
| Concorde (SST) | 110 |
| Lockheed L-1011 TriStar | 345 |
| McDonnell Douglas DC-8 | 189 |
| McDonnell Douglas DC-9 | |
| Series 20 | 119 |
| Series 30 & 40 | 125 |
| Series 50 | 139 |
| McDonnell Douglas DC-10 | 380 |
| Tupolev Tu-144 (Soviet SST) | 140 |

## What is the world's **fastest** and **highest flying jet** aircraft?

The Lockheed SR-71 holds three absolute world records for speed and altitude: speed in a straight line (2,193 mph, 3,529 kph); speed in a closed circuit (2,092 mph, 3,366 kph); and height in sustained horizontal flight (85,069 feet, 25,860 meters).

## What is the difference between an **amphibian plane** and a **seaplane**?

The primary difference is that an amphibian has retractable wheels that enable it to operate from land as well as water, while a seaplane is limited to water take-offs and landings, having only pontoons without wheels. Because its landing gear cannot retract, a seaplane is less aerodynamically efficient than an amphibian.

# MILITARY VEHICLES

## Where did the military **tank** get its name?

During World War I, when the British were developing the tank, they called these first armored fighting vehicles "water tanks" to keep their real purpose a secret. This code word has remained in spite of early efforts to call them "combat cars" or "assault carriages."

## Who invented the **culin device** on a tank?

In World War II, American tank man Sergeant Curtis G. Culin devised a crossbar welded across the front of the tank with four protruding metal tusks. This device made it possible to break through the German hedgerow defenses. In the hedgerow country of Normandy, France, countless rows or stands of bushes or trees surrounded the fields, limiting tank movement. The culin device, also known as the "Rhinoceros" because its steel angled teeth formed a tusk-like structure, cut into the base of the hedgerow and pushed a complete section ahead of it into the next field, burying any enemy troops dug in on the opposite side.

## What is a **Hummve**?

The U.S. Army originally developed the HMMWV (High Mobility Multipurpose Wheeled Vehicle), or Hummve, in 1979 as a possible replacement for the M-151 or Jeep. Today, the military uses more than 100,000 "Hummers," which can operate in all weather extremes and are designed as troop transports, light-weapon platforms, ambulances, and mobile shelters.

A civilian version is also available with such refinements as air conditioning, sound proofing, bucket seats, and a stereo sound system. It generally sells for around $50,000.

## Who was the **Red Baron**?

Manfred von Richthofen (1892–1918), a German fighter pilot during World War I, was nicknamed the "Red Baron" by the Allies because he flew a red-painted Albatros fighter. Although he became the top ace of the war by shooting down 80 Allied planes, only 60 of his kills were confirmed by both sides. The others are disputed and could have been joint kills by Richthofen and his squadron, the Flying Circus (so-named because of their brightly painted aircraft). Von Richthofen died on April 21, 1918, when he was attacked over the Semme River in France by Roy Brown, a Canadian ace, and Australian ground machine-gunners. Both parties claimed responsibility for his death.

## What is a Sopwith Camel and why is it so called?

The most successful British fighter plane of World War I, the Camel was a development of the earlier Sopwith Pup, with a much larger rotary engine. Its name "camel" was derived from the humped shape of the covering of its twin synchronized machine guns. The highly maneuverable Camel, credited with 1,294 enemy aircraft destroyed, proved far superior to all German types as dogfighters, until the introduction of the Fokker D. VII in 1918. Altogether, 5,490 Camels were built by Sopwith Aircraft. Its top speed was 118 miles per hour (189 kilometers per hour) and it had a ceiling of 24,000 feet (7,300 meters).

## When was the **B-17 Flying Fortress** introduced?

A Fortress prototype first flew on July 28, 1935, and the first Y1B-17 was delivered to the Air Corps in March 1937, followed by an experimental Y1B-17A fitted with turbo-supercharged engines in January 1939. An order for 39 planes was placed for this model under the designation B-17B. In addition to its bombing function, the B-17 was used for many experimental duties, including serving as a launching platform in the U.S.A.A.F. guided missile program and in radar and radio-control experiments. It was called a "Flying Fortress" because it was the best defended bomber of World War II. Altogether, it carried 13 50-caliber Browning M-2 machine guns, each having about 700 pounds (317.5 kilograms) of armor-piercing ammunition. Ironically, the weight of all its defensive armament and manpower severely restricted the space available for bombs.

## Who were the **Flying Tigers**?

They were members of the American Volunteer Group who were recruited early in 1941 by Major General Claire Lee Chennault (1890–1958) to serve in China as mercenaries. Some 90 veteran United States pilots and 150 support personnel served from December 1941 until June 1942 during World War II. The airplanes they flew were P-40 Warhawks, which had the mouths of tiger sharks painted on the planes' noses. It was from these painted-on images that the group got its nickname "Flying Tigers."

## Why was the designation **MiG** chosen for the Soviet fighter plane used in World War II?

The MiG designation, formed from the initials of the plane's designers, Artem I. Mikoyan and Mikhail I. Gurevich, sometimes is listed as the Mikoyan-Gurevich MiG. Appearing in 1940 with a maximum speed of 400 miles per hour (644 kilometers per hour), the MiG-3, a piston-engined fighter, was one of the few Soviet planes whose performance was comparable with Western types during World War II. One of the best-known fighters, the MiG-15, first flown in December 1947, was powered by a Soviet version of a Rolls-Royce turbo jet engine. This high performer saw action during the Korean Conflict (1950–1953). In 1955 the MiG-19 became the first Soviet fighter capable of supersonic speed in level flight.

## What is the name of the airplane that carried the **first atomic bomb**?

During World War II, the *Enola Gay*, a modified Boeing B-29 bomber, dropped the first atomic bomb on Hiroshima, Japan, at 8:15 a.m. on August 6, 1945. It was piloted by Col. Paul W. Tibbets Jr. of Miami, Florida. The bombardier was Maj. Thomas W. Ferebee of Mocksville, North Carolina. Bomb designer Capt. William S. Parsons was aboard as an observer.

Three days later, another B-29 called *Bockscar* dropped a second bomb on Nagasaki, Japan. The Japanese surrendered unconditionally on August 15, which confirmed the American belief that a costly and bloody invasion of Japan could be avoided at Japanese expense.

The *Enola Gay* was on display at the Smithsonian Institution's National Air and Space Museum in Washington, D.C., from 1995 to 1998. It will eventually be on permanent display at the National Air and Space Museum's Steven F. Udvar-Hazy Center scheduled to open in late 2003. *Bockscar* is on display at the U.S. Air Force Museum at Wright-Patterson Air Force Base, Dayton, Ohio.

## Which aircraft was the first to use **stealth** technology?

The F-117A *Nighthawk* was first deployed in 1982. The goal of stealth technology is to make an airplane invisible to radar. There are two different ways to achieve invisibility; one is for the airplane to be shaped so that any radar signal it reflects are reflected away from the radar equipment, and the other is to cover the airplane is materials that absorb radar signals. Stealth aircraft have completely flat surfaces and very sharp edges to reflect away radar signals. Stealth aircraft are also treated so they absorb radar energy.

## What is the **smallest spy plane**?

Miniature, unmanned micro air vehicles are the smallest spy planes. Weighing less than three ounces (100 grams) and no more than six inches (15.25 centimeters) across (small enough to fit in the palm of a hand), these tiny vehicles fly at low altitudes and are equipped with cameras and transmitters for a real-time video downlink.

# COMMUNICATIONS

## SYMBOLS, WRITING, AND CODES

### When was **papyrus** first used as a **writing medium**?

Papyrus is a plant (*Cyperus papyrus*) that grows in swamps and standing water. During ancient times it grew in the Nile Valley and Delta and along the Euphrates. It was used as a writing surface during the formative stages of civilization, but the origins of its first use are unknown. An unused roll was found in an Egyptian tomb from the First Dynasty (ca. 3100 B.C.E.). Papyrus was the main writing medium through the Roman Empire, but it was replaced by less expensive parchment in the third century C.E.

### What changes have occurred in the **paper manufacturing process** contributing to the **loss of historical writings**?

Most commercial cellulose paper manufactured within the last century is acidic. The acid makes paper brittle and eventually causes it to crumble with only minor use. The problem comes from two features of modern paper: the paper manufacturing process results in cellulose fibers that are very short and acid is introduced (or not removed by purification) during manufacture. Acid in the presence of moisture degrades the fibers, and the acidic hydrolysis reaction repeatedly splits the cellulose chains into smaller fragments. The reaction itself produces acid, accelerating the degradation. Ironically, the older the paper the longer it lasts. Paper manufactured up until about the mid-19th century was made from cotton and linen. These early papers had very long fibers, the key to their longevity. Today's newsprint paper is the weakest paper; it is unpurified and it has the shortest fibers. Consequently, newspapers generally fade and yellow within a few months.

Acidic paper can be de-acidified. Books, for example, can be dipped in or sprayed with an alkaline solution. However, this process does not reverse brittleness. Once the damage is done to paper fiber it is irreversible. Paper can crumble in as little as fifty years, which has placed many older manuscripts in danger.

The most durable modern papers are alkaline, where chalk is added to neutralize acid. Books that are marked in the frontmatter with the symbol for infinity (∞) generally indicate that the paper was specially prepared to meet the requirements of American National Standard for Information Science Permanence of Paper for Printed Library Materials.

## What is the **standard phonetic alphabet**?

| Letter | Phonetic equivalent |
|---|---|
| A | Alpha |
| B | Bravo |
| C | Charlie |
| D | Delta |
| E | Echo |
| F | Foxtrot |
| G | Golf |
| H | Hotel |
| I | India |
| J | Juliett |
| K | Kilo |
| L | Lima |
| M | Mike |
| N | November |
| O | Oscar |
| P | Papa |
| Q | Quebec |
| R | Romeo |
| S | Sierra |
| T | Tango |
| U | Uniform |
| V | Victor |
| W | Whiskey |
| X | X-ray |
| Y | Yankee |
| Z | Zulu |

## Which **animals** other than horses have been used to **deliver the mail**?

During the 19th century, cows hauled mail wagons in some German towns. In Texas, New Mexico, and Arizona, camels were used. In Russia and Scandinavia, reindeer pulled mail sleighs. The Belgian city of Liége even tried cats, but they proved to be unreliable.

### What is a hornbook?

Found in English and American classrooms from the 15th century to the 18th century, the hornbook was a flat board with a handle that beginning students used. On the board was pasted a sheet of paper usually containing the alphabet, the Benediction, the Lord's Prayer, and the Roman numerals. A thin, flat piece of clear horn covered the whole board to protect the paper, which was scarce and expensive at the time. Hornbooks were used as early as 1442 and became standard equipment in English schools by the 1500s. They were discontinued around 1800, when books became cheaper.

## Who invented the **Braille** alphabet?

The Braille system, used by the blind to read and write, consists of combinations of raised dots that form characters corresponding to the letters of the alphabet, punctuation marks, and common words such as "and" and "the." Louis Braille (1809–1852), blind himself since the age of three, began working on developing a practical alphabet for the blind shortly after he started a school for the blind in Paris. He experimented with a communication method called night-writing, which the French army used for nighttime battlefield missives. With the assistance of an army officer, Captain Charles Barbier, Braille pared the method's 12-dot configurations to a 6-dot one and devised a code of 63 characters. The system was not widely accepted for several years; even Braille's own Paris school did not adopt the system until 1854, two years after his death. In 1916, the United States sanctioned Louis Braille's original system of raised dots, and in 1932 a modification called "Standard English Braille, Grade 2" was adopted throughout the English-speaking world. The revised version changed the letter-by-letter codes into common letter combinations, such as "ow," "ing," and "ment," making reading and writing a faster activity.

Before Braille's system, one of the few effective alphabets for the blind was devised by another Frenchman, Valentin Haüy (1745–1822), who was the first to emboss paper to help the blind read. Haüy's letters in relief were actually a punched alphabet, and imitators immediately began to copy and improve on his system. Another letter-by-letter system of nine basic characters was devised by Dr. William Moon (1818–1894) in 1847, but it is less versatile in its applications.

## What is the **Morse Code**?

The success of any electrical communication system lies in its coding interpretation, for only series of electric impulses can be transmitted from one end of the system to the other. These impulses must be "translated" from and into words, numbers, etc. This problem plagued early telegraphy until American painter-turned-scientist Samuel F. B.

Samuel F. B. Morse, developed the Morse Code for use with the telegraph device.

Morse (1791–1872), with the help of Alfred Vail (1807–1859), devised in 1835 a code composed of dots and dashes to represent letters, numbers, and punctuation. Telegraphy uses an electromagnet—a device that becomes magnetic when activated and raps against a metal contact. A series of short electrical impulses repeatedly can make and break this magnetism, resulting in a tapped-out message.

Having secured a patent on the code in 1837, Morse and Vail established a communications company on May 24, 1844. The first long-distance telegraphed message was sent by Morse in Washington, D.C., to Vail in Baltimore, Maryland. This was the same year that Morse took out a patent on telegraphy; Morse never acknowledged the unpatented contributions of Joseph Henry (1797–1878), who invented the first electric motor and working electromagnet in 1829 and the electric telegraph in 1831.

The International Morse Code (shown below) uses sound or a flashing light to send messages. The dot is a very short sound or flash; a dash equals three dots. The pauses between sounds or flashes should equal one dot. An interval of the length of one dash is left between letters; an interval of two dashes is left between words.

| A .- | J .—- | S ... |
|---|---|---|
| B-... | K-.- | T- |
| C-.-. | L .-.. | U ..— |
| D-.. | M — | V ...- |
| E . | N-. | W .— |
| F ..-. | O —- | X-..- |

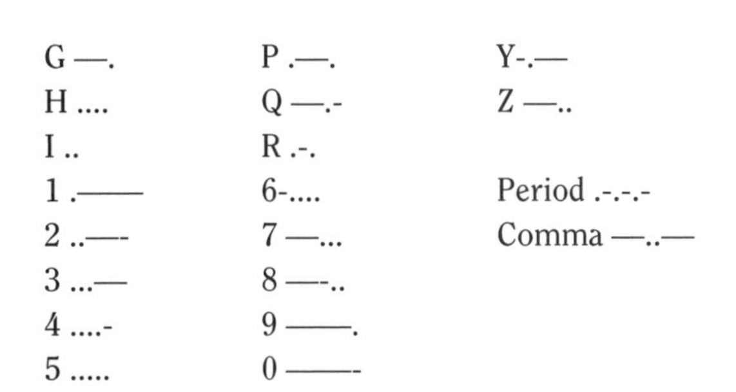

| | | |
|---|---|---|
| G —. | P .—. | Y-.— |
| H .... | Q —.- | Z —.. |
| I .. | R .-. | |
| 1 .—— | 6-.... | Period .-.-.- |
| 2 ..—- | 7 —... | Comma —..— |
| 3 ...— | 8 —-.. | |
| 4 ....- | 9 ——. | |
| 5 ..... | 0 ——- | |

## How does **ASCII** code work?

ASCII (pronounced "ask-eee") is the abbreviation for American Standard Code for Information Interchange. ASCII is a coding system that uses 128 different combinations of a group of seven bits to form common keyboard characters, including upper and lower case A to Z, numbers 0 to 9, and special symbols such as ! or @ or #. ASCII assigns each of these characters a number between 0 and 127. These assignments are graphed on a vertical line (00, 10, 20, through 70) and a horizontal line (00, 01, 02 to 09, 0A, 0B, 0C to 0F). Each character is then arranged with the vertical row position first followed by the horizontal row. For example, the name John would be coded as follows:

J=4A

o=70

h=68

n=6E

ASCII was created in 1963. The United States government and ANSI, the American National Standards Institute, officially adopted it in 1968. ASCII is unsuitable for most non-English languages and for complex computer applications because it is limited to only 128 different combinations of characters. More robust codes based on 16 and 24 bit, such as Unicode—which supports 65,356 different characters—are gradually becoming integrated into the newer operating systems and software applications.

## What were **Enigma** and **Purple** in **World War II**?

Enigma and Purple were the electric rotor cipher machines of the Germans and Japanese, respectively. The Enigma machine, used by the Nazis, was invented in the 1920s and was the best-known cipher machine in history. One of the greatest triumphs in the history of cryptanalysis was the Polish and British solution of the German Enigma ciphers. This played a major role in the Allies' conduct of World War II.

In 1939, the Japanese introduced a new cipher machine adapted from Enigma. Code-named Purple by U.S. cryptanalysts, the new machine used telephone stepping switches instead of rotors. U.S. cryptanalysts were able to solve this new system as well.

Cryptography—the art of sending messages in such a way that the real meaning is hidden from everyone but the sender and the recipient—is done in two ways: code and cipher. A code is like a dictionary in which all the words and phrases are replaced by code words or code numbers. A code book is used to read the code. A cipher works with single letters, rather than complete words or phrases. There are two kinds of ciphers: transposition and substitution. In a transposition cipher, the letters of the ordinary message (or plain text) are jumbled to form the cipher text. In substitution, the plain letters can be replaced by other letters, numbers, or symbols.

## Who was **Etaoin Shrdlu**?

The sequence of letters, Etaoin Shrdlu, appeared occasionally in newspapers many years ago, leading some to conclude that these letters were the name of some mysterious person. The reason for its appearance, however, was no mystery: Etaoin Shrdlu are the letters produced by running the finger down the first two vertical rows of the Linotype machine keyboard. The Linotype was the brand name for a printing device invented by Ottmar Mergenthaler in 1886 and first used in the *New York Tribune*. The text line was cast with molten lead when the operator keyed in and finished a line of text. Etaoin Shrdlu was used as a temporary marking slug or to indicate that a typographical mistake was made that required resetting of the cast. Linotype operators used this sequence because it was so easy to make on the keyboard. Sometimes the sequence inadvertently made it into print. Linotype typesetters were ubiquitous in the newspaper and industry (and were even used on the battlefields of WWI) up to about 1960, after which they were replaced with photocomposition. Etaoin Shrdlu likewise disappeared from print, except for the name on the occasional novel, comic strip, or science fact book.

## Which **code** went **undeciphered** during World War II?

The enemy never deciphered the secret code of the Navajo Code Talkers. Twenty-nine Navajo tribe members in the United States Marine Corps at the beginning of World War II went to San Diego and developed a code based on their native language. The code consisted of three parts: the Navajo word, its translation, and its military meaning. For instance, the Navajo word "ha-ih-des-ee" translates to "watchful." The military meaning is "alert." The original 29 Navajos grew to more than 400 by the end of the war, and the code book increased from 274 original words to 508 words by war's end. These cryptologist are the subject of the motion picture *Windtalkers*.

## What are the **10-codes**?

Almost as many different codes exist as agencies using codes in radio transmission. The following are officially suggested by the Associated Public Safety Communications Officers (APSCO):

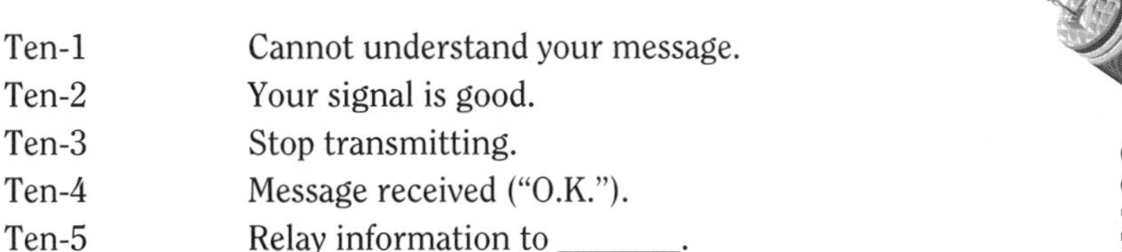

| | |
|---|---|
| Ten-1 | Cannot understand your message. |
| Ten-2 | Your signal is good. |
| Ten-3 | Stop transmitting. |
| Ten-4 | Message received ("O.K."). |
| Ten-5 | Relay information to _______. |
| Ten-6 | Station is busy. |
| Ten-7 | Out of service. |
| Ten-8 | In service. |
| Ten-9 | Repeat last message. |
| Ten-10 | Negative ("no"). |
| Ten-11 | ______ in service. |
| Ten-12 | Stand by. |
| Ten-13 | Report ______ conditions. |
| Ten-14 | Information. |
| Ten-15 | Message delivered. |
| Ten-16 | Reply to message. |
| Ten-17 | Enroute. |
| Ten-18 | Urgent. |
| Ten-19 | Contact ______. |
| Ten-20 | Unit location. |
| Ten-21 | Call ______ by telephone. |
| Ten-22 | Cancel last message. |
| Ten-23 | Arrived at scene. |
| Ten-24 | Assignment completed. |
| Ten-25 | Meet ______. |
| Ten-26 | Estimated time of arrival is ______. |
| Ten-27 | Request for information on license. |
| Ten-28 | Request vehicle registration information. |
| Ten-29 | Check records. |
| Ten-30 | Use caution. |
| Ten-31 | Pick up. |
| Ten-32 | Units requested. |
| Ten-33 | Emergency! Officer needs help. |
| Ten-34 | Correct time. |

## How was a **palindromic square** thought to be used as a **secret code**?

A palindrome is a sequence of characters (words, letters, or a combination of them) that reads the same from left to right and right to left. Examples include the girl's name Hannah and the year 2002. A "recurrent" palindrome occurs where reading from right to left and left to right creates separate meanings such as the words "trap" and "star." A palindromic square is a complex sequence of letters forming a square that is the same when read from all variations of left to right and top to bottom. Some historians believe that the following intricate square written on a Roman wall in England was a code message left by an early Christian while eluding persecution.

SATOR
AREPO
TENET
OPERA
ROTAS

## What do the lines in a **UPC bar code** mean?

A Universal Product Code (UPC) or bar code is a product description code designed to be read by a computerized scanner or cash register. It consists of 11 numbers in groups of "0"s (dark strips) and "1"s (white strips). A bar will be thin if it has only one strip or thicker if there are two or more strips side by side.

The first number describes the type of product. Most products begin with a "0"; exceptions are variable weight products such as meat and vegetables (2), health-care products (3), bulk-discounted goods (4), and coupons (5). Since it might be misread as a bar, the number 1 is not used.

The next five numbers describe the product's manufacturer. The five numbers after that describe the product itself, including its color, weight, size, and other distinguishing characteristics. The code does not include the price of the item. When the identifying code is read, the information is sent to the store's computer database, which checks it against a price list and returns the price to the cash register.

The last number is a check digit, which tells the scanner if there is an error in the other numbers. The preceding numbers, when added, multiplied, and subtracted in a certain way will equal this number. If they do not, a mistake exists somewhere.

## What does the code that follows the letters **ISBN** mean?

ISBN, or International Standard Book Number, is an ordering and identifying code for book products. It forms a unique number to identify that particular item. The first number of the series relates to the language the book is published in; for example, the zero is designated for the English language. The second set of numbers identifies the publisher, and the last set of numbers identifies the particular item. The very last number is a "check number," which mathematically makes certain that the previous numbers have been entered correctly.

Senator Guglielmo Marconi, inventor of the wireless radio.

# RADIO AND TELEVISION

## Who invented **radio**?

Guglielmo Marconi (1874–1937), of Bologna, Italy, was the first to prove that radio signals could be sent over long distances. Radio is the radiation and detection of signals propagated through space as electromagnetic waves to convey information. It was first called wireless telegraphy because it duplicated the effect of telegraphy without using wires. On December 12, 1901, Marconi successfully sent Morse code signals from Newfoundland to England.

In 1906, the American inventor Lee de Forest (1873–1961) built what he called "the Audion," which became the basis for the radio amplifying vacuum tube. This device made voice radio practical, because it magnified the weak signals without distorting them. The next year, de Forest began regular radio broadcasts from Manhattan, New York. As there were still no home radio receivers, de Forest's only audience was ship wireless operators in New York City Harbor.

## What was the first **radio broadcasting station**?

The identity of the "first" broadcasting station is a matter of debate since some pioneer AM broadcast stations developed from experimental operations begun before the institu-

With his invention of the FM receiver, Edward Howard Armstrong revolutionized the future of radio broadcasting.

tion of formal licensing practices. According to records of the Department of Commerce, which then supervised radio, WBZ in Springfield, Massachusetts, received the first regular broadcasting license on September 15, 1921. However, credit for the first radio broadcasting station has customarily gone to Westinghouse station KDKA in Pittsburgh for its broadcast of the Harding-Cox presidential election returns on November 2, 1920. Unlike most other earlier radio transmissions, KDKA used electron tube technology to generate the transmitted signal and hence to have what could be described as broadcast quality. It was the first corporate-sponsored radio station and the first to have a well-defined commercial purpose—it was not a hobby or a publicity stunt. It was the first broadcast station to be licensed on a frequency outside the amateur bands. Altogether, it was the direct ancestor of modern broadcasting.

## How are the **call letters** beginning with "K" or "W" assigned to **radio stations**?

These beginning call letters are assigned on a geographical basis. For the majority of radio stations located east of the Mississippi River, their call letters begin with the letter "W"; if the stations are west of the Mississippi, their first call letter is "K." There are exceptions to this rule. Stations founded before this rule went into effect kept their old letters. So, for example, KDKA in Pittsburgh has retained the first letter "K"; likewise, some western pioneer stations have retained the letter "W." Since many AM licensees also operate FM and TV stations, a common practice is to use the AM call letters followed by "–FM" or "–TV."

## Why do **FM radio stations** have a limited broadcast range?

Usually radio waves higher in frequency than approximately 50 to 60 megahertz are not reflected by the Earth's ionosphere, but are lost in space. Television, FM radio, and high frequency communications systems are therefore limited to approximately line-of-sight ranges. The line-of-sight distance depends on the terrain and antenna height, but is usually limited to from 50 to 100 miles (80 to 161 kilometers). FM (frequency–modulation) radio uses a wider band than AM (amplitude–modulation) radio

to give broadcasts high fidelity, especially noticeable in music—crystal clarity to high frequencies and rich resonance to base notes, all with a minimum of static and distortion. Invented by Edwin Howard Armstrong (1891–1954) in 1933, FM receivers became available in 1939.

## Why do **AM stations** have a wider **broadcast range at night**?

This variation is caused by the nature of the ionosphere of the Earth. The ionosphere consists of several different layers of rarefied gases in the upper atmosphere that have become conductive through the bombardment of the atoms of the atmosphere by solar radiation, by electrons and protons emitted by the sun, and by cosmic rays. These layers, sometimes called the Kennelly–Heaviside layer, reflect AM radio signals, enabling AM broadcasts to be received by radios that are great distances from the transmitting antenna. With the coming of night, the ionosphere layers partially dissipate and become an excellent reflector of the short waveband AM radio waves. This causes distant AM stations to be heard more clearly at night.

## Can **radio transmissions** between **space shuttles** and ground control be picked up by shortwave radio?

Amateur radio operators at Goddard Space Flight Center, Greenbelt, Maryland, retransmit shuttle space-to-ground radio conversations on shortwave frequencies. These retransmissions can be heard freely around the world. To hear astronauts talking with ground controllers during liftoff, flight, and landing, a shortwave radio capable of receiving single-sideband signals should be tuned to frequencies of 3.860, 7.185, 14.295, and 21.395 megahertz. British physics teacher Geoffrey Perry, at the Kettering Boys School, has taught his students how to obtain telemetry from orbiting Russian satellites. Since the early 1960s Perry's students have been monitoring Russian space signals using a simple taxicab radio, and using the data to calculate position and orbits of the spacecraft.

Philo T. Farnsworth played a key role in the development of television.

## Who was the founder of **television**?

The idea of television (or "seeing by electricity," as it was called in 1880) was offered by several people over the years, and several individuals contributed a multiplicity of partial inventions. For example, in 1897 Ferdinand Braun (1850–1918) constructed the first cathode ray oscilloscope, a fundamental component to all television receivers. In 1907, Boris Rosing proposed using Braun's tube to receive images, and in the following year Alan Campbell-Swinton likewise suggested using the tube, now called the cathode-ray tube, for both transmission and receiving. The figure most frequently called the father of television, however, was a Russian-born American named Vladimir K. Zworykin (1889–1982). A former pupil of Rosing, he produced a practical method of amplifying the electron beam so that the light/dark pattern would produce a good image. In 1923, he patented the iconoscope (which would become the television camera), and in 1924 he patented the kinoscope (television tube). Both inventions rely on streams of electrons for both scanning and creating the image on a fluorescent screen. By 1938, after adding new and more sensitive photo cells, Zworykin demonstrated his first practical model.

Another "father" of television is the American Philo T. Farnsworth (1906–1971). He was the first person to propose that pictures could be televised electronically. He came up with the basic design for an apparatus in 1922 and discussed his ideas with his high school teacher. This documented his ideas one year before Zworykin and was critical in settling a patent dispute between Farnsworth and his competitor at the Radio Corporation of America. Farnsworth eventually licensed his television patents to the growing industry and let others refine and develop his basic inventions.

During the early 20th century others worked on different approaches to television. The best-known is John Logie Baird (1888–1946), who in 1936 used a mechanized scanning device to transmit the first recognizable picture of a human face. Limitations in his designs made any further improvements in the picture quality impossible.

## How does **rain** affect **television reception** from a satellite?

The incoming microwave signals are absorbed by rain and moisture, and severe rainstorms can reduce signals by as much as 10 decibels (reduction by a factor of 10). If the installation cannot cope with this level of signal reduction, the picture may be momentarily lost. Even quite moderate rainfall can reduce signals enough to give noisy reception on some receivers. Another problem associated with rain is an increase in noise due to its inherent noise temperature. Any body above the temperature of absolute zero (0°K or –459°F or –273°C) has an inherent noise temperature generated by the release of wave packets from the body's molecular agitation (heat). These wave packets have a wide range of frequencies, some of which will be within the required bandwidth for satellite reception. The warm Earth has a high noise temperature, and consequently rain does as well.

## What name is used for a **satellite dish** that picks up **TV broadcasts**?

*Earth station* is the term used for the complete satellite receiving or transmitting station. It includes the antenna, the electronics, and all associated equipment necessary

to receive or transmit satellite signals. It can range from a simple, inexpensive, receive-only Earth station that can be purchased by the individual consumer, to elaborate, two-way communications stations that offer commercial access to the satellite's capacity. Signals are captured and focused by the antenna into a feedhorn and low noise amplifier. These are relayed by cable to a down converter and then into the satellite receiver/modulator.

Satellite television became widely available in the late 1970s when cable television stations, equipped with satellite dishes, received signals and sent them to their subscribers by coaxial cable. Taylor Howard designed the first satellite dish for personal use in 1976. By 1984 there were 500,000 installations, and in recent years that number has increased worldwide to 3.7 million.

## How do **flat-panel displays** differ from traditional screens?

Flat-panel screens are different because they do not use the cathode-ray tube. Cathode-ray tube monitors—the monitors that are nearly omnipresent around the world—work by bombarding a phosphorescent screen with a ray of electrons. The electrons illuminate "phosphors" on the screen into the reds, greens, and blues that form the picture. In contrast, flat-panel displays use a grid of electrodes, crystals, or vinyl polymers to create the small dots that make up the picture. Flat-panel screens are not a new idea: LCD (short for liquid crystal display) watches and calculators going back decades have relied on crystal-based flat-panel displays. The newest flat-panel screens found on laptop computers use plasma display panels (or "PDPs"). They can be very wide—over a meter—but only a few centimeters thick. A PDP screen is made of a layer of picture elements in the three primary colors. Electrodes from a grid behind produce a charge that create ultraviolet rays. These rays illuminate the various picture elements and form the picture.

## What is **high-definition television**?

The amount of detail shown in a television picture is limited by the number of lines that make it up and by the number of picture elements on each line. The latter is mostly determined by the width of the electron beam. To obtain pictures closer to the quality associated with 35-millimeter photography, a new television system, HDTV (high-definition television), will have more than twice the number of scan lines with a much smaller picture element. Currently, American and Japanese television has 525 scanning lines, while Europe uses 625 scanning lines. The Japanese are generally given credit for being pioneers in HDTV, ever since the Japanese broadcasting company NHK began research in 1968. In fact, the original pioneer was RCA's Otto Schade, who began his research after the end of World War II. Schade was ahead of his time, and decades passed before television pickup tubes and other components became available to take full advantage of his research.

### Can a TV satellite dish be painted a different color?

Although not recommended, TV satellite dishes can be painted a different color as long as the same standards adhered to in manufacture are maintained. The paint used should not be optically reflective. Metallic paints or gloss finishes may focus the sun's radiation on the head unit, causing performance problems. Only "vinyl matte" finish paints should be used. They exhibit lower solar reflection properties, and they cause a minimal amount of microwave absorption and reflection errors. Finally, the paint should be applied as smoothly as possible, as any bumps or drips may cause reflection errors.

HDTV cannot be used in the commercial broadcast bands until technical standards are approved by the United States Federal Communications Commission (FCC) or the various foreign regulating agencies. The more immediate problem, however, has been a technological one—HDTV needs to transmit five times more data than is currently assigned to each television channel. One approach is signal compression—squeezing the 30-megahertz bandwidth signal that HDTV requires into the six-megahertz bandwidth currently used for television broadcasting. The Japanese and Europeans have explored analog systems that use wavelike transmission, while the Americans based their HDTV development on digital transmission systems. In 1994, the television industry cleared this hurdle when it accepted a digital signal transmission system developed by Zenith.

The public can now enjoy HDTV: programming and television sets are available, albeit in limited broadcast areas and at a high cost. The programming is available through digital service for about 450 stations in 130 markets (almost every large city). Where digital HDTV service is not yet available, customers can order satellite service, but it requires a specialized dish and receiver. Television sets are also available, but as with any new technology prices are still high, starting at about $2,000 for an entry-level set with wide-screen models approaching $5,000.

## How do submerged **submarines communicate**?

Using frequencies from very high to extremely low, submarines can communicate by radio when submerged if certain conditions are met, and depending on whether or not detection is important. Submarines seldom transmit on long-range high-radio frequencies if detection is important, as in war. However, Super (SHF), Ultra (UHF), or Very High Frequency (VHF) two-way links with cooperating aircraft, surface ships, via satellite, or with the shore are fairly safe with high data rate, though they all require that the boat show an antenna above the water or send a buoy to the surface.

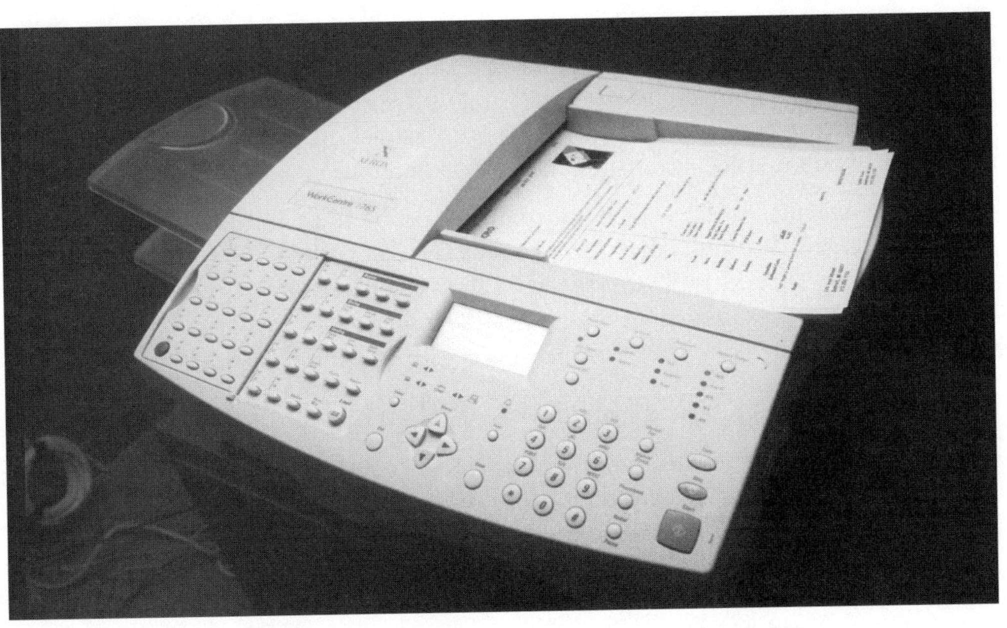

The telefacsimile, or fax machine, transmits information between locations by way of telephone lines.

# TELECOMMUNICATIONS, RECORDINGS, THE INTERNET, ETC.

## When was the first **commercial communications satellite** used?

In 1960 *Echo 1,* the first communications satellite, was launched. Two years later, on July 10, 1962, the first commercially funded satellite, *Telstar 1* (paid for by American Telephone and Telegraph), was launched into low Earth orbit. It was also the first true communications satellite, being able to relay not only data and voice, but television as well. The first broadcast, which was relayed from the United States to England, showed an American flag flapping in the breeze. The first commercial satellite (in which its operations are conducted like a business) was *Early Bird*, which went into regular service on June 10, 1965, with 240 telephone circuits. *Early Bird* was the first satellite launched for Intelsat (International Telecommunications Satellite Organization). Still in existence, the system is owned by member nations—each nation's contribution to the operating funds are based on its share of the system's annual traffic.

## How does a **fax** machine work?

Telefacsimile (also telefax or facsimile or fax) transmits graphic and textual information from one location to another through telephone lines. A transmitting machine uses either a digital or analog scanner to convert the black-and-white representations of the image into electrical signals that are transmitted through the telephone lines to a designated receiving machine. The receiving unit converts the transmission back to an image of the original and prints it. In its broadest definition, a facsimile terminal is simply a copier equipped to transmit and receive graphics images.

The fax was invented by Alexander Bain of Scotland in 1842. His crude device, along with scanning systems invented by Frederick Bakewell in 1848, evolved into several modern versions. In 1924, faxes were first used to transmit wire photos from Cleveland to New York, a boon to the newspaper industry.

## Can a **fax** and an **answering machine** be used on the **same telephone line**?

Most fax machines come with an interface that allows it to work with an answering machine. The fax "listens" for the incoming call and sends it to the answering machine if no one picks up the phone. As the message is being recorded, the fax machine listens for a fax tone. If it hears the tone, it sends the fax through. If not, the answering machine continues to function as it normally would. Fax machines with built-in answering machines are also available, making such an interface unnecessary.

## How does a **fiber-optic cable** work?

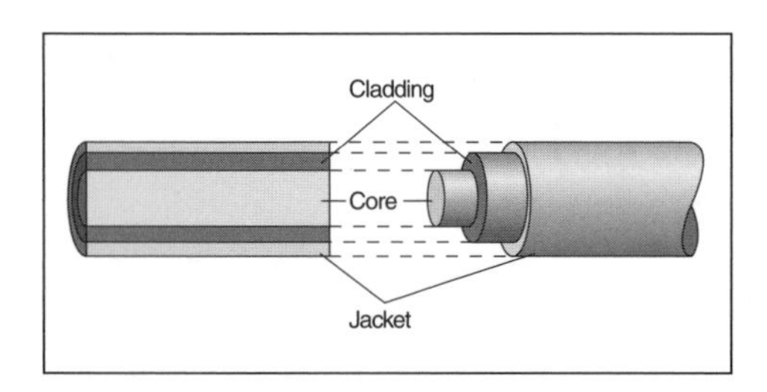


**A cross section of a fiber-optic cable.**

A fiber-optic cable is composed of many very thin strands of coated glass or plastic fibers that transmit light through the process of "cladding," in which total internal reflection of light is achieved by using material that has a lower refractive index. Once light enters the fiber, the cladding layer inside it prevents light loss as the beam of light zigzags inside the glass core. Glass fibers can transmit messages or images by directing beams of light inside itself over very short or very long distances up to 13,000 miles (20,917 kilometers) without significant distortion. The pattern of light waves forms a code that carries a message. At the receiving end, the light beams are converted back into electric current and decoded. Uses include telecommunications medical fiber-optic viewers, such as endoscopes and fiberscopes, to see internal organs; fiber-optic message devices in aircraft and space vehicles; and fiber-optic connections in automotive lighting systems.

Fiber-optic cables have greater "bandwidth": they can carry much more data than metal cable. Because fiber optics is based on light beams, the transmissions are more impervious to electrical noise and can also be carried greater distances before fading. The cables are thinner than metal wires. Fiber-optic cable delivers data in digital code

instead of an analog signal, the delivery method of metal cables; computers are structured for digital, so there is a natural symbiosis. The main disadvantage is cost: fiber optics are much more expensive than traditional metal cable.

## What is a **Clarke belt**?

Back in 1945, Arthur C. Clarke (b. 1917), the famous scientist and science fiction writer, predicted that an artificial satellite placed at a height of 22,248 miles (35,803 kilometers) directly above the equator would orbit the globe at the same speed with which the Earth was rotating. As a result, the satellite would remain stationary with respect to any point on the Earth's surface. This equatorial belt, rather like one of Saturn's rings, is affectionately known as the Clarke belt.

## What is the difference between "**analog**" and "**digital**"?

The word analog is derived from "analogous," meaning something corresponds to something else. A picture of a mountain is a representation of the mountain. If the picture is on traditional color film it is called an analog picture and is characterized by different colors and range of color on the paper. However, if the photographer took the picture with a digital camera, the image is digital: it is stored as a series of numbers in the camera's memory. Its colors are discrete. In communications and computers, a range of continuous variables characterizes analog. Digital, in contrast, is characterized by measurements that happen at discrete intervals. Digital representations are, therefore, more precise. Common examples of analog media include records (music) and VCR tapes (movies). Counterpart examples of digital media include CD-ROM discs and DVD movies.

## How have **cellular telephones** styles evolved?

The first and oldest variety of mobile cellular phones are permanently attached to an automobile and are powered by the car's battery. They also have an antenna that must be mounted outside the vehicle. The second type are transportable, or bag, cellular phones. These are essentially mobile phones with their own battery packs that allow owners to detach them from the car and carry them in a pouch. However, most weigh about five pounds (2.25 kilograms) and are not very practical when used this way. The third type is a portable cellular phone. Similar in appearance to a cordless phone handset, a portable generally weighs less than a pound and is the most versatile type of cellular phone. It is also the most expensive and has a transmitter of less power than a mobile or transportable cellular phone. Kits are available for some models, however, that boosts the transmitter's power. Despite claims that they may cause health problems or that they may distract drivers, cell phone use is more popular than ever: in 2000 about 100 million Americans routinely used mobile phones.

## What is virtual reality?

Virtual reality combines state-of-the-art imaging with computer technology to allow users to experience a simulated environment as reality. Several different technologies are integrated into a virtual reality system, including holography, which uses lasers to create three-dimensional images; liquid crystal displays; high-definition television; and multimedia techniques that combine various types of displays in a single computer terminal.

## What is the latest type of **wireless** telephone?

The latest wireless telephone introduced in 1995 is called a PCS (personal communications services) phone. These are digital cellular telephones; they are digital because they convert the speech into a chain of numbers. This makes the transmission much clearer and enhances security by making eavesdropping more difficult. It also allows for computer integration. Users can send and receive e-mail and browse the web. They are cellular because antennas on metal towers—visible on the horizon of nearly every reasonably developed location in America—send and receive all calls within a geographical area. Transmissions are carried over microwave radiation. Wireless companies can use the same frequency over again in different cells (except for cells that are nearby) because each cell is small and therefore requires only a limited range signal. When users move across cells the signal is "handed off" to the next cell tower.

## What is the **Dolby** noise reduction system?

The magnetic action of a tape produces a background hiss—a drawback in sound reproduction on tape. A noise reduction system known as Dolby—named after R. M. Dolby (b. 1933), its American inventor—is widely used to deal with the hiss. In quiet passages, electronic circuits automatically boost the signals before they reach the recording head, drowning out the hiss. On playback, the signals are reduced to their correct levels. The hiss is reduced at the same time, becoming inaudible.

## What is **digital audio tape** (DAT)?

DAT, a new concept in magnetic recording, produces a mathematical value for each sound based on the binary code. When the values are reconstructed during playback, the reconstructed sound is so much like the original that the human ear cannot distinguish the difference. The reproduction is so good that American record companies have lobbied lawmakers to prevent DAT from being sold in the United States, claiming that it could encourage illegal CD copying.

On average, the expected life span of a CD-ROM disc is between three and five years.

## How are **compact discs** (CDs) made?

The master disc for a CD is an optically flat glass disc coated with a resist. The resist is a chemical that is impervious to an etchant that dissolves glass. The master is placed on a turntable. The digital signal to be recorded is fed to the laser, turning the laser off and on in response to the binary on-off signal. When the laser is on, it burns away a small amount of the resist on the disc. While the disc turns, the recording head moves across the disc, leaving a spiral track of elongated "burns" in the resist surface. After the recording is complete, the glass master is placed in the chemical etchant bath. This developing removes the glass only where the resist is burned away. The spiral track now contains a series of small pits of varying length and constant depth. To play a recorded CD, a laser beam scans the three miles (five kilometers) of playing track and converts the "pits" and "lands" of the CD into binary codes. A photodiode converts these into a coded string of electrical impulses. In October 1982, the first CDs were marketed; they were invented by Phillips (Netherlands) Company and Sony in Japan in 1978.

## What is the life span of a **CD-ROM** disc?

Although manufacturers claim that a CD-ROM disc will last 20 years, recent statements by the United States National Archives and Records Administration suggest that a life span of three to five years is more accurate. The main problem is that the aluminum substratum on which the data is recorded is vulnerable to oxidation.

## What is the **information highway**?

A term originally coined by former Vice President Al Gore (1948–), the information highway is envisioned as an electronic communications network of the near future that would easily connect all users to one another and provide every type of electronic service possible, including shopping, electronic banking, education, medical diagnosis, video conferencing, and game playing. Initially implemented on a national scale, it would eventually become a global network.

The exact form of the information highway is a matter of some debate. Two principle views currently exist. One visualizes the highway as a more elaborate form of the Internet, the principle purpose of which would be to gather and exchange written information via a global electronic mail network. The other possibility centers around plans to create an enhanced interactive television network that would provide video services on demand.

## What is the **Internet**?

The Internet is the world's largest computer network. It links computer terminals together via wires or telephone lines in a web of networks and shared software. With the proper equipment, an individual can access vast amounts of information and search databases on various computers connected to the Internet, or communicate with someone located anywhere in the world as long as he or she has the proper equipment.

Originally created in the late 1960s by the U.S. Department of Defense Advanced Research Projects Agency to share information with other researchers, the Internet mushroomed when scientists and academics using the network discovered its great value. Despite its origin, however, the Internet is not owned or funded by the U.S. government or any other organization or institution. A group of volunteers, the Internet Society, addresses such issues as daily operations and technical standards.

## Which technology holds the most promise for keeping information sent over the **Internet secure**?

Public-key cryptography is a means for authenticating information sent over the Internet. The system works by encrypting and decrypting information through the use of a combination of "keys." One key is a published "public key"; the second is a "private key," which is kept secret. An algorithm is used to decipher each of the keys. The method is for the sender to encrypt the information using the public key, and the recipient to decrypt the information using the secret private key.

The strength of the system depends on the size of the key: a 128-bit encryption is about 3 × 1,026 times stronger than 40-bit encryption. No matter how complex the encryption, as with any code, keeping the secret aspects secret is the important part to safeguarding the information.

### Is there a patron saint of the Internet?

A patron saint in the Roman Catholic Church is a special protector over all manner of life. There are patron saints for everything from animals (Saint Francis of Assisi) to librarians (Saint Jerome) to the former Yugoslavia (Saints Cyril and Methodius). The proposed patron saint of the Internet is Saint Isidore of Seville. Saint Isidore of Seville is well-matched for the task of Internet saint because he loved information. He wrote an encyclopedia of knowledge titled *Etymologies*, and also books on grammar, astronomy, geography, history, and theology.

### How many people **use the Internet**?

According to *Newsweek*, there are as of the year 2000 about 259 to 300 million people in 56 countries who are in some way using the Internet. About 111 million users are in the United States, followed by Japan with about 18 million, and then Britain with 14 million. Users are projected to reach one billion by 2005.

### What is the **Netplex**?

The Netplex is the name given to an area around Washington, D.C., that has become the world center of the data communications industry and the focal point of the Internet. The businesses in the Netplex include companies that build and manage optical fiber networks, sell Internet connections to companies and individuals, or offer other services. It has been compared to such other "technology centers" as California's Silicon Valley or the Research Triangle of North Carolina, where a large concentration of corporations involved in the telecommunication and computer industries are located.

## COMPUTERS

### What is an **algorithm**?

An algorithm is a set of clearly defined rules and instructions for the solution of a problem. It is not necessarily applied only in computers, but can be a step-by-step procedure for solving any particular kind of problem. A nearly 4,000-year-old Babylonian banking calculation inscribed on a tablet is an algorithm, as is a computer program that consists of step-by-step procedures for solving a problem.

The term is derived from the name of Muhammad ibn Musa al Kharizmi (ca. 780–ca. 850), a Baghdad mathematician who introduced Hindu numerals (including 0) and decimal calculation to the West. When his treatise was translated into Latin in the 12th century, the art of computation with Arabic (Hindu) numerals became known as algorism.

Charles Babbage conceived his "analytical engine"—building upon a science that has produced today's modern computer.

## Who invented the **computer**?

Computers developed from calculating machines. One of the earliest mechanical devices for calculating, still widely used today, is the abacus—a frame carrying parallel rods on which beads or counters are strung. The abacus originated in Egypt in 2000 B.C.E.; it reached the Orient about a thousand years later, and arrived in Europe in about the year 300 C.E. In 1617, John Napier (1550–1617) invented "Napier's Bones"—marked pieces of ivory for multiples of numbers. In the middle of the same century, Blaise Pascal (1623–1662) produced a simple mechanism for adding and subtracting. Multiplication by repeated addition was a feature of a stepped drum or wheel machine of 1694 invented by Gottfried Wilhelm Leibniz (1646–1716). In 1823, the English visionary Charles Babbage (1792–1871) persuaded the British government to finance an "analytical engine." This would have been a machine that could undertake any kind of calculation. It would have been driven by steam, but the most important innovation was that the entire program of operations was stored on a punched tape. Babbage's machine was not completed in his lifetime because the technology available to him was not sufficient to support his design. However, in 1991 a team led by Doron Swade at London's Science Museum built the "analytical engine" (sometimes called the "difference engine") based on Babbage's work. Measuring 10 feet (3 meters) wide by 6.5 feet (2 meters) tall, it weighed three tons and could calculate equations down to 31 digits. The feat proved that Babbage was way ahead of his time, even though the device was impractical because one had to turn a crank hundreds of times in order to generate a single calculation. Modern computers use electrons, which travel at nearly the speed of light.

Based on the concepts of British mathematician Alan M. Turing (1912–1954), the earliest programmable electronic computer was the 1,500-valve "Colossus," formulated

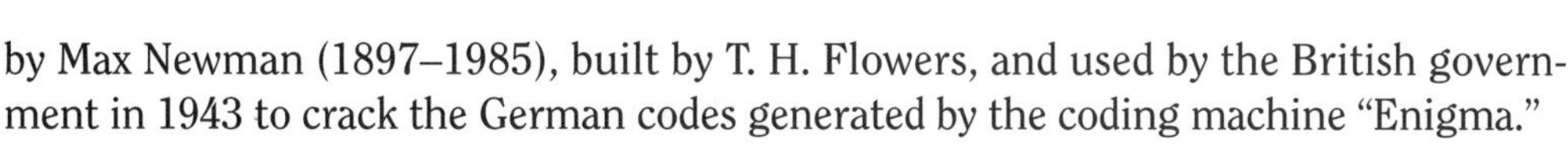

by Max Newman (1897–1985), built by T. H. Flowers, and used by the British government in 1943 to crack the German codes generated by the coding machine "Enigma."

## What was the first major use for **punched cards**?

Punched cards were a way of programming, or giving instructions to, a machine. In 1801, Joseph Marie Jacquard (1752–1834) built a device that could do automated pattern weaving. Cards with holes were used to direct threads in the loom, creating predefined patterns in the cloth. The pattern was determined by the arrangement of holes in the cards, with wire hooks passing through the holes to grab and pull through specific threads to be woven into the cloth.

By the 1880s, Herman Hollerith (1860–1929) was using the idea of punched cards to give machines instructions. He built a punched card tabulator that processed the data gathered for the 1890 United States Census in six weeks (three times the speed of previous compilations). Metal pins in the machine's reader passed through holes punched in cards the size of dollar bills, momentarily closing electric circuits. The resulting pulses advanced counters assigned to details such as income and family size. A sorter could also be programmed to pigeonhole cards according to pattern of holes, an important aid in analyzing census statistics. Later, Hollerith founded Tabulating Machines Co., which in 1924 became IBM. When IBM adopted the 80-column punched card (measuring 73/8 × 31/4 inches [18.7 × 8.25 centimeters] and 0.007 inches [0.17 millimeters] thick), the de facto industry standard was set, which has endured for decades.

## What is meant by **fifth-generation computers**? What are the other four generations?

The evolution of computers has advanced so much in the past few decades that "generations" are used to describe these important advances:

*First-generation computer*—a mammoth computer using vacuum tubes, drum memories, and programming in machine code as its basic technology. Univax 1, used in 1951, was one of the earliest of these vacuum-tube based electronic computers. The generation starts at the end of World War II and ends about 1957.

*Second-generation computer*—a computer using discrete transistors as its basic technology. Solid-state components replaced the vacuum tubes during this period from 1958 to 1963. Magnetic core memories store information. This era includes the development of high-level computer languages.

*Third-generation computer*—a computer having integrated circuits, semiconductor memories, and magnetic disk storage. New operating systems, minicomputer systems, virtual memory, and timesharing are the advancements of this period from 1963 to 1971.

*Fourth-generation computer*—a computer using microprocessors and large-scale integrated chips as its basic technology, which made computers accessible to a

### What is the origin of the expression "Do not fold, spindle, or mutilate"?

This is the inscription on an IBM punched card. Frequently, office workers organize papers and forms by stapling or folding them together, or by impaling them on a spindle. Because Hollerith (punched) card readers scan uniform rectangular holes in a precise arrangement, any damage to the physical card makes it unusable. In the 1950s and 1960s, when punched cards became widespread, manufacturers printed a warning on each card; IBM's "Do not fold, spindle, or mutilate" was the best known. In 1964, the student revolution at the University of California, Berkeley, used the phrase as a symbol of authority and regimentation.

large segment of the population. Networking, improved memory, database management systems, and advanced languages mark the period from 1971 to the end of the 1980s.

*Fifth-generation computer*—a computer that uses inference to draw reasoned conclusions from a knowledge base and interacts with its users via an intelligent user interface to perform such functions as speech recognition, machine translation of natural languages, and robotic operations. These computers using artificial intelligence have been under development since the early 1980s, especially in Japan, as well as in the United States and Europe. In 1991, however, Japan began a new 10-year initiative to investigate neural networks, which will probably divert resources from development of the fifth generation as traditionally defined.

## A lot of people have heard of **ENIAC**, the first large electronic computer. What was **MANIAC**?

MANIAC (mathematical analyzer, numerator, integrator, and computer) was built at the Los Alamos Scientific Laboratory under the direction of Nicholas C. Metropolis (1915–1999) between 1948 and 1952. It was one of several different copies of the high-speed computer built by John von Neumann (1903–1957) for the Institute for Advanced Studies (IAS). It was constructed primarily for use in the development of atomic energy applications, specifically the hydrogen bomb.

It was originated with the work on ENIAC (electronic numerical integrator and computer), the first fully operational, large-scale, electronic digital computer. ENIAC was built at the Moore School of Electrical Engineering at the University of Pennsylvania between 1943 and 1946. Its builders, John Presper Eckert Jr. (1919–1995) and John William Mauchly (1907–1980), virtually launched the modern era of the computer with ENIAC.

## What is an **expert system**?

An expert system is a type of software that analyzes a complex problem in a particular field and recommends possible solutions based on information previously programmed into it. The person who develops an expert system first analyzes the behavior of a human expert in a given field, then inputs all the explicit rules resulting from their study into the system. Expert systems are used in equipment repair, insurance planning, training, medical diagnosis, and other areas.

## What was the **first computer game**?

Despite that computers were not invented for playing games, the idea that they could be used for games did not take long to emerge. Alan Turing proposed a famous game called "Imitation Game" in 1950. In 1952, Rand Air Defense Lab in Santa Monica created the first military simulation games. In 1953, Arthur Samuel created a checkers program on the new IBM 701. From these beginnings computer games have today become a multi-billion-dollar industry.

## What was the first successful **video-arcade game**?

Pong, a simple electronic version of a tennis game, was the first successful video-arcade game. Although it was first marketed in 1972, Pong was actually invented 14 years earlier in 1958 by William Higinbotham, who headed instrumentation design at Brookhaven National Laboratory at the time. Invented to amuse visitors touring the laboratory, the game was so popular that visitors would stand in line for hours to play it. Higinbotham dismantled the system two years later, and, considering it a trifle, did not patent it. In 1972, Atari released Pong, an arcade version of Higinbotham's game, and Magnavox released Odyssey, a version that could be played on home televisions.

## What was **"The Turk"**?

The Turk was the name for a famous chess-playing automaton. An automaton, such as a robot, is a mechanical figure constructed to act as if it moves by its own power. On a dare in 1770, a civil servant in the Vienna Imperial Court named Wolfgang von Kempelen created a chess-playing machine. This mustached, man-sized figure carved from wood wore a turban, trousers, and robe, and sat behind a desk. In one hand it held a long Turkish pipe, implying that it had just finished a pre-game smoke, and its innards were filled with gears, pulleys, and cams. The machine seemed a keen chess player and dumbfounded onlookers by defeating all the best human chess players. It was a farce, however: its moves were surreptitiously made by a man hiding inside. The Turk, so dubbed because of the audacious outfit, is regarded as a forerunner to the industrial revolution because it created a commotion over devices that could complete complex tasks. Historians argue that it inspired people to invent other early devices such as the

### What was the name of the microcomputer introduced by Apple in the early 1980s?

Lisa was the name of the microcomputer that Apple introduced. The forerunner of the Macintosh microcomputer, Lisa had a graphical user interface and a mouse.

power loom and the telephone, and it even was a precursor to concepts such as artificial intelligence and computerization. Today, however, computer chess games are so sophisticated that they can defeat even the world's best chess masters. In May 1997, the "Deep Blue" chess computer, an IBM RS/6000 SP that relied on 32 processors functioning as 512 to calculate 200 million chess moves per second, defeated World Champion Garry Kasparov.

## What is a **silicon chip**?

A silicon chip is an almost pure piece of silicon, usually less than one centimeter square and about half a millimeter thick. It contains hundreds of thousands of microminiature electronic circuit components, mainly transistors, packed and interconnected in layers beneath the surface. These components can perform control, logic, and memory functions. There is a grid of thin metallic strips on the surface of the chip; these wires are used for electrical connections to other devices. The silicon chip was developed independently by two researchers: Jack Kilby of Texas Instruments in 1958, and Robert Noyce (b. 1927) of Fairchild Semiconductor in 1959.

While silicon chips are essential to most computer operation today, a myriad of other devices depend on them as well, including calculators, microwave ovens, automobile diagnostic equipment, and VCRs.

## What are the **sizes of silicon chips**?

SSI—small-scale integration
MSI—medium-scale integration
LSI—large-scale integration
VLSI—very-large-scale integration
ULSI—ultra-large-scale integration
GSI—gigascale integration

The number of components packed into a single chip is very loosely defined. ULSI units can pack millions of components on a chip. GSI, a long-term target for the industry, will potentially hold billions of components on a chip.

## Are any devices being developed to **replace** silicon chips?

When transistors were introduced in 1948, they demanded less power than fragile, high-temperature vacuum tubes; allowed electronic equipment to become smaller, faster, and more dependable; and generated less heat. These developments made computers much more economical and accessible; they also made portable radios practical. However, the smaller components were harder to wire together, and hand wiring was both expensive and error-prone.

In the early 1960s, circuits on silicon chips allowed manufacturers to build increased power, speed, and memory storage into smaller packages, which required less electricity to operate and generated even less heat. While through most of the 1970s manufacturers could count on doubling the components on a chip every year without increasing the size of the chip, the size limitations of silicon chips are becoming more restrictive. Though components continue to grow smaller, the same rate of shrinking cannot be maintained.

Researchers are investigating different materials to use in making circuit chips. Gallium arsenide is harder to handle in manufacturing, but it has potential for greatly increased switching speed. Organic polymers are potentially cheaper to manufacture and could be used for liquid-crystal and other flat-screen displays, which need to have their electronic circuits spread over a wide area. Unfortunately, organic polymers do not allow electricity to pass through as well as the silicons do. Several researchers are working on hybrid chips, which could combine the benefits of organic polymers with those of silicon. Researchers are also in the initial stages of developing integrated optical chips, which would use light rather than electric current. Optical chips would generate little or no heat, would allow faster switching, and would be immune to electrical noise.

## What are **carbon nanotubes**?

Carbon nanotubes are structures of microscopic cylinders that scientists can create when electricity arcs between two carbon electrodes. The carbon atoms become arranged in a lattice that forms a tiny tube a few nanometers long (a nanometer is a billionth of a meter). Scientists can then position them to act as transmitters. The benefit of such a structure is that they are a hundred times smaller than transistors now used on computer chips. Carbon nanotubes may eventually replace the standard silicon computer chip, allowing increased performance in a smaller size.

## What is **Moore's Law**?

Gordon Moore, cofounder of Intel, a top microchip manufacturer, observed in 1965 that the number of transistors per microchip—and hence a chip's processing power—would double about every year and a half. The press dubbed this Moore's Law. Despite claims that this ever-increasing trend cannot perpetuate, history has shown that microchip advances are, indeed, keeping pace with Moore's prediction.

## How is a **hard disk** different from a **floppy disk**?

Both types of disk use a magnetic recording surface to record, access, and erase data, in much the same way as magnetic tape records, plays, and erases sound or images. A read/write head, suspended over a spinning disk, is directed by the central processing unit (CPU) to the sector where the requested data is stored, or where the data is to be recorded. A hard disk uses rigid aluminum disks coated with iron oxide to store data. It has much greater storage capacity than several floppy disks (from 10 to hundreds of megabytes). While most hard disks used in microcomputers are "fixed" (built into the computer), some are removable. Minicomputer and mainframe hard disks include both fixed and removable hard disks (in modules called disk packs or disk cartridges). A floppy disk, also called a diskette, is made of plastic film covered with a magnetic coating, which is enclosed in a nonremovable plastic protective envelope. Floppy disks vary in storage capacity from 100 thousand bytes to more than two megabytes. Floppy disks are generally used in minicomputers and microcomputers.

In addition to storing more data, a hard disk can provide much faster access to storage than a floppy disk. A hard disk rotates from 2,400 to 3,600 revolutions per minute (rpm) and is constantly spinning (except in laptops, which conserve battery life by spinning the hard disk only when in use). An ultra-fast hard disk has a separate read/write head over each track on the disk, so that no time is lost in positioning the head over the desired track; accessing the desired sector takes only milliseconds, the time it takes for the disk to spin to the sector. A floppy disk does not spin until a data transfer is requested, and the rotation speed is only about 300 rpm.

## How much **data** can a **floppy disk** hold?

The three common floppy disk (diskette) sizes vary widely in storage capacity.

| Envelope size (inches) | Storage capacity |
|---|---|
| 8 | 100,000–500,000 bytes |
| 5.25 | 100 kilobytes–1.2 megabytes |
| 3.5 | 400 kilobytes–more than 2 megabytes |

An 8-inch or 5-inch diskette is enclosed in a plastic protective envelope, which does not protect the disk from bending or folding; parts of the disk surface are also exposed and can be contaminated by fingerprints or dust. The casing on a 3.5-inch floppy disk is rigid plastic, and includes a sliding disk guard that protects the disk surface, but allows it to be exposed when the disk is inserted in the disk drive. This protection, along with the increased data storage capacity, makes the 3.5-inch disk currently the most popular. With computer file sizes increasing at the pace of increasing computing power and memory, the venerable floppy disk will eventually give way to more expansive storage media. The latest portable storage is Zip® disks: they can store up to 250 megabytes, and re-writable CD and DVD disks can hold 600 or more

megabytes of data. In the coming years people will find useless floppy disks in cupboards and drawers—perhaps filled with important information—because there will be no disk drives left to read them. Archivists call this "obsolescence," and people are encouraged to "migrate" important information onto new storage media.

## Why does a computer **floppy disk** have to be **"formatted"**?

A disk must first be organized so that data can be stored on and retrieved from it. The data on a floppy disk or a hard disk is arranged in concentric tracks. Sectors, which can hold blocks of data, occupy arc-shaped segments of the tracks. Most floppy disks are soft-sectored, and formatting is necessary to record sector identification so that data blocks can be labeled for retrieval. Hard-sectored floppy disks use physical marks to identify sectors; these marks cannot be changed, so the disks cannot be reformatted. The way that sectors are organized and labeled dictates system compatibility: disks formatted for DOS computers can only be used in other DOS machines; those formatted for Macintoshes can only be used in other Macintoshes. Formatting erases any pre-existing data on the disk. Hard-disk drives are also formatted before being initialized, and should be protected so that they are not reformatted unintentionally. Today most disks come pre-formatted.

## Who invented the computer **mouse**?

A computer "mouse" is a hand-held input device that, when rolled across a flat surface, causes a cursor to move in a corresponding way on a display screen. A prototype mouse was part of an input console demonstrated by Douglas C. Englehart in 1968 at the Fall Joint Computer Conference in San Francisco. Popularized in 1984 by the Macintosh from Apple Computer, the mouse was the result of 15 years devoted to exploring ways to make communicating with computers simpler and more flexible.

The physical appearance of the small box with the dangling, tail-like wire suggested the name of "mouse."

## What is **Hopper's rule**?

Electricity travels one foot in a nanosecond (a billionth of a second). This is one of a number of rules compiled for the convenience of computer programmers; it is also considered to be a fundamental limitation on the possible speed of a computer—signals in an electrical circuit cannot move any faster.

## Is an **assembly language** the same thing as a **machine language**?

While the two terms are often used interchangeably, an assembly language is a more "user friendly" translation of a machine language. A machine language is the collec-

### Who was the first programmer?

According to historical accounts, Lord Byron's daughter, Augusta Ada Byron, the Countess of Lovelace, was the first person to write a computer program for Charles Babbage's (1792–1871) "analytical engine." This machine was to work by means of punched cards that could store partial answers that could later be retrieved for additional operations and would then print the results. Her work with Babbage and the essays she wrote about the possibilities of the "engine" established her as a "patron saint," if not a founding parent, of the art and science of programming. The programming language called "Ada" was named in her honor by the United States Department of Defense. In modern times the honor goes to Commodore Grace Murray Hopper (1906–1992) of the United States Navy. She wrote the first program for the Mark I computer.

tion of patterns of bits recognized by a central processing unit (CPU) as instructions. Each particular CPU design has its own machine language. The machine language of the CPU of a microcomputer generally includes about 75 instructions; the machine language of the CPU of a large mainframe computer may include hundreds of instructions. Each of these instructions is a pattern of 1's and 0's that tells the CPU to perform a specific operation.

An assembly language is a collection of symbolic, mnemonic names for each instruction in the machine language of its CPU. Like the machine language, the assembly language is tied to a particular CPU design. Programming in assembly language requires intimate familiarity with the CPU's architecture, and assembly language programs are difficult to maintain and require extensive documentation.

The computer language C, developed in the late 1980s, is now frequently used instead of assembly language. It is a high-level programming language that can be compiled into machine languages for almost all computers, from microcomputers to mainframes, because of its functional structure.

## What are the **generations** of **programming languages**?

Computer scientists use the following abbreviations to grade computer languages in steps or evolutions:

*1GL*—A first-generation language is called a "machine language," the set of instructions that the programmer writes for the processor to perform. It appears in binary form. It is written in 0s and 1s.

*2GL*—A second-generation language is termed an "assembly" language, because an assembler converts it into machine language for the processor.

*3GL*—A third-generation language is called a "high-level" programming language. Java and C++ are 3GL languages. A compiler then converts it into machine language, typically written like this:

```
if (chLetter ≥ 'B')
Console.WriteLine ("Usage: one argument");
return 1; // sample code
```

*4GL*—A fourth-generation language resembles plain language. Relational databases use this language. An example is the following:

```
FIND All Titles FROM Books WHERE Title begins with "Handy"
```

*5GL*—A fifth-generation language uses a graphical design interface that allows a 3GL or 4GL compiler to translate the language. This is similar to HTML text editors because it allows drag and drop of icons and visual display of hierarchies.

## Who invented the **COBOL** computer language?

COBOL (common business oriented language) is a prominent computer language designed specifically for commercial uses, created in 1960 by a team drawn from several computer makers and the Pentagon. The best-known individual associated with COBOL was then-Lieutenant Grace Murray Hopper (1906–1992) who made fundamental contributions to the United States Navy standardization of COBOL. COBOL excels at the most common kinds of data processing for business—simple arithmetic operations performed on huge files of data. The language endures because its syntax is very much like English and because a program written in COBOL for one kind of computer can run on many others without alteration.

## Who invented the **PASCAL** computer language?

Niklaus Wirth (b. 1934), a Swiss computer programmer, created the PASCAL computer language.

## What's the difference between a **"bit"** and a **"byte"**?

A byte, a common unit of computer storage, holds the equivalent of a single character, such as a letter ("A"), a number ("2"), a symbol ("$"), a decimal point, or a space. It is usually equivalent to eight "data bits" and one "parity bit." A bit (a binary digit), the smallest unit of information in a digital computer, is equivalent to a single "0" or "1". The parity bit is used to check for errors in the bits making up the byte. Although eight data bits per byte is the most common size, computer manufacturers are free to define a differing number of bits as a byte. Six data bits per byte is another common size.

## What does it mean to "**boot**" a computer?

Booting a computer is starting it, in the sense of turning control over to the operating system. The term comes from bootstrap, because bootstraps allow an individual to pull on boots without help from anyone else. Some people prefer to think of the process in terms of using bootstraps to lift oneself off the ground, impossible in the physical sense, but a reasonable image for representing the process of searching for the operating system, loading it, and passing control to it. The commands to do this are embedded in a read-only memory (ROM) chip that is automatically executed when a microcomputer is turned on or reset. In mainframe or minicomputers, the process usually involves a great deal of operator input. A cold boot powers on the computer and passes control to the operating system; a warm boot resets the operating system without powering off the computer.

## Should a PC be **turned off** when not in use?

Personal computers (PCs) consume a lot of electricity. The average use for the computer is about 120 watts and the average monitor consumption is 150 watts. If the computer runs for four hours a day at the average cost of 8.5 cents per kilowatt hour, you can calculate as follows to see the cost to run a PC:

(120 + 150) watts × 4 hours per day × 365 days per year/1000

= 394 kilowatt hours × 8.5 cents per hour

= $33.51 per year to run the computer.

It's no wonder that many locations are feeling an energy crunch, especially if this figure is multiplied by the millions of PCs across America. While shutting off personal computers for one or two hours during the work day is not a cost-effective practice, turning off the monitor and leaving on the central processing unit (CPU) for the same amount of time saves a substantial fraction of the PC's energy use. However, to save energy and extend the computer's lifetime, both the CPU and the monitor should be shut off at the end of the day and before the weekend.

## What is the correct way to **face a computer** screen?

Correct positioning of the body at a computer is essential to preventing physical problems such as carpal tunnel syndrome and back pain. You should sit so that your eyes are 18 to 24 inches (45 to 61 centimeters) from the screen, and at a height so that they are six to eight inches (15 to 20 centimeters) above the center of the screen. Your hands should be level with or slightly below the arms.

Correct posture is also necessary. You should sit upright, keeping the spine straight. Sit all the way back in the chair with the knees level with or below the thighs. Both feet should be on the floor. The arms may rest on the desk or chair arms, but make sure you do not slouch. If you need to bend or lean forward, do so from the waist.

## Where did the term **bug** originate?

The slang term bug is used to describe problems and errors occurring in computer programs. The term may have originated during the early 1940s at Harvard University, when computer pioneer Grace Murray Hopper (1906–1992) discovered that a dead moth had caused the breakdown of a machine on which she was working. When asked what she was doing while removing the corpse with tweezers, she replied, "I'm debugging the machine." The moth's carcass, taped to a page of notes, is preserved with the trouble log notebook at the Virginia Naval Museum.

## How did the term **glitch** originate?

A glitch is a sudden interruption or fracture in a chain of events, such as in commands to a processor. The stability may or may not be salvageable. The word is thought to have evolved from the German *glitschen*, meaning "to slip," or by the Yiddish *glitshen*, to slide or skid.

One small glitch can lead to a cascade of failure along a network. For instance, in 1997 a small Internet service provider in Virginia unintentionally provided incorrect router (a router is the method by which the network determines the next location for information) information to a backbone operator (a backbone is a major network thoroughfare in which local and regional networks patch into for lengthy interconnections). Because many other Internet service providers rely on the backbone providers, the error echoed around the globe, causing temporary network failures.

## What is a computer **virus** and how is it spread?

Taken from the obvious analogy with biological viruses, a computer "virus" is a program that searches out other programs and "infects" them by replicating itself in them. When the programs are executed, the embedded virus is executed too, thus propagating the "infection." This normally happens invisibly to the user. A virus cannot infect other computers, however, without assistance. It is spread when users communicate by computer, often when they trade programs. The virus might do nothing but propagate itself and then allow the program to run normally. Usually, however, after propagating silently for a while, it starts doing other things—possibly inserting "cute" messages or destroying all of the user's files. Computer "worms" and "logic bombs" are similar to viruses, but they do not replicate themselves within programs as viruses do. A logic bomb does its damage immediately—destroying data, inserting garbage into data files, or reformatting the hard disk; a worm can alter the program and database either immediately or over a period of time.

In the 1990s, viruses, worms, and logic bombs have become such a serious problem, especially among IBM PC and Macintosh users, that the production of special detection and "inoculation" software has become an industry.

## What is a **fuzzy search**?

Fuzzy search is a feature of some software programs that allows a user to search for text that is similar to but not exactly the same as what he or she specifies. It can produce results when the exact spelling is unknown, or it can help users obtain information that is loosely related to a topic.

## What is **GIGO**?

GIGO is not a computer language despite its similarity to the names of computer languages; instead it is an acronym for the truism that one gets out of something what one puts into it: GIGO stands for the phrase Garbage In, Garbage Out. This is a computer hacker term for when people input imprecise information and get imprecise results.

## What is a **pixel**?

A pixel (from the words pix, for picture, and element) is the smallest element on a video display screen. A screen contains thousands of pixels, each of which can be made up of one or more dots or a cluster of dots. On a simple monochrome screen, a pixel is one dot; the two colors of image and background are created when the pixel is switched either on or off. Some monochrome screen pixels can be energized to create different light intensities to allow a range of shades from light to dark. On color screens, three dot colors are included in each pixel—red, green, and blue. The simplest screens have just one dot of each color, but more elaborate screens have pixels with clusters of each color. These more elaborate displays can show a large number of colors and intensities. On color screens, black is created by leaving all three colors off; white by all three colors on; and a range of grays by equal intensities of all the colors.

The most economical displays are monochrome, with one bit per pixel, with settings limited to on and off. High-resolution color screens, which can use a million pixels, with each color dot using four bytes of memory, would need to reserve many megabytes just to display the image.

## What does **DOS** stand for?

DOS stands for "disk operating system," a program that controls the computer's transfer of data to and from a hard or floppy disk. Frequently, it is combined with the main operating system. The operating system was originally developed at Seattle Computer Products as SCP-DOS. When IBM decided to build a personal computer and needed an operating system, it chose the SCP-DOS after reaching an agreement with the Microsoft Corporation to produce the actual operating system. Under Microsoft, SCP-DOS became MS-DOS, which IBM referred to as PC-DOS (personal computer), and which everyone eventually simply called DOS.

## What is **e-mail**?

Electronic mail, also known as E mail or e-mail, uses communication facilities to transmit messages. Many systems use computers as transmitting and receiving interfaces, but fax communication is also a form of E mail. A user can send a message to a single recipient, or to many. Different systems offer different options for sending, receiving, manipulating text, and addressing. For example, a message can be "registered," so that the sender is notified when the recipient looks at the message (though there is no way to tell if the recipient has actually read the message). Many systems allow messages to be forwarded. Usually messages are stored in a simulated "mailbox" in the network server or host computer; some systems announce incoming mail if the recipient is logged onto the system. An organization (such as a corporation, university, or professional organization) can provide electronic mail facilities; national and international networks can provide them as well. In order to use e-mail, both sender and receiver must have accounts on the same system or on systems connected by a network.

## Who sent the **first e-mail**?

In the early 1970s computer engineer Ray Tomlinson noticed that people working at the same mainframe computer could leave one another messages. He imagined great utility of this communication system if messages could be sent to different mainframes. So he wrote a software program over the period of about a week that used file-transfer protocols and send-and-receive features. It enabled people to send messages from one mainframe to another over the Arpanet, the network that became the Internet. To make sure the messages went to the right system he adopted the @ symbol because it was the least ambiguous keyboard symbol and because it was brief. As of April 2001, Hotmail®—just one of several e-mail providers—had over 70 million users and added about 100,000 e-mail accounts every day.

## What is a **hacker**?

A hacker is a skilled computer user. The term originally denoted a skilled programmer, particularly one skilled in machine code and with a good knowledge of the machine and its operating system. The name arose from the fact that a good programmer could always hack an unsatisfactory system around until it worked.

The term later came to denote a user whose main interest is in defeating password systems. The term has thus acquired a pejorative sense, with the meaning of one who deliberately and sometimes criminally interferes with data available through telephone lines. The activities of such hackers have led to considerable efforts to tighten security of transmitted data. The "hacker ethic" is that information-sharing is the proper way of human dealing, and, indeed, it is the responsibility of hackers to liberally impart their wisdom to the software world by distributing information. Nefarious hacker attacks by people outside the company costs companies on average about $56,000 per attack.

## What is a **kludge**?

A kludge (also spelled kluge) is a sloppy, crude, cumbersome solution to a problem. It refers to a makeshift solution as well as to any poorly designed product, or a product that becomes unmanageable over time.

## Who coined the term **technobabble**?

John A. Barry used the term "technobabble" to mean the pervasive and indiscriminate use of computer terminology, especially as it is applied to situations that have nothing at all to do with technology. He first used it in the early 1980s.

## How did the **Linux operating system** get its name?

The name Linux is combination of the first name of its principal programmer, Finland's Linus Torvalds (1970–), and the UNIX operating system. Linux (pronounced with a short "i") is an open source computer operating system that is comparable to more powerful, expansive, and usually costly UNIX systems, of which it resembles in form and function. Linux allows users to run an amalgam of reliable and hearty open-source software tools and interfaces, including powerful Web utilities such as the popular Apache server, on their home computers. Anyone can download Linux for free or can obtain it on disk for only a marginal fee. Torvalds created the kernel—or heart of the system—"just for fun," and released it freely to the world, where other programmers helped further its development. The world, in turn, has embraced Linux and made Torvalds into a folk hero.

## What is the idea behind **"open-source-software"**?

Open-source software is computer software where the code (the rules governing its operation) is available for users to modify. This is in contrast to proprietary code, where the software vendor veils the code so users cannot view and, hence, manipulate (or steal) it. The software termed open source is not necessarily free—that is, without charge; authors can charge for its use, and some do, albeit nominal fees. According to the Free Software Foundation, "Free software" is a matter of liberty, not price. To understand the concept, you should think of "free" as in "free speech," not as in "free beer." Free software is a matter of the users' freedom to run, copy, distribute, study, change, and improve the software." Despite this statement, most of it is available without charge. Open-source software is usually protected under the notion of "copyleft," instead of "copyright" law. Copyleft does not mean releasing material to the public domain, nor does it mean near absolute prohibition from copying, like the federal copyright law. Instead, according to the Free Software Foundation, copyleft is a form of protection guaranteeing that whoever redistributes software, whether modified or not, "must pass along the freedom to further copy and share it." Open source has

evolved into a movement of sharing, cooperation, and mutual innovation, ideas that many believe are necessary in today's cutthroat corporatization of software.



## Who invented the **World Wide Web**?

Tim Berners-Lee (1955–) is considered the creator of the World Wide Web (WWW). The WWW is a massive collection of interlinked hypertext documents that travel over the Internet and are viewed through a browser. The Internet is a global network of computers developed in the 1960s and 1970s by the U.S. Department of Defense's Advanced Research Project Agency (hence the term "Arpanet"). The idea of the Internet was to provide redundancy of communications in case of a catastrophic event (like a nuclear blast), which might destroy a single connection or computer but not the entire network. The browser is used to translate the hypertext, usually written in Hypertext Markup Language (HTML), so it is human-readable on a computer screen. Along with Gutenberg's invention of the printing press, the inception of the WWW in 1990 and 1991, when Berners-Lee released the tools and protocols onto the Internet, is considered one of humanity's greatest communications achievements.

## How does a **search engine** work?

Internet search engines are akin to computerized card catalogs at libraries. Viewed through a web browser with an Internet connection, they provide a hyperlinked listing of locations on the World Wide Web according to the requested keyword or pattern of words submitted by the searcher. A Web directory service provides the same service, but uses different methods to gather its source information. A Web directory service employs humans to "surf" Web sites and organize them into a hierarchical index by subject or some other category. A search engine, however, uses computer software called "spiders" or "bots" to search out, inventory, and index Web pages automatically. The spiders scan each Web page's content for words and the frequency of the words, then stores that information in a database. When the user submits words or terms, the search engine returns a list of sites from the database and ranks them according to the relevancy of the search terms.

## How can a grocery store customer become involved in **data mining**?

Many grocery stores today offer a free card that allows instant price reductions that function like a plastic, reusable coupon. The only catch is the customer must sign up for the card and reveal demographic information, such as age and gender. The particulars of each sale are stored in a database each time the customer uses the card, including facts such as day and time of purchase and items purchased. The grocery store then uses "data mining" to better target its sales campaigns and position its in-store displays. Data mining, as its name implies, is a computer and statistical technology where computers sift through extensive data to extract patterns, identify relationships, and allow for predictions. For instance, if patterns show that diapers are purchased before 3 p.m. on weekends by men aged 26 to 35, and most wine is purchased after 3 p.m. by men aged 46 to 55, then the store can move diapers to a prominent, high-traffic location during the day and place items close by that men of the same age group also purchase. Then after 3 p.m. the store can move the diapers out and the wine into the same prominent location, along with items that men in the 46 to 55 age range also purchase.

# GENERAL SCIENCE AND TECHNOLOGY

## NUMBERS

### When and where did the concept of **"numbers"** and **counting** first develop?

The human adult (including some of the higher animals) can discern the numbers one through four without any training. After that people must learn to count. To count requires a system of number manipulation skills, a scheme to name the numbers, and some way to record the numbers. Early people began with fingers and toes, and progressed to shells and pebbles. In the fourth millennium B.C.E. in Elam (near what is today Iran along the Persian Gulf), accountants began using unbaked clay tokens instead of pebbles. Each represented one order in a numbering system: a stick shape for the number one, a pellet for ten, a ball for 100, and so on. During the same period, another clay-based civilization in Sumer in lower Mesopotamia invented the same system.

### When was a symbol for the concept **zero** first used?

Surprisingly, the symbol for zero emerged later than the concept for the other numbers. The ancient Greeks, for instance, conceived of logic and geometry, concepts providing the foundation for all mathematics, yet they never had a symbol for zero. Hindu mathematicians are usually given credit for developing a symbol for the concept "zero"; it appears in an inscription at Gwalior dated 870 C.E. It is certainly older than that; it is found in inscriptions dating from the seventh century in Cambodia, Sumatra, and Bangka Island (off Sumatra). While there is no documented evidence for the zero in China before 1247, some historians believe that it originated there and arrived in India via Indochina.

## What is a perfect number?

A perfect number is a number equal to the sum of all its proper divisors (divisors smaller than the number) including 1. The number 6 is the smallest perfect number; the sum of its divisors 1, 2, and 3 equals 6. The next three perfect numbers are 28, 496, and 8,126. No odd perfect numbers are known. The largest known perfect number is

$$2^{13466916}(2^{13466917}-1).$$

It was discovered in 2001.

## What are **Roman numerals**?

Roman numerals are symbols that stand for numbers. They are written using seven basic symbols: I (1), V (5), × (10), L (50), C (100), D (500), and M (1,000). Sometimes a bar is place over a numeral to multiply it by 1,000. A smaller numeral appearing before a larger numeral indicates that the smaller numeral is subtracted from the larger one. This notation is generally used for 4s and 9s; for example, 4 is written IV, 9 is IX, 40 is XL, and 90 is XC.

## What are **Fibonacci numbers**?

Fibonacci numbers are a series of numbers where each, after the second term, is the sum of the two preceding numbers—for example, 1, 1, 2, 3, 5, 8, 13, 21 .... (1+1=2, 1+2=3, 2+3=5, 3+5=8, 5+8=13, 8+13=21, and so on). They were first described by Leonardo Fibonacci (ca. 1180–ca. 1250), also known as Leonard of Pisa, as part of a thesis on series in his most famous book *Liber abaci* (*The Book of the Calculator*), published in 1202 and later revised by him.

## What is the **largest prime number** presently known?

A prime number is one that is evenly divisible only by itself and 1. The integers 1, 2, 3, 5, 7, 11, 13, 17, and 19 are prime numbers. Euclid (ca. 300 B.C.E.) proved that there is no "largest prime number," because any attempt to define the largest results in a paradox. If there is a largest prime number (P), adding 1 to the product of all primes up to and including P, $1 + (1 \times 2 \times 3 \times 5 \times \ldots \times P)$, yields a number that is itself a prime number, because it cannot be divided evenly by any of the known primes. In 2001, Michael Cameron from Canada discovered the largest known (and the 39th) prime number: $2^{13466917} - 1$. This is over 4 million digits long and would take about three weeks to write out by hand. The 39th prime is part of a special class of prime numbers called Mersenne Primes (named after Marin Mersenne,

1588–1648, a French monk who did the first work in this area). Mersenne primes occur where $2^{n-1}$ is prime.

There is no apparent pattern to the sequence of primes. Mathematicians have been trying to find a formula since the days of Euclid, without success. The 39th prime was discovered on a personal computer as part of the GIMPS effort (the Great Internet Mersenne Prime Search), which was formed in January 1996 to discover new world-record-size prime numbers. GIMPS relies on the computing efforts of thousands of small, personal computers across the planet. Interested participants can become involved in the search for primes by pointing their browsers to: http://www.mersenne.org/prime.htm.

## What is the **largest number** mentioned in the **Bible**?

The largest number specifically named in the Bible is a thousand thousand; i.e., a million. It is found in 2 Chronicles 14:9.

## How are names for large and small **quantities constructed** in the **metric system**?

Each prefix listed below can be used in the metric system and with some customary units. For example, centi + meter = centimeter, meaning one-hundredth of a meter.

| Prefix | Power | Numerals |
|---|---|---|
| Exa- | $10^{18}$ | 1,000,000,000,000,000,000 |
| Peta- | $10^{15}$ | 1,000,000,000,000,000 |
| Tera- | $10^{12}$ | 1,000,000,000,000 |
| Giga- | $10^{9}$ | 1,000,000,000 |
| Mega- | $10^{6}$ | 1,000,000 |
| Myria- | $10^{5}$ | 100,000 |
| Kilo- | $10^{3}$ | 1,000 |
| Hecto- | $10^{2}$ | 100 |
| Deca- | $10^{1}$ | 10 |
| Deci- | $10^{-1}$ | 0.1 |
| Centi- | $10^{-2}$ | 0.01 |
| Milli- | $10^{-3}$ | 0.001 |
| Micro- | $10^{-6}$ | 0.000001 |
| Nano- | $10^{-9}$ | 0.000000001 |
| Pico- | $10^{-12}$ | 0.000000000001 |
| Femto- | $10^{-15}$ | 0.000000000000001 |
| Atto- | $10^{-18}$ | 0.000000000000000001 |

## Why is the number **ten** considered **important**?

One reason is that the metric system is based on the number ten. The metric system emerged in the late-eighteenth century out of a need to bring standardization to measurement, which had up to then been fickle, depending upon the preference of the ruler of the day. But ten was important well before the metric system. Nicomachus of Gerasa, a second-century neo-Pythagorean from Judea, considered ten a "perfect" number, the figure of divinity present in creation with mankind's fingers and toes. Pythagoreans believed ten to be "the first-born of the numbers, the mother of them all, the one that never wavers and gives the key to all things." And shepherds of West Africa counted sheep in their flocks by colored shells based on ten, and ten had evolved as a "base" of most numbering schemes. Some scholars believe the reason ten developed as a base number had more to do with ease: ten is easily counted on fingers and the rules of addition, subtraction, multiplication, and division for the number ten are easily memorized.

## What are some **very large numbers**?

| Name | Value in powers of 10 | Number of 0's | Number of groups of three 0's after 1,000 |
|---|---|---|---|
| Billion | $10^{9}$ | 9 | 2 |
| Trillion | $10^{12}$ | 12 | 3 |
| Quadrillion | $10^{15}$ | 15 | 4 |
| Quintillion | $10^{18}$ | 18 | 5 |
| Sextillion | $10^{21}$ | 21 | 6 |
| Septillion | $10^{24}$ | 24 | 7 |
| Octillion | $10^{27}$ | 27 | 8 |
| Nonillion | $10^{30}$ | 30 | 9 |
| Decillion | $10^{33}$ | 33 | 10 |
| Undecillion | $10^{36}$ | 36 | 11 |
| Duodecillion | $10^{39}$ | 39 | 12 |
| Tredecillion | $10^{42}$ | 42 | 13 |
| Quattuor-decillion | $10^{45}$ | 45 | 14 |
| Quindecillion | $10^{48}$ | 48 | 15 |
| Sexdecillion | $10^{51}$ | 51 | 16 |
| Septen-decillion | $10^{54}$ | 54 | 17 |
| Octodecillion | $10^{57}$ | 57 | 18 |
| Novemdecillion | $10^{60}$ | 60 | 19 |
| Vigintillion | $10^{63}$ | 63 | 20 |
| Centillion | $10^{303}$ | 303 | 100 |

The British, French, and Germans use a different system for naming denominations above one million. The googol and googolplex are rarely used outside the United States.

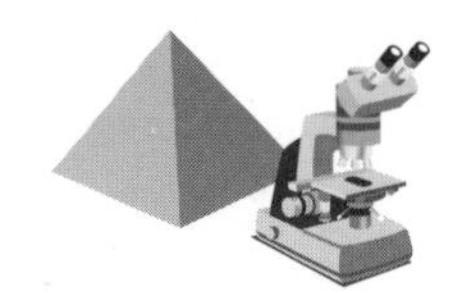



## How large is a **googol**?

A googol is $10^{100}$ (the number 1 followed by 100 zeros). Unlike most other names for numbers, it does not relate to any other numbering scale. The American mathematician Edward Kasner first used the term in 1938; when searching for a term for this large number, Kasner asked his nephew, Milton Sirotta, then about nine years old, to suggest a name. The googolplex is 10 followed by a googol of zeros, represented as $10^{googol}$. The popular Web search engine Google.com is named after the concept of a googol.

## What is an **irrational number**?

Numbers that cannot be expressed as an exact ratio are called irrational numbers; numbers that can be expressed as an exact ratio are called rational numbers. For instance, 1/2 (one half, or 50 percent of something) is rational. 1.61803 ($\theta$), 3.14159 ($\pi$), 1.41421 ($\sqrt{2}$), are irrational. History claims that Pythagoras in the sixth century B.C.E. first used the term when he discovered that the square root of 2 could not be expressed as a fraction.

## What are **imaginary numbers**?

Imaginary numbers are the square roots of negative numbers. Since the square is the product of two equal numbers with like signs it is always positive. Therefore, no number multiplied by itself can give a negative real number. The symbol $i$ is used to indicate an imaginary number.

## What is the **value of pi** out to 30 digits past the decimal point?

Pi ($\pi$) represents the ratio of the circumference of a circle to its diameter, used in calculating the area of a circle ($\pi r^2$) and the volume of a cylinder ($\pi r^2 h$) or cone. It is a "transcendental number," an irrational number with an exact value that can be measured to any degree of accuracy, but that can't be expressed as the ratio of two integers. In theory, the decimal extends into infinity, though it is generally rounded to 3.1416. The Welsh-born mathematician William Jones selected the Greek symbol ($\pi$) for Pi. Rounded to 30 digits past the decimal point, it equals 3.141592653589793238462643383279. In 1989, Gregory and David Chudnovsky at Columbia University in New York City calculated the value of pi to 1,011,961,691 decimal places. They performed the calculation twice on an IBM 3090 mainframe and on a CRAY-2 supercomputer with matching results. In 1991, they calculated pi to 2,260,321,336 decimal places. In 1999, Yasumasa Kanada and Daisuke Takahashi of the University of Tokyo calculated pi out to 206,158,430,000 digits. Mathematicians have also calculated pi in binary format (i.e. 0s and 1s). The five trillionth binary digit of pi was computed by Colin Percival and 25 others at Simon Fraser University. The computation took over 13,500 hours of computer time.

## Why is **seven** considered a **supernatural number**?

In magical lore and mysticism, all numbers are ascribed certain properties and energies. Seven is a number of great power, a magical number, a lucky number, a number of psychic and mystical powers, of secrecy and the search for inner truth. The origin of belief in seven's power lies in the lunar cycle. Each of the moon's four phases lasts about seven days. The Sumerians, who based their calendar on the moon, gave the week seven days and declared the seventh and last day of each week to be uncanny. Life cycles on Earth also have phases demarcated by seven. Furthermore, there are said to be seven years to each stage of human growth; there are seven colors to the rainbow; and seven notes are in the musical scale. The seventh son of a seventh son is said to be born with formidable magical and psychic powers. The number seven is widely held to be a lucky number, especially in matters of love and money, and it also carries great prominence in the old and new testaments. Here are a few examples: the Lord rested on the seventh day; there were seven years of plenty and seven years of famine in Egypt in the story of Joseph; God commanded Joshua to have seven priests carry trumpets, and on the seventh day they were to march around Jericho seven times; Solomon built the temple in seven years; and there are seven petitions of the Lord's Prayer.

## What are some examples of **numbers** and **mathematical concepts** in **nature**?

The world can be articulated with numbers and mathematics. Some numbers are especially prominent. The number six is ubiquitous: every normal snowflake has six sides; every honeybee colony's combs are six-sided hexagons. The curved, gradually decreasing chambers of a nautilus shell are propagating spirals of the Golden Section and the Fibonacci sequence of numbers. Pine cones also rely on the Fibonacci sequence, as do many plants and flowers in their seed and stem arrangements. Fractals are evident in shorelines, blood vessels, and mountains.

# MATHEMATICS

## How is **arithmetic** different from **mathematics**?

Arithmetic is the study of positive integers (i.e. 1, 2, 3, 4, 5, ...) manipulated with addition, subtraction, multiplication, and division, and the use of the results in daily life. Mathematics is the study of shape, arrangement, and quantity. It is traditionally viewed as consisting of three fields: algebra, analysis, and geometry. But any lines of division have evaporated because the fields are now so interrelated.

## Who invented **calculus**?

The German mathematician Gottfried Wilhelm Leibniz published the first paper on calculus in 1684. Most historians agree that Isaac Newton invented calculus eight to ten years earlier, but he was typically very late in publishing his works. The invention of calculus marked the beginning of higher mathematics. It provided scientists and mathematicians with a tool to solve problems that had been too complicated to attempt previously.

Gottfried Wilhelm Leibniz made important contributions to the science of calculus and designed early manifestations of the computer.

## What is the most **enduring mathematical work** of all time?

*The Elements of Euclid* (fl. about 300 B.C.E.) has been the most enduring and influential mathematical work of all time. In it, the ancient Greek mathematician presented the work of earlier mathematicians and included many of his own innovations. The *Elements* is divided into 13 books: the first six cover plane geometry; seven to nine address arithmetic and number theory; 10 treats irrational numbers; and 11 to 13 discuss solid geometry. In presenting his theorems, Euclid used the synthetic approach, in which one proceeds from the known to the unknown by logical steps. This method became the standard procedure for scientific investigation for many centuries, and the *Elements* probably had a greater influence on scientific thinking than any other work.

## Is it possible to **count** to **infinity**?

No. Very large finite numbers are not the same as infinite numbers. Infinite numbers are defined as being unbounded, or without limit. Any number that can be reached by counting or by representation of a number followed by billions of zeros is a finite number.

## How long has the **abacus** been used? Is it still used?

The abacus grew out of early counting boards, with hollows in a board holding pebbles or beads used to calculate. It has been documented in Mesopotamia back to around 3500 B.C.E.. The current form, with beads sliding on rods, dates back at least to 15th-century China. Before the use of decimal number systems, which allowed the familiar paper-and-pencil methods of calculation, the abacus was essential for almost all multiplication and division. The abacus is still used in many countries where modern calculators are not available. It is also still used in countries, such as Japan and China, that

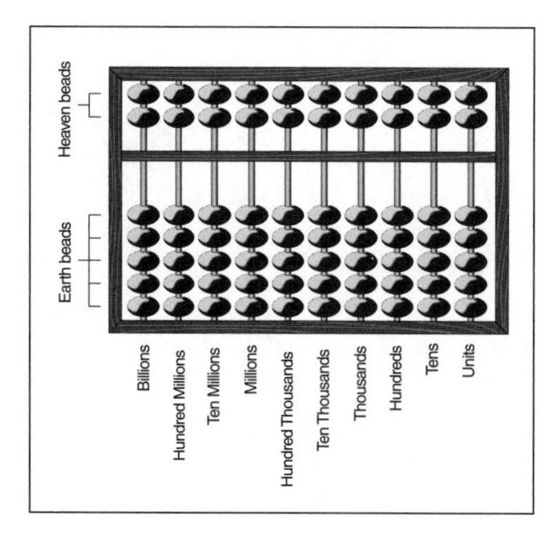


A Chinese abacus called a suan pan.

John Napier, the Scottish mathematician known for a device called "Napier's bones," devised the use of the decimal point when writing numbers.

have long traditions of abacus use. As recently as the mid-1970s, most Japanese shopkeepers used abaci for totaling customers' bills. While the calculator is now more widely used, many people still prefer to check the results on an abacus. At least one manufacturer offers calculators with small, built-in abaci.

## Can a person using an **abacus calculate more rapidly** than someone using a calculator?

In 1946, the Tokyo staff of *Stars and Stripes* sponsored a contest between a Japanese abacus expert and an American accountant using the best electric adding machine then available. The abacus operator proved faster in all calculations except the multiplication of very large numbers. While today's electronic calculators are much faster and easier to use than the adding machines used in 1946, undocumented tests still show that an expert can add and subtract faster on an abacus than someone using an electronic calculator. It also allows long division and multiplication problems with more digits than a hand calculator can accommodate.

## What are **Napier's bones**?

In the 16th century, the Scottish mathematician John Napier (1550–1617), Baron of Merchiston, developed a method of simplifying the processes of multiplication and division, using exponents of 10, which Napier called logarithms (commonly abbreviated as logs). Using this system, multiplication is reduced to addition and division to subtraction. For example, the log of 100 ($10^2$) is 2; the log

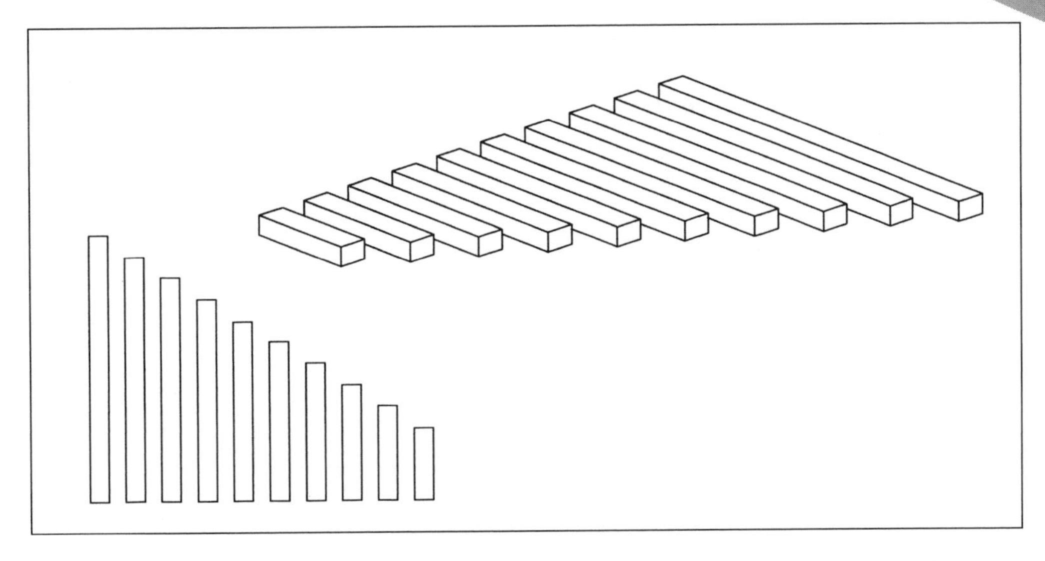

Cuisenaire rods.

of 1000 ($10^3$) is 3; the multiplication of 100 by 1000, $100 \times 1000 = 100{,}000$, can be accomplished by adding their logs: $\log[(100)(1000)] = \log(100) + \log(1000) = 2 + 3 = 5 = \log(100{,}000)$. Napier published his methodology in *A Description of the Admirable Table of Logarithms* in 1614. In 1617 he published a method of using a device, made up of a series of rods in a frame, marked with the digits 1 through 9, to multiply and divide using the principles of logarithms. This device was commonly called "Napier's bones" or "Napier's rods."

## What are **Cuisenaire rods**?

The Cuisenaire method is a teaching system used to help young students independently discover basic mathematical principles. Developed by Emile-Georges Cuisenaire, a Belgian schoolteacher, the method uses rods of 10 different colors and lengths that are easy to handle. The rods help students understand mathematical principles rather than merely memorizing them. They are also used to teach elementary arithmetic properties such as associative, commutative, and distributive properties.

## What is a **slide rule**, and who invented it?

Up until about 1974, most engineering and design calculations for buildings, bridges, automobiles, airplanes, and roads were done on a slide rule. A slide rule is an apparatus with moveable scales based on logarithms, which were invented by John Napier (1550–1617), Baron of Merchiston, and published in 1614. The slide rule can, among other things, quickly multiply, divide, square root, or find the logarithm of a number.

In 1620, Edmund Gunter (1581–1626) of Gresham College, London, England, described an immediate forerunner of the slide rule, his "logarithmic line of numbers." William Oughtred (1574–1660), rector of Aldbury, England, made the first rectilinear slide rule in 1621. This slide rule consisted of two logarithmic scales that could be manipulated together for calculation. His former pupil, Richard Delamain, published a description of a circular slide rule in 1630 (and received a patent about that time for it), three years before Oughtred published a description of his invention (at least one source says that Delamain published in 1620). Oughtred accused Delamain of stealing his idea, but evidence indicates that the inventions were probably arrived at independently.

The earliest existing straight slide rule using the modern design of a slider moving in a fixed stock dates from 1654. A wide variety of specialized slide rules were developed by the end of the 17th century for trades such as masonry, carpentry, and excise tax collecting. Peter Mark Roget (1779–1869), best known for his *Thesaurus of English Words and Phrases,* invented a log-log slide rule for calculating the roots and powers of numbers in 1814. In 1967, Hewlett-Packard (now known as HP) produced the first pocket calculators. Within a decade, slide rules became the subject of science trivia and collector's books.

## How is **casting out nines** used to check the results of addition or multiplication?

The method of "casting out nines" is based on the excess of nines in digits of whole numbers (the remainder when a sum of digits is divided by 9). Illustrating this process in the multiplication example below, the method begins by adding the digits in both the multiplicand (one of the terms that is being multiplied) and the multiplier (the other term being multiplied). In the example below, this operation leads to the results of "13" and "12," respectively. If these results are greater than 9 (>9), then the operation is repeated until the resulting figures are less than 9 (<9). In the example below, the repeated calculation gives the results as "4" and "3," respectively. Multiply the resulting "excess" from the multiplicand by the excess from the multiplier (4 × 3 below). Add the digits of the result to eventually yield a number equal to or less than 9 (≤). Repeat the process of casting out nines in the multiplication product (the result of the multiplication process). The result must equal the result of the previous set of transactions, in this case "3." If the two figures disagree, then the original multiplication procedure was done incorrectly. "Casting out nines" can also be applied to check the accuracy of the results of addition.

328 → 13 → 4<br>
*624* → 12 → *3*<br>
1312      12 → 3<br>
656<br>
*1968*<br>
204672 → 21 → 3

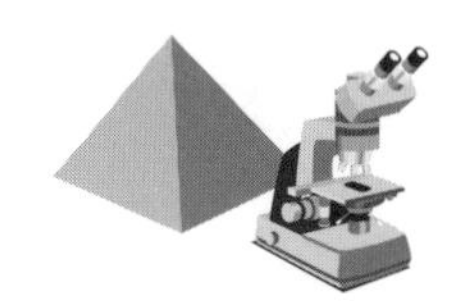



## What is the difference between a **median** and a **mean**?

If a string of numbers is arranged in numerical order, the median is the middle value of the string. If there is an even number of values in the string, the median is found by adding the two middle values and dividing by two. The arithmetic mean, also known as the simple average, is found by taking the sum of the numbers in the string and dividing by the number of items in the string. While easy to calculate for relatively short strings, the arithmetic mean can be misleading, as very large or very small values in the string can distort it. For example, the mean of the salaries of a professional football team would be skewed if one of the players was a high-earning superstar; it could be well above the salaries of any of the other players. The mode is the number in a string that appears most often.

For the string 111222234455667, for example, the median is the middle number of the series: 3. The arithmetic mean is the sum of numbers divided by the number of numbers in the series, $51 \div 15 = 3.4$. The mode is the number that occurs most often, 2.

## When does $\mathbf{0 \times 0 = 1}$?

Factorials are the product of a given number and all the factors less than that number. The notation n! is used to express this idea. For example, 5! (five factorial) is $5 \times 4 \times 3 \times 2 \times 1 = 120$. For completeness, 0! is assigned the value 1, so $0 \times 0 = 1$.

## When did the concept of **square root** originate?

A square root of a number is a number that, when multiplied by itself, equals the given number. For instance, the square root of 25 is 5 ($5 \times 5 = 25$). The concept of the square root has been in existence for many thousands of years. Exactly how it was discovered is not known, but several different methods of exacting square roots were used by early mathematicians. Babylonian clay tablets from 1900 to 1600 B.C.E. contain the squares and cubes of integers 1–30. The early Egyptians used square roots around 1700 B.C.E., and during the Greek Classical Period (600 to 300 B.C.E.) better arithmetic methods improved square root operations. In the 16th century, French mathematician René Descartes was the first to use the square root symbol, called "the radical sign," $\sqrt{\phantom{x}}$.

## What are **Venn diagrams**?

Venn diagrams are graphical representations of set theory, which use circles to show the logical relationships of the elements of different sets, using the logical operators (also called in computer parlance "Boolean Operators") *and, or,* and *not*. John Venn (1834–1923) first used them in his 1881 Symbolic Logic, in which he interpreted and corrected the work of George Boole (1815–1864) and Augustus de Morgan (1806–

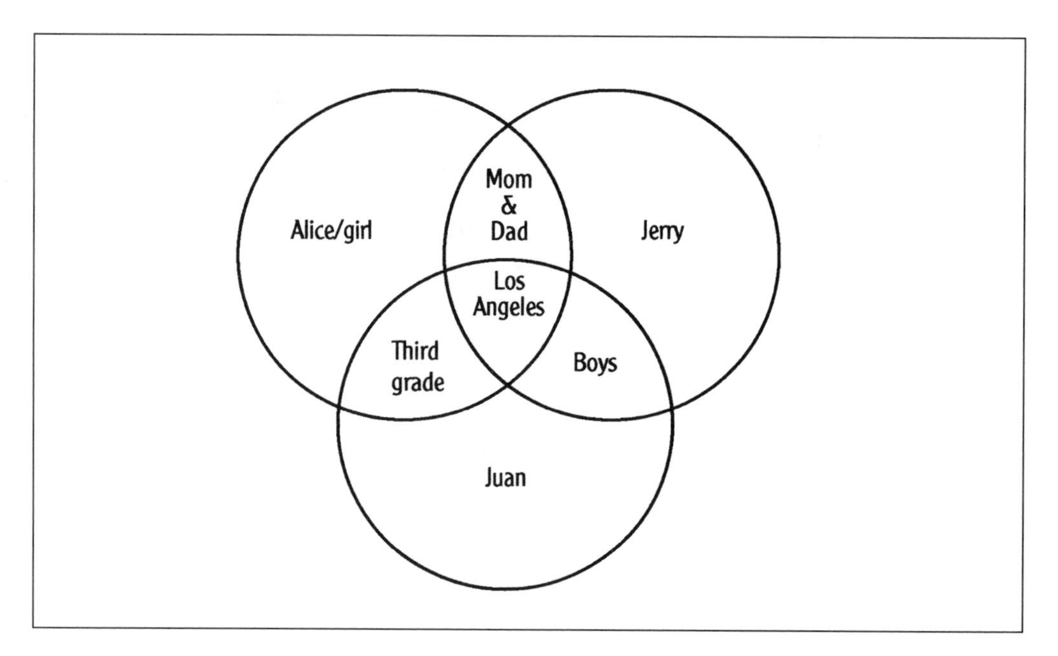


**Example of a Venn diagram.**

1871). While his attempts to clarify perceived inconsistencies and ambiguities in Boole's work are not widely accepted, the new method of diagraming is considered to be an improvement. Venn used shading to better illustrate inclusion and exclusion. Charles Dodgson (1832–1898), better known as Lewis Carroll, refined Venn's system, in particular by enclosing the diagram to represent the universal set.

## What are the common **mathematical formulas** for **volume**?

*Volume of a sphere*:

Volume = 4/3 times pi times the cube of the radius

$V = 4/3 \times \pi r^3$

*Volume of a pyramid*:

Volume = 1/3 times the area of the base times the height

$V = 1/3bh$

*Volume of a cylinder*:

Volume = area of the base times the height

$V = Ah$

*Volume of a circular cylinder* (with circular base):

Volume = pi times the square of the radius of the base times the height

$V = \pi r^2 h$

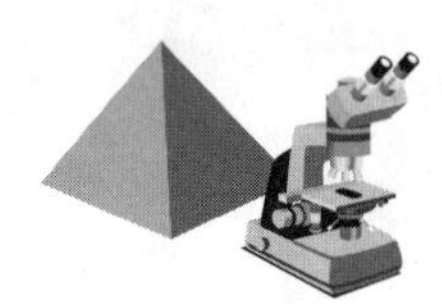

*Volume of a cube*:

Volume = the length of one side cubed

$V = S^3$

*Volume of a cone*:

Volume = 1/3 times pi times the square of the radius of the base times the height.

$V = 1/3\ \pi r^2 h$

*Volume of a rectangular solid*:

Volume = length times width times height

$V = lwh$

## How many feet are on each side of an **acre** that is square?

An acre that is square in shape has about 208.7 feet (64 meters) on each side.

## What are the common mathematical formulas for **area**?

*Area of a rectangle*:

Area = length times width

$A = lw$

Area = altitude times base

$A = ab$

*Area of a circle*:

Area = pi times the radius squared

$A = \pi r^2$ or $A = 1/4\pi d^2$

*Area of a triangle*:

Area = one half the altitude times the base

$A = 1/2ab$

*Area of the surface of a sphere*:

Area = four times pi times the radius squared

$A = 4\pi r^2$ or $A = \pi d^2$

*Area of a square*:

Area = length times width, or length of one side squared

$A = s^2$

*Area of a cube*:

Area = square of the length of one side times 6

$A = 6s^2$

*Area of an ellipse*:

Area = long diameter times short diameter times 0.7854.

## Who discovered the formula for the **area of a triangle**?

Heron (or Hero) of Alexandria (first century B.C.E.) is best known in the history of mathematics for the formula that bears his name. This formula calculates the area of a triangle with sides a, b, and c, with s = half the perimeter: $A = \surd[s\ (2 - a)(s - b)(s - c)]$. The Arab mathematicians who preserved and transmitted the mathematics of the Greeks reported that this formula was known earlier to Archimedes (ca. 287–212 B.C.E.), but the earliest proof now known is that appearing in Heron's Metrica.

## How is **Pascal's triangle** used?

Pascal's triangle is an array of numbers, arranged so that every number is equal to the sum of the two numbers above it on either side. It can be represented in several slightly different triangles, but this is the most common form:

```
                    1
                 1     1
              1     2     1
           1     3     3     1
        1     4     6     4     1
     1     5    10    10     5     1
```

The triangle is used to determine the numerical coefficients resulting from the computation of higher powers of a binomial (two numbers added together). When a binomial is raised to a higher power, the result is expanded, using the numbers in that row of the triangle. For example, $(a + b)^1 = a^1 + b^1$, using the coefficients in the second line of the triangle. $(a + b)^2 = a^2 + 2ab + b^2$, using the coefficients in the next line of the triangle. (The first line of the triangle correlates to $(a + b)^0$.) While the calculation of coefficients is fairly straightforward, the triangle is useful in calculating them for the higher powers without needing to multiply them out. Binomial coefficients are useful in calculating probabilities; Blaise Pascal was one of the pioneers in developing laws of probability.

French mathematician and scientist Blaise Pascal (1623–1662) uncovered a relationship between numbers in Pascal's triangle.

As with many other mathematical developments, there is some evidence of a previous appearance of the triangle in China. Around 1100 C.E., the Chinese mathematician Chia Hsien wrote about "the tabulation system for unlocking binomial coefficients"; the first publication of the triangle was probably in a book called *Piling-Up Powers and Unlocking Coefficients,* by Liu Ju-Hsieh.

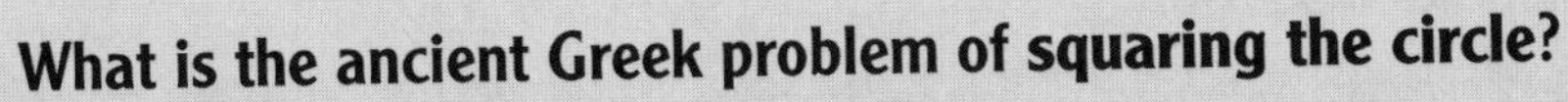

### What is the ancient Greek problem of squaring the circle?

This problem was to construct, with a straight-edge and compass, a square having the same area as a given circle. The Greeks were unable to solve the problem because the task is impossible, as was shown by the German mathematician Ferdinand von Lindemann (1852–1939) in 1882.

## What is the **Pythagorean theorem**?

In a right triangle (one where two of the sides meet in a 90-degree angle), the hypotenuse is the side opposite the right angle. The Pythagorean theorem, also known as the rule of Pythagoras, states that the square of the length of the hypotenuse is equal to the sum of the squares of the other two sides ($h^2 = a^2 + b^2$). If the lengths of the sides are: h = 5 inches, a = 4 inches, and b = 3 inches, then

$$h = \sqrt{a^2 + b^2} = \sqrt{4^2 + 3^2} = \sqrt{16 + 9} = \sqrt{25} = 5$$

The theorem is named for the Greek philosopher and mathematician Pythagoras (ca. 580–ca. 500 B.C.E.). Pythagoras is credited with the theory of the functional significance of numbers in the objective world and numerical theories of musical pitch. As he left no writings, the Pythagorean theorem may actually have been formulated by one of his disciples.

## What are the **Platonic solids**?

The Platonic solids are the five regular polyhedra: the four-sided tetrahedron, the six-sided cube or hexahedron, the eight-sided octahedron, the twelve-sided dodecahedron, and the twenty-sided icosahedron. While they had been studied as long ago as the time of Pythagorus (around 500 B.C.E.), they are called the Platonic solids because they were first described in detail by Plato around 400 B.C.E. The ancient Greeks gave mystical significance to the Platonic solids: the tetrahedron represented fire, the icosahedron represented water, the stable cube represented the Earth, the octahedron represented the air. The twelve faces of the dodecahedron corresponded to the twelve signs of the zodiac, and this figure represented the entire universe.

## What does the expression **tiling the plane** mean?

It is a mathematical expression describing the process of forming a mosaic pattern (a "tessellation") by fitting together an infinite number of polygons so that they cover an entire plane. Tesselations are the familiar patterns that can be seen in designs for quilts, floor coverings, and bathroom tilework.

## What is a **golden section**?

Golden section, also called the divine proportion, is the division of a line segment so that the ratio of the whole segment to the larger part is equal to the ratio of the larger part to the smaller part. The ratio is approximately 1.61803 to 1. The number 1.61803 is called the Golden Number (also called Phi [with a capital P]). The golden number is the limit of the ratios of consecutive Fibonacci numbers, such as, for instance, 21/13, and 34/21. A golden rectangle is one whose length and width correspond to this ratio. The ancient Greeks thought this shape had the most pleasing proportions. Many famous painters have used the Golden Rectangle in their paintings, and architects have used it in their design of buildings, the most famous example being the Greek Parthenon.

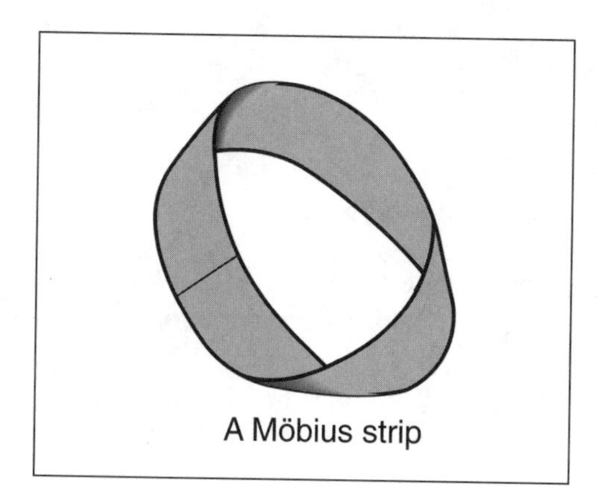


A Möbius strip.

## What is a **Möbius strip**?

A Möbius strip is a surface with only one side, usually made by connecting the two ends of a rectangular strip of paper after putting a half-twist (180 degrees relative to the opposite side) in the strip. Cutting a Möbius strip in half down the center of the length of the strip results in a single band with four half-twists. Devised by the German mathematician August Ferdinand Möbius (1790–1868) to illustrate the properties of one-sided surfaces, it was presented in a paper that was not discovered or published until after his death. Another 19th-century German mathematician, Johann Benedict Listing, developed the idea independently at the same time.

## How is the **rule of 70** used?

This rule is a quick way of estimating the period of time it will take a quantity to double given the percentage of increase. Divide the percentage of increase into 70. For example, if a sum of money is invested at 6 percent interest, the money will double in value in $70 \div 6 = 11.7$ years.

## What are **fractals**?

A fractal is a set of points that is too irregular to be described by traditional geometric terms, but that often have some degree of self-similarity; that is, they are made of parts that resemble the whole. They are used in image processing to compress data and to depict apparently chaotic objects in nature such as mountains or coastlines.

**How many different bridge games are possible?**

Roughly 54 octillion different bridge games are possible.

Scientists also use fractals to better comprehend rainfall trends, patterns formed by clouds and waves, and the distribution of vegetation. Fractals are also used to create computer-generated art.

### How is **percent of increase** calculated?

To find the percent of increase, divide the amount of increase by the base amount. Multiply the result by 100 percent. For example, a raise in salary from $10,000 to $12,000 would have percent of increase = (2,000 ÷ 10,000) × 100% = 20%.

### What is the difference between **simple interest** and **compound interest**?

Simple interest is calculated on the amount of principal only. Compound interest is calculated on the amount of principal plus any previous interest already earned. For example, $100 invested at a rate of 5 percent for 1 year will earn $5.00 after one year earning simple interest. The same $100 will earn $5.12 if compounded monthly.

### If 30 people are chosen at random, what is the probability that at least two of them have their **birthday on the same day**?

The probability that at least two people in a group of 30 share the same birthday is about 70 percent.

### What is the **law of very large numbers**?

Formulated by Persi Diaconis (1945–) and Frederick Mosteller (1916–) of Harvard University, this long-understood law of statistics states that "with a large enough sample, any outrageous thing is apt to happen." Therefore, seemingly amazing coincidences can actually be expected if given sufficient time or a large enough pool of subjects. For example, when a New Jersey woman won the lottery twice in four months, the media publicized it as an incredible long shot of 1 in 17 trillion. However, when statisticians looked beyond this individual's chances and asked what were the odds of the same happening to any person buying a lottery ticket in the United States over a six-month period, the number dropped dramatically to 1 in 30. According to researchers, coincidences arise often in statistical work, but some have hid-

Representation of the Königsberg Bridge Problem.

den causes and therefore are not coincidences at all. Many are simply chance events reflecting the luck of the draw.

## What is the **Königsberg Bridge Problem**?

The city of Königsberg was located in Prussia on the Pregel River. Two islands in the river were connected by seven bridges. By the eighteenth century, it had become a tradition for the citizens of Königsberg to go for a walk through the town trying to cross each bridge only once. No one was able to succeed, and the question was asked whether it was possible to do so. In 1736, Leonhard Euler (1707–1783) proved that it was not possible to cross the Königsberg bridges only once. Euler's solution led to the development of two new areas of mathematics: graph theory, which deals with questions about networks of points that are connected by lines, and topology, which is the study of those aspects of the shape of an object that do not depend on length measurements.

## What is **Zeno's paradox**?

Zeno of Elea (ca. 490–ca. 425 B.C.E.), a Greek philosopher and mathematician, is famous for his paradoxes, which deal with the continuity of motion. One form of the paradox is: If an object moves with constant speed along a straight line from point 0 to point 1, the object must first cover half the distance (1/2), then half the remaining distance (1/4), then half the remaining distance (1/8), and so on without end. The con-

**What is the probability of a successful triple play occurring in a single baseball game?**

The odds against a triple play in a game of baseball are 1,400 to 1.

clusion is that the object never reaches point 1. Because there is always some distance to be covered, motion is impossible. In another approach to this paradox, Zeno used an allegory telling of a race between a tortoise and Achilles (who could run 100 times as fast), where the tortoise started running 10 rods in front of Achilles. Because the tortoise always advanced 1/100 of the distance that Achilles advanced in the same time period, it was theoretically impossible for Achilles to pass him. The English mathematician and writer Charles Dodgson (1832–1898), better known as Lewis Carroll, used the characters of Achilles and the tortoise to illustrate his paradox of infinity.

## TERMS AND THEORIES

### What is the difference between **science** and **technology**?

Science is the process by which humans go about understanding the world. What is the nature of energy, space, and matter, are some of the issues. Engineering is the use of this knowledge in the creation of plans to reach a destination. Technology is the tools or processes to carry out the plans.

### What book is considered the **greatest** and **most influential scientific** work?

Isaac Newton's 1687 book, *Philosophiae Naturalis Principia Mathematica* (known for short as *Principia*). Newton wrote *Principia* in 18 months, summarizing his work and covering almost every aspect of modern science. Newton introduced gravity as a universal force, explaining that the motion of a planet responds to gravitational forces in inverse proportion to the planet's mass. Newton was able to explain tides, and the motion of planets, moons, and comets using gravity. He also showed that spinning bodies such as earth are flattened at the poles. The first printing of *Principia* produced only 500 copies.

### What was the **first technical report** written in **English**?

Geoffrey Chaucer's *Treatise on the Astrolabe* was written in 1391.

## What scientific article has the **most authors**?

The article "First Measurement of the Left-Right Cross Section Asymmetry in Z-Boson Production by $e^+e^-$ Collisions," published in *Physical Review Letters* listed 406 authors on two pages.

## What is the **most frequently cited** scientific journal article?

The most frequently cited scientific article is "Protein Measurement with the Folin Phenol Reagent" by O. H. Lowry and coworkers, published in 1951 in the *Journal of Biological Chemistry*. By the mid 1990s, this article had been cited more than 245,000 times since it first appeared.

## What was the first important **scientific society** in the United States?

The first significant scientific society in the United States was the American Philosophical Society, organized in 1743 in Philadelphia, Pennsylvania, by Benjamin Franklin.

## What was the **first national science institute**?

On March 3, 1863, President Abraham Lincoln signed a congressional charter creating the National Academy of Sciences, which stipulated that "the Academy shall, whenever called upon by any department of the government, investigate, examine, experiment, and report upon any subject of science of art, the actual expense of such investigations, examinations, experiments, and reports to be paid from appropriations which may be made for the purpose, but the Academy shall receive no compensation whatever for any services to the Government of the United States." The Academy's first president was Alexander Dallas Backe. Today, the Academy and its sister organizations—the National Academy of Engineering, establish in 1964, and the Institute of Medicine, established in 1970—serve as the country's preeminent sources of advice on science and technology and their bearing on the nation's welfare.

The National Research Council was established in 1916 by the National Academy of Sciences at the request of President Woodrow Wilson "to bring into cooperation existing governmental, educational, industrial and other research organizations, with the object of encouraging the investigation of natural phenomena, the increased use of scientific research in the development of American industries, the employment of scientific methods in strengthening the national defense, and such other applications of science as will promote the national security and welfare."

The National Academy of Sciences, the National Academy of Engineering, and the Institute of Medicine work through the National Research Council of the United States, one of world's most important advisory bodies. More than 10,000 scientists, engineers, industrialists, and health and other professionals participating in more than 1,000 committees comprise the National Research Council.



## What was the first **national scientific society** organized in the United States?

The first national scientific society organized in the United States was the American Association for the Advancement of Science (AAAS), organized on September 20, 1848, in Philadelphia, Pennsylvania, for the purpose of "advancing science in every way." The first president was William Charles Redfield.

## Who are the **youngest** and **oldest Nobel Laureates**?

The youngest Nobel Laureate was Sir William Lawrence Bragg, who shared the physics prize with his father, Sir William Henry Bragg, in 1915 at the age of 25. Three individuals were in their eighties when they became Nobel Laureates—Pyotr Leonidovich Kapitsa, who won the physics prize in 1978 at age 84; Charles J. Pedersen, who won the chemistry prize in 1987 at age 83, and Georg Witting, who won the chemistry prize in 1979 at age 82.

## Are there any **multiple Nobel Prize winners**?

Four individuals have received multiple Nobel prizes. They are Marie Curie (physics in 1903 and chemistry in 1911); John Bardeen (physics in 1956 and 1972); Linus Pauling (chemistry in 1954 and peace in 1962); and Frederick Sanger (chemistry in 1958 and 1980).

## What is **high technology** or **high tech**?

This buzz term used mainly by the lay media (as opposed to scientific, medical, or technological media) appeared in the late 1970s. It was initially used to identify the newest, "hottest" application of technology to fields such as medical research, genetics, automation, communication systems, and computers. It usually implied a distinction between technology to meet the information needs of society and traditional heavy industry, which met more material needs. By the mid-1980s, the term had become a catch-all applying primarily to the use of electronics (especially computers) to accomplish everyday tasks.

## How is a **therblig** defined?

Frank Bunker Gilbreth (1868–1924), the founder of modern motion study technique, called the fundamental motions of the hands of a worker "therbligs" (Gilbreth roughly spelled backwards). He concluded that any and all operations are made up of series of

### What is the new science of chaos?

Chaos or chaotic behavior is the behavior of a system whose final state depends very sensitively on the initial conditions. The behavior is unpredictable and cannot be distinguished from a random process, even though it is strictly determinate in a mathematical sense. Chaos studies the complex and irregular behavior of many systems in nature, such as changing weather patterns, flow of turbulent fluids, and swinging pendulums. Scientists once thought they could make exact predictions about such systems, but found that a tiny difference in starting conditions can lead to greatly different results. Chaotic systems do obey certain rules, which mathematicians have described with equations, but the science of chaos demonstrates the difficulty of predicting the long-range behavior of chaotic systems.

these 17 divisions. The 17 divisions are search, select, grasp, reach, move, hold, release, position, pre-position, inspect, assemble, disassemble, use, unavoidable delay, avoidable delay, plan, and test to overcome fatigue.

Gilbreth developed many of the concepts that became part of modern management techniques, and he also patented many inventions useful in the construction industry. One device he used in time and motion studies was his cyclograph or "motion recorder." An ordinary camera and a small electric bulb showed the path of movement. The light patterns reveal all hesitation or poor habits interfering with a worker's dexterity.

## How did a **total solar eclipse** confirm Einstein's **theory of general relativity**?

When formulating his theory of general relativity, Albert Einstein proposed that the curvature of space near a massive object like the Sun would bend light that passed close by. For example, a star seen near the edge of the sun during an eclipse would appear to have shifted by 1.75 arc seconds from its usual place. The British astronomer Arthur Eddington confirmed Einstein's hypothesis during an eclipse on May 29, 1919. The subsequent attention given Eddington's findings helped establish Einstein's reputation as one of science's greatest figures.

## What is **Occam's Razor**?

Occam's Razor is the scientific doctrine that states that "entities must not be multiplied beyond what is necessary;" it proposes that a problem should be stated in its basic and simplest terms. In scientific terms, it states that the simplest theory that fits the facts of a problem should be the one selected. Credit for outlining the law is usually given to William Occam (1284?–1347?), an English philosopher and theologian. This concept is also known as the principle of parsimony or the economy principle.

# Further Reading

## Books


Allaby, Michael. *Encyclopedia of Weather and Climate.* New York: Facts on File, 2002.

Allaby, Michael, and Derek Gjertsen. *Makers of Science.* New York: Oxford University Press, 2002.

Alsop, Fred. *Birds of North America.* New York: DK, 2001.

American Kennel Club. *The Complete Dog Book.* New York: Howell Book House, 1997.

*Animal.* Washington, DC: Smithsonian Institution; New York: DK Publishing, 2001.

Anzovin, Steven, and Janet Podell. *Famous First Facts: International Edition.* New York: H. W. Wilson Company, 2000.

Attenborough, David. *The Life of Birds.* Princeton, NJ: Princeton University Press, 1998.

Audesirk, Teresa, and Gerald Audesirk. *Life on Earth.* Upper Saddle River, NJ: Prentice Hall, 1997.

Bakich, Michael E. *The Cambridge Planetary Handbook.* Cambridge, UK; New York: Cambridge University Press, 2000.

Barnes-Svarney, Patricia. *The New York Public Library Science Desk Reference.* New York: Macmillan USA, 1995.

Barrow, John D. *The Book of Nothing: Vacuums, Voids, and the Latest Ideas about the Origins of the Universe.* New York: Pantheon, 2000.

*Beacham's Guide to the Endangered Species of North America.* Detroit: Gale Group, 2001.

Berinstein, Paula. *Alternative Energy.* Westport, CT: Oryx Press, 2001.

Berlow, Lawrence H. *The Reference Guide to Famous Engineering Landmarks of the World.* Phoenix, AZ: Oryx Press, 1998.

*Biographical Dictionary of Scientists.* 3rd ed. New York: Oxford University Press, 2000.

*Biological and Chemical Weapons.* San Diego, CA: Greenhaven Press, 2001.

*Biology Data Book.* 2nd ed. Bethesda, MD: Federation of American Societies for Experimental Biology, 1972-1974.

Bless, R. C. *Discovering the Cosmos.* Sausalito, CA: University Science Books, 1996.

Blocksma, Mary. *Reading the Numbers.* New York: Penguin Books, 1989.

Bloomfield, Louis A. *How Things Work: The Physics of Everyday Life.* New York: J. Wiley, 1997.

Bothamley, Jennifer. *Dictionary of Theories.* Detroit: Visible Ink Press, 2002.

Bowler, Peter J. *Evolution: The History of an Idea.* Revised ed. Berkeley, CA: University of California Press, 1989.

Brady, George S. *Materials Handbook.* 13th ed. New York: McGraw-Hill Book Co., 1991.

*The Brooklyn Botanic Garden Gardener's Desk Reference.* New York: Henry Holt, 1998.

Bruno, Leonard C. *On the Move.* Collingdale, PA: Diane Publishing Co., 1998.

Bruno, Leonard C. *Science & Technology Firsts.* Detroit: Gale Research, 1997.

Bush, Mark B. *Ecology of a Changing Planet.* Upper Saddle River, NJ: Prentice Hall, 1997.

Bynum, W. F., E. J. Browne, and Roy Porter. *Dictionary of the History of Science.* Princeton, NJ: Princeton University Press, 1981.

Campbell, Ann, and Ronald N. Rood. *New York Public Library Incredible Earth.* New York: Wiley, 1996.

Cazeau, Charles J. *Science Trivia: From Anteaters to Zeppelins.* New York: Plenum Press, 1986.

Cazeau, Charles J. *Test Your Science IQ.* Amherst, NY: Prometheus Books, 2000.

Claiborne, Ray C. *The New York Times Book of Science Questions & Answers.* New York: Anchor Books, 1997.

Collin, S. M. H. *Dictionary of Information Technology.* London: Peter Collin Pub., 2002.
*The Complete Family Health Book.* New York: St. Martin's Press, 2001.
*The Complete Garden Guide.* Alexandria, VA: Time-Life Books, 2000.
*Concise Dictionary of Scientific Biography.* 2nd ed. New York: Scribner's, 2000.
*Concise Encyclopedia Biology.* New York: Walter de Gruyter Berlin, 1996.
Cone, Robert J. *How the New Technology Works.* Phoenix: Oryx Press, 1998.
*Consumer Drug Reference.* 2002 ed. Yonkers, NY: Consumer Reports, 2001.
Cook, Theodore. *The Curves of Life: Being an Account of Spiral Formations and Their Application to Growth in Nature, to Science, and to Art.* New York: Dover Publications, 1979.
Cooper, Paulette, and Paul Noble. *277 Secrets Your Dog Wants You to Know.* Revised ed. Berkeley, CA: Ten Speed Press, 1999.
Couper, Heather, and Nigel Henbest. *DK Space Encyclopedia.* New York: DK Publishing, 1999.
Cowing, Renee D. *The Complete Book of Pet Names.* San Mateo, CA: Fireplug Press, 1990.
Cox, Jeff. *Landscape with Roses.* Newton, CT: Taunton Press, 2002.
Cox, John D. *Weather for Dummies.* Foster City, CA: IDG Books Worldwide, 2000.
*CRC Handbook of Chemistry and Physics.* 83rd ed. Boca Raton, FL: CRC Press, 2003.
Croddy, Eric. *Chemical and Biological Warfare.* New York: Copernicus Books, 2002.
Cunningham, Sally Jean. *Great Garden Companions.* Emmaus, PA: Rodale Press, 1998.
*Current Medical Diagnosis & Treatment.* 41st ed. New York: Lange Medical Books/McGraw-Hill, 2002.
*Current Pediatric Diagnosis & Treatment.* 12th ed. Norwalk, CT: Appleton & Lange, 1995.
*The Cutting Edge: An Encyclopedia of Advanced Technologies.* New York: Oxford University Press, 2000.
DeJauregui, Ruth. *100 Medical Milestones That Shaped World History.* San Mateo, CA: Bluewood Books, 1998.
Dennis, Carina, and Richard Gallagher, eds. *The Human Genome.* New York: Nature/Palgrave, 2001.
Diagram Group. *Space and Astronomy on File.* New York: Facts on File, 2001.
Diagram Group. *The Facts on File Chemistry Handbook.* New York: Facts on File, Inc., 2000.
Diagram Group. *Weather and Climate on File.* New York: Facts on File, 2001.
*Dictionary of Computer and Internet Words.* Boston: Houghton Mifflin, 2001.
*Dictionary of Computer Science, Engineering, and Technology.* Boca Raton, FL: CRC Press, 2001.
*Dictionary of Mathematics.* Chicago: Fitzroy Dearborn Publishers, 1999.
*Dictionary of Scientific Biography.* New York: Charles Scribner's Sons, 1973.
*Dogs: The Ultimate Care Guide.* Emmaus, PA: Rodale Press, 1998.
*Dorland's Illustrated Medical Dictionary.* 29th ed. Philadelphia: W.B. Saunders Co., 2000.
*Earth Sciences for Students.* New York: Macmillan Reference USA, 1999.
*Encyclopedia of Astronomy and Astrophysics.* Philadelphia: Institute of Physics Pub.; London; New York: Nature Publishing Group, 2001.
*Encyclopedia of Climate and Weather.* New York: Oxford University Press, 1996.
*Encyclopedia of Computer Science.* 4th ed. London: Nature Pub. Group, 2000.
*Encyclopedia of Environmental Issues.* Pasadena, CA: Salem Press, 2000.
*Encyclopedia of Genetics.* Pasadena, CA: Salem Press, 1999.
*Encyclopedia of Genetics.* Chicago: Fitzroy Dearborn, 2001.
*Encyclopedia of Genetics.* San Diego, CA: Academic Press, 2002.
*Encyclopedia of Global Change.* Oxford; New York: Oxford University Press, 2002.
*The Encyclopedia of Mammals.* New York: Facts on File, 2001.
*Encyclopedia of Science and Technology.* New York: Routledge, 2001.
*Encyclopedia of Stress.* San Diego: Academic Press, 2000.
*Encyclopedia of the Biosphere.* Detroit: Gale Group, 2000.
*Endangered Animals.* Danbury, CT: Grolier Educational, 2002.
Engelbert, Phillis. *Astronomy & Space.* Detroit, MI: U-X-L, 1997.
Engelbert, Phillis. *The Complete Weather Resource.* Detroit: U-X-L, 1997.
Engelbert, Phillis, and Diane L. Dupuis. *The Handy Space Answer Book.* Detroit, MI: Visible Ink Press, 1998.
*The Facts on File Dictionary of Astronomy.* 4th ed. New York: Facts on File, 2000.
*The Facts on File Dictionary of Biology.* New York: Facts on File, 1999.
*The Facts on File Dictionary of Computer Science.* 4th ed. New York: Facts on File, 2001.
*The Facts on File Dictionary of Earth Science.* New York: Facts on File, 2000.
*The Facts on File Dictionary of Weather and Climate.* New York: Facts on File, 2001.
*The Facts on File Encyclopedia of Science.* New York: Facts on File, 1999.
Farndon, John. *Dictionary of the Earth.* London; New York: Dorling Kindersley, 1994.

Ferguson, Nicola. *Take Two Plants: The Gardener's Complete Guide to Companion Planting*. Lincolnwood, IL: Contemporary Books, 1999.
Fitzpatrick, Patrick J. *Natural Disasters: Hurricanes*. Santa Barbara, CA: ABC-CLIO, 1999.
Fogle, Bruce. *The New Encyclopedia of the Cat*. New York: DK Publishing, 2001.
Fogle, Bruce. *The New Encyclopedia of the Dog*. New York: DK Publishing, 2000.
Freedman, Alan. *The Computer Glossary*. 9th ed. New York: AMACOM, 2001.
*Fundamentals of Complementary and Alternative Medicine*. 2nd ed. New York: Churchill Livingstone, 2001.
*Gaiam Real Goods Solar Living Sourcebook*. Hopland, CA: Gaiam Real Goods, 2001.
*The Gale Encyclopedia of Alternative Medicine*. Detroit: Gale Group, 2001.
*Gale Encyclopedia of Medicine*. 2nd ed. Detroit: Gale Group, 2002.
*The Gale Encyclopedia of Science*. 2nd ed. Detroit: Gale Group, 2001.
Gillespie, James R. *Modern Livestock & Poultry Production*. 5th ed. Albany, NY: Delmar Publishers, 1995.
Gingras, Pierre. *The Secret Lives of Birds*. Willowdale, Ont.; Buffalo, NY: Firefly Books, 1997.
Goldwyn, Martin. *How Does a Bee Make Honey & Other Curious Facts*. Secaucus, NJ: Carol Publishing Group, 1995.
Greenfield, Sheldon L. *ASPCA Complete Guide to Dogs*. San Francisco, CA: Chronicle Books, 1999.
*Grzimek's Encyclopedia of Evolution*. New York: Van Nostrand Reinhold Company, 1976.
*Grzimek's Encyclopedia of Mammals*. 2nd ed. New York: McGraw-Hill, 1990.
Guiley, Rosemary. *Moonscapes*. New York: Prentice Hall Press, 1991.
*Guinness World Records*. Enfield, England: Guinness Publishing, 2002.
Gundersen, P. Erik. *The Handy Physics Answer Book*. Detroit: Visible Ink Press, 1999.
Harding, Anne S. *Milestones in Health and Medicine*. Phoenix: Oryx Press, 2000.
Hargrave, Frank. *Hargrave's Communications Dictionary*. New York: IEEE Press, 2001.
Harrison, Tinsley Randolph, and Eugene Braunwald. *Harrison's Principles of Internal Medicine*. 15th ed. New York: McGraw-Hill, 2001.
*Harvard Medical School Family Health Guide*. New York: Simon & Schuster, 1999.
*Hawley's Condensed Chemical Dictionary*. 14th ed. New York: John Wiley & Sons, 2002.
*Health Issues*. Pasadena, CA: Salem Press, 2001.
Hopkins, Nigel J., John W. Mayne, and John R. Hudson. *The Numbers You Need*. Detroit: Gale Research Inc., 1992.
*The Human Body*. New York: Dorling Kindersley, 1995.
*Human Diseases and Conditions*. New York: Charles Scribner's Sons, 2000.
Hunt, Andrew. *Dictionary of Chemistry*. Chicago: Fitzroy Dearborn Publishers, 1999.
Ifrah, Georges. *The Universal History of Numbers: From Prehistory to the Invention of the Computer*. New York: John Wiley & Sons, 2000.
*The Illustrated Encyclopedia of Birds*. New York: Prentice Hall Editions, 1990.
*The Illustrated Encyclopedia of Wildlife*. Lakeville, CT: Grey Castle Press, 1991.
Indge, Bill. *Dictionary of Biology*. Chicago: Fitzroy Dearborn Publishers, 1999.
*Innovations in Earth Sciences*. Santa Barbara, CA: ABC-CLIO, 1999.
Jackson, Donald C. *Great American Bridges and Dams*. New York: John Wiley & Sons, 1995.
*The International Encyclopedia of Science and Technology*. New York: Oxford University Press, 1999.
James, Glenn, and Robert James. *Mathematics Dictionary*. New York: Van Nostrand Reinhold, 1992.
Jargodzki, Christopher, and Franklin Porter. *Mad about Physics*. New York: John Wiley & Sons, 2001.
*Johns Hopkins Consumer Guide to Drugs*. New York: Rebus Books, 2002.
*Johns Hopkins Family Health Book*. New York: HarperCollins, 1999.
Johnson, George B. *Biology: Visualizing Life*. New York: Holt, Rinehart and Winston, 1994.
Johnson, George B. *The Living World*. Dubuque, IA: William C. Brown Publishers, 1997.
Johnson, Ingrid. *Why Can't You Tickle Yourself?* New York: Warner Books, 1993.
Jonas, Ann Rae. *Museum of Science Book of Answers and Questions*. Holbrook, MA: Adams Media, 1996.
Kahn, Ada P. *Stress A–Z*. New York: Facts on File, 1998.
Kane, Joseph Nathan. *Famous First Facts*. 5th ed. New York: H. W. Wilson Company, 1997.
King, Robert C., and William D. Stansfield. *A Dictionary of Genetics*. Oxford; New York: Oxford University Press, 2002.
Krebs, Robert E. *The History and Use of Our Earth's Chemical Elements*. Westport, CT: Greenwood Press, 1998.
Kuttner, Paul. *Science's Trickiest Questions: 401 Questions That Will Stump, Amuse and Surprise*. New York: Henry Holt, 1994.
Langone, John. *National Geographic's How Things Work*. Washington, DC: National Geographic Society. 1999.
*Larousse Dictionary of Science and Technology*. Edinburgh; New York: Larousse, 1995.
Le Vine, Harry. *Genetic Engineering*. Santa Barbara, CA: ABC-CLIO, 1999.
Levin, Simon A. *Encyclopedia of Biodiversity*. San Diego: Academic Press, 2001.

Lewis, Grace Ross. *1001 Chemicals in Everyday Products.* 2nd ed. New York: Wiley, 1999.

Lewis, Ricki. *Life.* 3rd ed. Boston: WCB/McGraw-Hill, 1998.

Lewis, William M. *Wetlands Explained.* Oxford; New York: Oxford University Press, 2001.

Loewer, Peter. *Solving Weed Problems.* Guilford, CT: The Lyons Press, 2001.

Long, Kim. *The Moon Book.* Boulder, CO: Johnson Books, 1998.

Lyons, Walter A. *The Handy Weather Answer Book.* Detroit: Visible Ink Press, 1997.

Macaulay, David. *The New Way Things Work.* Boston: Houghton Mifflin, 1998.

*Macmillan Encyclopedia of Energy.* New York: Macmillan Reference USA, 2001.

*Magill's Medical Guide.* 2nd revised ed. Pasadena, CA: Salem Press, 2002.

Margolis, Philip E. *Random House Webster's Computer & Internet Dictionary.* 3rd ed. New York: Random House, 1999.

*Math & Mathematicians.* Detroit: U-X-L, 1999.

*Mathematics Dictionary.* New York: Van Nostrand Reinhold, 1992.

*McGraw-Hill Dictionary of Scientific and Technical Terms.* 5th ed. New York: McGraw-Hill, 1994.

*McGraw-Hill Encyclopedia of Science and Technology.* 9th ed. New York: McGraw-Hill, 2002.

McGrayne, Sharon Bertsch. *365 Surprising Scientific Facts, Breakthroughs, and Discoveries.* New York: Wiley, 1994.

McGrayne, Sharon Bertsch. *Blue Genes and Polyester Plants: 365 More Surprising Scientific Facts, Breakthroughs, and Discoveries.* New York: John Wiley & Sons, 1997.

Meinesz, Alexandre. *Killer Algae.* Chicago: The University of Chicago Press, 1999.

*Merck Index.* 13th ed. Whitehouse Station, NJ: Merck, 2001.

*Merck Manual of Diagnosis and Therapy.* 17th ed. Rahway, NJ: Merck, 1999.

*Merck Manual of Medical Information.* New York: Pocket Books, 2000.

Mertz, Leslie A. *Recent Advances and Issues in Biology.* Phoenix, AZ: Oryx Press, 2000.

Mitton, Jacqueline. *Cambridge Dictionary of Astronomy.* Cambridge, UK; New York: Cambridge University Press, 2001.

Mongillo, John, and Linda Zierdt-Warshaw. *Encyclopedia of Environmental Science.* Phoenix, AZ: Oryx Press, 2000.

Nagel, Rob. *Endangered Species.* Detroit: U-X-L, 1999.

National Safety Council. *Injury Facts.* Itasca, IL: National Safety Council, 2000.

*Natural Disasters.* Pasadena, CA: Salem Press, 2001.

*The New Book of Popular Science.* Danbury, CT: Grolier, 2000.

*New York Times Almanac.* New York: New York Times, 2002.

Newton, David E. *Chemical Elements.* Detroit: U-X-L, 1999.

Newton, David E. *Chemistry.* Phoenix, AZ: Oryx Press, 1999.

*The Nobel Prize Winners: Chemistry.* Pasadena, CA: Salem Press, 1990.

*The Nobel Prize Winners: Physics.* Pasadena, CA: Salem Press, 1989.

*The Nobel Prize Winners: Physiology or Medicine.* Pasadena, CA: Salem Press, 1991.

*Notable Scientists: From 1900 to the Present.* Detroit: Gale Group, 2001.

*Numbers: How Many, How Far, How Long, How Much.* New York: HarperPerennial, 1996.

Ochoa, George, and Melinda Corey. *The Wilson Chronology of Science and Technology.* New York: H. W. Wilson, 1997.

Odenwald, Sten F. *The Astronomy Cafe.* New York: W. H. Freeman, 1998.

Olson, Todd R. *PDR Atlas of Anatomy.* Montvale, NJ: Medical Economics Co., 1996.

*The Oxford Companion to the Body.* Oxford; New York: Oxford University Press, 2001.

*The Oxford Companion to the Earth.* Oxford; New York: Oxford University Press, 2000.

*The Oxford Illustrated Companion to Medicine.* Oxford; New York: Oxford University Press, 2001.

*The PDR Family Guide to Natural Medicines and Healing Therapies.* New York: Three Rivers Press, 1999.

*Plant Sciences.* New York: Macmillan Reference USA, 2001.

*Science Explained: The World of Science in Everyday Life.* New York: Henry Holt and Company, 1993.

Purves, William K., et al. *Life, the Science of Biology.* 5th ed. Sunderland, MA: Sinauer Associates; Salt Lake City, UT: W. H. Freeman, 1998.

Quadbeck-Seeger, Hans-Jürgen. *World Records in Chemistry.* Weinheim: Wiley, 1999.

*Reader's Guide to the History of Science.* London; Chicago: Fitzroy Dearborn Publishers, 2000.

Reading, Richard P., and Brian Miller. *Endangered Animals.* Westport, CT: Greenwood Press, 2000.

Richards, James R. *ASPCA Complete Guide to Cats.* San Francisco, CA: Chronicle Books, 1999.

Ridpath, Ian. *The Illustrated Encyclopedia of the Universe.* New York: Watson-Guptill Publications, 2001.

Ritchie, David, and Alexander E. Gates. *Encyclopedia of Earthquakes and Volcanoes.* New ed. New York: Facts on File, 2001.

*Rodale's All-New Encyclopedia of Organic Gardening.* Emmaus, PA: Rodale Press, 1992.

Royston, Angela. *You and Your Body.* New York: Facts on File, 1995.

Ryrie, Charlie. *Garden Folklore That Works.* New York: Readers Digest, 2001.

Sankaran, Neeraja. *Microbes and People*. Phoenix, AZ: Oryx Press, 2000.

Schappert, Phillip. *A World for Butterflies*. Buffalo, NY: Firefly Books, 2000.

Schlager, Neil. *When Technology Fails*. Detroit: Gale Research, 1994.

Schlosberg, Suzanne, and Liz Neporent. *Fitness for Dummies*. Foster City, CA: IDG Books Worldwide, 2000.

*Science and Technology Almanac*. Phoenix, AZ: Oryx Press, 2001.

*Science & Technology Encyclopedia*. Chicago: University of Chicago Press, 2000.

*Scientific American: How Things Work Today*. New York: Crown Publishers, 2000.

*Scientific American Science Desk Reference*. New York: J. Wiley & Sons, 1999.

Serway, Raymond. *Physics for Scientists and Engineers*. 3rd ed. Philadelphia: Saunders College Publishing, 1990.

Shafritz, Jay M., et al. *The Facts on File Dictionary of Military Science*. New York: Facts on File, 1989.

*Sharks*. Pleasantville, NY: The Reader's Digest Association, 1998.

Shearer, Benjamin F., and Barbara S. Shearer. *State Names, Seals, Flags, and Symbols*. Westport, CT: Greenwood Press, 1994.

*The Sibley Guide to Bird Life & Behavior*. New York: Alfred A. Knopf, 2001.

*The Simon & Schuster Encyclopedia of Animals*. New York: Simon & Schuster Editions, 1998.

*The Simon & Schuster Encyclopedia of Dinosaurs & Prehistoric Creatures*. New York: Simon & Schuster, 1999.

Smith, Roger. *The Solar System*. Pasadena, CA: Salem Press, 1998.

Spencer, Donald D. *The Timetable of Computers*. 2nd ed. Ormond Beach, FL: Camelot Publishing Co., 1999.

Stary, Frantisek. *Poisonous Plants*. Wigston, Leicester: Magna Books, 1995.

Stashower, Daniel. *The Boy Genius and the Mogul*. New York: Broadway Books, 2002.

*Statistical Abstract of the United States*. 121st ed. Washington, DC: U.S. Government Printing Office, 2001.

*Stedman's Medical Dictionary*. 27th ed. Philadelphia: Lippincott Williams & Wilkins, 2000.

Stein, Paul. *The Macmillan Encyclopedia of Weather*. New York: Macmillan Reference USA, 2001.

Strong, Debra L. *Recycling in America: A Reference Handbook*. 2nd ed. Santa Barbara, CA: ABC-CLIO, 1997.

Svarney, Thomas E., and Patricia Barnes-Svarney. *The Handy Dinosaur Answer Book*. Detroit: Visible Ink Press, 2000.

Svarney, Thomas E., and Patricia Barnes-Svarney. *The Handy Ocean Answer Book*. Detroit: Visible Ink Press, 2000.

Tapson, Frank. *Barron's Mathematics Study Dictionary*. Hauppauge, NY: Barron's Educational Series, 1998.

Taylor, Norman. *1001 Questions Answered about Flowers*. New York: Dover Publications, 1999

*The Time Almanac*. Boston: Information Please LLC, 2002.

Tobias, Russell R. *USA in Space*. 2nd ed. Pasadena, CA: 2001.

Tomajczyk, S. F. *Dictionary of the Modern United States Military*. Jefferson, NC: McFarland & Company, Inc. 1996.

*Top 10 of Everything*. American ed. New York: DK Publishing, 2002.

Trefil, James. *1001 Things Everyone Should Know about Science*. New York: Doubleday, 1992.

Troshynski-Thomas, Karen. *The Handy Garden Answer Book*. Detroit, MI: Visible Ink Press, 1999.

Turkington, Carol. *The Poisons and Antidotes Sourcebook*. 2nd ed. New York: Facts on File, 1999.

Tyson, Neil De Grasse. *Just Visiting This Planet*. New York: Doubleday, 1998.

Van Dulken, Stephen. *Inventing the 19th Century*. New York: New York University Press, 2001.

Van Dulken, Stephen. *Inventing the 20th Century*. New York: New York University Press, 2000.

*Van Nostrand's Scientific Encyclopedia*. 9th ed. New York: John Wiley & Sons, 2002.

Volti, Rudi. *The Facts on File Encyclopedia of Science, Technology, and Society*. New York: Facts on File, 1999.

von Jezierski, Dieter. *Slide Rules: A Journey through Three Centuries*. Astragal Press, 2000.

Waldbauer, Gilbert. *The Birder's Bug Book*. Cambridge, MA: Harvard University Press, 1998.

Walker, Richard D., and C. D. Hurt. *Scientific and Technical Literature*. Chicago: American Library Association, 1990.

*Ward's Motor Vehicle Facts & Figures*. Southfield, MI: Ward's Communications, 1999.

Watson, Lyall. *Jacobson's Organ and the Remarkable Nature of Smell*. New York: W.W. Norton, 2000.

*Weapons and Warfare*. Pasadena, CA: Salem Press, 2002.

Weigel, Marlene. *U-X-L Encyclopedia of Biomes*. Detroit: U-X-L, 2000.

Wells, Edward R., and Alan M. Schwartz. *Historical Dictionary of North American Environmentalism*. Lanham, MD; London: Scarecrow Press, 1997.

Wilmut, Ian, Keith Campbell, and Colin Tudge. *The Second Creation: Dolly and the Age of Biological Control*. New York: Farrar, Straus & Giroux, 2000.

*Wood Engineering Handbook*. 2nd ed. Englewood Cliffs, NJ: Prentice Hall, 1990.

Woodham, Anne, and David Peters. *Encyclopedia of Natural Healing*. 2nd American ed. New York: DK Publishing, 2000.

*World Almanac and Book of Facts*. New York: World Almanac, 2002.

*World of Biology*. Detroit: Gale Group, 1999.

*World of Chemistry*. Detroit: Gale Group, 2000.

*World of Computer Science*. Detroit: Gale Group, 2002.

*World of Genetics.* Detroit: Gale Group, 2002.
*World of Health.* Detroit: Gale Group, 2000.
*World of Mathematics.* Detroit: Gale Group, 2001.
*World of Physics.* Detroit: Gale Group, 2001.
Xenakis, Alan P. *Why Doesn't My Funny Bone Make Me Laugh?* New York: Villard Books, 1993.

## Journals and Periodicals

*American Forests*
*American Health*
*American Heritage of Invention and Technology*
*American Scientist*
*Astronomy*
*Audubon Magazine*
*Automotive Industries*
*Automotive News*
*Aviation Week and Space Technology*
*Biocycle*
*Bioscience*
*Buzzworm: The Environmental Journal*
*Cat Fancy*
*Cats Magazine*
*Chemical & Engineering News*
*Chilton's Automotive Industries*
*Country Journal*
*Current Health*
*Discover*
*Environment*
*E: The Environmental Magazine*
*FDA Consumer*
*Facts on File World News Digest*
*Fine Gardening*
*Fine Woodworking*
*Flower and Garden*
*Harvard Health Letter*
*Health*
*Home Mechanix*
*Horticulture*
*International Wildlife*
*Journal of the American Medical Association*
*Motor Trend*
*National Geographic*
*National Geographic World*
*National Wildlife*
*Natural History*
*Nature*
*New Scientist*
*New York Times*
*Nuclear News*
*Organic Gardening*
*Physics Today*
*Planetary Report*
*Popular Mechanics*
*Popular Science*
*Recycling Today*

*Road & Track*
*Safety and Health*
*Science*
*Science News*
*Science Teacher*
*Scientific American*
*Sky and Telescope*
*Smithsonian*
*Technology and Culture*
*Weatherwise*

## Web Sites

### General

http://scienceworld.wolfram.com
http://www.ars.usda.gov
http://www.exploratorium.edu
http://www-groups.dcs.st-and.ac.uk/~history/HistTopics/Perfect_numbers.html
http://www.guinnessworldrecords.com
http://www.howstuffworks.com
http://www.infoplease.com
http://www.lacim.uqam.ca/pi/records.html
http://www.madsci.org
http://www.mersenne.org/prime.htm
http://www.nature.com
http://www.nsf.gov
http://www.pbs.org
http://www.sciam.com
http://www.tc.cornell.edu/Edu/MathSciGateway/
http://www.thinkquest.org/library/IC_index.html

### Animal World

http://www.acfacat.com
http://www.cfainc.org
http://www.dnr.state.sc.us/water/envaff/aquatic/zebra.html
http://www.geobop.com/Symbols/
http://www.geog.ouc.bc.ca/physgeog/home.html
http://www.lam.mus.ca.us/cats/main.htm
http://www.lrp.usace.army.mil/
http://www.mans-best-friend.org/bites.html
http://www.oar.noaa.gov/spotlite/archive/spot_corals.html
http://www.spotsociety.org/breeds.htm
http://www.vet.cornell.edu

### Biology

http://anthro.palomar.edu/animal/default.htm
http://micro.magnet.fsu.edu
http://pubs.usgs.gov/gip/dinosaurs/
http://tolweb.org/tree/phylogeny.html
http://waynesword.palomar.edu/ww0504.htm
http://www.biology.arizona.edu
http://www.bt.cdc.gov
http://www.herb.lsa.umich.edu/kidpage/factindx1.htm
http://www.ncbi.nlm.nih.gov/About/primer/index.html

http://www.nhgri.nih.gov
http://www.ornl.gov/hgmis/
http://www.ucmp.berkeley.edu

## Boats, Trains, Cars, and Planes

http://www.abs-education.org
http://www.aerovironment.com/news/news-archive/mav99.html
http://www.af.mil/news/factsheets/F_117A_Nighthawk.html
http://www.airforce-technology.com/projects/f117/index.html
http://www.harristechnical.com/skid11.htm
http://www.janes.com
http://www.nhtsa.dot.gov
http://www.nsc.org/lrs/statstop.htm
http://www.sprucegoose.org

## Buildings, Bridges, and Other Structures

http://www.aisc.org
http://www.cntower.ca
http://www.iti.northwestern.edu
http://www.lehigh.edu/ctbuh/credits.html
http://www.nps.gov/jeff/ar-facts.htm

## Chemistry

http://pearl1.lanl.gov/periodic/default.htm
http://www.acs.org
http://www.chemfinder.camsoft.com
http://www.chemicool.com
http://www.lbl.gov/Science-Articles/Archive/118-retraction.html
http://www.iupac.org

## Climate and Weather

http://www.almanac.com
http://www.nhc.noaa.gov
http://www.noaa.gov

## Communications

http://webopedia.internet.com
http://www.computer.org
http://www.computerhistory.org
http://www.obsoletecomputermuseum.org

## Earth

http://earthquake.usgs.gov/faq
http://image.gsfc.nasa.gov/poetry/ask/amag.html
http://neic.usgs.gov/neis/eqlists/10maps_usa.html
http://nisee.berkeley.edu
http://nsidc.org/glaciers/questions/located.html
http://volcanoes.usgs.gov/update.html
http://www2.nature.nps.gov/stats
http://www.aero.org
http://www.avalanche.org/
http://www.bbc.co.uk/education/rocks
http://www.fi.edu/earth/index.html
http://www.geo.cornell.edu/geology/Galapagos.html

http://www.geology.about.com
http://www.hartrao.ac.za/geodesy/tectonics.html
http://www.lib.noaa.gov/docs/windandsea.html
http://www.neic.cr.usgs.gov/neis/plate_tectonics/rift_man.html
http://www.phys.ocean.dal.ca/other-sites.html
http://www.uc.edu/geology/geologylist/
http://www.ucmp.berkeley.edu/geology/tectonics.html
http://www.whoi.edu/Resources/oceanography.html
http://www.yoto98.noaa.gov/oceanl.htm

## Environment

http://www.biodiversity.org
http://www.epa.gov
http://www.fws.gov
http://www.nmfs.noaa.gov/prot_res/PR3/Turtles/turtles.html
http://www.nps.gov
http://www.nrc.gov
http://www.nrdc.org/health/pesticides/hcarson.asp
http://www.panda.org
http://www.rainforest-alliance.org
http://www.smokeybear.com

## Energy

http://epic.er.doe.gov/epic
http://www.ans.org
http://www.eia.doe.gov
http://www.epa.gov/swerust1/mtbe/index.htm
http://www.eren.doe.gov
http://www-formal.stanford.edu/jmc/progress/nuclear-faq.html
http://www.ieer.org
http://www.nea.fr/
http://www.nei.org
http://www.nirs.org
http://www.nuc.umr.edu/nuclear_facts/nuclearfacts.html

## Health and Medicine and Human Body

http://eire.census.gov/popest/archives/national/nation2.php
http://nccam.nih.gov
http://orthoinfo.aaos.org
http://sln.fi.edu/biosci/blood/types.html
http://www.alz.org
http://www.alzheimers.org
http://www.ama-assn.org
http://www.cancer.org
http://www.cdc.gov/nchs/fastats
http://www.cryonics.org
http://www.daawat.com/resources/calorieburn.htm
http://www.diabetes.org
http://www.fda.gov
http://www.ncbi.nlm.nih.gov
http://www.nhlbi.nih.gov
http://www.nhlbisupport.com/bmi/bmicalc.htm
http://www.nidcd.nih.gov
http://www.nsc.org

http://www.redcross.org
http://www.surgeongeneral.gov
http://www.vestibular.org

## Minerals and Other Materials

http://mineral.galleries.com
http://www.czplatinum.com
http://www.diasource.com
http://www.usgs.gov

## Physics

http://www.aip.org
http://www.aps.org
http://www.colorado.edu/physics/2000/index.pl
http://www.physicsweb.org
http://www.physlink.com

## Plant World

http://www.ashs.org
http://www.botanical.com
http://www.chestnut.acf.org/Chestnut_history.htm
http://www.citygardening.net/complant/
http://www.co.mo.md.us/services/dep/Landscape/shakespeare.htm
http://www.ea.pvt.k12.pa.us/medant/hemlock.htm
http://www.greendesign.net/understory/sumfall97/notsotrv.htm
http://www.plainfield.com/bardgard/links.html
http://www.safnet.org/archive/sherman201.htm

## Space

http://antwrp.gsfc.nasa.gov/apod/astropix.html
http://cfa-www.harvard.edu/iau/info/OldDesDoc.html
http://hubble.stsci.edu
http://www.aas.org
http://www.amsmeteors.org
http://www.astronomy.com
http://www.cnn.com/TECH/space/
http://www.jpl.nasa.gov
http://www.ksc.nasa.gov
http://www.nasa.gov
http://www.nasm.si.edu/nasm/dsh/
http://www.noao.edu
http://www.nrao.edu
http://www.planetary.org
http://www.setileague.org
http://www.space.com

## Weights, Measure, Time, Tools, and Weapons

http://aa.usno.navy.mil/
http://www.af.mil/news/indexpages/fs_index.shtml
http://www.atomicmuseum.com
http://www.chinfo.navy.mil/navpalib/factfile/ffiletop.html
http://www.nist.gov
http://www.time.gov

# Index

Note: (ill.) indicates photos and illustrations.


## A

Abacus, 578, 601–2, 602 (ill.)
"ABCD" survey, 417
ABO system of typing blood, 394
Abortion pill, 456
ABS (antilock braking systems), 540
Absolute zero, 1
Accelerated Mass Spectronometer, 265
Accidents, men vs. women, 409
Acid rain, 229, 229 (chart), 230, 230 (chart)
Acidity, 290
Acquired immunodeficiency syndrome (AIDS), 423–25, 424 (chart)
Acre, 607
Activated charcoal, 419
Active solar energy systems, 182
Ada computer programming language, 586
Adams, John Quincy, 365
Adiabatic process, 6
Adrenals, 391
Advisory (weather), 143
Aesculapius, 440–41
Aflatoxin, 497
African Americans in space, 76
African elephants, 223, 344, 351–52, 352 (chart)
African honeybees, 321–22
Agent Orange, 232
Aging, 405, 437
Agricultural tools, 489
AIDS, 423–25, 424 (chart)
Air. *See also* Air phenomena
- blue sky, 83
- breathed in lifetime, 388
- composition of Earth's atmosphere, 81
- density of, 24, 24 (chart)
- layers of Earth's atmosphere, 81–83
- Van Allen belts, 82–83

Air bags, 544–45
Air conditioner, 200, 538
Air phenomena. *See also* Air
- aurora, 123, 123 (ill.)
- ball lightning, 126
- bishop's ring, 122
- Brown Mountain lights, 127
- clouds, 123–25
- fulgurites, 126–27
- green flash phenomenon, 123
- lightning and thunder, 126
- lightning, color, 125
- lightning, heat from, 125
- lightning strike, length of, 125–26
- lightning, striking twice, 126
- lightning, volts in, 125
- rainbow, 127
- Saint Elmo's fire, 127

Air pollution, 228
Aircraft
- amphibian plane, 552
- around-the-world balloon flight, first, 549
- avionics, 549
- bird shot test, 550
- black box, 549–50
- fastest aircraft, 551
- fastest jet aircraft, 552
- fuel consumption, 204
- highest flying jet aircraft, 552
- *Hindenburg,* 546, 547 (ill.)
- hovercraft, 547
- Mach number, 551
- nonstop transatlantic flight, first, 548
- nonstop, unrefueled, around-the-world, first, 548–49
- parachute jump, first, 549
- seaplane, 552
- seating capacity in jets, 552 (chart)
- *Spruce Goose,* 550–51, 551 (ill.)
- supersonic flight, first, 548
- tires, 550
- wind tunnel, 550
- wings, 547
- Wright Brothers, 547–48

Al Aziziyah, Libya, 120
Alarm clock, 488
Albatross, 338
Alberta clipper, 130
Alchemical symbols, 157, 157 (chart)
Alcock, John W., 548

Alcohol, 415
Aldrin, Edwin E., Jr., 72, 75
Algae, 207
Algorithm, 577–78
Alkali metals, 19
Alkaline Earth metals, 19
Alkalinity, 290
Alligators, 334
Alpha Centauri, 37, 43
Alpha particles, 11
Alpine black salamander, 305
ALS (amyotrophic lateral sclerosis), 431
Alternative fuel vehicle, 535
Aluminum, 23, 156–57, 203
Alvarez, Luis, 218
Alvarez, Walter, 218
Alzheimer's disease, 437–38
a.m. and p.m., 486
AM stations, 567
*Amanita phalloides* mushroom, 421
Amazon River, 94
Amber, 166, 317
Ambergris, 167
American Association for the Advancement of Science, 615
American Chemical Society, 27
American chestnut tree, 276
American foxhound, 357
American National Standard for Information Science Permanence of Paper for Printed Library Materials, 558
American Philosophical Society, 614
American Physical Society, 27
American shorthair cat, 362
American Standard Code for Information Interchange (ASCII), 561
Amino acid, 264
Ammann, Othmar H., 518
Ammonia, 169–70
Amor asteroids, 60
Ampere, 31
Ampère, André Marie, 31
Amphibian plane, 552
Amundsen, Roald, 98
Amyand, Claudries, 390
Amyotrophic lateral sclerosis (ALS), 431
Anabolic steroids, 454–55
Analog, 573
Analytical chemistry, 27
Andromeda galaxy, 36
Angel, Jimmy, 94
Ångström, Anders, 8
Animal and plant classification system, 267–68, 268 (chart)
Animal cloning, 259
Animalia kingdom of biology, 267
Animals. *See also* Aquatic life; Birds; Cats; Dogs; Insects; Mammals; Reptiles and amphibians
  bears, hibernation, 309, 309 (ill.)
  blood color, 311
  blood types, 311
  color vision, 310
  female names, 313–14 (chart)
  fingerprints, 31
  group names, 316–17 (chart)
  hearing, 311
  juvenile names, 314–16 (chart)
  largest, 308 (chart)
  life expectancy, 305–8 (chart)
  longest gestation, 305
  male names, 313–14 (chart)
  most intelligent, 310
  organs, 461
  regeneration, 310–11
  running speed, 312 (chart)
  smallest, 308–9 (chart)
  smells, 309
  snoring, 312
  in space, 73–74
  zoo, first, 305
Anise, 287
Anorexia, 436
Answering machines, 572
Antarctic ozone hole, 208
Antarctica, 98, 99 (ill.)
Antelopes, 350
Anthracite, 184
Anthrax, 269, 497
Antibody, 370
Antigen, 370
Antilock braking systems (ABS), 540
Antimatter, 28
Antimony, 26
Antiquark, 14
Ants, 324, 324 (chart)
Aorta, 394
Apollo asteroids, 60
*Apollo* moon flights, 72
Apothecary, 449
Appendix, 390
Appleseed, Johnny, 298 (ill.), 299
Appliance energy, 200–202, 201 (chart)
April (month), 481
Aqua regia, 169
Aquatic life. *See also* Animals
  coral and coral reef, 329, 330
  dolphins, 331
  electric eel, 330–31
  fish, age of, 328
  fish, swimming direction, 328
  fish, swimming speed, 328–29
  giant tube worms, 330
  lobsters, 329
  mermaid's purse, 330
  mollusk shells, 331–32
  pearls, 329
  salmon, 331
  sharks and shark attacks, 332, 332 (chart)
  zebra mussels, 328
Aquifers, 88–89
Arboretum, 294
Arch bridge, 516
Arctic tern, 337–38
Area, 471–72 (charts), 607, 608
Argon, 22–23
Arithmetic, 600
Armadillos, 355, 355 (ill.)
Armstrong, Edward Howard, 566 (ill.), 567
Armstrong, Neil A., 72, 75
Aromatherapy, 463–64
Arrhenius, Svante, 208, 260
Arrow poisons, 420–21
Arsenic, 26
Arteriosclerosis, 433, 433 (ill.)
Arthritis, 420
Artificial hearts, 461
Asbestos, 411–12
ASCII (American Standard Code for Information Interchange), 561
Aspdin, Joseph, 174
Assembly computer language, 585–86
Assisted living facility, 444
Asteroids, 60, 61, 65
Asthenosphere, 84
Astrolabe, 65
Astrological symbols for planets, 157, 157 (chart)
Astronauts, 72
Astronomer Royal, 64
Astronomical unit (AU), 64

Aten asteroids, 60
Atom, 6, 12
Atomic bomb, 555
Atomic time, 474, 486
Atomic weapons, 501–2
AU (astronomical unit), 64
August (month), 482
Aurora, 123, 123 (ill.)
Aurora Australis, 123
Aurora Borealis, 8, 123, 123 (ill.)
Australian eucalyptus tree, 277
Automatic transmission, 538
Automaton, 581
Automobiles
air bags, 544–45
air conditioner, 538
alternative fuel vehicle, 535
American automobile company, first, 535–36
antilock braking systems (ABS), 540
automatic transmission, 538
braking distance, 540, 540 (chart)
colors, most popular, 542 (chart)
colors, safest, 541
Corvair, 541–42
cost to operate, 539 (chart)
crashes, 410
electric, 535
energy economics, 203
engine, 534
fatal accidents, 541
horsepower, 533
inventor of, 534–35
laser speed guns, 544
license plate, 539
manufacturing materials, 537 (chart)
Michelin tire, 537
Nader, Ralph, 541–42
nuclear-powered, 538
numbers on tires, 538, 538 (chart)
parking meters, 545
police radar, 543–44
registration, 539
rumble seat, 537
safety recalls, 540
seatbelts, 541
skidding, 540 (chart)
speed trap, 543
taxicab, 546
theft, 545
time taken to build, 536 (chart)
tires, 243–44, 537, 538, 538 (chart)
vs. trucks, 545–46
tubeless tires, 537
turns on red light, 542–43
VASCAR (Visual Average Speed Computer and Recorder), 543
vehicle identification number (VIN), 539
Autumn, 482–83
Avalanches, 100
Avdeyev, Sergei Vasilyevich, 77
Avionics, 549
Avogadro, Amedeo, 32
Avogadro's law, 25
Avoirdupois measurements, 472–73, 473 (chart)
"Aztec money," 150

## B

B. F. Goodrich Company, 537
B lymphocytes, 371–72
B-17 Flying Fortress, 554
Babbage, Charles, 578, 578 (ill.)
Baboon heart, 461
Baby boomers, 410
Babylonian calendar, 476
Backe, Alexander Dallas, 614
Bacon, Francis Thomas, 190
Bacon, Roger, 178
Bacteria, reproduction of, 269
Bacteriology, 271
"Bad cholesterol," 415
Badgers, 144–45
Baekeland, Leo Hendrik, 179
Bailly (crater), 59
Baily's beads, 45
Bain, Alexander, 572
Baird, John Logie, 568
Bakelite, 179
Bakewell, Frederick, 572
Balance, 404–5
Bald eagle, 341, 341 (ill.)
Ball lightning, 126
Ballard, Robert, 529
Balled-and-burlapped plants, 293–94
Balloons, 230, 549
Baltimore, David, 423
Bamboo plant, 279
Banned or strictly controlled chemicals, 226
Banyan tree, 278
Barbier, Charles, 559
Bardeen, John, 2, 615
Bare-rooted plants, 294
Barnard, Christiaan, 460
Barnum, P. T., 381
Barometric pressure, 144
Barrel of oil, 195
Barry, John A., 592
Bartholdi, Fréderic-Auguste, 522
Bartram, John, 295
Baseball, 3–4, 613
Basenji, 360
Bates, Henry Walter, 255
Batesian mimicry, 255
Bateson, William, 253
Bats, 347
Battista, Giovanni, 447
Bay of Fundy, 91
Bazooka, 500
Bean, Alan L., 72
Bean moth, 327
Bears, 309, 309 (ill.), 354
Beaufort, Francis, 133
Beaufort scale, 133
Beaumont, William, 374
Beckett, Frederick, 161
Becquerel, Antoine Henri, 11, 15
Bednorz, J. Georg, 2
Bees, 321–23, 421
Belgian block, 175
Bell, Jocelyn, 38
Bell, Lawrence, 548
Bench mark, 473–74
Bends, 437
Benedictus, Edouard, 172
Benz, Karl, 534, 534 (ill.)
Bernard, Claude, 376, 376 (ill.), 377
Berners-Lee, Tim, 593
Beryl, 155
Berzelius, Jöns Jakob, 15, 17
Beta particles, 12
Beverage deposits, 241
Bible, 597
Biblical shekel, 465–66 (chart)
Bifocal lenses, 402
Big Bang theory, 33
Big Bertha, 501
Big Bore theory, 33
Big Crunch theory, 33
Big Dipper, 39–40
Binary stars, 37
Biochemistry, 264
Biodiversity, 205
Bioinformatics, 264

Biological clock, 263
Biology. *See also* Cells; Evolution and genetics; Fungi, bacteria, and algae; Life processes and structures
animal and plant classification system, 267–68, 268 (chart)
bioinformatics, 264
five kingdoms of, 266–67
gnotobiotics, 266
molecular, 266
number of organisms, 268
origins of, 265–66
radiocarbon dating, 265
Bioluminescence, 327
Biomass energy, 182–83
Biome, 205, 206 (ill.)
Bioremediation, 225
Biorhythms, 263
Bio-terrorism, 497–98
Bird houses, 343
Bird shot test, 550
Birds. *See also* Animals
arctic tern, 337–38
bald eagle, 341, 341 (ill.)
bird houses, 343
bluebirds, 344
canary, 341
electrocution, 343
emperor penguin, 340–41
European starling, 342
flightless, 338
geese, 338 (ill.), 338–39
hearing of, 336
homing pigeons, 342
human touch, 343
hummingbirds, feeding, 339–40
hummingbirds, speed of, 339, 339 (chart)
hummingbirds, wing speed, 340, 340 (ill.)
largest eggs, 337
migration, 337–38
names of, 335–36 (chart)
orphaned, 343
oxpecker on rhinoceros's back, 342
penguin, predator of, 341
as pets, 365 (chart)
singing, 336
smallest eggs, 337
state birds, 342
swallows of Capistrano, 338
"V" flight formation, 338 (ill.), 338–39
wing span, 338
woodpeckers, 343
Birth defects caused by Thalidomide, 456
Bishop's ring, 122
Bismuth, 15
Bison, 354
Bits, 587
Bituminous coal, 184
Black bile, 392
Black box, 549–50
Black damp, 163
Black Death (1347–1351), 423
Black hole, 37–38
Black mamba snake, 334
Black widow spider bite, 421
"Blackout Bomb," 503
Blackwell, Elizabeth, 442
Bleeding to death, 417
"Blind as a bat," 347
Blinking, 402
Blood alcohol level, 415, 415 (chart)
Blood and body fluids. *See also* Human body
ABO system of typing blood, 394
animals, blood in, 311
aorta, 394
blood, amount of, 394
blood flow in heart, 393 (ill.)
blood grouping, 396, 396 (chart)
blood types, 394 (chart), 394–95, 396
blood vessels, 394
carbon dioxide in blood, 395
donating blood, 395
eyes tearing around onions, 396
"falling asleep" (arms and legs) sensation, 394
humors of the body, 392
oxygen in blood, 392
pH of blood, urine, and saliva, 392
plateletpheresis, 395
Rh factor, 395
runny nose while eating, 397
saliva, 393
seawater vs. blood, 392 (chart)
sneezing and blood, 397
sweating when eating spicy foods, 397
universal donor/recipient, 395
water, percentage of human body weight, 396
wounds, 396
Blood banks, 442–42
Blood clots, 396
Blood pressure, 445
Blood-clotting factor VIII, 257–58, 432
Blue moon, 57–58, 58 (chart)
"Blue People," 429
Blue sea, 89
Blue sky, 83
Bluebirds, 344
Bluford, Guion S., Jr., 76, 79
BOCA code, 507
Body fluids. *See* Blood and body fluids
Body temperature, 391
Body typing, 380, 380 (ill.)
Bohr, Niels, 12
Bombay blood type, 396
Bones, muscles, and nerves. *See also* Human body
bone that does not touch another bone, 383
bones, number of, 383
broken bones, 383
cracking knuckles, 383–84
ecorche, 385, 385 (ill.)
frowning, 386
funny bone, 384
hamstring muscles, 385
human bite, 386
molars and premolars, 384
muscle, longest, 385
muscle, most variable 385
muscles, number of, 384–85
nerve, largest, 386
primary teeth, 384
smiling, 386
stiff and sore muscles, 386
teeth, 384
uvula, 386
Bonsai trees, 299–300
Boole, George, 605–6
Boomerang, 4
"Boomeritis," 410
Booting computers, 588
Borchers, W., 161
Borel, Jean-François, 460
Borglum, Gutzon, 105
Borsch, J. L., 402

Bosch, Karl, 170
Bosenberg, Henry F., 303
Bosons, 14
Botanical garden, 294–95
Botany, 280
Botulism, 419, 497
Bovine spongiform encephalopathy, 426
Bowie, Jim, 498
Bowie knife, 498
Bowie, Rezin, 498
Bowser, Sylanus, 188
Boyle, Robert, 17, 18
Boyle's law, 25
"B.P." date designation, 484
*Brachiosaurus,* 217
Brackish water, 91
Bragg, William Henry, 16, 615
Bragg, William Lawrence, 16, 615
Braille alphabet, 559
Braille, Louis, 559
Brain, 387, 388 (ill.)
Braking distance, 540, 540 (chart)
Brand, Hennig, 18
Braun, Ferdinand, 568
Brearly, Harry, 161
Breathalyzer, 495
Brewster, David, 16
Bridge games, 611
Bridges
- arch, 516
- bridge spans, longest, 517 (chart), 517–18
- bridge-tunnel, 516
- Brooklyn Bridge, 518–19
- cantilever, 516
- causeway, longest, 519
- city with most bridges, 518
- covered, 516–17
- floating, 518
- Golden Gate Bridge, 519
- "kissing," 516–17
- Mississippi River, first bridge across, 519
- rigid beam, 516
- suspension, 516
- suspension, longest, 518
- types of, 516

Bridges, Robert, 160
Briggs, Lyman, 3
Brillouin, Léon, 4
Bristlecone pine tree, 275
British thermal unit (BTU), 196
Bromium, 22
Brompton's cocktail, 456
Brooke, Robert, 357
Brooklyn Bridge, 518–19
Brown, Adolphe F., 508
Brown, Arthur W., 548
Brown, Louise, 462
Brown Mountain lights, 127
Brown, Robert, 245
Brown, Roy, 553
Brown, Samuel, 534
Brownfields, 236
Bruhn, Wilhelm, 546
Brunel, Isambard K., 530
Bryant, Gridley, 531
BTU (British thermal unit), 196
Buckminsterfullerene, 170–71
"Bucky balls," 171
"Bug" (computer), 589
Building Officials and Code Administrators (BOCA) International, 507
Buildings and building parts, 505. *See also* Structures
- BOCA code, 507
- chimney, 505
- crown molding, 506
- doorjamb, 505
- Empire State Building, 509
- flue, 505
- geodesic dome, 511, 512 (ill.)
- Leaning Tower of Pisa, 509
- mail-order homes, 513
- non-air-supported clear span roof, 512
- "penny" as nail size, 507–8
- Pentagon, 509–10
- Phillips screws, 506–7
- *pisé de terre,* 508
- R-value, 506, 506 (chart)
- rammed earth, 508
- retractable roof stadium, 511–12
- shopping center, first, 509
- skyscraper, first, 508–9
- STC rating, 507
- tallest building criteria, 510
- tallest buildings in U.S., 510, 510 (chart)
- tallest buildings in world, 510
- tallest self-supporting structure in world, 511, 511 (ill.)
- tallest structure in world, 511
- "topping out" party, 512
- wood used in house construction, 505
- yurt, 508

Bulbs, 280–81
Bull, Henryk, 98
Bulletproof glass, 172
Bunker, Chang and Eng, 381, 381 (ill.)
Burdach, Karl, 265
Burns, 439 (chart)
Burns, Bob, 500
Burroughs, John, 214, 215 (ill.)
Burton, William, 186
Bushnell, David, 498
Butcher's blocks, 164
Butterflies, 254 (ill.), 255, 296, 319 (ill.), 321 (chart)
Butterfly gardens, 320–21
"By and large," 525–26
Byrd, Richard, 130–31
Byron, Augusta Ada (countess of Lovelace), 586
Bytes, 587

## C

Cable, 572–73
Cable car, 532–33, 533
Cabooses, 531–32
Cabriolet, 546
Caduceus, 440–41
Calculator, 602
Calculus, 601
Calendar year, 475
Call letters, 566
Calmet, Jeanne Louise, 381
Calories, 373, 373–74 (chart)
Camels, 354
Cameron, Michael, 596
Canary, 185, 341
Cancer, 434
Cannel coal, 185
Cannon King, 500
Cannon, Walter Bradford, 377
Cantilever bridge, 516
Cape May diamonds, 152
Cape Morris K. Jesup, 97
Captured rotation, 57
Capybara, 355–56
Carbon, 25
Carbon atoms, even vs. odd, 23–24
Carbon black, 176
Carbon dating, 106
Carbon dioxide, 226 (chart), 395
Carbon nanotubes, 583
Carbon-carbon composites (CCCs), 180
Carcinogens, 413, 414 (chart)
Cardano, Gerolamo, 447

Cardiac pacemakers, 461–62
Cardiopulmonary resuscitation (CPR), 416–17
Cargo Quilt, 200
Carlsbad Caverns National Park, 13
Carnemolla, Rosie, 380
Carnivorous plants, 281, 282 (ill.)
Carpal tunnel syndrome, 430–31
Carroll, Lewis, 255, 422, 606, 613
Cars. *See* Automobiles
Carson, Rachel, 226
Carter, Jimmy, 71
Carver, George Washington, 302, 302 (ill.)
Cashmere, 168
Cassini, Jean Domenique, 51
"Casting out nines," 604
CAT scans, 446
Catgut, 460
Catnip, 289, 301
Cats. *See also* Animals
  catnip, 289, 301
  desert, 352–53
  eyes shining in the dark, 363
  favorite names, 364
  memories of, 361–62
  original domestic, 362
  poisonous plants, 364
  purring, 363–64
  Siamese, 363
  tabby, 362
  whiskers, 364
Cattley, William, 285
Causeway, 519
Caves, 102, 103
"cc" designation in engine sizes, 490
CD (compact disc), 575, 575 (ill.)
CD-ROM disc, 575, 575 (ill.)
Celestial objects, naming of, 65
Cell fusion, 256
Cells. *See also* Biology; Evolution and genetics; Fungi, bacteria, and algae; Life processes and structures
  cell theory, 245
  chloroplasts, 248, 248 (ill.)
  eukaryotic, 246, 246 (chart), 247, 247 (chart)
  in human body, 369, 369 (chart)
  mitochondria, 246–47
  mitosis, 248, 249 (ill.)
  nucleus of, 248
  organelles, 247 (chart)
  plant, 248, 248 (ill.)
  prokaryotic, 246, 246 (ill.), 246 (chart), 247 (chart)
  scientists associated with, 245
Cellular telephones, 573
Celluloid, 179
Celsius, Anders, 30
Celsius temperature scale, 30
Celsius-Fahrenheit conversion, 30, 30 (chart)
Cement, 174
Cenozoic Era, 250 (chart)
Centenarians, 382
Centipede, 326
Central processing unit (CPU), 586
Century, 475–76
Ceramics, 168
Ceres, 60
Cernan, Eugene A., 72
Cesium, 22, 23
CFCs (chlorofluorocarbons), 209, 226, 227–28
Chadwick, James, 13
Chaffee, Roger, 78
Chain, Ernest, 453
*Challenger,* 78–79
Chamois, 356
Chaos (science), 616
Charcoal, 163
Chardack, William M., 445
Charles, Jules, 302
Charles's law, 25
Charon, 54–55
Chaucer, Geoffrey, 613
Chemical elements, 17–26
  alkali metals, 19
  alkaline Earth metals, 19
  boiling points, highest and lowest, 24
  carbon atoms, even vs. odd, 23–24
  conductors of electricity, 23
  density, highest, 24–25
  density of air, 24
  embalming fluids, 26
  first element discovered, 18
  founders of modern chemistry, 17
  gas laws, 25
  gold, 21
  hardest, 25
  Harkin's rule, 21
  heavy water, 25–26
  isomers, 25
  isotopes, 23
  lead-acid batteries, 23
  Lewis acid, 26
  liquid at room temperature, 22
  most abundant in universe, 22
  most abundant on Earth, 22
  most used chemical, 26
  named after women, 20
  noble metals, 21
  periodic table, 17–18
  philosopher's stone, 21
  platinum group metals, 21
  rare gases and rare Earth elements, 22–23
  silver, 21
  softest, 25
  sweetest chemical compound, 18, 18 (chart)
  symbols not derived from their English names, 22 (chart)
  transition elements, 19–20
  transuranic elements, 20
Chemical warfare, 496
Chemistry, 27
Chemistry measurement and methodology. *See* Physics and chemistry measurement and methodology
Chemotherapy, 454
Chemotropism, 274
Chennault, Claire Lee, 554
Chernobyl accident, 193–94, 228–29
Chess, 581–82
Chest cavity, 389
Chestnut blight, 276–77, 277
Cheyenne, Wyoming, 130
Chia Hsien, 608
Chia seeds, 300
Chicago, Illinois, 130
Childproof containers, 420
Child-safe plants, 281, 282 (chart)
Chimney, 505
Chimpanzees in space, 73–74
"China Syndrome," 194
China White, 456
Chinese calendar, 476
Chinese desert cat, 353
Chinese earthquake detector, 110
Chinese years, 479, 480 (chart)
Chinook, 130
Chiron, 60
Chiropractic medicine, 443
Chlorofluorocarbons (CFCs), 209, 226, 227–28
Chloroplasts, 248, 248 (ill.)

Cholesterol, 414–15, 446 (chart)
"Christmas factor," 432 (ill.), 432–33
Christy, James, 54
Chromatography, 31
Chromium, 160–61
Chromosomes, 367–69, 368 (ill.)
Chudnovsky, Gregory and David, 599
Cigarettes, 415–16, 416 (ill.)
Cinder cones, 106
Cinnabar, 150
Circle of Fire, 106–7, 107 (ill.)
Civil engineering, 513
Civil War innovations, 498
Clams, 329
Clark, Barney B., 461
Clarke, Arthur C., 573
Clarke belt, 573
Clavicle, 383
Clay, 163
Clay soil, 289
Clean Air Act (1995), 187, 229
Clemens, Samuel L., 527
Clockwise movement, 485
Cloning, 258–60
"Close encounters," 69–70
*Clostridium perfringens,* 498
Clouds, 123–25
Clouds with Vertical Development, 125
Clover, 288
Clydesdale horses, 351
CN Tower, 511, 511 (ill.)
Coal, 162, 163, 184–85
Coastal redwood tree, 277
Coast-to-coast highway, 514
COBOL computer language, 587
Cocaine, 457, 458 (ill.)
Cochran, Jacqueline, 9
Cocklebur, 180
Cockroaches, 326
Codes
- ASCII, 561
- Braille alphabet, 559
- Enigma and Purple machines, 561–62
- "Etaoin Shrdlu," 562
- ISBN, 564
- Morse Code, 559–60, 560–61 (chart)
- palindromic square, 563–64
- phonetic alphabet, 558
- 10-codes, 562, 563 (chart)
- undeciphered, 562
- UPC bar code, 564

Cogeneration, 189
Collar bone, 383
College of Philadelphia Department of Medicine, 442
Colles fracture, 383
Colligative properties, 14
Collins, Eileen, 76
Collins, Michael, 72
Color, 8, 403
Colt revolver, 498
Colt, Samuel, 498
Coltan, 156
Combustion, 5
Comets, 61–62, 65
Common cold, 423
Compact disc (CD), 575, 575 (ill.)
Compact fluorescent light bulbs, 202
Companion planting, 294, 294 (chart)
Compass, 113, 490
Composite cones, 106
Compound interest, 611
Compound microscope, 491–92
*Compsognathus,* 217
Computerized axial tomography (CAT) scans, 446
Computers
- algorithm, 577–78
- assembly language, 585–86
- bit vs. byte, 587
- booting, 588
- "bug," 589
- carbon nanotubes, 583
- COBOL language, 587
- data mining, 593–94
- data storage, 584, 584 (chart)
- "do not fold, spindle, or mutilate," 580
- DOS (disk operating system), 590
- e-mail, 591
- energy usage of, 588
- ENIAC (electronic numerical integrator and computer), 580
- expert system, 581
- floppy disk, 584–85
- formatted disk, 585
- fuzzy search, 590
- games, first, 581
- generations of computers, 579–80
- generations of programming languages, 586–87
- GIGO ("garbage in, garbage out"), 590
- glitch, 589
- hacker, 591
- hard disk, 584
- Hopper's rule, 585
- Internet, 576–77, 593
- kludge, 592
- languages, 585–87
- Linux operating system, 592
- Lisa (microcomputer), 582
- machine language, 585–86
- MANIAC (mathematical analyzer, numerator, integrator, and computer), 580
- Moore's Law, 583
- mouse, 585
- open source software, 592–93
- origins of, 578–79
- PASCAL language, 587
- physical positioning of, 588
- pixel, 590
- programmer, first, 586
- punched cards, 579, 580
- silicon chips, 582–83
- technobabble, 592
- "the Turk," 581–82
- video-arcade games, first, 581
- virtual reality, 574
- virus, 589

Condom, 462
Cones, 401–2
Conifers, 278–79, 299
Conrad, Charles P., 72
Conservation. *See* Recycling, conservation, and waste
Constellations, 41, 41–43 (chart), 43
Container-grown plants, 293, 294
Continental divide, 104
Continents, 95–96, 96 (ill.)
Contraception, 462
Controlled substances, 456, 457 (chart)
Cook, James, 98
Coolidge, Calvin, 366
Cooling capacity, 198
Cooling degree day, 197
Cooper, Leon N., 2
Copper, 23, 26
Coptic calendar, 476–77
"Copyleft," 592
Coral and coral reef, 329, 330

Cord of wood, 198
Coriolis effect, 6, 128
Coriolis, Gaspard C., 128
Corm, 281
Corning Glass Works, 173
Corrosive materials, 224
Corsages, 285
Corvair, 541–42
Cougar, 353
Counting, 595
Covered bridge, 516–17
Cows, 352
CPR (cardiopulmonary resuscitation), 416–17
CPU (central processing unit), 586
Cracking knuckles, 383–84
Craters, 59, 102
Creationism, 255
Cremation, 382
Creosote, 176
Creutzfeldt-Jakob disease, 426
Crick, Francis, 261
Cricket chirps, 122
Crocodiles, 334
Cross-pollination, 273–74
Crown glass, 171–72
Crown molding, 506
Cruise missile, 502
Crust (Earth's), 86, 86 (chart)
Cryonic suspension, 382
Cryptography, 562, 576
Crystal ball, 501
Crystal garden, 16
Crystal polarimetry, 16
Crystallography, 15–16
Cubic boron nitride, 151
Cubic zirconium, 154–55
Cugnot, Nicolas-Joseph, 534
Cuisenaire, Emile-Georges, 603
Cuisenaire rods, 603, 603 (ill.)
Culin, Curtis, 553
Culin device, 553
Cullinan diamond, 154, 154 (ill.)
Cullinan, Thomas M., 154
Curare, 420–21
Curbside recycling programs, 241
Curie, Jacques, 16
Curie, Marie, 12, 20, 59, 151, 615
Curie, Pierre, 12, 16, 20, 59, 151
Curium, 20
Currents, 87
Curve ball, 3–4
Cyclones, 131–32
Cyclotron, 6
Cytomegalovirus, 428

## D

"Daddy longlegs," 324–25
Daffodils, 297
Daimler, Gottlieb, 534
Dalmatians, 353
Dalton, John, 17
Damadian, Raymond, 446
Damp, 163
Dams, 521–22
"Dance of the bees," 322
Dark nebulae, 36
Darval, Denise Ann, 460
Darwin, Charles, 249, 253, 253–54
Darwin's finches, 249
Data mining, 593–94
Data storage, 584, 584 (chart)
Davis, Jan, 76
Davis, John, 98
Dawn redwood trees, 278–79
Day, time differences among planets, 52, 52 (chart)
Daylight Savings time, 484–85
Days of the week, 481 (chart)
DDT, 226–27
De Forest, Lee, 565
Dead reckoning, 525
Dead Sea, 92
Deadweight tonnage, 527
Death Valley, California, 87, 120
December, 482
Decibel, 10, 10 (chart)
"Deep Blue" chess computer, 582
Deforestation, 211
De-icing roads, 170
Delamain, Richard, 604
Delamare-Deboutteville, Edouard, 535
DeMaestral, George, 180
Denali National Park, 86
Dental X-ray radiation, 413
Dentistry, 444, 444 (chart)
Dermabrasion, 399
Dermis, 397
Descartes, René, 605
Deserts, 100–101, 101 (chart)
Designer drugs, 456
Deuterium, 23
Deuterium oxide, 25
Deville, Sainte-Claire, 156
DeVries, William, 461
Dew point, 138–39
Dexter, Thomas, 160
Diabetes, 430
Diaconis, Persi, 611
Diagnostic equipment and tests. *See also* Surgery and non-drug treatments
  blood pressure, 445
  cholesterol test ranges, 446 (chart)
  hearing aids, 447
  heart-monitoring machine, 445
  nuclear magnetic resonance imaging, 446–47
  "NYD" ("not yet diagnosed"), 445
  pacemaker, 445–46
  reflexes, 447
Diamonds
  Cape May, 152
  carbon, 25
  and cubic zirconium, 155
  four "C"s, 153
  identification of genuine, 152–53
  mines, 152
  value of, 153
  weighing of, 153–54
  world's largest, 154
Diaphorase, 429
Diatomite, 162–63
Diatoms, 269
Dietary supplement, 452–53
Dietary Supplement Health and Education Act (1994), 452
Digestion, 374, 375 (ill.), 376
Digges, Leonard, 474
Digital, 573
Digital audio tape, 574
Dimples on golf balls, 3
Dinosaurs, 216–19
Dirac, Paul, 28
"Dirty bomb," 503
Diseases. *See also* Health conditions
  AIDS, 423–25, 424 (chart)
  Alzheimer's disease, 437–38
  Amyotrophic lateral sclerosis (ALS), 431
  anorexia, 436
  arteriosclerosis, 433, 433 (ill.)
  "Blue People," 429
  cancer, 434
  "Christmas factor," 432 (ill.), 432–33
  deadliest, 423
  death of Napoleon, 440
  diabetes, 430

factor VIII, 432
"flesh-eating bacteria," 428
Hansen's disease, 427
heart attack, 433
herpes viruses, 427–28
HIV, 424–25
iatrogenic illness, 437
kidney stones, 438–39
Legionnaire's disease, 427
leprosy, 427
Lou Gehrig's disease, 431
Lyme disease, 426
"mad cow disease," 426
malaria, 425
Marfan's syndrome, 430
mosquitoes, 425
most common, 423
necrotizing fasciitis, 428
Panama Canal, 426
pelvic inflammatory disease (PID), 438
progeria, 437
retrovirus, 423
Typhoid Mary, 427
vectors, 425
virus, 423
West Nile Virus, 426
yellow fever, 425
zoonosis, 425
Dishwasher, 243
Disk operating system (DOS), 590
Displacement tonnage, 527
DNA, 255, 256, 261, 262, 369–70
DNA fingerprinting, 369–70
"Do not fold, spindle, or mutilate," 580
Dobrovolsky, Georgi, 78
Dodgson, Charles, 606, 613
Dodo, 219
"Dog days," 122
Dogs. *See also* Animals
basenji, 360
bones, 358 (chart)
Canary Islands, 341
classifications of, 358 (chart)
computation of age, 361
computerized method of tagging, 361
dalmatians, 353
easiest to train, 359
favorite food odors, 360
favorite names, 359
hearing, 359
howling at sirens, 359–60
kept alive by insulin, 361
memories of, 361
most dangerous, 358–59 (chart)
oldest breed, 356–57
pug dog, 360–61
rarest breed, 361
shar-pei, 360
shedless, 360
voiceless, 360
wrinkled, 360
and young children, 357
Dogwood trees, 300
Dolby noise reduction system, 574
Dolby, R. M., 574
Dolly (sheep), 259–60, 260 (ill.)
Dolphin-safe tuna, 223–24
Dolphins, 223–24, 331, 349
Donating blood, 395
Donkey engine, 490
Doomsday clock, 488
Doorjamb, 505
Doppler, Christian, 9, 543–44
Doppler effect, 9–10, 10 (ill.)
Doppler radar, 144
DOS (disk operating system), 590
Dots and dashes, 560
Double dormant plants, 292
Double-blind study, 450
Double-digging, 292
Dox, Paul, 115
Drake, Edwin L., 185
Drake, Frank, 68, 70
Drake well, 185
Draper, John William, 27
Dreams, 373
Drew, Richard C., 442
Drug tests, 457–59
Drugs and medicines. *See also* First aid
abortion pill, 456
anabolic steroids, 454–55
birth defects caused by Thalidomide, 456
Brompton's cocktail, 456
chemotherapy, 454
China White, 456
cocaine, 457, 458 (ill.)
controlled substance, 456, 457 (chart)
derived from plants, 450–51
designer drugs, 456
dietary supplement, 452–53
Dietary Supplement Health and Education Act (1994), 452
double-blind study, 450
drug tests, 457–59
drugs, most frequently prescribed, 449 (chart)
herbal medicine, 451, 452 (chart)
marijuana, 459
monoclonal antibodies, 454
natural drugs, 447–48
new drugs, development of, 450
orphan drugs, 450
patient-controlled analgesia, 455–56
penicillin, 453
pharmacognosy, 447–48
polio vaccine, 454
prescriptions, abbreviations, 448 (chart)
prescriptions, measurements, 449 (chart)
prescriptions, shelf life, 448 (chart)
rain forest plant extracts, 451
RU-486, 456
Rx, 448
streptomycin, 453–54
taxol, 451
tetanus shot, 453
tumors, 455 (ill.)
Dry ice, 168–69
Dry measures, 470–71 (charts)
Dubus-Bonnel, 173
Duck-billed platypus, 348
Duhamel, J. P. F., 5
Duke, Charles M., Jr. 72
Duryea, Charles, 535
Duryea, Frank, 535
Duryea Motor Wagon Company, 535–36
Dust mites, 435
Dutrochet, Henri, 245
Dwarf conifers, 299
Dynamic time, 474
Dynamite, 178–79
Dyslexia, 435–36, 436 (ill.)


## E


Eads, James B., 519
Earhart, Amelia, 548
Ears
balance, 404–5
human, 404 (ill.)
protruding, 381
range of sound frequency, 405

three bones, 405
Earth. *See also* Air; Earth observation and measurement; Earthquakes; Land; Moon; Planets; Volcanoes; Water
center of, 85
changes in distance to sun, 53
circular movement of, 53–54, 54 (ill.)
circumference of, 53
composition of Earth's atmosphere, 81
crust, 86, 86 (chart)
distance from planets, 55
as a giant magnet, 4
highest elevation, 86–87
highest point, 85–86
interior of, 84–85, 85 (ill.)
land vs. water coverage, 87
layers of atmosphere, 81–83
lowest elevation, 86–87
lowest point, 86
mass of, 83
precision of the equinoxes, 53–54, 54 (ill.)
rotation of, 52
sinkholes, 85
temperature, 85
variant rotation speed of, 52–53
Earth Day, 216
Earth observation and measurement
compass, at North Pole, 113
Foucault pendulum, 113
founder of American geology, 115
Gaia Hypothesis, 113
geologic timeframes, 114, 114 (chart)
Global Positioning System (GPS), 117
Gulf stream mapping, 116
Landsat maps, 116–17
magnetic declination, 113
Mercator's projection, 115
Piri Re'is map, 114
prime meridian, 115
relief maps, 115
satellite photographs, 116
Earth Resources Technology Satellite (ERTS), 116
Earth station, 568–69
Earthquakes
Chinese earthquake detector, 110
faults, 108–9, 109 (ill.)
modified Mercalli Scale, 110–12
most severe, 112
Richter scale, 110, 110 (chart)
San Andreas Fault, 109
San Francisco, 112–13
seismograph, 109–10, 111 (ill.)
tsunami, 108
Earthworm, 326
Easter, 483, 483 (chart)
Easter Island, 107
Eastern native cat, 344
Eating and runny noses, 397
Echidna, 348
Eckert, John Presper, Jr., 580
Eclipses, 45, 46 (ill.), 47, 47 (chart), 58, 616
Ecliptic, 44
Ecology and resources
Antarctic ozone hole, 208
biodiversity, 205
biome, 205, 206 (ill.)
deforestation, 211
Earth Day, 216
El Niño, 210
Environmental Protection Agency, 215
eutrophication, 207, 207 (ill.)
fire prevention, 212–13
food chain, 206
food web, 206
forest fires, 212
"green product," 215–16
greenhouse effect, 208–9, 209 (ill.)
greenhouse gases, 209–10, 210 (chart)
Hawk Mountain Sanctuary, 214
killer algae, 207
limnology, 205–6
modern conservation, founder of , 214
national parks, 213–14, 214 (charts)
ozone, 208, 209
protected areas, 212, 212 (chart)
red tide, 211
Smokey the Bear, 212–13, 213 (ill.)
"Spaceship Earth," 215
species extinction in tropical rain forests, 210
tropical rain forests, 210–12
wetlands, 208
Ecorche, 385, 385 (ill.)
Ectomorph, 380
Eddington, Arthur, 13
Edwards, Robert, 462
Eels, 330–31
Egg-laying mammals, 348–49
Egyptian calendar, 476
Ehrlich, Paul, 454
Eiffel Tower, 521 (chart)
Einstein, Albert, 8, 13, 27 (ill.), 27–28, 616
El Niño, 119, 210
Eldredge, Niles, 249
Electric automobile, 535
Electric current unit, 31
Electric eel, 330–31
Electric power, 32
Electrical charge, 6–7, 31–32
Electricity, 23
Electrocardiograph, 445
Electrocution of birds, 343
Electromagnetic fields, 411
Electron, 13
Electron microscope, 492
Electronic numerical integrator and computer (ENIAC), 580
*Elements of Euclid,* 601
Elephants, 223, 346
Elevation, highest and lowest, 86–87
Elizabeth I (queen), 468
Ellis, Charles, 519
E-mail, 591
Embalming fluids, 26
Embryology, 262
Emerald, 155
Emission nebulae, 36
Empedocles of Agrigentum, 392
Emperor penguin, 340–41
Empire State Building, 126, 509
Endangered Species Act (1973), 220
Endocrine glands, 390–91
Endomorph, 380
Endorphins, 372
Energy consumption and conservation. *See also* Recycling, conservation, and waste
air conditioner, 200
aircraft fuel consumption, 204

appliance energy, 200–202, 201 (chart)
automobile energy economics, 203
Cargo Quilt, 200
compact fluorescent light bulbs, 202
energy efficiency, 199–200
Energy Star program, 199–200
fluorescent light, 202
furnace, 200
gas lighting, 202
gas mileage, 202–3
incandescent light bulbs, 202
insulation, 200
light bulbs, 202
recycling aluminum, 203
reserves, 199
transportation energy, 204
underinflated tires, 203–4
U.S. consumption, 199 (chart)
U.S. production, 198
windows, 202
worldwide consumption, 198 (chart)

Energy Department (U.S.), 236

Energy measures and measurement
barrel of oil, gallons in, 195
barrel of oil, weight of, 195
common fuels, weight per gallon of, 195 (chart)
cooling capacity, 198
cooling degree day, 197
cord of wood, 198
energy source comparisons, 195–96 (chart)
heating degree day, 197
heating values of fuels, 196 (chart)
natural gas, heat from, 197
quad of energy, 196–97
utility meters, 197

Energy, motion, force, and heat
absolute zero, 1
adiabatic process, 6
boomerang, 4
curve ball, 3–4
cyclotron, 6
freezing hot water vs. cold water, 1
golf balls, dimples on, 3
inertia, 2–3
kindling point of paper, 6
Leyden jar, 6–7
magnetism, 4–5
Maxwell's demon, 4
phlogiston, 5–6
spontaneous combustion, 5
string theory, 2
superconductivity, 2
water, rotation in drain, 6

Energy Star program, 199–200
EnergyGuide labels, 201–2
Engines, 489–90, 491 (ill.)
Englehart, Douglas C., 585
English channel, 515
ENIAC (electronic numerical integrator and computer), 580
Enigma and Purple machines, 561–62
*Enola Gay,* 555
Environmental Protection Agency (U.S.), 215, 235, 236–37
Enzymes, 256
Epicondylitis, 431
Epidermis, 397
Epstein-Barr virus, 428
Equinoxes, 53–54
Ergonomics, 370
Error theories, 405
ERTS (Earth Resources Technology Satellite), 116
Espaliering, 298
Essential oils, 166–67
"Etaoin Shrdlu," 562
Ethanol, 189, 534
Euclid, 601
Eukaryotic cells, 246, 246 (chart), 247, 247 (chart)
Euler, Leonhard, 612
European starling, 342
Eutrophication, 207, 207 (ill.)
Evans, Ronald E., 72
Evergreen tree, 513

Evolution and genetics. *See also* Biology; Cells; Fungi, bacteria, and algae; Life processes and structures
Batesian mimicry, 255
Cenozoic Era, 250 (chart)
cloned animal, first, 259
cloned mammal, first, 259–60
cloning, 258–60
Darwin, Charles, 249, 253, 253–54
Darwin's finches, 249
DNA, 261, 262
embryology, founder of, 262
genetic engineering, 256–58
genetics, founder of, 252–53
geologic time divisions, 250–52 (charts)
human cloning, 259
human evolution, 252
Human Genome Project, 258
largest protein from genetic engineering, 257–58
Mendelian inheritance, 253
Mesozoic Era, 250–51 (chart)
*On the Origin of Species,* 253–54
"ontogeny recapitulates phylogeny," 262
organisms with most chromosones, 262
p53 (gene), 261–62
Paleozic Era, 251 (chart)
panspermia, 260
plasmid vs. prion, 255–56
polymerase chain reaction, 256
Precambrian Era, 252 (chart)
punctuated equilibrium, 249–50
Red Queen hypothesis, 255
RNA, 261
Scopes (monkey) trial, 255
stem cells, 260
"survival of the fittest," 254

Ewing, William Maurice, 96
Excelsior, 167
Exhaust, 189
Exosphere, 82
Expert system, 581
*Explorer* (satellite), 77

Extinct and endangered plants and animals
conditions for becoming endangered, 220
dinosaurs, 216–19
dodo, 219
dolphin-safe tuna, 223–24
elephants in Africa, 223
great whales, 222–23 (chart)
Jurassic mammal, 216
list of threatened or endangered species, 221 (chart)
mastodon vs. mammoth, 217
passenger pigeon, 218, 219 (ill.)
recent extinct species, 220 (chart)
recent species removed from endangered list, 221 (chart)
terminology, 219

in tropical rain forest, 210
turtles, 223
Extraterrestrial life, 70
Eyes
bifocal lenses, 402
blinking, 402
color perception, 403
floating specks in eyes, 400–401
human, 401 (ill.)
nearsightedness vs. farsightedness, 402
newborn babies and blue eyes, 403
phosphenes, 402
red dots in photographs, 403
rods and cones, 401–2
"sand" in eyes, 403
tearing around onions, 396
transplants, 462
20/20 vision, 402

## F

Factor VIII, 257–58, 432
Factorials, 605
Fahrenheit, Daniel, 28–29
Fahrenheit-Celsius conversion, 30, 30 (chart)
Fairy ring, 269–70
Fall, 482–83
"Fall back, spring forward," 484
Fall of the Roman Empire, 422
"Falling asleep" (arms and legs) sensation, 394
Fallopius, Gabriel, 462
Fantus, Bernard, 443
Farming. *See* Gardening and farming
Farnsworth, Philo T., 567 (ill.), 568
Farsightedness, 402
"Fat Man," 502
Fatalities in space, 78, 78 (chart)
Fathoms, 526
Faults, 108–9, 109 (ill.)
Fax machines, 571 (ill.), 571–72
February, 481
Feiser, Louis, 496
Feldman, Holly, 402
Female astronauts, 74, 75–76
Fences and deer, 301
Ferebee, Thomas W., 555
Fern, 262
Ferris, George Washington Gale, 523
Ferris Wheel, 523
Fertilizer, 290
Feynman, Richard, 14
Fibonacci, Leonardo, 596
Fibonacci numbers, 596
Fiberglass, 173–74
Fiber-optic cable, 572 (ill.), 572–73
Fifth generation computer, 580
Figure head, 525
Finches, 249
Fingernails, 400
Fingerprints, 31, 398
Fir trees, 278 (chart)
Fire hydrant, 494
Fire prevention, 212–13
Fireflies, 327
"Firehouse dogs," 353
Fireworks, 178
First aid. *See also* Drugs and medicines; Poisons and poisonings
"ABCD" survey, 417
bleeding to death, 417
cardiopulmonary resuscitation (CPR), 416–17
Heimlich maneuver, 417
kit, 418
thunder storm safety rules, 418
First generation computer, 579
"First Measurement of the Left-Right Cross Section Asymmetry ...," 614
Fish, age of, 328–29
Fisher, Carl G., 514
Fitzgerald, William, 463
Five-in-one tree, 298
Flail, 496
Flamsteed, John, 64
Flat-panel display televisions, 569
Fleas, 326–27
Fleming, Alexander, 453, 453 (ill.)
Flemming, Walther, 245
"Flesh-eating bacteria," 428
Flightless birds, 338
Float glass, 173
Floating bridge, 518
Floppy disk, 584–85
Floral clock, 487
Florey, Howard, 453
Flower arrangement, 295
Flowers. *See also* Gardening and farming; Plants; Trees
attracting hummingbirds, 296
botany, founder of, 280
bulb, 280–81
colors, most common, 285 (chart)
corm, 281
daffodils, 297
geraniums, 297
Ikebana, 295
"imperfect," 280
national, 285, 286 (chart)
orchid, 285
parts of, 280
passionflower, 286–87
poinsettias, 283
rhizome, 281
roses, 285 (chart)
Shakespeare garden, 295
symbolic of each month, 284 (chart)
tuber, 281
tuberous root, 281
water-lily Victoria amazonica, 289
wildflowers, 287
wildflowers, Native Americans and red dye, 287
wormwood, 287
Flowers, T. H., 579
Flue, 505
Fluorescent light, 8, 202
Fly ash, 163
*Flyer,* 547
Flying mammals, 344–45
Flying Tigers, 554
FM stations, 566–67
Food chain, 206
Food poisoning, 498
Food web, 206
Fool's gold, 153
Force. *See* Energy, motion, force, and heat
Ford, Henry, 535, 536, 536 (ill.)
Ford Motor Company, 544
Forel, F. A., 206
Forensic science and hair, 400
Forest death, 229
Forest fires, 212
Formaldehyde, 26, 232–33
Formatted disk, 585
Fortrel EcoSpun, 242
Fossett, Steve, 549
Fossil fuels, 183–84
Fossils, 148 (ill.), 148–49, 149 (ill.)
Foucault, Jean, 113
Foucault pendulum, 113
Four-horned antelope, 350
Four-stroke engine, 489–90

Fourth generation computer, 579–80
Fourth state of matter, 12
Fractals, 610–11
Frankincense, 167
Franklin, Benjamin, 116, 341, 402, 614
Freckles, 398
Friday, 481
Fried, Henry, 485
Frog cloning, 259
Frogs, 145, 333 (ill.)
Frowning, 386
Fruit fly, 327
Fruit trees, 301
Fuel cell, 190
Fujita and Pearson Tornado Scale, 132, 132 (chart)
Fujita, T. Theodore, 132
Fulgurites, 126–27
Full moons, 57, 57 (chart), 408
Fuller, R. Buckminster, 170, 215, 511, 512 (ill.)
Fullerenes, 171
Fuller's earth, 163
Fungi, bacteria, and algae. *See also* Biology; Cells; Evolution and genetics; Life processes and structures
  anthrax, 269
  bacteria, reproduction of, 269
  bacteria, visible, 269
  bacteriology, founder of, 271
  diatoms, 269
  fairy ring, 269–70
  fungi kingdom of biology, 267
  Koch's postulates, 271
  lichens, 270
  mushrooms, 289
  mycology, 270
  virus, 270–71
Fungicide, 301–2
Funicular railway, 533
Funny bone, 384
Furnace, 200
Fuzzy search, 590


## G


Gabor, Dennis, 492
Gadolinium, 23
Gagarin, Yuri, 70, 71 (ill.), 77
Gage, Andres A., 445
Gaia Hypothesis, 113
Galaxies, 36, 39
Galena, 151
Galileo, 28, 51, 56, 66
*Galileo* spacecraft, 77
Gallium, 22
Gallo, Robert, 423
Galton, Francis, 398
Galveston, Texas, 138
Gamma rays, 12
Gandy dancer, 532
Garbage, 238
Garbage barge, 238
"Garbage in, garbage out" (GIGO), 590
Garbage incinerator, 238
Gardening and farming. *See also* Flowers; Plants; Trees
  acidity vs. alkalinity, 290
  Appleseed, Johnny, 298 (ill.), 299
  arboretum, 294
  balled-and-burlapped plants, 293–94
  bare-rooted plants, 294
  best time to work the soil, 291
  bonsai trees, 299–300
  botanical garden, 294–95
  Carver, George Washington, 302, 302 (ill.)
  cats and catnip, 301
  chia seeds, 300
  chilling requirement for fruit trees, 297–98
  clay soil, 289
  companion planting, 294, 294 (chart)
  container-grown plants, 293, 294
  daffodils, 297
  dogwood trees, 300
  double dormant plants, 292
  double-digging, 292
  dwarf conifers, 299
  espaliering fruit trees, 298
  fence and deer, 301
  fertilizer, 290
  five-in-one tree, 298
  fruit trees and mice, 301
  fungicide, 301–2
  geraniums, 297
  greenhouse, first, 302
  hardening off seedlings, 293
  hydroponics, 292–93
  Ikebana, 295
  insects and strawberries, 301
  lawn clippings, 297
  loam soil, 290
  lunar gardening, 291
  method of early growth, 293–94
  pH, 290
  plant patent, first, 303
  pleaching, 299
  poison ivy, 300
  poison oak, 300
  poison sumac, 300
  potting soil, 291
  railroad worm, 300–301
  rain shadow, 292
  sandy soil, 290
  seedless watermelon, 299
  seeds, 293, 293 (chart)
  Shakespeare garden, 295
  snowmold, 297
  soil, 289–91
  squirrels, 301
  synthetic soil, 291
  tomato plants, 296
  vegetable garden, 296
  victory garden, 295
  weeding, 296
  xeriscaping, 292
Gardner, Richard, 295
Garnerin, Jacques, 549
Gas, 12
Gas laws, 25
Gas lighting, 202
Gasohol, 188–89, 534
Gasoline, 186–88, 188 (chart)
Gasoline mileage, 202–3
Gasoline stations, 187–88
Gateway Arch, 519–20, 520 (charts)
Gatlefosse, Rene Maurice, 463
Gatling, Richard J., 499
Gatty, Harold, 549
Geese, 338 (ill.), 338–39
Gell-Mann, Murray, 13 (ill.), 13–14
Gemstones, 154, 180
Gene cloning, 256
Gene splicing, 256
General Motors, 541–42, 544
Genesis rock, 59
Genetic engineering, 256–58
Genetic fingerprinting, 369–70
Genetics. *See* Evolution and genetics
Geodesic dome, 511, 512 (ill.)
Geologic timeframes, 114, 114 (chart), 250–52 (charts)
Geology, 115
Geothermal energy, 181

Geotropism, 274
Geraniums, 297
Gericke, William, 292
German silver, 160
Ghawar field, 185
Giant tube worms, 330
GIGO ("garbage in, garbage out"), 590
Gilbert, William, 4, 5 (ill.)
Gilbreth, Frank Bunker, 615–16
GIMPS (Great Internet Mersenne Prime Search), 597
Giraffe, 351
Glaciers, 97
Glare, 7
Glass, 171–73
Glass blocks, 172–73
Glass in movie stunts, 173
Glazing, 202
Glenn, John H., Jr., 71
Glitch, 589
Global Positioning System (GPS), 117
Global warming, 119–20
Glow-in-the-dark chemical, 177
Gluteus maximus, 385
Glycerine, as leaf preservative, 279
Gnotobiotics, 266
Goddard, Robert H., 68
Gold, 21, 23, 87, 158–59
Gold leaf, 158, 159 (ill.)
Golden Gate Bridge, 519
Golden section, 610
Golf balls, 3
Golf on the moon, 75
Gondwanaland, 96, 96 (ill.)
"Good cholesterol," 415
Googol, 599
Goose-bumps, 398
Gopher plant, 183, 184 (ill.)
Gordon, John B., 259
Gordon, Richard F., Jr., 72
Gore, Al, 576
Gorgas, William C., 426
Gorillas, 310
Gould, Stephen J., 249
Gram atomic weight, 32
Gram formula weight, 32
Grand Canyon, 104
"Grand tour" of the planets, 71
Grandfather clock, 487
Granite Railway, 531
Grant, Ulysses S., 213
Graphite, 25
Gravitational force, relative to Earth, 51, 51 (chart)
Gravity, 68
"Gravity assist," 71
Gray (radiation measurement), 413
Gray wolf, 353–54
Great Falls, Montana, 130
Great Ice Age, 98–99
Great Internet Mersenne Prime Search (GIMPS), 597
Great Lakes, 93–94
Great whales, 222–23 (chart), 349 (chart)
Greatbatch, Wilson, 445
Green flash phenomenon, 123
"Green product," 215–16
Greenhouse effect, 208–9, 209 (ill.)
Greenhouse gases, 209–10, 210 (chart)
Greenhouses, 302
"Greening of the galaxy," 69
Greenwich Mean Time, 485
Gregoire, Marc, 180
Gregorian calendar, 477
Gregory XII (pope), 475, 477, 480
Greiner, Friedrich, 546
Grissom, Virgil "Gus," 78
Grocery bags, 243
Grocery stores, 593–94
Gross tonnage, 527
Groundhog Day, 144–45
Grove, William, 190
Groves, Leslie R., 501
"Guardian Angel of the Genome," 261–62
Gulf stream mapping, 116
Gunpowder, 178
Gunter, Edmund, 604
Gurevich, Mikhail I., 554
Gutenberg, Beno, 84
Gutenberg discontinuity, 84
Gutta percha, 167
Gypsy moth, 318
Gypsy moth caterpillars, 318–19

## H

Haber, Fritz, 169
Haboobs, 127
Hacker, 591
*Hadrocodium wui,* 216
Hahneman, Christian F. S., 443
Hailstones, 141, 141 (ill.)
Hair. *See* Skin, hair, and nails
Halcyon days, 129–30
Hale, Gerald A., 545
Half-life, 15
Hall, Charles Martin, 156
Halley, Edmund, 61–62, 62 (ill.)
Halley's comet, 61–62
Hallidie, Andrew S., 533
Halo, 145
Hamstring muscles, 385
Hand motions, 615–16
Hansen's disease, 427
Hard disk, 584
Hardening off seedlings, 293
Hares, 352
Harkin's rule, 21
Harrison, William Henry, 366
Harvestmen, 324–25
Haüy, René-Just, 15–16
Haüy, Valentin, 559
Haven, C. D., 173
Hawk Mountain Sanctuary, 214
Hawking, Stephen, 35, 35 (ill.)
Hayes, Denis, 216
Haynes, Elwood, 161
Hazardous waste, 224–25
Hazardous waste sites, 236
Head injuries on bicycles, 410
Health and medicine. *See* Diagnostic equipment and tests; Diseases; Drugs and medicines; First aid; Health conditions; Health hazards; Health industry; Poisons and poisonings; Surgery and non-drug treatments
Health conditions. *See also* Diseases
  bends, 437
  burns, 439 (chart)
  carpal tunnel syndrome, 430–31
  dust mites, 435
  dyslexia, 435–36, 436 (ill.)
  heat exhaustion, 434 (chart)
  heat stroke, 434 (chart)
  HeLa cells, 435
  jet lag, 431–32
  lactose intolerance, 429–30
  narcolepsy, 431
  phobias, 439, 439–40 (chart)
  pica, 438
  poison ivy, 435
  stomach ulcer, 430
  sulfites, 434
  tennis elbow, 431
  warts, 429
Health hazards
  accidents, men vs. women, 409

asbestos, 411–12
automobile crashes, 410
blood alcohol level, 415, 415 (chart)
"boomeritis," 410
carcinogens, 413, 414 (chart)
cholesterol, 414–15
cigarettes, 415–16, 416 (ill.)
dental X-ray radiation, 413
electromagnetic fields, 411
head injuries on bicycles, 410
increase in speed limit, 409
leading causes of death, 409, 409 (chart)
lightning strikes, 408
motorcycle deaths, 408
owning pets, 409
ozone, 412
radiation, 412–13
radon, 412
risk factors, 407
smoking, 415
sports injuries, 409, 410 (chart), 410–11
stress, 407, 408 (chart)
violence and full moon, 408

Health industry
assisted living facility, 444
blood bank, first, 442–42
chiropractic medicine, 443
dentistry, 444, 444 (chart)
father of medicine, 442
Hippocratic Oath, 441
homeopathy, 443
intermediate care facility, 444
medical college, first, 442
modern medicine, founder of, 441–42
naturopathy, 443
nursing homes, number of, 443
ophthalmologists, 444
opticians, 444
optometrists, 444
osteopathic medicine, 443
physicians, number of, 443
skilled nursing facility, 443–44
symbol representing medicine, 440–41
woman physician, first, 442

Hearing aids, 447
Hearing in animals, 311
Heart, 393 (ill.)
Heart attack, 433
Heart transplants, 460
Heartbeat, 387
Heart-monitoring machine, 445
Heat. *See* Energy, motion, force, and heat
Heat exhaustion, 434 (chart)
Heat index, 121, 121 (chart)
Heat stroke, 434 (chart)
Heat wave, 120–21
Heating degree day, 197
Heating values of fuels, 196 (chart)
Heaviest humans, 379–80
Heavy water, 25–26
Heimlich, Henry J., 417
Heimlich maneuver, 417
Heineken, Albert, 244
Heisenberg, Werner Karl, 28
HeLa cells, 435
Helium, 22, 22–23, 24
Hell's Canyon, 104
Hemlock, 275
Hemophilia B, 432
Henry I (king), 467
Henry, Edward, 398
Henry, Joseph, 560
Herbal medicine, 451, 452 (chart)
Herbicides, 232
Hermit crab, 365
Hero of Alexandria, 28, 608
Heroult, Paul, 156
Herpes simplex type 1 virus, 427–28
Herpes simplex type 2 virus, 428
Herpes viruses, 427–28
Hess, Harry Hammond, 96
Hetrick, John W., 544
Hibernation, 309
Hiccups, 390
High clouds, 124
High definition television (HDTV), 569
High density lipoproteins, 446 (chart)
High Mobility Multipurpose Wheeled Vehicle (HMMVE), 553
High speed steel, 160
"High tech," 615
Highway barriers, 515
Higinbotham, William, 581
Hillier, James, 492
*Hindenburg,* 546, 547 (ill.)
Hindu calendars, 477
Hipparchus, 64
Hippocrates, 441, 442
Hippocratic Oath, 441
Hiroshima, Japan, 555
Hisinger, Wilhelm, 17
HIV, 424–25
HMMVE (High Mobility Multipurpose Wheeled Vehicle), 553
Hogs, 352
Hollerith, Herman, 579, 580
Holmes, Thomas H., 407
Holography, 492–93
Holter, J. J., 445
Holter monitor, 445
Homeopathy, 443
Homeostasis, 377
Homing pigeons, 342
Honeybees, 321–22, 322
Hoodoo, 100
Hooke, Robert, 245, 492
Hooker, R. D., 417
Hoover Dam, 521–22
Hopper, Grace Murray, 586, 587, 589
Hopper's rule, 585
Hoppe-Seyler, F., 264
Horizon, 473
Hornbook, 559
Horse latitudes, 129
Horsepower, 533
Horses, 351
Horus's eye, 448
Hospital ship, 528
Houston, Texas, 277
Hovercraft, 547
Howard, Luke, 123
Howard, Taylor, 569
HTML (Hypertext Markup Language), 593
Hubble, Edwin, 33, 67
Hubble telescope, 67–68
Hubble's Constant, 34
Hubble's Law, 67
Hughes, Howard, 550–51
Human bite, 386
Human body. *See also* Blood and body fluids; Bones, muscles, and nerves; Ears; Eyes; Organs and glands; Skin, hair, and nails; Taste
B lymphocytes, 371–72
body typing, 380, 380 (ill.)
calories, 373, 373–74 (chart)
cells, 369, 369 (chart)
centenarians, 382
chemicals in, 367 (chart)
chromosomes, 367–69, 368 (ill.)
cremation, 382

cryonic suspension, 382
digestion, 374, 375 (ill.), 376
DNA fingerprinting, 369–70
dreams, 373
endorphins, 372
ergonomics, 370
heat lost through head, 377
heaviest humans, 379–80
height, 377
homeostasis, 377
immune system, 370, 371 (ill.)
intestines, 376
left-right preferences, 382
oldest person, 381
physiology, founder of, 376–77
protruding ears, 381
REM sleep, 372–73
Siamese twins, 381, 381 (ill.)
sleep needs, 372, 372 (chart)
snoring, 373
swallowing, 376
T cells, 371–72
weight, 377, 378–79 (charts), 379–80
Human cloning, 259
Human evolution, 252
Human Genome Project, 258
Human immunodeficiency virus (HIV), 424–25
Humidity, 122
"Hummers," 553
Hummingbirds, 296, 339 (chart), 339–40, 340 (ill.)
Hummve, 553
Humors of the body, 392
Hunter's moon vs. harvest moon, 58
Hurricane Andrew, 138
Hurricanes, 131–37
Hutchins, Levi, 488
Hutchinson-Gilford syndrome, 437
Hutchison, Miller Reese, 447
Huxley, Thomas Henry, 253
Huygens, Christiaan, 49, 51, 487
Hyatt, John Wesley, 179
Hydramatic automatic transmission, 538
Hydrocarbon cracking, 186
Hydroelectric power plant, 190
Hydrogen, 22, 23, 24
Hydrogen isotopes, 24 (ill.)
Hydrogen peroxide, 170
Hydroponics, 292–93
Hydrotropism, 274
Hynek, J. Allen, 69
Hyoid bone, 383
Hypertext Markup Language (HTML), 593
Hypothalamus, 391


## I


Iatrogenic illness, 437
Ice, 88, 92, 92 (chart), 96–97
Ice Age, 98–99
Ice caps, 97
Ice sheets, 96–97
Icebergs, 88
Icebreakers, 530
Identical twins, 398
Igneous rocks, 147
Ignitable materials, 224
Ikebana, 295
Imaginary numbers, 599
Immune system, 370, 371 (ill.)
"Imperfect" flowers, 280
Incandescent light bulbs, 202
Incus, 405
Indian dollars, 150
Indian elephants, 351–52, 352 (chart)
Indian summer, 122
Indoor air pollution, 233, 233–34 (chart)
Inertia, 2–3
Inferior planets, 51–52
Infinity, 601
Information highway, 576
Ingham, Eunice D., 463
Inorganic chemistry, 27
Insects. *See also* Animals
amber, 317
ant weight, 324
ants vs. termites, 324, 324 (chart)
bee colony, 323
bees, 321–23
beneficial, 319–20
butterflies, 319 (ill.), 321 (chart)
butterfly gardens, 320–21
centipede, 326
cockroaches, 326
"daddy longlegs," 324–25
"dance of the bees," 322
earthworm, 326
fireflies, 327
fleas, 326–27
fruit fly, 327
gypsy moth, 318
gypsy moth caterpillars, 318–19
honey production, 323
"killer bees," 321–22
lack of national insect, 321
locust, 318, 319 (ill.)
metamorphosis, 319 (ill.), 320
Mexican jumping bean, 327
migratory beekeepers, 322–23
mosquitoes, 324
most destructive, 317
moths, 321 (chart)
number of species, 317
sense of smell, 324
spider eggs, 325
spider web, 325, 325 (ill.)
state insects, 321
and strawberries, 301
termites, 323
Institute of Medicine, 614
Insulation, 200, 506
Insulin, 257
Intelligent life on other planets, 68–69
Interest, simple vs. compound, 611
Intermediate care facility, 444
Internal combustion engines, 490, 491 (ill.)
International Astronomical Union (IAU), 65
International date line, 485
International Morse Code, 560–61
International Standard Book Number (ISBN) code, 564
Internet. *See also* Computers
cryptography, 576
definition of, 576
information highway, 576
Netplex, 577
patron saint of, 577
search engine, 593
users of, 577
World Wide Web, 593
Interstate highways, 514
Intestines, 376
In-vitro fertilization, 462
Ionosphere, 82
Iridium, 24–25
Iridology, 463
Iron pyrite, 153
Ironwood, 165–66
Ironworks, 160
Irrational numbers, 599
Irwin, James B., 72

ISBN (International Standard Book Number) code, 564
Isinglass, 168
Isomers, 25
Isotopes, 23, 24 (ill.)
Ivy, 288

## J

Jackhammer, 494
Jacquard, Joseph Marie, 579
Jacquel-Droz, Henri-Louis, 493
Jacquel-Droz, Pierre, 493
*Jahre Viking,* 529
Janssen, Hans, 492
Janssen, Zacharias, 66, 492
January, 476, 481
Japanese calendar, 477
Jarvik, Robert, 461, 461 (ill.)
Jarvis, Gregory, 78
Jefferson, Thomas, 105
Jeffreys, Alec, 369
Jemison, Mae Carol, 75, 76
Jenner, Edward, 270
Jenney, William Le Baron, 508
Jersey barrier, 515
Jewish calendar, 476
Jet lag, 431–32
Jet stream, 128–29, 129 (ill.)
Jimson weed, 288
Johnson, Andrew, 366
Johnson, Robert Underwood, 214
Joliot-Curie, Frederic, 59
Jones, Meredith, 330
Jones, William, 599
Joshua tree, 277
Jovian planets, 52
Julian calendar, 477
Julian day calendar, 477
Julian day count, 478
Julius Caesar, 476, 477, 480
July, 482
June, 482
Jupiter (planet), 48, 48–49, 51–52, 52
*Jupiter* (space flight), 73
Jurassic mammal, 216

## K

Ka Lae, Hawaii, 98
Kaiser, Henry J., 528, 551
Kanada, Yasumasa, 599
Kantrowitz, Adrian, 460
Kapitsa, Leonidovich, 615
Karats of gold, 158, 158 (chart)
Kasner, Edward, 599
Kasparov, Garry, 582
Kasperak, Mike, 460
Katydid chirps, 122
KDKA radio station (Pittsburgh, PA), 566
Kelvin, Lord, 260
Kelvin temperature scale, 29–30, 30 (chart)
Kennedy, John F., 70
Keratinocytes, 400
Kerr black hole, 38
Kerr-Newman black hole, 38
Kevlar, 179
Key West, Florida, 98
Kharizmi, Muhammad ibn Musa al, 578
Kidney stones, 438–39
Kilby, Jack, 582
Killer algae, 207
"Killer bees," 321–22
Kindling point of paper, 6
Kingdoms of, 266–67
Kings Canyon, 104
"Kissing bridge," 516–17
Kludge, 592
Knight, Nancy, 142
Knoll, Max, 492
Knoop hardness scale, 25
Knuckles, 383–84
Koala bears, 310
Koch, Robert, 271
Koch's postulates, 271
Komarov, Vladimir, 78
Königsberg Bridge Problem, 612, 612 (ill.)
Korolev, Sergei P., 77
Kouwenhoven, William R., 417
Kowal, Charles, 60
Krupp, Alfred, 500
Krupp, Bertha, 501
Krupp, Friedrich A., 501
Krypton, 22–23, 177
Kwolek, Stephanie, 179

## L

La Niña, 119
LaBrea Tar Pits, 105
Lacks, Henriette, 435
Lactose intolerance, 429–30
Lafayette, Marquis de, 365
Laika, 73
Lake Baikal, 93
Lake Huron, 93–94
Lake Michigan, 93–94
Lake Pontchartrain Causeway, 519
Lake Tanganyika, 93
Lakes, 93 (chart)
Lamarck, Jean, 123, 265, 265 (ill.)
Land
- Antarctica, 98, 99 (ill.)
- avalanches, 100
- caves, 102, 103
- continental divide, 104
- continents, 95–96, 96 (ill.)
- coverage on Earth, 87
- craters, 102
- deserts, 100–101, 101 (chart)
- glaciers, 97
- Grand Canyon, 104
- hoodoo, 100
- ice, 96–97
- Ice Age, 98–99
- LaBrea Tar Pits, 105
- moraine, 99
- Mount Rushmore National Monument, 105
- national parks, 104
- northernmost point, 97–98
- permanently frozen, 97
- quicksand, 101–2
- Rock of Gibralter, 105
- sand dunes, 100, 101 (ill.)
- southernmost point, 97–98
- speleothem, 102
- spelunking vs. speleology, 102
- stalactite vs. stalagmite, 103 (ill.), 103–4
- tides, 95
- tufa, 103

Landfills, 239
Landsat maps, 116–17
Landsteiner, Karl, 394, 395
Lanes, 514
Langworthy, O. R., 416
Laparoscopy, 459
Large intestine, 376
Large numbers, 598 (chart)
Laser speed guns, 544
Laser surgery, 398–99
Laurasia, 96, 96 (ill.)
Lauterbar, Paul, 446
Lava domes, 106
Lavoisier, Antoine Laurent, 6, 17
Law of constant extinction, 255
Law of thermodynamics, 4
Law of very large numbers, 611–12
Lawn clippings, 297
Lawrence, Ernest, 6
Laws of motion, 3
Lawson, Andrew, 109

Lazarus, Emma, 522
Lead, 26, 421–22
Lead poisoning, 421–22
Lead-acid batteries, 23
Leaded vs. lead-free gasoline, 186
Leaning Tower of Pisa, 509
Leap second, 480
Leap year, 480
Leaves, brightness of, 276–77
Lebon, Philippe, 202
Lechuguilla Cave, 103
Lee, Mark, 76
Left turn on red, 543
Left-right preferences, 382
Legionnaire's disease, 427
Leibniz, Gottfried Wilhelm, 578, 601, 601 (ill.)
Leith, Emmet, 492
Lemoine, Louis, 494
Lenoire, J. J. Etienne, 534–35
Leonardo da Vinci, 276
Leonov, Alexei, 74
Leopard seal, 341
Leprosy, 427
Leptons, 14
Levine, Philip, 395
Lewis acid, 26
Lewis, Gilbert Newton, 26
Leyden jar, 6–7
Libbey Glass Company, 173
Libby, Willard F., 265
Liberty ship, 528
License plate, 539
Lichens, 270
Lichtenberg figures, 495
Lichtenberg, George Christopher, 495
Life expectancy, 305–8 (chart)
Life processes and structures. *See also* Biology; Cells; Evolution and genetics; Fungi, bacteria, and algae
  amino acid, 264
  biochemistry, founder of, 264
  biological clock, 263
  biorhythms, 263
  "Spiegelman monster," 263–64
Light, 7–8
  Ångström, Anders, 8
  color difference between sunlight and fluorescent light, 8
  Michelson-Morley experiment, 8
  polarized sunglasses, 7
  primary colors, 7
  spectroscopy, 8
  speed of light, 7
  waves, 8
Light bulbs, 202
Light year, 64
Lightning, 125–26, 140, 408
Lightning bugs, 327
Lignite, 184
Limnology, 205–6
Lincoln, Abraham, 105, 366, 614
Lincoln Highway, 514
Lindbergh, Charles, 548
Linnaeus, Carolus, 30, 266, 266 (ill.), 268
Linotype machines, 562
Linux operating system, 592
Lipoproteins, 414–15
Lippershey, Hans, 65–66, 492
Liquid measures, 470–71 (charts)
Liquids, 12
Lisa (microcomputer), 582
Listing, Johann Benedict, 610
Lithotripsy, 462
"Little Boy," 501–2
Liver, 391
Living stones, 289
Loadline, 527–28
Loam soil, 290
Lobsters, 329
"Local Noon," 484
Lockheed SR-71, 552
Locks, 521
Locust, 318, 319 (ill.)
Lodestone, 162
Logarithms, 602–4
Long ton, 471
Long-nosed echidna, 348
Lost wax process, 495
Lou Gehrig's disease, 431
Lovelock, James, 113
Low clouds, 124
Lucid, Shannon W., 75
Low density lipoproteins, 446 (chart)
Lowry, O. H., 614
Luffa sponges, 288
Lunar eclipses, 58, 59 (ill.)
Lunar gardening, 291
Lungs, 389
Lyme disease, 426

## M

Macadam, 174
MacAdam, John Louden, 174
Mach number, 551
Machine computer language, 585–86
Machine gun, 499 (ill.), 499–500
Machines
  breathalyzer, 495
  "cc" designation in engine sizes, 490
  donkey engine, 490
  fire hydrant, 494
  four-stroke vs. two-stroke engine, 489–90
  internal combustion engines, 489–90, 491 (ill.)
  power take-off, 490
  road roller, 494
  robots, 493
  simple, 489
Maclure, William, 115
"Mad as a hatter," 422
"Mad cow disease," 426
Magee, Carlton C., 545
MAGLEV trains, 530–31
Magnetic declination, 113
Magnetic recording, 574
Magnetism, 4–5
Magnets, 162
Magnitude, 39
Magnus, H. G., 3
Maiden-hair tree, 275
Mail delivery, 558
Mail-order homes, 513
"Main Street of the United States," 514
Malaria, 425
Malayan sun bear, 354
Malleus, 405
Mallon, Mary, 427
Mammals. *See also* Animals
  African elephants, 351–52, 352 (chart)
  armadillos, 355, 355 (ill.)
  bats, 347
  bear living in tropical rain forest, 354
  bison, 354
  "blind as a bat," 347
  breath-holding capability, 346–47 (chart)
  camels, 354
  capybara, 355–56
  chamois, 356
  cloning, 259–60
  Clydesdale horses, 351
  cougar, 353
  cows, 352
  Dalmatians, 353
  desert cats, 352–53

dolphins, 349
egg-layers, 348–49
flying, 344–45
four-horned antelope, 350
gestation period, 344
giraffe, 351
gray wolf, 353–54
great whales, 349 (chart)
group names, 345–46 (chart)
hares, 352
heartbeat, 347 (chart)
hogs, 352
horses, 351
inability to jump, 346
Indian elephants, 351–52, 352 (chart)
largest terrestrial, 354
marine mammals, diving of, 349 (chart)
marsupials, 347–48
pigs, 352
porcupines, 354–55
porpoises, 349
with pouches, 347–48
Przewalski's horse, 350 (ill.), 351
rabbits, 352
skunk odor, 356, 363
squirrels, 356
thoroughbred horses, 351
tree-climbing canine, 353
venomous, 348
West Indian manatee, 350
wombats, 348
Mammoth, 217
Manatees, 350
Mandslay, H., 494
Manganese, 23
Manhattan Project, 501
Manhole covers, 516
MANIAC (mathematical analyzer, numerator, integrator, and computer), 580
Man-made products. *See also* Metals; Minerals; Natural substances; Rocks
ammonia, 169–70
aqua regia, 169
Belgian block, 175
buckminsterfullerene, 170–71
bulletproof glass, 172
carbon black, 176
cement, 174
ceramics, 168
creosote, 176
crown glass, 171–72
de-icing roads, 170
dry ice, 168–69
dynamite, 178–79
fiberglass, 173–74
fireworks, 178
float glass, 173
glass, 171–73
glass blocks, 172–73
glass in movie stunts, 173
glow-in-the-dark chemical, 177
gunpowder, 178
hydrogen peroxide, 170
Kevlar, 179
lightest known solid, 170
macadam, 174
metal objects in microwaves, 179
neatsfoot oil, 176
parchment paper, 177
plastic, 179
salt, 170
sandpaper, 177
slag, 176
solder, 175 (ill.), 175–76
space vehicle material, 180
Styrofoam, 168
sulfuric acid, 169
synthetic gemstone, 180
teflon, 180
tempered glass, 172
thermopane glass, 173
titanium dioxide, 177–78
TNT, 178
Velcro, 180
white pigment, 177–78
Mantle, 84
Maps, 116
March (month), 481
Marconi, Guglielmo, 565, 565 (ill.)
Marcus, Siegfried, 535
Marfan's syndrome, 430
Margulis, Lynn, 113
Mariana Trench, 90
Marijuana, 459
Marine mammals, 349 (chart)
"Mark twain," 526–27
Married couple in space, 76
Mars, 48, 51–52, 52, 54
Marsh, George Perkins, 214
Marshall, Barry, 430
Marshall, James C., 501
Marsupials, 347–48
Mastodon, 217
Mathematical analyzer, numerator, integrator, and computer (MANIAC), 580
Mathematics. *See also* Numbers
abacus, 601–2, 602 (ill.)
acre, 607
area formulas, 607
area of triangle, 608
arithmetic, 600
bridge games, 611
calculator, 602
calculus, 601
"casting out nines," 604
Cuisenaire rods, 603, 603 (ill.)
*Elements of Euclid,* 601
factorials, 605
fractals, 610–11
golden section, 610
infinity, 601
interest, simple vs. compound, 611
Königsberg Bridge Problem, 612, 612 (ill.)
law of very large numbers, 611–12
median vs. mean, 605
Möbius strip, 610, 610 (ill.)
Napier's bones, 602–3
Pascal's triangle, 608
percent of increase, 611
Platonic solids, 609
Pythagorean theorem, 609
rule of 70, 610
shared birthday, probability of, 611
slide rule, 603–4
square root, 605
squaring the circle, 609
"tilling the plane," 609
triple play, probability of, 613
Venn diagrams, 605–6, 606 (ill.)
volume formulas, 606–7
Zeno's paradox, 612–13
Matter, 12–16
atom, theory of, 12
bismuth, density of, 15
colligative properties, 14
crystal garden, 16
crystallography, 15–16
discoverers of electron, proton, and neutron, 13
first organic compound synthesized from inorganic ingredients, 15

fourth state, 12
half-life, 15
nuclear fission vs. nuclear fusion, 13
quantum electrodynamics, 14
quark, 13–14
solar eclipse, 13
subatomic particles, 14
theory of general relativity, 13
water, density of, 15
Mattingly, Thomas K., 72
Mauchly, John William, 580
Maxim, Hiram S., 500
Maxwell, James C., 4
Maxwell's demon, 4
May, Gene, 548
May (month), 482
McAdoo, William, 543
McAuliffe, Christa, 78
McCandless, Bruce, 74, 79
McKinley, William, 85
McNair, Ronald, 78
Mean, 605
Mechanical watches, 487–88
Median, 605
Medical colleges, 442
Medicine. *See* Drugs and medicine
Meitner, Lise, 20
Meitnerium, 20
Melanin, 399–400
Melanocyte, 399–400
Meltdown, 194
Mendel, Gregor, 252 (ill.), 252–53, 253
Mendeleyev, Dmitry Ivanovich, 17–18, 18 (ill.)
Mendelian inheritance, 253
Mercalli, Guiseppe, 111
Mercator, Gerardus, 115
Mercator's projection, 115
Mercury (chemical element), 22, 26, 422
Mercury (planet), 48, 51–52, 52
*Mercury* (space flight), 74
Meridians, 115
Mermaid, 350
Mermaid's purse, 330
Mersenne, Marin, 596
Mersenne Primes, 596–97
Mesomorph, 380
Meson, 14
Mesosphere, 81
Mesozoic Era, 250–51 (chart)
Metal objects in microwaves, 179
Metal recycling, 238
Metals. *See also* Man-made products; Minerals; Natural substances; Rocks
alchemical symbols, 157, 157 (chart)
astrological symbols for planets, 157, 157 (chart)
coltan, 156
German silver, 160
gold leaf, 158, 159 (ill.)
gold-producing countries, 159, 159 (chart)
high speed steel, 160
ironworks, 160
karats of gold, 158, 158 (chart)
most abundant, 156–57
noble, 157
pewter, 160
precious, 157
stainless steel, 160–61
sterling silver, 159
technetium, 161
troy ounce of gold, 158
tuning fork, 161
uranium, 161
white gold, 158
Metamorphic rocks, 147
Metamorphosis, 319 (ill.), 320
Meteor crater, 102
Meteorites
landing on Earth, 63
largest, 63, 63 (chart)
lunar, 64
meteor showers, 62–63, 63 (chart)
vs. meteoroids, 62
meteors, 62
Meter, 467
Methane fuel, 239
Methodology. *See* Measurement and methodology
Methyl tertiary butyl ether (MTBE), 187
Metius, Jacob, 66
Metric conversions, 468–70 (chart)
Metric system, 466, 470, 597 (chart)
Metric ton, 471
Metropolis, Nicholas C., 580
Mexican jumping bean, 327
Mice, 301
Michelin, André, 537
Michelin, Edouard, 537
Michelin tire, 537
Michelson, Albert A., 8
Michelson-Morley experiment, 8
Microburst, 128
Microclimates, 128
Microgravity, 68
Microscopes, 491–92
Microwaves, 179
Middle clouds, 124
MiG, 554–55
Migration, 337–38
Migratory beekeepers, 322–23
Mikoyan, Artem I., 554
Military time, 488
Military vehicles
atomic bomb, first, 555
B-17 Flying Fortress, 554
culin device, 553
*Enola Gay,* 555
Flying Tigers, 554
Hummve, 553
MiG, 554–55
Red Baron, 553
Sopwith Camel, 554
spy plane, smallest, 555
stealth technology, 555
tank, 553
Milky Way galaxy, 36, 39
Mimosa ("sensitive plant"), 279
Mine barrage, 498
Minerals. *See also* Diamonds; Man-made products; Metals; Natural substances; Rocks
color standardization, 151
cubic boron nitride, 151
cubic zirconium, 154–55
emerald, 155
fool's gold, 153
galena, 151
gemstones, common cuts, 154
Mohs scale, 150, 150 (chart)
pitchblende, 151
precious stones, largest, 155
and rocks, 150
star sapphires, 155
stibnite, 152
strategic, 151
tiger's eye, 156
Miner's canary, 185
Minimally invasive surgery, 459
Minnoch, John Brower, 379
*Mir,* 73
Missiles, 502
Mississippi River, 519
Mr. Yuk, 419
Mistletoe, 420
Mitchell, Edgar D., 72

Mitochondria, 246–47
Mitosis, 248, 249 (ill.)
Möbius, August Ferdinand, 610
Möbius strip, 610, 610 (ill.)
*Mobro* (ship), 238
Mode, 605
Modified Mercalli Scale, 110–12
Mohorovicic, Andrija, 84
Mohorovicic discontinuity, 84
Mohs, Friedrich, 150
Mohs scale, 150, 150 (chart)
Moissieff, Leon, 519
Mojave Desert, 101
Molars and premolars, 384
Mole, 32
Mole Day, 32
Molecular biology, 266
Molecular cloning, 256
Molecules, 4, 170–71, 263–64
Mollusk shells, 331–32
Molybdenum, 160
Monarch butterfly, 254 (ill.), 255
Moncky, Charles, 490
Monday, 481
Monera kingdom of biology, 267
Monkey ball tree, 279
Monkey wrench, 490
Monkeys in space, 73–74
Monnartz, P., 161
Monoclonal antibodies, 454
Monosodium glutamate (MSG), 438
Months of the year, 481–82
Montreal Protocol (1987), 228
Monuments, 519–20, 520 (charts), 522–23
Moon. *See also* Planets
  atmosphere, 56
  blue moon, 57–58, 58 (chart)
  craters, 59
  Curie family, craters named after, 59
  diameter and circumference, 56
  full moons, names of, 57, 57 (chart)
  Genesis rock, 59
  halo, 145
  hunter's moon vs. harvest moon, 58
  lunar eclipses, 58, 59 (ill.)
  moonquakes, 57
  phases of, 56–57
  side facing Earth, 57
  tail, 58–59
Moonquakes, 57
Moore, Gordon, 583
Moore's Law, 583
Moraine, 99
Moreno, Jacob L., 464
Morgan, Augustus de, 605
Morley, E. W., 8
Morse Code, 559–60, 560–61 (chart)
Morse, Samuel F. B., 559–60, 560 (ill.)
Mosaic patterns, 609
Mosquitoes, 324, 425
Mosteller, Frederick, 611
Moths, 321 (chart)
Motion. *See* Energy, motion, force, and heat
Motor vehicle exhaust, 189
Motorcycle deaths, 408
Mount Erebus, 99 (ill.)
Mt. Everest, 85
Mt. McKinley, 86
Mount Rushmore National Monument, 105
Mount St. Helens, 107–8
Mount Tambora, 121–22
Mt. Vesuvius, 105–6
Mount Washington, 130
Mt. Whitney, 86
Mouse (computer), 585
MSG (monosodium glutamate), 438
MTBE (methyl tertiary butyl ether), 187
Muir, John, 214, 215 (ill.)
Müller, Fritz, 255
Müller, K. Alex, 2
Müller, Paul, 226
Müllerian mimicry, 255
Mullis, Kary, 256
Mundy, Peter, 523
Municipal waste recycling, 241
Murdock, William, 202
Muscles, 384–86
Musgrave, Story, 79
Mushrooms, 289, 421
Music of the spheres, 55
Musical scale, 11, 11 (chart)
Muslim calendar, 476
Mustard gas, 497
Mustard plaster, 463
Mutke, Hans Guido, 548
Mycology, 270
Mycotoxin, 497
Myocardial infarction, 433
Myrrh, 167–68

## N

Nader, Ralph, 541–42
Nagaoka, Hantaro, 12
Nagasaki, Japan, 555
Nails. *See* Skin, hair, and nails
Napalm, 496
Napier, John, 578, 602 (ill.), 602–3, 603
Napier's bones, 578, 602–3
Napoleon III, 156
Napoleon Bonaparte, 440
Narcolepsy, 431
National Academy of Engineering, 614
National Academy of Sciences, 614
National flowers, 285, 286 (chart)
National Heart, Lung, and Blood Institute, 377
National Institute of Diabetes and Digestive and Kidney Diseases, 377
National monuments, 519–20, 520 (charts), 522–23
National parks, 104, 213–14, 214 (charts)
National Research Council, 614–15
National Weather Service, 143–44
Native Americans and wildflowers, 287
Natural drugs, 447–48
Natural gas, 197
Natural substances. *See also* Man-made products; Metals; Minerals; Rocks
  amber, 166
  ambergris, 167
  cashmere, 168
  charcoal, 163
  coal, 162, 163
  damp, 163
  diatomite, 162–63
  essential oils, 166–67
  excelsior, 167
  fly ash, 163
  frankincense, 167
  fuller's earth, 163
  gutta percha, 167
  isinglass, 168
  lodestone, 162
  myrrh, 167–68
  naval stores, 166
  obsidian, 162
  petrified wood, 165 (ill.), 166
  red dog, 162
  rosin, 166

trees, yield of an acre of, 165
tropical forest products, 164 (chart)
wood, sinking, 165–66
wood used in butcher's blocks, 164
wood used in paper, 163
wood used in railroad ties, 165
wood used in telephone poles, 165
Nature and numbers, 600
Naturopathy, 443
Nautical mile, 468
Navaho Code Talkers, 562
Naval stores, 166
Neanderthal, 388, 489
Nearsightedness, 402
Neatsfoot oil, 176
Nebulae, 36
Necrotizing fasciitis, 428
Nectar, 280, 323
Negative ion generator, 463
Nelson, Gaylord, 216
Nelson, Pinky, 79
Neon, 22–23
Neptune, 48–49, 51–52, 52
Nerves, 386
Net registered tonnage, 527
Netplex, 577
Neumann, Frank, 111
Neurons, 387
Neutron, 13
New drugs, development of, 450
New Madrid earthquakes, 112
Newborn babies and blue eyes, 403
Newman, Max, 579
Newspaper recycling and waste, 240
Newton, Isaac, 2–3, 3 (ill.), 601, 613
Niagara Falls, 94–95
*Nighthawk,* 555
Nile River, 94
NIMBY ("Not in My Back Yard") syndrome, 236
Nixon, Richard M., 215
Noah's ark, 526
Nobel, Alfred, 178–79
Nobel Laureates, 615
Nobel Prize, 615
Noble metals, 21, 157
Non-air-supported clear span roof, 512
Non-nuclear fuels. *See also* Nuclear power plants
biomass energy, 182–83
cannel coal, 185
coal, formation of, 184
coal mining, 185
coal, types of, 184
cogeneration, 189
fossil fuels, 183–84
fuel cell, 190
gasohol, 188–89
gasoline additives, 187
gasoline alternatives, 188, 188 (chart)
gasoline stations, 187–88
geothermal energy, 181
hydrocarbon cracking, 186
hydroelectric power plant, first, 190
leaded vs. lead-free gasoline, 186
miner's canary, 185
motor vehicle exhaust, 189
octane numbers, 187
offshore drilling, 186
oil and gas fields, largest, 185
oil well, first, 185
passive vs. active solar energy systems, 182
plants as source of petroleum, 183, 184 (ill.)
Pennsylvania crude oil, 186
primary energy, 181–82
reformulated gasoline, 186–87
solar cell, 182
solar radiation, 181–82
Tesla, Nikola, 190
tidal and wave energy, 182
wind energy, 190
wind farms, 189–90
wood-burning stove, 183
Nonstop flights, 548, 548–49
North American X-15, 551
North Pole, 97, 113, 120, 121
North Star, 40
Northern Lights, 123, 123 (ill.)
Northwest Angle, Minnesota, 98
"Not in My Back Yard" (NIMBY) syndrome, 236
"Not yet diagnosed" ("NYD"), 445
November, 482
Noyce, Robert, 582
Nuclear annihilation, 488
Nuclear autumn, 503
Nuclear fission, 13
Nuclear fusion, 13
Nuclear magnetic resonance, 31
Nuclear power plants, 192 (ill.). *See also* Non-nuclear fuels
accidents at, 191 (chart)
Chernobyl accident, 193–94
"China Syndrome," 194
life of, 190
meltdown, 194
number of, 192–93 (chart)
oldest operational, 191
Rasmussen report, 193
Three Mile Island, 191–92
Nuclear magnetic resonance imaging, 446–47
Nuclear Regulatory Commission (U.S.), 236–37
Nuclear transplantation, 259
Nuclear Waste Policy Act (1982), 237
Nuclear waste storage, 236–37
Nuclear winter, 502–3
Nuclear-powered automobile, 538
Nuclear-powered vessels, 529–30
Nucleotides, 258
Numbered highways, 514
Numbers. *See also* Mathematics
in Bible, 597
counting, origins of, 595
Fibonacci, 596
googol, 599
imaginary, 599
irrational, 599
metric system, 597 (chart)
in nature, 600
perfect, 596
pi, 599
prime number, largest, 596–97
Roman numerals, 596
seven, as supernatural number, 600
ten, importance of, 598
very large, 598 (chart)
zero, 595
Nursing homes, 443
"NYD" ("not yet diagnosed"), 445

## O

Oberth, Herman, 68
Oblique faults, 109
Observatories, 66–67
Obsidian, 162
Occam, William, 616
Occam's Razor, 616
Ocean, 87, 89 (chart), 89–90, 90, 91
Ocean debris, 232 (chart)

Octane numbers, 187
October, 482
O'Dwyer Variable Lethality Law Enforcement pistol, 500
Offshore drilling, 186
Ogallala Aquifer, 88–89
Oil, 185–86, 195
Oil pollution, 230–32, 231 (chart)
Oil spills, 230–31, 231 (chart)
Oil well, 185
Oldest person, 381
Olds, Ransom Eli, 536
Omnes, Heinke Kamerlingh, 2
*On the Origin of Species,* 253–54
Onager, 496
Onions and teary eyes, 396
Onizuka, Ellison, 78
"Ontogeny recapitulates phylogeny," 262
Oodaq, 97
Oort, Jan, 61
"Open sesame," 287
Open source software, 592–93
Operation Ranch Hand, 232
Ophthalmologists, 444
Opossum, 344
Oppenheimer, J. Robert, 501–2
Opposition, 55
Optical mineralogy, 16
Opticians, 444
Optometrists, 444
Orchid, 285
Organelles, 247 (chart)
Organic chemistry, 27
Organic compound, first to be synthesized from inorganic ingredients, 15
Organisms, 262, 268
Organs and glands. *See also* Human body
- air breathed in lifetime, 388
- appendix, 390
- body temperature, 391
- brain, 387, 388 (ill.)
- brain, Neanderthal vs. modern human, 388
- brain weight, 387
- chest cavity, 389
- endocrine glands, 390–91
- gland, largest, 391
- heartbeat, 387
- hiccups, 390
- liver, 391
- lungs, 389
- organ, largest, 386
- respiratory system, 389 (ill.)
- skin, 386, 387 (ill.)

*Orient Express,* 532
Oroville Dam, 521
Orphan drugs, 450
Orphaned birds, 343
Osmium, 21, 24–25
Osteopathic medicine, 443
Otis, Elisha, 509
Oughtred, William, 604
Outer Space Treaty, 69
Ovaries, 391
Owen, Richard, 216
Owens Illinois Glass Company, 173
Oxidation, 5
Oxpeckers, 342
Oxygen, 22, 392
Oymyakon, Siberia, 120
Oyster Creek nuclear plant, 191
Oysters, 329
Ozone, 208, 209, 228, 412

## P

p53 (gene), 261–62
Pacemaker, 445–46, 461–62
Packard Motor Car Company, 538
Painted Desert National Monument, 165 (ill.)
Paleozic Era, 251 (chart)
Palindromic square, 563–64
Palitzsch, Johann, 62
Palladium, 21
Pallas, 60
Palmer, Daniel David, 443
Palmer, Nathaniel, 98
Panama Canal, 426, 521
Pancreas, 391
Pangaea, 96, 96 (ill.)
Panspermia, 260
Paper, 163
- bags, 243
- manufacturing process, 557–58
- mills, 240
- recycling, 239–40

Papyrus, 557
Parachute jumps, 549
Paraheliotropism, 274
Parathyroids, 391
Parchment paper, 177
Parker, James, 174
Parkes, Alexander, 179
Parking meters, 545
Parsec, 64
Parsons, William S., 555
Particle physics, 2
Particles, 14
Pascal, Blaise, 578, 608, 608 (ill.)
PASCAL computer language, 587
Pascal's triangle, 608
Passenger pigeon, 218, 219 (ill.)
Passionflower, 286–87
Passive solar energy systems, 182
Passover, 483–84
Pasteur, Louis, 16, 271, 271 (ill.)
Pasteurization, 271
Patient-controlled analgesia, 455–56
Patsayev, Viktor, 78
Pauling, Linus, 615
PC (personal computer), 588
PCBs, 226, 227
Pearl, 155, 329
Pearl, Mrs. Percy, 380
Pearson, Allen, 132
Pedersen, Charles J., 615
Pelvic inflammatory disease (PID), 438
Penguins, 341
Penicillin, 453
Pennsylvania crude oil, 186
"Penny" as nail size, 507–8
Pentagon, 509–10
Penumbral eclipse, 58
Penzias, Arno A., 33
Percent of increase, 611
Percival, Colin, 599
Perfect numbers, 596
Periodic table, 17–18, 19 (ill.)
Periodontal disease, 423
Permafrost, 97, 143
Permian Basin, 185
Perpetual calendar, 477–78
Perrier, C., 161
Perry, Geoffrey, 567
Pershing missile, 502
Personal computer (PC), 588
Petals, 280
Peterson, Donald, 79
Petrified wood, 165 (ill.), 166
Petroleum nut, 183
Petrology, 148
Petronas towers, 510
Pets. *See* Cats; Dogs
Pewter, 160
pH scale, 29, 29 (chart), 290
Pharmaceutical research, 450
Pharmacognosy, 447–48
Philadelphia Zoological Garden, 305
Phillips screws, 506–7

Philo of Byzantium, 28
Philosopher's stone, 21
*Philosophiae Naturalis Principia Mathematica,* 613
Phlegm, 392
Phlogiston, 5–6
Phobias, 439, 439–40 (chart)
Phonetic alphabet, 558
Phosphenes, 402
Phosphorus, 18
Phototropism, 274
Physical chemistry, 27
Physics, national society, first, 27
Physics and chemistry measurement and methodology
  ampere, 31
  antimatter, 28
  Celsius temperature scale, 30
  Celsius-Fahrenheit conversion, 30, 30 (chart)
  chromatography, 31
  divisions of chemistry, 27
  Einstein, Albert, 27 (ill.), 27–28
  gram atomic weight vs. gram formula weight, 32
  Kelvin temperature scale, 29–30, 30 (chart)
  mole, 32
  Mole Day, 32
  national chemical society, first, 27
  national physics society, first, 27
  nuclear magnetic resonance, 31
  pH scale, 29, 29 (chart)
  quantum mechanics, founder of, 28
  standard temperature and pressure (STP), 31
  thermometer, inventor of, 28–29
  volt, 31–32
  watt, 32
Physiology, founder of, 376–77
Pi, 599
Piazzi, Giuseppe, 60
Pica, 438
Piccard, Jacques, 90
Pickering, William, 82 (ill.)
PID (pelvic inflammatory disease), 438
Piezoelectricity, 16
Pigeons, 342
Pigs, 352
Pilkington, Alastair, 173
Pinchot, Gifford, 214
Pine trees, 278 (chart)
Pinhole camera, 47 (ill.)
"Pioneer dollars," 150
Piri Re'is map, 114
Pisano, Bonanno, 509
*Pisé de terre,* 508
Pistil, 280
Pistol Star, 39
Pitchblende, 151
Pittsburgh, Pennsylvania, 518
Pituitary gland, 390
Pixel, 590
Placebo, 450
Plague, 423
Planetary nebulae, 36
Planet X, 55
Planets. *See also* Earth; Moon
  colors of, 50 (chart)
  "day," differences among, 52, 52 (chart)
  diameters of, 50 (chart)
  distance from Earth, 55
  distance from sun, 48, 48 (chart)
  circular movement of, 53–54, 54 (ill.)
  gravitational force, relative to Earth, 51, 51 (chart)
  inferior vs. superior, 51–52
  Jovian, 52
  Mars, life on, 54
  music of the spheres, 55
  naming of surface features of, 65
  number of moons per planet, 56, 56 (chart)
  "in opposition," 55
  Planet X, 55
  Pluto, as outermost planet, 54–55
  with rings, 48–49
  Saturn's rings, 49, 51
  sidereal time, 51
  solar system, age of, 48
  vs. stars, 55
  terrestrial, 52
  time of revolution around sun, 50 (chart)
  Venus, rotation of, 52
  visibility of, 48
Plant cells, 248, 248 (ill.)
Plant classification system, 267–68, 268 (chart)
  bioinformatics, 264
  five kingdoms of, 266–67
  gnotobiotics, 266
  molecular, 266
  number of organisms, 268
  origins of, 265–66
  radiocarbon dating, 265
Plantae kingdom of biology, 267
Plants. *See also* Flowers; Gardening and farming; Trees
  anise, 287
  attracting butterflies, 296
  carnivorous, 281, 282 (ill.)
  catnip, 289
  child-safe plants, 281, 282 (chart)
  double dormant, 292
  and drugs, 450–51
  evolution, 273 (chart)
  fastest growing, 279
  first plant patent, 303
  hemlock, 275
  hydroponics, 292–93
  ivy, 288
  Jimson weed, 288
  living stones, 289
  luffa sponges, 288
  mimosa ("sensitive plant"), 279
  poison ivy, 300
  pollination, 273–74
  sesame seeds ("open sesame"), 287
  as source of petroleum, 183, 184 (ill.)
  symbolic meanings of, 282–84 (chart)
  toxins, 420
  tropism, 274
  weeds, 288
Plasma, 12
Plasmid, 255–56, 256
Plastic, 179, 241, 242, 242 (chart)
Plastic bags, 243
Plateau theory, 33
Plateletpheresis, 395
Platinum, 21
Platinum group metals, 21
Platonic solids, 609
Platypus, 348
Platysma muscle, 385
Pleaching, 299
Plessor, 447
Plimsoll line, 527–28
Plimsoll, Samuel, 527
Pliny, 436

Plunkett, Roy J., 180
Pluto, 52
p.m. and a.m., 486
Poinsettias, 283
Point Barrow, Alaska, 98
Points, highest and lowest, 85–86
Poison ivy, 300, 435
Poison oak, 300
Poison sumac, 300
Poisonous snakes, 334 (chart)
Poisons and poisonings. *See also* First aid
  activated charcoal, 419
  arrow poisons, 420–21
  bee stings, 421
  black widow spider bite, 421
  childproof containers, 420
  common causes of, 420
  deadliest natural toxin, 419
  lead and the fall of the Roman Empire, 422
  lead poisoning, 421–22
  "mad as a hatter," 422
  mistletoe, 420
  mushroom, 421
  plant toxins, 420
  stitches, 422
  strychnine, 420
  symbol for poisonous products, 419
Polar jet stream, 128–29
Polaris, 40
Polarized sunglasses, 7
Police radar, 543–44
Polio vaccine, 454
Pollination, 273–74
Pollutant Standard Index, 225
Pollution
  acid rain, 229, 229 (chart), 230, 230 (chart)
  Agent Orange, 232
  alleviating air pollution, 228
  balloons, 230
  banned or strictly controlled chemicals, 226
  bioremediation, 225
  carbon dioxide, 226 (chart)
  Chernobyl accident, 228–29
  chlorofluorocarbons (CFCs), 227–28
  DDT, 226–27
  formaldehyde contamination, 232–33
  hazardous waste, 224–25
  indoor air pollution, 233, 233–34 (chart)
  ocean debris, 232 (chart)
  oil pollution, 230–32, 231 (chart)
  oil spills, 230–31, 231 (chart)
  Operation Ranch Hand, 232
  PCBs, 227
  Pollutant Standard Index, 225
  smog, 228
  toxic chemicals, 227
  Toxic Release Inventory, 225–26
Polyakov, Valerij, 73
Polymerase chain reaction, 256
Pong (computer game), 581
Porcupines, 354–55
Porpoises, 349
Post, Wiley, 549
Potting soil, 291
Power take-off, 490
Prandial rhinorrhea, 397
Precambrian Era, 252 (chart)
Precious metals, 157
Precious stones, 155
Precision of the equinoxes, 53–54, 54 (ill.)
Prescriptions, 448 (charts), 449 (chart)
Primary colors, 7
Primary energy, 181–82
Primary teeth, 384
Prime meridian, 115
Prime numbers, 596–97
Primordial black hole, 37
Prion, 255–56
Progeria, 437
Programmed theories, 405
Project BETA, 70
Project META, 70
Project Sentinel, 70
Prokaryotic cells, 246, 246 (ill.), 246 (chart), 247 (chart)
Promethium, 177
Pronghorn sheep, 346
Prospect Creek, Alaska, 120
Protected areas, 212, 212 (chart)
Protein, 257–58
"Protein Measurement with the Folin Phenol Reagent," 614
Protista kingdom of biology, 267
Protium, 23
Proton, 13, 14
Protruding ears, 381
Przewalski, Nikolai, 351
Przewalski's horse, 350 (ill.), 351
Psychodrama, 464
Pug dog, 360–61
Pulsar, 38
Punched cards, 579, 580
Punctuated equilibrium, 249–50
Purring, 363–64
PVC plastics, 241
Pythagoras, 55, 599, 609
Pythagorean theorem, 609

## Q

Quad of energy, 196–97
Quantum electrodynamics, 14
Quantum mechanics, 28
Quark, 13–14
Quartz, 156
Quartz watches, 487–88
Quasars, 35
Quicksand, 101–2
Quills, 354–55

## R

R-value, 506, 506 (chart)
Rabbits, 352
Rabies, 423
Rad, 412–13
Radar, 543–44
Radial tires, 537
Radiation, 11–12, 412–13
Radio
  AM stations, 567
  broadcasting station, first, 565–66
  call letters, 566
  inventor of, 565
  FM stations, 566–67
  space shuttle transmissions on shortwave, 567
  submarines' communication, 570
Radioactive nuclei, 15
Radiocarbon dating, 265
Radiological dispersal device, 503
Radium, 177
Radon, 412
Rahe, Richard H., 407
Railroad ties, 165
Railroad worm, 300–301
Railroads. *See* Trains and railroads
Rain
  freezing, 141
  frogs and toads, 145
  hailstones, 141, 141 (ill.)
  lightning, 125–26, 140

raindrops, 139
rainiest place on Earth, 139
record rainfall, 139
sleet, 141
speed of rainfall, 140
and television, 568
thunder and thunderstorms, 140
Rain forest plant extracts, 451
Rain forests, 210–12
Rain shadow, 292
Rainbow, 127
Raindrops, 139
Rammed earth, 508
Rare Earth elements, 22–23
Rare gases, 22–23
Rasmussen report, 193
Reactive materials, 224
Recombinant DNA (RNA), 256, 257, 261
Recordings
CD-ROM disc, 575, 575 (ill.)
compact disc (CD), 575, 575 (ill.)
digital audio tape, 574
Dolby noise reduction system, 574
Recycling, conservation, and waste. *See also* Energy consumption and conservation
aluminum, 203
automobile tires, 243–44
beverage deposits, 241
Brownfields, 236
curbside recycling programs, 241
garbage, amount generated, 238
garbage incinerator, 238
hand washing vs. dishwasher, 243
hazardous waste sites, 236
landfills, 239
metal recycling, 238
methane fuel, 239
*Mobro* (ship), 238
municipal waste recycling, 241
newspaper recycling, 240
newspaper waste, 240
NIMBY ("Not in My Back Yard") syndrome, 236
nuclear waste storage, 236–37
paper mills, 240
paper recycling, 239–40
plastic vs. paper bags, 243
plastics, biodegradable, 241
PVC plastics, burning of, 241
recycled plastic products, 242, 242 (chart)
recycling symbols, 242, 242 (chart)
Resource Conservation and Recovery Act (1976), 235
Superfund Act (1980), 235
"throw-away society," 237–38
Toxic Substances Control Act (1976), 235
WOBO (world bottle), 244
Red Baron, 553
Red dog, 162
Red dots in photographs, 403
Red dye, Native Americans and wildflowers, 287
Red Queen hypothesis, 255
Red tide, 211
Redfield, William Charles, 615
Redier, Antoine, 488
Reed, Ezekiel, 508
Reflecting telescope, 66
Reflection nebulae, 36
Reflexes, 447
Reflexology, 463
Reformulated gasoline, 186–87
Refracting telescopes, 66
Regeneration, 310–11
Reissner-Nordstrom black hole, 38
Relativity, 8, 616
Relief maps, 115
Rem, 412–13
REM sleep, 372–73
Remak, Robert, 245
Replicating molecules, 263–64
Reptiles and amphibians. *See also* Animals
alligators, 334
crocodiles, 334
difference between, 333–34
fastest snake, 334
frog, life cycle of, 333 (ill.)
poisonous snakes, 334 (chart)
Surinam toad, 335
turtle, 335
Resin, 317
Resnik, Judith, 78
Resource Conservation and Recovery Act (1976), 235
Respiratory system, 389 (ill.)
Retractable roof stadium, 511–12
Retrovirus, 423
Reverse faults, 109
Revetment, 521
Rh factor, 395
Rhenium, 24
Rhinoceros, 342, 346
Rhizome, 281
Rhodium, 21
Ribicoff, Abraham, 542
Richter, Charles W., 110
Richter scale, 110, 110 (chart)
Ricin, 497
Ride, Sally K., 74, 75, 76 (ill.), 79
Right turn on red, 542
Rigid beam bridge, 516
Ring of Fire, 106–7, 108
Ringed planets, 48–49
Rings (trees), 275–76, 276 (ill.)
Rip currents, 92
Rip tides, 92
Rivers, 94
RNA (recombinant DNA), 256, 257, 261
Road roller, 494
Roads
coast-to-coast highway, first, 514
Jersey barrier, 515
lanes, most, 514
manhole covers, 516
numbered highways, 514
state with most miles of roads, 513–14
traffic light, first, 514–15
Robots, 493
Rock of Gibralter, 105
Rocks. *See also* Man-made products; Metals; Minerals; Natural substances
cinnabar, 150
differences among, 147
fossils, 148 (ill.), 148–49, 149 (ill.)
igneous, 147
Indian dollars, 150
metamorphic, 147
and minerals, 150
petrology, 148
sedimentary, 147
tektite, 149
Rods, 401–2
Roebling, John A., 518
Roebling, Washington A., 518
Rogers Commission, 78–79
Rogers Pass, Montana, 120
Roget, Peter Mark, 604

Roller coasters, 523 (chart), 523–24
Roman calendar, 477
Roman Empire, 422
Roman numerals, 486, 596
Roosa, Stuart A., 72
Roosevelt, Theodore, 105, 366
Rose family and trees, 279
Roses, 285 (chart)
Rosin, 166
Rosing, Boris, 568
Ross, Sir James Clark, 98
Rotational time, 474
Rowland, Henry Augustus, 27
RU-486, 456
Ruby, 155, 180
Rule of 70, 610
Rumble seat, 537
Runny nose while eating, 397
Ruska, Ernst, 492
Rutan, Dick, 548
Rutherford, Ernest, 12, 13
Rx, 448


## S


Sabin, Albert, 454
Saffir, Herbert, 134
Saffir/Simpson Damage-Potential scale, 134
Sahara Desert, 101
Saint Elmo's fire, 127
St. Isidore of Seville (patron saint of the Internet), 577
St. Lawrence Seaway, 93
St. Martin, Alexis, 374
St. Petersburg, Florida, 121
Saliva, 392, 393
Salk, Jonas E., 454
Salmon, 331
Salt, 26, 91–92, 170
Saluki, 357
Samuel, Arthur, 581
San Andreas Fault, 109
San Francisco earthquake (1906), 112–13
San Juan Capistrano, California, 338
Sand, 100–102
Sand cat, 352–53
Sand dunes, 100, 101 (ill.)
"Sand" in eyes, 403
Sandpaper, 177
Sandy soil, 290
Sanger, Frederick, 615
Santorio, Santorio, 28
Sapphire, 155
Sargent, Charles Sprague, 214
Sartorius, 385
Satellite dishes, 568–69, 570
Satellite photographs, 116
Satellite television, 569
Satellites, 77, 571
Saturday, 481
Saturn, 48, 48–49, 49, 51–52, 52
*Savannah,* 529–30
Savitskaya, Svetlana, 74
Scaliger, Joseph Justus, 478
Schade, Otto, 569
Schafer, Edward, 416–17
Schieffelin, Eugene, 342
Schleiden, Matthias, 245
Schmitt, Harrison H., 72
Schrieffer, John Robert, 2
Schroeder, William, 461
Schwann, Theodor, 245
Schwartz, Berthold, 178
Schwarzschild black hole, 38
Sciatic nerve, 386
Science vs. technology, 613
Scobee, Francis, 78
Scopes, John T., 255
Scopes (monkey) trial, 255
Scott, David R., 72
Scott, Robert Falcon, 98
Screw thread standardization, 494
Screws, 506–7
Sea level, 91
Sea, time at, 487, 487 (chart)
Seaplane, 552
Search engine, 593
Search for Extraterrestrial Intelligence (SETI), 70
Sears, Roebuck, 513
Seashell sounds, 9
Seasons, 53, 53 (ill.), 482, 482–83, 483 (chart)
Seatbelts, 541
Seawall, 520–21
Seawater, 87, 91–92, 392 (chart)
Second generation computer, 579
Sedimentary rocks, 147
Seedless watermelon, 299
Seeds, 293, 293 (chart)
Segrè, Emilio, 161
Seismograph, 109–10, 111 (ill.)
Semper, Karl, 206
"Sensitive plant," 279
Sepal, 280
September, 482
Serpent, 440–41
Serpollet, Leon, 539
Service stations, 187–88
Sesame seeds ("open sesame"), 287
SETI (Search for Extraterrestrial Intelligence), 70
Seven, as supernatural number, 600
Shakespeare garden, 295
Shakespeare, William, 295
Sharks and shark attacks, 332, 332 (chart)
Shar-pei, 360
Shedless dogs, 360
Sheep cloning, 259–60, 260 (ill.)
Shekel, 465, 465–66 (charts)
Sheldon, William Herbert, 380
Shepard, Alan B., Jr., 71, 72, 75
Shield volcanoes, 106
Shining planets vs. twinkling stars, 55
Ships
- "by and large," 525–26
- dead reckoning, 525
- deadweight tonnage, 527
- displacement tonnage, 527
- gross tonnage, 527
- hospital ship, 528
- largest, 529
- Liberty, 528
- loadline, 527–28
- "mark twain," 526–27
- net registered tonnage, 527
- Noah's ark, 526
- nuclear-powered vessels, first, 529–30
- "starboard," 526
- *Titanic,* 528–29
- wooden figure of woman, 525

Shopping center, first, 509
Short ton, 471
Short-nosed echidna, 348
Shortwave radio, 567
Shot, 499
Shumway, Norman, 460
SI system, 466
Siamese cats, 363
Siamese twins, 381, 381 (ill.)
Siberia, 60
Siberian express, 130
Sidereal time, 51
Sievert, 413
*Silent Spring,* 226
Silica aerogels, 170
Silicon, 22
Silicon chips, 582–83

Silver, 21, 23, 159–60
Simple, interest, 611
Simple machines, 489
Simpson, Robert, 134
Sinkholes, 85
Sintering, 495
Siple, Paul A., 13
Sirens, 359–60
Sirius, 37
Skidding, 540 (chart)
Skilled nursing facility, 443–44
Skin, hair, and nails. *See also* Human body
amount of, 397
fingernails after death, 400
fingernails, speed of growth, 400
fingerprints, 398
forensic science and hair, 400
freckles, 398
goose-bumps, 398
graying hair, 399–400
hair, amount of, 399
hair, curly vs. straight, 400
hair, speed of growth, 399
identical twins and fingerprints, 398
identical twins and rolling their tongue, 398
tattoos, 398–99
warts, 429
Skunk odor, 356, 363
Sky, 83
Skydome, 511–12
Skyscrapers, 508–9
Slag, 176
Slate, Thomas Benton, 169
Sleep, 372, 372 (chart), 431
Sleet, 141
Slide rule, 603–4
Small intestine, 376
Smalley, Richard, 171
Smart Bomb, 502
Smeaton, John, 174
Smiling, 386
Smith, Delford, 551
Smith, Michael, 78
Smith, Robert Angus, 229
Smith, William "Uncle Billy," 185
Smog, 228
Smokey the Bear, 212–13, 213 (ill.)
Smoking, 415
Snakes, 334, 334 (chart)
Sneezing, 397
Snoring, 312, 373
Snow
air temperature, 142
amount of water in, 143
forming of, 142
record snowfall, 143
sleet, 141
snowflakes, 142
"white-out," 138
Snowflakes, 142
Snowmold, 297
Sobriero, Ascanio, 178
Sodium chloride, 26
Soil, 289–91
Solar cell, 182
Solar cycle, 45
Solar eclipse, 13, 45–46, 47, 47 (chart), 616
Solar radiation, 181–82
Solar system, 34 (ill.), 34–35, 48, 49 (ill.)
Solar wind, 46–47
Solder, 175 (ill.), 175–76
Solids, 12
Somatic cell nuclear transfer, 259
Sommelier, German, 494
Sonic boom, 9
Sopwith Camel, 554
Sound, 8–11
breaking sound barrier, 8–9
decibel, 10, 10 (chart)
Doppler effect, 9–10
double sonic boom, 9
musical scale, 11, 11 (chart)
seashell sounds, 9
speed of sound, 11
Sound barrier, 8–9
Sound frequency, 405
Sound Transmission Class (STC), 507
South Africa, 159
South Cape, Hawaii, 98
South Pole, 97, 120, 121
Soviet space program, 77
Space. *See* Earth; Moon; Planets; Space exploration; Space observation and measurement; Stars; Sun; Universe
Space exploration
African American in space, first, 76
animal in space, first, 73
astronauts on the moon, 72
*Challenger,* 78–79
"close encounters," 69–70
extraterrestrial life, 70
fatalities, 78, 78 (chart)
female astronauts, 75–76
*Galileo* spacecraft, 77
golf shot on the moon, first, 75
"grand tour" of the planets, 71
"greening of the galaxy," 69
intelligent life on other planets, 68–69
man and woman to walk in space, first, 74
man in space, first, 70–71
manned flight, longest, 73
married couple in space, first, 76
meal on the moon, first, 75
monkeys and chimpanzees in space, first, 73–74
most time in space, 77
Outer Space Treaty, 69
rocket proposals, 68
Soviet space program, founder of, 77
space shuttles, 78–79
successful space flights in 2000, 76 (chart), 76–77
U.S. flag on the moon, 75
U.S. satellite, first, 77
*Voyager,* 71–72
*Voyager* messages, 71–72
woman in space, first, 74
words spoken on the moon, first, 75
zero gravity vs. microgravity, 68
Space observation and measurement
astrolabe, 65
Astronomer Royal, 64
celestial objects, naming of, 65
Hubble telescope, 67–68
light year, 64
reflecting vs. refracting telescopes, 66
Stonehenge, 66
systematic astronomy, 64
telescopes, 65–66, 67–68
units of astronomical distances, 64
Very Large Array (VLA), 66–67, 67 (ill.)
Space shuttles, 78–79, 567
Space vehicle material, 180
"Spaceship Earth," 215
Spectroscopy, 8
Speed guns, 544

Speed limit, 409
Speed of light, 7, 8
Speed of sound, 11
Speed trap, 543
Speleothem, 102
Spelunking vs. speleology, 102
Spencer, Herbert, 254
Sphygmomanometer, 445
Spicy foods and sweating, 397
Spiders, 325, 325 (ill.), 421
"Spiegelman monster," 263–64
Spiegelman, Sol, 263
Spiny anteater, 348
Spontaneous combustion, 5
Sports injuries, 409, 410 (chart), 410–11
Spring, 482–83
*Spruce Goose,* 550–51, 551 (ill.)
Spruce trees, 278 (chart)
*Sputnik,* 73
Spy plane, smallest, 555
Square root, 605
Squaring the circle, 609
Squirrels, 301, 356
Stack, John, 548
Staehle, Albert, 212
Stahl, Georg Ernst, 5
Stainless steel, 160–61
Stalactite, 103,103 (ill.)
Stalagmite, 103 (ill.), 104
Stamen, 280
Standard gauge railroad, 530
Standard temperature and pressure (STP), 31
Stapes, 405
Star sapphires, 155
"Starboard," 526
Stardust, 37
Stars, 36–47. *See also* Sun
 Big Dipper, 39–40
 binary, 37
 black hole, 37–38
 brightest, 39, 39 (chart)
 closest to Earth, 43
 colors of, 38, 38 (chart)
 constellations, 41, 41–43 (chart), 43
 largest, 39
 Milky Way, 39
 naming of, 65
 nebulae, 36
 North Star, 40
 vs. planets, 55
 Polaris, 40
 pulsar, 38
 stardust, 37
 summer triangle, 40
 supernova, 36
 twinkling, 36–37
State birds, 342
State insects, 321
Statement (weather), 143
States of matter, 12
Statue of Liberty, 522–23
Statute mile, 468
STC rating, 507
Stealth technology, 555
Stem cells, 260
Stephenson, George, 530
Steptoe, Patrick, 462
Sterling silver, 159
Steroids, 454–55
Stetson, R. E., 395
Stevens, John, 531
Stewart, Robert, 79
Stibnite, 152
Still, Andrew Taylor, 443
Stitches, 422
Stomach, 376
Stomach ulcer, 430
Stonehenge, 66
Storm chasers, 133–34
STP, 31
Stratosphere, 81
Strauss, Joseph B., 519
Strawberries, 301
Streptomycin, 453–54
Stress, 407, 408 (chart)
Strike-slip faults, 109
String theory, 2
Structures. *See also* Buildings and building parts
 dams, highest, 521–22
 Eiffel Tower, 521 (chart)
 Ferris Wheel, 523
 Gateway Arch, 519–20, 520 (charts)
 Hoover Dam, 521–22
 national monument, tallest, 519–20, 520 (charts)
 Panama Canal, 521
 roller coasters, 523 (chart), 523–24
 seawall, 520–21
 Statue of Liberty, 522–23
 Texas tower, 520
Strychnine, 420
Styrofoam, 168
Subatomic particles, 14
Subbituminous coal, 184
Submarines, 570
Subtropical jet streams, 129
Sulfites, 434
Sulfuric acid, 169
Sullivan, Kathryn D., 74, 75, 79
Sullivan, Roy, 125
Summer, 482–83
Summer triangle, 40
Sun. *See also* Stars
 colors of, 44–45
 distance of planets from, 48, 48 (chart)
 eclipses, 45, 46 (ill.), 47, 47 (chart)
 ecliptic, 44
 elements of, 44, 44 (chart)
 halo, 145
 heat of, 43–44
 predicted death of, 44
 safe viewing of solar eclipse, 47, 47 (ill.)
 solar cycle, 45
 solar wind, 46–47
 star closest to Earth, 43
 sun dog, 46
 sunlight, 8, 45
 sunspot cycle, 45
Suncoast Dome, 512
Sunday, 481
Sundial, 485, 486–87
Sunshine, 121
Sunspot cycle, 45
Superconductivity, 2
Superfund Act (1980), 235
Supergiant, 38
Superior planets, 51–52
Supernatural numbers, 600
Supernova, 36, 38
Supersonic flight, 548
Surgery and non-drug treatments
 animal organs, 461
 aromatherapy, 463–64
 artificial hearts, 461
 cardiac pacemakers, 461–62
 catgut, 460
 contraception, 462
 eye transplants, 462
 first African American to perform heart surgery, 460
 heart transplants, 460
 iridology, 463
 lithotripsy, 462
 minimally invasive surgery, 459
 mustard plaster, 463

negative ion generator, 463
psychodrama, 464
reflexology, 463
surgical procedures, number of, 46
synthetic skin, 463
test-tube baby, first, 462
Surgical excision, 399
Surinam toad, 335
"Survival of the fittest," 254
Suspended animation, 382
Suspension bridge, 516, 518
Swade, Doron, 578
Swallowing, 376
Swallows of Capistrano, 338
Sweating when eating spicy foods, 397
Sydenham, Thomas, 441–42
Synesthesia, 403
Synthetic gemstone, 180
Synthetic skin, 463
Synthetic soil, 291
Systematic astronomy, 64
Syzygy, 36

## T

T cells, 371–72
Tabby cat, 362
Tadpoles, 365
Taft, William Howard, 366
Tahitian bear dog, 361
Tail of the moon, 58–59
Takahashi, Daisuke, 599
Tallest buildings and structures, 510, 510 (chart), 511, 511 (ill.)
Tamarand, 302
Tamayo-Mendez, Arnaldo, 76
Tank, 553
Tantalum, 156
Tar pits, 105
Tarquinius Priscus, 477
Taste, 406, 406 (ill.)
Tattoos, 398–99
Taxicab, 546
Taximeter, 546
Taxol, 451
Taylor, J. G., 524
Technetium, 161
Technobabble, 592
Technology vs. science, 613
Teeth, 384
Teflon, 180
Tektite, 149
Telecommunications
analog, 573
answering machines, 572
cellular telephones, 573
Clarke belt, 573
digital, 573
fax machines, 571 (ill.), 571–72
fiber-optic calbe, 572 (ill.), 572–73
satellite, 571
wireless telephones, 574
Telefacsimile machines, 571 (ill.), 571–72
Telephones
answering machines, 572
cellular, 573
fax machines, 571–72
poles, 165
wireless, 574
Telescopes, 65–66, 67–68
Television
flat-panel displays, 569
founder of, 567–68
high definition (HDTV), 569
rain's effect on, 568
satellite dishes, 568–69, 570
Telford, Thomas, 513
Temin, Howard, 423
Temperature
cricket chirps, 122
"dog days," 122
El Niño, 119, 210
global warming, 119–20
heat index, 121, 121 (chart)
heat wave, 120–21
highest recorded, 120
humidity, 122
Indian summer, 122
La Niña, 119
lowest recorded, 120
North Pole, 120
South Pole, 120
sunshine, 121
"year without a summer," 121–22
Tempered glass, 172
Ten, importance of, 598
10-codes, 562, 563 (chart)
Tennis elbow, 431
Terbium, 23
Tereshkova-Nikolaeva, Valentina V., 74
Termites, 323
Terrestrial planets, 52
Tesla, Nikola, 190
Tesselations, 609
Testes, 391
Test-tube baby, 462
Tetanus shot, 453
Tetrahydrocannabinol (THC), 459
Texas tower, 520
TGV *(Train à Grande Vitesse),* 530
Thalidomide, 456
Thallium, 177
Theodolite, 474
Theophrastus, 280
Theory of evolution, 254, 255
Theory of relativity, 8, 616
Theory of uncertainty, 28
Therblig, 615–16
Thermal cracking, 186
Thermodynamic temperature, 29
Thermodynamics, 4
Thermometer, 28–29
Thermopane glass, 173
Thermosphere, 82
Thermotropism, 274
Thigmotropism, 274
Third generation computer, 579
Thompson, John Taliaferro, 500
Thompson, LaMarcus, 524
Thomson, Sir Joseph John, 13
Thomson, William (Lord Kelvin), 30
Thorium, 23
Thoroughbred horses, 351
Three Mile Island, 191–92
"Throw-away society," 237–38
Thunder and thunderstorms, 126, 140, 418
Thursday, 481
Thyroid gland, 391
Tibbetts, Paul W., Jr., 555
Tidal bore, 91
Tidal energy, 182
Tides, 91, 95
Tiger's eye, 156
Tight building syndrome, 233, 233–34 (chart)
"Tilling the plane," 609
Time
alarm clock, 488
a.m. and p.m., 486
atomic time, 486
Babylonian calendar, 476
"B.P." date designation, 484
calendar year, 475
century, 475–76
Chinese calendar, 476
Chinese years, 479, 480 (chart)
clockwise movement, 485
Coptic calendar, 476–77

Daylight Savings time, 484–85
days of the week, 481 (chart)
doomsday clock, 488
Easter, 483, 483 (chart)
Egyptian calendar, 476
floral clock, 487
grandfather clock, 487
Gregorian calendar, 477
Hindu calendars, 477
international date line, 485
January 1 as first day of year, 476
Japanese calendar, 477
Jewish calendar, 476
Julian calendar, 477
Julian day calendar, 477
Julian day count, 478
leap second, 480
leap year, 480
"Local Noon," 484
longest years, 480–81
measurement of, 474
military time, 488
modern timekeeping, 475
months of the year, 481–82
Muslim calendar, 476
Passover, 483–84
perpetual calendar, 477–78
quartz vs. mechanical watches, 487–88
Roman calendar, 477
Roman numerals on clocks, 486
sea, time at, 487, 487 (chart)
seasons, beginning dates of, 482–83, 483 (chart)
seasons, length of, 482
shortest years, 480–81
sundial, 486–87
time zones, 485
twelve-month calendars, 479, 479 (charts)
United States Time Standard signal, 486
universal time, 485–86
week, 481
world calendar, 478
Time zones, 485
Timekeeping system, 475
Tire-derived fuel, 243
Tires, 203–4, 243–44, 537, 538, 538 (chart), 550
*Titanic,* 528–29
Titanium dioxide, 177–78
TNT, 178
Toads, 145
Toadstools, 289
Tomato plants, 296
Tombaugh, Clyde, 54
Tomlinson, Ray, 591
"Tommy gun," 500
Tongue, 398, 406, 406 (ill.)
Tons, 471, 471 (charts)
Tools
compass, 490
compound microscope, 491–92
earliest agricultural, 489
electron microscope, 492
jackhammer, 494
monkey wrench, 490
Neanderthal, 489
screw thread standardization, 494
"Topping out" party, 512
Tornadoes, 131–32, 133, 134
Torvalds, Linus, 592
Tossach, William, 416
Total solar eclipse, 616
Toxic chemicals, 227
Toxic materials, 225
Toxic Release Inventory, 225–26
Toxic Substances Control Act (1976), 235
Tradescant, John, 167
Traffic light, 514–15
*Train à Grande Vitesse (TGV),* 530
Trains and railroads
cable car, 532–33, 533
cabooses, 531–32
fastest train, 530–31
funicular railway, 533
gandy dancer, 532
longest railway, 531
*Orient Express,* 532
standard gauge railroad, 530
U.S. railway, first, 531
velocipede, 532
Transition elements, 19–20
Transportation energy, 204
Trans-Siberian Railway, 531
Transuranic elements, 20
*Treatist on the Astrolabe,* 613
Tree-climbing canine, 353
Trees. *See also* Flowers; Gardening and farming; Plants
banyan, 278
bonsai, 299–300
chestnut blight, 276–77, 277
city with most trees, 277
clover, 288
conifers, 278–79
dogwood, 300
fir, 278 (chart)
five-in-one, 298
fruit, 301
glycerine, as leaf preservative, 279
Joshua, 277
leaves, brightness of, 276
leaves, number of, 276
leaves, turning color, 277
monkey ball, 279
navel orange, 298
oldest, 275, 275 (chart)
pine, 278 (chart)
rose family, 279
spruce, 278 (chart)
tallest, 277
tree rings, 275–76, 276 (ill.)
trunk, largest, 278
as weather predictor, 145
yield of one acre of, 165
Trevithick, Richard, 534
Triangles, 608
Triglycerides, 446 (chart)
Triple play, 613
Tritium, 23, 177
Trojan asteroids, 60
Tropical forest products, 164 (chart)
Tropical rain forests, 210–12
Tropism, 274
Troposphere, 81
Trouvelot, Leopold, 318
Troy measurements, 472–73, 473 (chart)
Troy ounce of gold, 158
Trucks, 545–46
Trunks (trees), 278
Tsiolkovsky, Konstantin E., 68
Tsunami, 108
Tswett, Mikhail, 31
Tubeless tires, 537
Tuberculosis, 453–54
Tuberous root, 281
Tubers, 281
Tuesday, 481
Tufa, 103
Tumors, 455 (ill.)
Tuna, 223–24
Tungsten, 24, 160
Tunguska Event, 60–61
Tuning fork, 161
Tunnels, 515–16
Turing, Alan M., 578, 581

"The Turk," 581–82
Turko, Richard P., 502
Turtles, 223, 335
Tutunendo, Colombia, 139
TV. *See* Television
Twelve-month calendars, 479, 479 (charts)
20/20 vision, 402
Twinkling stars, 36–37, 55
Two-stroke engine, 489–90
Tyndall, John, 208
Typhoid fever, 427
Typhoid Mary, 427

## U

Ulcers, 430
Undeciphered codes, 562
Underinflated tires, 203–4
United States Time Standard signal, 486
Universal donor/recipient, 395
Universal Product Code (UPC) bar code, 564
Universal time, 485–86
Universe, 33–36, 34 (ill.)
- age of, 34
- Big Bang theory, 33
- Big Crunch theory, 33
- closest galaxy, 36
- Hawking, Stephen, 35, 35 (ill.)
- quasars, 35
- solar system size, 34–35
- syzygy, 36

University of Pennsylvania School of Medicine, 442
UNIX operating system, 592
Upatnieks, Juris, 492
UPC (Universal Product Code) bar code, 564
Uranium, 161
Uranus, 48–49, 51–52, 52
Urey, Harold C., 26
Urine, 392
Ursa Major, 39, 39 (ill.)
U.S. flag on the moon, 75
U.S. money, 469
USS *Enterprise,* 529
USS *Long Beach,* 529
USS *Nautilus,* 529
Utility meters, 197
Uvula, 386

## V

"V" flight formation, 338 (ill.), 338–39
Vail, Alfred, 560
Valerius Maximus, 436
Van Allen belts, 82–83
Van Allen, James A., 77, 82, 82 (ill.)
Van Buren, Martin, 366
Van Helmont, Jan Baptista, 264
Van Hoften, Ox, 79
Van Kleist, E. Georg, 6
Van Musschenbroek, Pieter, 6
Van Vleck, John H., 5
Van Vleck paramagnetism, 5
Varicella-zoster (Herpes zoster) virus, 428
VASCAR (Visual Average Speed Computer and Recorder), 543
Vauquelin, Nicolas-Louis, 264
Vectors, 425
Vegetable garden, 296
Vehicle identification number (VIN), 539
Velcro, 180
Velocipede, 532
Venn diagrams, 605–6, 606 (ill.)
Venn, John, 605–6
Venus, 48, 51–52, 52
Venus fly trap, 281
Verkhoyansk, Siberia, 120
Verneuil, Auguste Victor Louis, 180
Verrazano, Giovanni da, 518
Verrazano-Narrows Bridge, 518
Very Large Array (VLA), 66–67, 67 (ill.), 611–12
Very large numbers, 598 (chart)
Viceroy butterfly, 254 (ill.), 255
Victory garden, 295
Video-arcade games, first, 581
Violence and full moon, 408
Virtual reality, 574
Virus (computer), 589
Virus (human), 270–71, 423, 497–98
Viscosity, 171
Visual Average Speed Computer and Recorder (VASCAR), 543
Voiceless dog, 360
Volcanoes
- active, 107
- ancient eruptions, 106
- Circle of Fire, 106–7, 107 (ill.)
- greatest concentration of, 107
- kinds of, 106
- most destructive, 108 (chart)
- Mount St. Helens, 107–8
- Mt. Vesuvius, 105–6
- "year without a summer," 121–22

Volkov, Vladislav, 78
Volt, 31–32
Volta, Alessandro, 31
Volume formulas, 606–7
Von Bellingshausen, Fabian Gottlieb, 98
Von Braun, Wernher, 82 (ill.)
Von Frisch, Karl, 322
Von Kempelen, Wolfgang, 581
Von Lindemann, Ferdinand, 609
Von Neumann, John, 580
Von Richthofen, Manfred, 553
Von Sachs, Julius, 292
Von Waldeyen-Hartz, Wilhelm, 245
Vostok Station, Antarctica, 120
*Voyager,* 71–72
*Voyager of the Seas,* 529

## W

Waksman, Selman A., 453
Waldheim, Kurt, 71
"Waldsterben," 229
Walsh, David, 90
Warning (weather), 143–44
Warts, 429
Washansky, Louis, 460
Washington, George, 105
Washington (state) bridges, 518
Waste. *See* Recycling, conservation, and waste
Watch (weather), 143
Water
- amount of, on Earth, 88
- aquifers, 88–89
- blue sea, 89
- brackish, 91
- coverage on Earth, 87
- density of, 15
- Dead Sea, 92
- erosion, 102
- freezing, 1
- gold in seawater, 87
- Great Lakes, 93–94
- ice, bearing capacity of, 92, 92 (chart)
- icebergs, 88
- lake, deepest, 93
- lakes, largest, 93, 93 (chart)
- melting ice, 88
- Niagara Falls, 94–95
- ocean, chemical composition of, 89, 89 (chart)
- ocean, circulation of, 87

ocean, depth of, 90
ocean, penetration of sunlight, 90
ocean vs. sea, 91
percentage of human body weight, 396
rip tides, 92
rivers, longest, 94
rotation in drain, 6
saltiness in seawater, 91–92
sea level, 91
tidal bore, 91
tides, 91
waterfall, highest, 94
waves, 89–90
weight, 472 (charts)
yazoo, 93
Waterfalls, 94
Water-lily Victoria amazonica, 289
Watermelons, 299
Watson, James, 261
Watt, 32
Watt, James, 32, 533
Watts, William, 499
Wave energy, 182
Wavelength, 9–10
Waves, 89–90
WBZ radio station (Springfield, MA), 566
Weapons
anthrax, 497
atomic weapons, 501–2
bazooka, 500
Big Bertha, 501
bio-terrorism, 497–98
"Blackout Bomb," 503
Bowie knife, 498
Cannon King, 500
chemical warfare, 496
Civil War innovations, 498
Colt revolver, 498
cruise missile, 502
crystal ball, 501
"Fat Man," 502
flail, 496
handgun, most technologically advanced, 500
"Little Boy," 501–2
machine gun, 499 (ill.), 499–500
Manhattan Project, 501
mine barrage, 498
mustard gas, 497
napalm, 496
non-lethal, 503
nuclear winter, 502–3
onager, 496
Pershing missile, 502
radiological dispersal device, 503
Smart Bomb, 502
viruses, 497–98
Weather prediction
barometric pressure, 144
Doppler radar, 144
frogs and toads, 145
Groundhog Day, 144–45
National Weather Service, 143–44
origins of, 144
sun/moon halo, 145
trees, 145
wooly-bear caterpillar superstition, 145
Weaver, Warren, 266
Weddell, James, 98
Wednesday, 481
Weeds and weeding, 288, 296
Week, 481
Wegener, Alfred Lothar, 95
Weights and measures
area, 471–72 (charts)
avoirdupois measurements, 472–73, 473 (chart)
bench mark, 473–74
biblical shekel, 465–66 (chart)
dry and liquid measures, 470–71 (charts)
horizon, 473
meter, original and current, 467
metric conversions, 468–70 (chart)
metric system, 466
metric system, countries not using, 470
nautical mile, 468
SI system, 466
statute mile, 468
theodolite, 474
tons, types of, 471, 471 (charts)
troy measurements, 472–73, 473 (chart)
U.S. money, 469
water weight, 472 (charts)
yard, 467, 467–68 (charts)
Weiner, A. S., 395
Welch, George, 548
Werner, Abraham Gottlob, 151
Werner's syndrome, 437
West Indian manatee, 350
West Nile virus, 426
Wetlands, 208
Wheeler, John, 37
Whiskers, cat, 364
White blood cells, 370, 371–72
White damp, 163
White, Edward, 74, 78
White gold, 158
White House pets, 365–66
White pigment, 177–78
"White-out," 138
Whittaker, R. H., 267
Wickard, Claude R., 295
Wiener, Norbert, 493
Wilbrind, J., 178
Wildflowers, 287
Wilkes, Charles, 98
Wilkins, Maurice, 261
Williams, Daniel Hale, 460
Wilmut, Ian, 259
Wilson, Edward O., 310
Wilson, Robert W. 33
Wilson, Woodrow, 213, 614
Wind
Alberta clipper, 130
Beaufort scale, 133
chinook, 130
Coriolis effect, 128
cyclone, 131–32
Fujita and Pearson Tornado Scale, 132, 132 (chart)
haboobs, 127
halcyon days, 129–30
horse latitudes, 129
hurricanes, 131–32
hurricanes, classification of, 134 (chart), 134–35
hurricanes, deadliest, 137–38
hurricanes, most destructive, 138
hurricanes, names of, 135–36 (chart), 135–37
jet stream, 128–29, 129 (ill.)
microburst, 128
microclimates, 128
Siberian express, 130
solar, 46–47
storm chasers, 133–34
tornadoes, 131–32, 133, 134
wind chill, 130–31, 131 (chart)
wind shear, 128
windiest cities, 130
Wind chill, 130–31, 131 (chart)
Wind energy, 190

Wind farms, 189–90
Wind shear, 128
Wind tunnel, 550
Windows, 202
Wings, 547
Winter, 482–83
Winthrop, John, Jr., 160
Wireless telephones, 574
Wirth, Niklaus, 587
Withworth, J., 494
Witting, Georg, 615
WOBO (world bottle), 244
Wöhler, Friedrich, 15
Wolff, Kaspar Friedrich, 262
Wolves, 357 (ill.)
Wombats, 348
Women in space, 74, 75–76
Wood, 163–66, 198, 505
Wood, Harry, 111
Wood-burning stove, 183
Wooden figure of woman, 525
Woodpeckers, 343
Wooly-bear caterpillar superstition, 145
Worden, Alfred M., 72
Work, Henry Clay, 487
World bottle (WOBO), 244
World calendar, 478
World Trade Center, 509, 510
World Wide Web, 593
Wormwood, 287
Wounds, 396
Wright Brothers, 547–48
Wrinkled dog, 360
Writing
  hornbook, 559
  paper manufacturing process, 557–58
  papyrus, 557

## X

Xenon, 22–23, 23
Xeriscaping, 292
X-ray crystallography, 16
X-rays, 446

## Y

Yannas, Ioannis V., 463
Yard, 467, 467–68 (charts)
Yazoo, 93
Yeager, Chuck, 8, 548
Yeager, Jeana, 548
"Year without a summer," 121–22
Yellow bile, 392
Yellow fever, 425
Yellowstone National Park, 213
Yosemite Falls, 94
Young, John W., 72
Yttrium, 23
Yucca Mountain, Nevada, 237
Yudin, Sergei, 443
Yuma, Arizona, 121
Yurt, 508

## Z

Zebra mussels, 328
Zeidler, Othmar, 226
Zeno of Elea, 612–13
Zeno's paradox, 612–13
Zero, 595
Zero gravity, 68
Zhang Heng, 110
Zoll, Paul, 445
Zoonosis, 425
Zoos, 305
Zworykin, Vladimir, 492, 568